Molecular Design of Opto-Electronic Materials

Molecular Design of Opto-Electronic Materials

From Single Molecules to Molecular Aggregates

Edited by Zhen Li

WILEY-VCH

Editor

Prof. Zhen Li
Hubei Key Lab on Organic and Polymeric Optoelectronic Materials
Department of Chemistry
Wuhan University
Wuhan 430072
China

College of Chemistry and Chemical Engineering
Hubei University
Wuhan 430062
China

Cover Design: Wiley
Cover Image: © O-IAHI/Shutterstock

Typesetting: Lumina Datamatics, Madurai, India

Printing and Binding: CPI Group (UK) Ltd, Croydon, CR0 4YY

Library of Congress Card No.: applied for

British Library Cataloguing-in-Publication Data
A catalogue record for this book is available from the British Library.

Bibliographic information published by the Deutsche Nationalbibliothek TheDeutsche Nationalbibliothek lists this publication in the Deutsche Nationalbibliografie; detailed bibliographic data are available on the Internet at <http://dnb.d-nb.de>

Print ISBN: 9783527349395
ePDF ISBN: 9783527833429
ePUB ISBN: 9783527833443
o-Book ISBN: 9783527833436

C9783527349395_250626

Contents

Preface

Thanks to the great efforts of scientists, the development of functional organic and polymeric materials has made huge progress, with labels of organic light-emitting diodes, solar cells, biomedical materials, etc. However, besides many successful cases, some examples indicated unexpected phenomena and uncontrollable performance. Accordingly, more and more attention has been focused on molecular packing and intermolecular interactions in molecular aggregates. While molecular chemistry focuses on the covalent bond, the properties of organic and polymeric materials in the solid state are governed largely by the non-covalent, intermolecular interactions that arise when molecules come together. Thus, understanding and controlling these interactions lies at the heart of molecular aggregation science, a field that bridges chemistry, physics, materials science, biology, engineering, etc. The manner in which molecules arrange themselves in space, their aggregated states, has been a subject of fascination and intense research.

This book aims to provide a comprehensive overview of the principles and practices of molecular aggregation, based on the systematical interesting research results of the book authors coupled with related literature, from fundamental forces to cutting-edge applications. It is intended for graduate students, researchers, and professionals who seek to understand how the collective behavior of molecules, that is, Molecular Uniting Set Identified Characteristic (MUSIC), gives rise to emergent properties not present in isolated species.

Throughout the writing process, our goal has been to present a coherent picture that unifies synthetic design, supramolecular control, inherent mechanism understanding, computational prediction, and device engineering. We believe that a deep understanding of molecular aggregation will continue to drive innovation in flexible electronics, biomedical imaging, sensing, and beyond.

We are grateful to our colleagues and students whose insights and critiques have shaped this book. It is our hope that this book will inspire readers not only to appreciate the beauty of molecular aggregate science but also to actively design them for future technologies.

1

A Brief Introduction to Molecular Aggregates

Jiaqiang Wang, Juqing Gu, Guigui Ye, Wei Cao, Meng Wang, Changzun Jiang, Shuhui Li, Arui Huang, Qianqian Li, and Zhen Li**

Department of Chemistry, Wuhan University, Wuhan, China

*Corresponding authors: liqianqian@whu.edu.cn, lizhen@whu.edu.cn

1.1 Introduction

1.1.1 Motivation

The opto-electronic materials act as the key components in many functional devices, as well as the detection and imaging systems, which can permeate every aspect of modern life. Their diverse applications and potential for innovation have garnered increasing attention, and the diverse properties in aggregated states enrich their applications, including data storage and encryption,[1,2] photovoltaic devices,[3–5] organic light-emitting diodes (OLEDs),[6–8] organic light-emitting transistors (OLETs) devices,[9,10] biological sensing and imaging,[11–13] etc. In these applications, the macroscopic performance of molecular aggregates is not just the simple overlay of single molecules. In many cases, the emergence of new behaviors is enabled by specific molecular packing arrangements. Likewise, the adjustability and uncertainty of molecular aggregation bring opportunities and challenges for designing high-performance opto-electronic materials. Until now, there is still no comprehensive theoretical system to illustrate the detailed structure-performance relationship. However, certain advantageous molecular arrangements and packing modes have been discovered in specific scenarios to guide the construction of efficient opto-electronic materials. As the investigation goes deeper and deeper, it is essential to study their properties from the molecular aggregation levels, in addition to the individual molecules. The purpose of this book is to explore the influence of single molecules and molecular aggregates on the properties of opto-electronic materials, providing feasible ideas for molecular design, thereby broadening the application scenarios of opto-electronic materials, and improving the performance of corresponding detection or devices.

In the past several decades, the opto-electronic properties of organic materials have been primarily explored through the molecular design of functional groups and structural motifs. Molecules, as the smallest identifiable unit of the substance, have been the foundation for extensive research. Research based on molecular science has flourished, and a large number of organic opto-electronic materials have been developed.[14,15] However, some phenomena, such as the different luminescent behaviors observed in different polymorphic forms, remain difficult to be explained within a single-molecule framework. Molecular aggregates bridge the gap between single-molecule properties and the macroscopic opto-electronic performance of materials, providing more objective and perfect analysis methods. Accordingly, molecular aggregates are highlighted in these cases, thereby breaking the original single-molecule framework and gradually pushing the development of molecular aggregate science.[16–18]

Generally, organic molecules in extremely dilute solutions can be approximately considered as the single-molecule state by ignoring the influence of solvent molecules. As the concentrations increase along with the formation of molecular aggregates or precipitation, the overall properties in aggregated states change in most cases, compared to those in single-molecule states. The macroscopic properties in aggregated states often exhibit remarkable diversity and complexity.[19,20] Key factors such as intermolecular interactions and stacking modes are critical in determining these properties, necessitating a departure from a purely molecular perspective to a comprehensive framework for understanding structure-packing-performance relationships. This shift has profound implications for the development of opto-electronic materials, particularly in fields where luminescence properties play a central role (Figure 1.1).

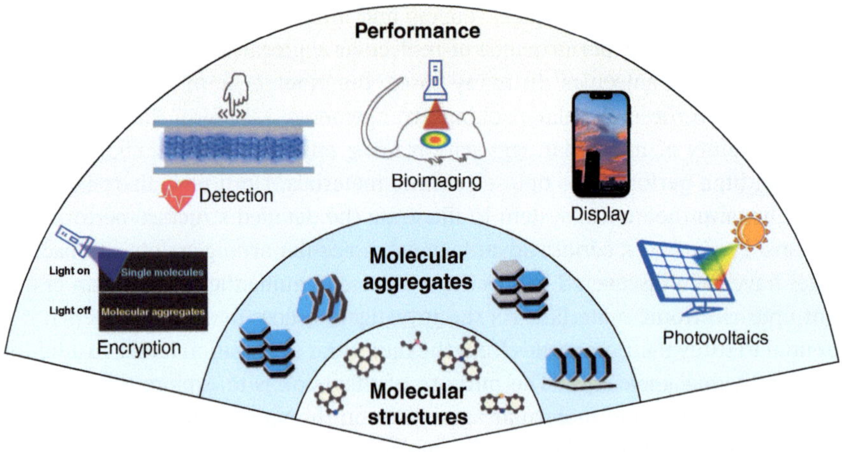

Figure 1.1 Schematic diagram of the relationship among molecular structures, molecular packing, and opto-electronic performance.

Luminescence properties can provide powerful and reliable information for exploring aggregated structures according to their changes from single molecules to aggregated states in some cases. Even for the same molecule, the tiny changes in aggregated structures can result in totally different luminescence properties, including emission wavelength, intensity, and lifetime. Moreover, specific phenomena such as mechanoluminescence (ML), mechanochromism, and room-temperature phosphorescence (RTP) often emerge from aggregated states, heavily influenced by molecular packing arrangements. Thus, in-depth exploration of the relationship between luminescence properties and aggregated structures represents a critical pathway to achieving excellent and diverse opto-electronic properties.[21–25]

The study of molecular aggregation and the related structure-performance relationship promotes the optimization of the photophysical properties and effectively improves the controllability of their opto-electronic property in the corresponding devices. For instance, in photovoltaic devices, the π–π coupling between two adjacent molecules is considered to be an efficient and important channel for charge transfer.[26–28] The research on the spatial structure of aggregated states can provide insights into the design of next-generation high-performance photovoltaic materials. For dye-sensitized solar cells, dye aggregation, which relates to the concentration of dye bath, ionic strength, temperature, solvents, and so on, usually leads to ineffective light-harvesting and charge recombination.[29] Accordingly, the efficiency of electron injections decreases, resulting in poor photovoltaic performance in most cases. Thus, the suppression of dye aggregation on the semiconductor surface is essential to improve the device performance. In perovskite solar cells, hole-transporting materials (HTMs) extract photogenerated holes and transport charges, thereby affecting the overall photovoltaic conversion efficiencies. The corresponding hole mobility and conductivity can be improved based on the optimization of morphology and crystallinity of hole-transporting layers (HTLs).[30] Although many effective strategies have been proposed, the relationship between molecular structure, aggregation behavior, and photovoltaic performance remains insufficiently understood. This limitation may be related to the complex carrier transfer processes and diverse molecular stacking in HTLs. Thus, precise molecular design guided by a deep understanding of aggregated structures is of great significance for improving device performance.

Similarly, the bio-applications of organic luminescence materials also depend on their aggregation behaviors under physiological environments, encompassing fields including biosensing, bioimaging, and therapy. For instance, aggregation-induced emission (AIE) luminogens and RTP nanoparticles demonstrate remarkable potential in biomedical imaging and therapeutic applications in aggregated states.[31,32] RTP materials can keep emissive states after turning off the light source, which shields the autofluorescence of tissues *in vivo*, thus achieving afterglow imaging of targeted tissues with high signal-to-noise ratios. In some cases, the reactive oxygen species (ROS) can be generated under external stimuli, including light and ultrasound, which can effectively kill tumor cells through photodynamic therapy or sonodynamic therapy.[33,34]

However, a lack of comprehensive analysis of molecular packing in opto-electronic materials still results in an increased gap between molecular design and practical applications. Although some dominant molecular packing modes have been studied in past research, several factors still hinder a systematic and deep understanding. On the one hand, the methods for characterizing and testing molecular aggregates are still limited, making it hard to monitor molecular aggregation behavior in practical applications. So far, the analysis of aggregated structures is heavily dependent on single crystals. However, in specific applications, molecular aggregation often takes more complex forms, including films, nanoparticles, powders, etc. At present, the intermolecular interactions and molecular stacking can be predicted through theoretical calculation, but discrepancies frequently arise between simulations and experimental results. On the other hand, the relationship among molecular structures, aggregation behaviors, and opto-electronic properties is still not clear enough to guide the practical application. Although some useful strategies have been proposed and applied in relatively simple systems, extending these approaches to complex molecular structures and multi-component systems across diverse opto-electronic fields remains a significant challenge. Thus, research on the relationship among molecular structures, aggregated modes, and opto-electronic properties is essential, and a mesoscale perspective bridging the gap between molecular design and material application is urgently needed.

1.1.2 A Brief History

The research and exploration of aggregated modes in organic opto-electronic materials has been one of the most advanced fields. The journey from molecular monomers to aggregated states has witnessed groundbreaking milestones since the first introduction of these organic opto-electronic materials, involving photoluminescence (PL), ML, OLEDs, OLETs, nonlinear optics (NLO), organic photovoltaics (OPV), perovskite solar cells, photocatalytic hydrogen evolution, and so on.

Luminescent materials, in particular, exhibit highly sensitive optical signals that enable rapid and pronounced responses from monomer to aggregate states, significantly influencing the development of modern opto-electronic devices and imaging technologies. It can be divided into fluorescence and phosphorescence based on the pathways of irradiative decay from singlet or triplet excited states. Since the nineteenth century, Sir John Herschel first observed that the colorless solution of quinine displayed a beautiful celestial blue color under light,[35] various organic fluorogens have been developed.[36] In 2001, Tang et al. proposed the concept of AIE from an unusual phenomenon that the non-emissive solution of 1-methyl-1,2,3,4,5-pentaphenylsilole became fluorescent upon aggregation on a thin-layer chromatography plate.[37] The emergence of the intriguing AIE effect established a new era of aggregated luminescent molecules, and AIE materials have since been widely applied in bioimaging and opto-electronics,[38,39] supported by a growing understanding of their underlying mechanisms.[40] Currently, a universal database for aggregate science has been established to facilitate a data-driven research paradigm within the field.[41]

Compared to fluorescence, phosphorescence refers to the emission of photons from the triplet excited states. Due to the inherent instability of the triplet states, phosphorescence is highly sensitive to the external environment and strongly dependent on its own aggregated modes. In 1939, the first observation of organic room-temperature phosphorescence in the solid state was reported by Clapp using the crystalline tetraphenylmethane and tetraphenylsilane derivatives.[42] The continuous discovery of the phosphorescence phenomenon advanced the corresponding theories. In the 1970s, Parker et al. adsorbed sodium 4-biphenylcarboxylate and sodium 1-naphthoate on paper, sucrose, starch, glass fiber paper, and silanized paper, and proposed the prototype of the concept that the restriction of molecular motions was favorable to RTP performance.[43] More recently, the emergence of crystallization-induced phosphorescence (CIP), reported by Sun and Tang et al. in 2010, demonstrated that ordered and compact packing with multiple intermolecular interactions could inhibit the possible quenching of the triplet state.[44] Since then, organic RTP materials with high efficiency in aggregated states have been gradually developed, and the relationship between molecular packing and RTP properties has been underscored.[45] In 2018, Pu and Li et al. studied the effects of the molecular packing on the organic RTP of the 10-phenyl-10H-phenothiazine-5,5-dioxide-based derivatives,[46] and elucidated the important role of π–π coupling in the stabilization of triplet excitons. Until now, the effect of molecular packing on RTP has been systematically investigated, highlighting the close relationship between aggregation modes and phosphorescent properties.[47]

ML materials, characterized by light emission induced by mechanical stimuli such as rubbing, scratching, pressing, or shaking, also show a strong dependence on their aggregated state. The first discovery of ML traces back to 1605, when Bacon observed light emission by scraping hard sugar crystals.[48] Over the next several centuries, various ML materials were reported, such as inorganic crystals, minerals, and organic crystals, including phenanthrene, 9-anthrylethanol, coumarin, acenaphthene, carbazole derivatives, *N*-phenyl imides, and so on.[49] With the development of AIEgens, a series of ML-AIE active molecules were reported in the 2010s,[50] which facilitated the understanding of ML mechanism and the exploration of the relationship between aggregated structures and ML. However, these ML materials usually exhibited only fluorescence as short-lived luminescence. In 2017, Li et al. reported a mechanoluminescent material based on phenothiazine derivatives with dual emission of fluorescence and phosphorescence.[51] More recently, Liu et al. reported ML materials by isostructural doping strategy, making the ML emission with an extended duration of up to 0.4 seconds.[52]

The advancement in aggregate science also facilitates the development of electroluminescence (EL) materials, which emit light in response to the passage of an electric current or a strong electric field. OLED devices based on EL materials have become a cornerstone in modern opto-electronics. The origins of OLED technology trace back to the early 1960s, with the discovery of delayed fluorescence in eosin and EL in anthracene crystals under high voltage.[53] While traditional fluorescent materials suffered from aggregation-caused quenching effect with limited efficiencies in OLEDs, in 2001, Tang et al. found the AIE effect in compound

1-methyl-1,2,3,4,5-pentaphenylsilole.[37] Since then, AIEgens with twisted molecular structures like triphenylamine, tetraphenylethylene, 10H-spiro[acridine-9,9'-fluoren] have been widely used in the non-doped OLEDs.[54,55] Meanwhile, researchers continuously investigated new OLED materials based on phosphorescence and thermally activated delayed fluorescence (TADF).[56] The aggregation-induced delayed fluorescence (AIDF) principle proposed by Chi and Tang et al., with the combination of AIE and TADF,[57,58] has further enhanced OLED performance. AIDF-based OLEDs have achieved unprecedented external quantum efficiencies (EQEs) by maximizing exciton utilization through aggregation.

Extending beyond luminescent materials to other critical applications, the research of aggregate science could be traced back to some other opto-electronic materials, without the consideration focusing on aggregate science at that time.[59–62] For example, as a significant branch of modern optics, nonlinear optics also exhibits a strong dependence on molecular packing modes. Indeed, early breakthroughs in this field originated from crystalline materials. In 1961, Franken et al.[63] made a landmark discovery and observed the occurrence of second harmonic generation (SHG) in a ruby crystal. In 1964, Rentzepis et al.[64] detected SHG in organic crystals such as benzopyrene, marking the early steps toward organic nonlinear optical materials. By 1970, Davydov et al.[65] reported that organic materials with electronic push-pull structures exhibited second-order nonlinear optical responses, utilizing intramolecular charge transfer to trigger nonlinear optical activity. In 1999, Dalton et al.[66] introduced the "site isolation principle," where chemical modifications were used to change chromophore spatial conformation toward a spherical shape, thus weakening strong dipole–dipole interactions and improving macroscopic NLO effects. Guided by this principle, dendrimers have emerged as promising NLO materials since they can protect chromophores and reduce dipole–dipole interactions among them through dendritic isolation groups.[67,68] The transition from single molecules to macroscopic systems, such as thin films, reveals several challenges and opportunities related to molecular packing and aggregation states. A clear discrepancy exists between the microscopic NLO effects of chromophores and the macroscopic second-order NLO effects of thin films. At the microscopic level, nonlinear polarization arises from electron transitions induced by photon absorption. However, in macroscopic systems, dipole–dipole interactions and the orientation of molecules significantly influence the overall NLO response. Significant progress has been made in fine-tuning molecular aggregation to optimize NLO materials by the incorporation of site isolation principle. From 2006, Li et al.[69,70] introduced various isolation groups to facilitate chromophore alignment during polarization, and the concept of "suitable isolation group" was further proposed to achieve spatial arrangements that facilitated orderly molecular alignment and enhanced NLO performance.[71–73]

The regulation of molecular aggregation to the improved performance of photovoltaic devices is equally groundbreaking.[74–77] For perovskite solar cells, the aggregation characteristics of organic semiconductors have long been recognized as a critical factor in enhancing the carrier transport capabilities of organic semiconductor-based hole transport materials (HTMs).[78,79] Photocatalytic hydrogen evolution has also benefited from the strategic manipulation of aggregation behavior. Upon rational aggregation of chromophores, intermolecular interactions promote the emergence of new coupled states that mitigate the Coulombic attraction between electrons and holes, thus enhancing charge separation and the subsequent hydrogen evolution.[80,81] The size of aggregates is a critical determinant in photocatalytic performance. It dictates the specific surface area of active sites and electron transport pathways. Meanwhile, aggregates with higher crystallinity and fewer recombination centers provide an ideal environment for photogenerated carriers to diffuse more effectively from the bulk to the surface, where redox reactions take place.[82]

In conclusion, the above discussions demonstrate how aggregate science has reshaped the design and applications of opto-electronic materials. By understanding and controlling molecular packing and aggregation states, researchers have unlocked unprecedented levels of performance in opto-electronic materials. These advancements not only highlight the versatility of aggregation phenomena but also provide a robust foundation for future innovations in molecular design and opto-electronic technologies.

1.1.3 Basic Knowledge of Organic Compounds in Aggregated States

The previous sections introduced the motivation and development of molecular aggregated states. The following content explains some common elements of molecular aggregate science, focusing on the driving forces behind molecular aggregation and their profound influence on the macroscopic properties of functional materials. These aspects are critical for understanding and regulating the aggregated states.

1.1.3.1 The Driving Force of Molecular Aggregates

The aggregated behaviors of organic molecules are derived from various intermolecular noncovalent interactions, involving hydrogen bonds, π–π interactions, electrostatic interactions, hydrophobic associations, and so on. In contrast to single covalent bonds with energies higher than 200 kJ mol^{-1}, noncovalent interactions are relatively weaker but highly significant in driving molecular aggregation. For example, hydrogen bonds exhibit energies of 25–40 kJ mol^{-1}, while the energies of other noncovalent interactions are lower than 10 kJ mol^{-1}.[83–86] These interactions act as the molecular linker, combining individual molecules to form macroscopic materials with diverse functional properties (Figure 1.2a), where molecular aggregates influence charge transport, luminescence, energy conversion efficiency, etc.[40,45,87] Broadly, these noncovalent interactions can be categorized as follows.

1.1.3.1.1 Hydrogen Bond

Hydrogen bond, one of the most extensively studied noncovalent interactions, involves a hydrogen atom bonded to a highly electronegative atom that interacts with a nearby electronegative atom with lone electron pairs, resulting in a special noncovalent interaction in the form of X—H···Y (Figure 1.2b). X and Y represent non-metallic atoms with high electronegativity and small atomic radii, such as

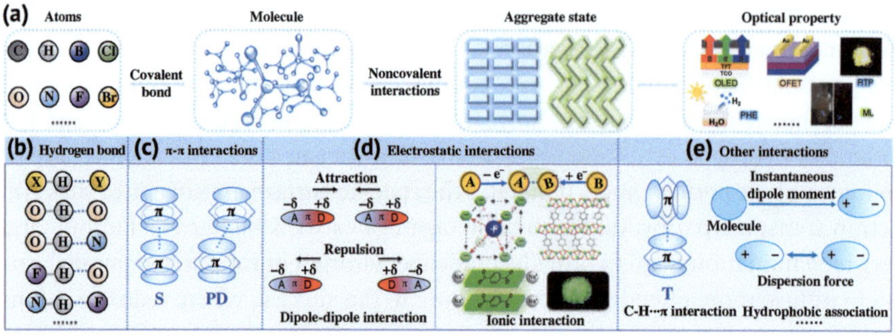

Figure 1.2 (a) Schematic diagram of molecular aggregated state formation. (b) Hydrogen bonding and the representation. (c) π–π interactions and the representation. (d) Electrostatic interactions and the representation including dipole–dipole interactions and ionic interactions. (e) Other interactions and the corresponding representation.

F, O, N, etc.[88,89] Hydrogen bonds can form between identical atoms, such as O—H···O in water, or between different atoms, like N—H···O in aqueous ammonia. Moreover, a hydrogen bond is characterized by specificity, saturability, and directionality, with optimal strength achieved when X—H and Y are aligned linearly, although it can also be bent or bifurcated. No matter the binding energy of a single hydrogen bond is relatively lower than that of covalent bonds, the cumulative effect of hydrogen bonding is significant.[90–92] In molecular aggregates, multiple hydrogen bonds often work synergistically to stabilize the aggregated structures.[93] This is particularly relevant in organic opto-electronic materials like hydrogen-bonded organic frameworks as chemical sensing materials and molecular devices, where strong hydrogen-bonding networks contribute to the thermal and mechanical stability of materials, improving their device performance.[94–96]

1.1.3.1.2 π–π Interaction

The π–π interaction is another pivotal one among aromatic systems, which is driven by dispersion forces between π-electron clouds. It is crucial for efficient charge transport and exciton diffusion in devices such as OLEDs and OPVs.[97–100] Depending on the molecular geometry, π–π interactions can occur in different stacking modes, such as S-shaped (sandwich) and PD-shaped (parallel-displaced) stacking (Figure 1.2c).[101–103] Understanding the nature of π–π interactions is essential for regulating the opto-electronic properties of organic materials, such as light absorption, charge transport, emission properties, etc. For instance, in pentacene derivatives, S-shaped stacking enables high charge-carrier mobility by creating continuous π-conduction pathways.[104,105] PD-shaped stacking, often observed in conjugated polymers, enhances the optical and electronic properties of these materials.[106]

1.1.3.1.3 Electrostatic Interactions

Electrostatic interactions add another dimension to the stabilization and functionality of molecular aggregates. These forces arise from the attraction between opposite dipoles or charged ions, including dipole–dipole and ionic interactions.[107–109] (Figure 1.2d). Ionic interactions occur between cations and anions, playing an important role in improving structural rigidity and tuning the solubility and processing conditions of organic materials. Polar molecules exhibit uneven charge distributions with permanent dipole moments, resulting in the generation of dipole–dipole interactions between them. Therefore, the interactions are created between partial positive and negative charges that attract one another. In opto-electronic materials, electrostatic interactions play a crucial role in improving the overall performance of opto-electronic devices by fine-tuning their arrangements. For example, in organic materials used in OPVs, electrostatic forces can facilitate the formation of phase-separated domains that optimize exciton dissociation and charge collection efficiency.[110,111]

1.1.3.1.4 Other Interactions

In addition to π–π and electrostatic interactions, several other types of interactions contribute to the aggregation modes of molecular systems in opto-electronic materials. C—H···π interactions, for example, are essential in π-stacking systems, where a hydrogen atom with slight electron deficiency on one aromatic ring interacts with the π-electron cloud of another aromatic ring, beneficial to the formation of T-shaped stacking (Figure 1.2e).[112] These interactions can also facilitate the formation of tightly packed and stabilize the aggregated structures. For example, the herringbone packing mode in ML materials, driven by edge-to-face molecular arrangements with multiple C—H···π interactions, prevents non-radiative relaxation by minimizing molecular slippage under mechanical force, thus facilitating the generation of ML.[113,114] Furthermore, hydrophobic associations, though less intuitive, are vital in the self-assembly of nonpolar molecules in polar environments like water. These interactions arise because polar water molecules preferentially interact with each other, forcing nonpolar molecules to aggregate into hydrophobic domains. This phenomenon contributes to the formation of micelles and vesicles and plays a significant role in stabilizing organic molecules in aqueous dispersions.[115–118]

While permanent dipole moments of polar molecules result in the generation of dipole–dipole interactions, instantaneous dipole moments can lead to the generation of dispersion forces due to the fluctuating distribution of electrons within a molecule. They exist between all molecules and result in weak attractions when molecules come sufficiently close to each other (Figure 1.2e).[119,120] Quantum mechanical calculations reveal that the strength of dispersion forces is strongly related to the polarizability of molecules, and the more deformable or polarizable molecules exhibit stronger dispersion interactions.[121] In opto-electronic materials, dispersion forces play a pivotal role in governing the stability of molecular assemblies. These interactions help stabilize the molecular packing within crystals and influence the morphology of aggregated structures, which in turn affect the macroscopic properties. For instance, halogen bonds can significantly limit the non-radiative transitions of molecules and enhance SOC, contributing to the RTP properties of the material.[122,123]

The aforementioned interactions often act synergistically to determine the overall properties of molecular aggregates. For example, in organic thin films used in light-emitting materials, hydrogen bonds can stabilize the molecular framework, while π–π stacking governs the packing density and electronic coupling between molecules. Electrostatic interactions further refine molecular orientation, enhancing the optical properties. In the design of materials for OFETs, precise tuning of π-π stacking and hydrogen bonding ensures high charge mobility and stability. Similarly, in organic bioimaging systems, hydrophobic interactions can help maintain the integrity of dye molecules in aqueous media, while hydrogen bonds, π-stacking, and electrostatic interactions can stabilize excitons and suppress non-radiative transitions, thereby enhancing the bioimaging performance.

1.1.3.2 The Adjustment of Molecular Interactions in Molecular Aggregates

Actually, the interplay of the driving forces of molecular aggregates, such as hydrogen bonding, π-stacking, electrostatic interactions, and hydrophobic associations, not only governs the formation and stability of molecular assemblies but also tailors their optical and electronic properties. The macroscopic functions of opto-electronic materials in aggregated states are usually different from those of single molecules, such as in dilute solutions, which are related to molecular packing in most cases.[20,124–126] In opto-electronic applications, the molecular packing of organic compounds is a critical factor influencing their properties, such as luminescence, charge mobility, and energy efficiency. Thus, controlling molecular packing is an essential issue for achieving the optimized optical and electronic properties. A comprehensive understanding of these interactions and the efficient strategies to regulate them will provide a powerful toolkit for designing innovative materials.

Though great challenges exist because of the unclear structures and complicated intermolecular interactions, advancements in molecular design have enabled the adjustment of molecular interactions and packing through internal and external forces (Figure 1.3). The internal forces include electronic and steric effects, the self-assembly effect, etc. The electronic properties of molecules can be manipulated to influence their packing behavior. Adjustments to the electron density of π-conjugated backbones or the electronic nature of substituents can regulate the degree of electronic coupling in aggregates, enabling diverse packing arrangements. For example, fine-tuning π-electron coupling through the regulation of substituents has been proven to be an effective strategy to regulate the RTP emission.[46] Similarly, substituents with electron-donating or electron-withdrawing characteristics can modulate the alignment of molecules and the electron affinities of materials, tailoring the electron mobility and charge transport properties of opto-electronic materials.[127]

The steric effect provides another strategy to adjust molecular packing by influencing spatial arrangement. When substituent groups with various sizes or shapes are integrated into molecular backbones, molecules will be dictated to adopt a compact or loose arrangement, leading to different optical and electronic properties.[128,22] Accordingly, researchers often maintain the main structures of molecules and change the substitutes, such as introducing specific functional groups or alkyl chains, to realize the subtle adjustments in molecular packing modes and associated performance.[129,130] The introduction of bulky substituents has been demonstrated to effectively disrupt π–π stacking, and the odd-even effect of alkanes has been widely utilized for the regulation of the molecular packing and intermolecular interactions.[131]

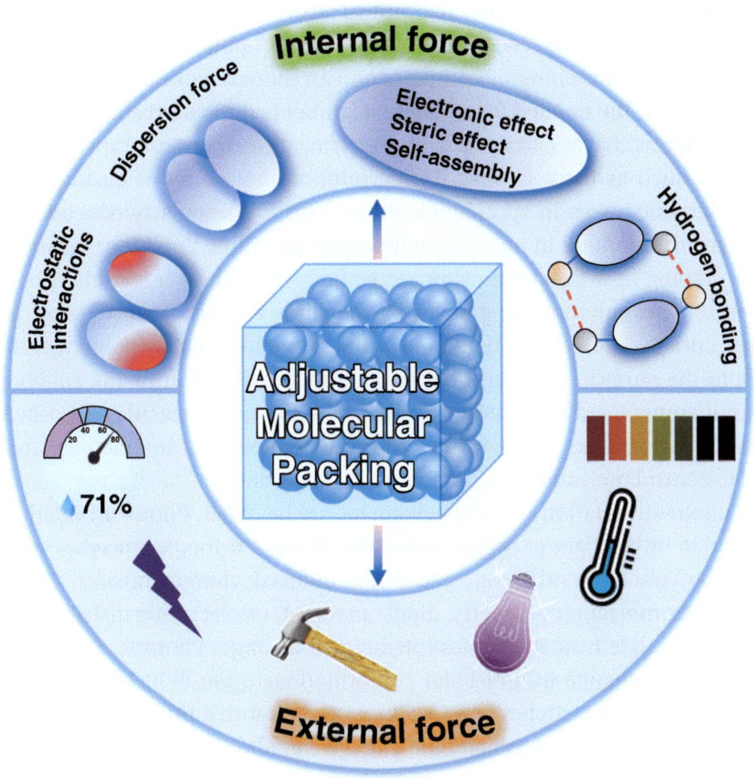

Figure 1.3 Schematic diagram of adjusting molecular packing through internal and external forces.

Self-assembly is an intrinsic mechanism that enables molecules to organize into well-defined aggregates without external guidance. This process is largely governed by noncovalent interactions, which can be modulated to achieve desired packing modes. It arises from balances of intermolecular forces, including hydrogen bonding, van der Waals forces, π–π interactions, dipole–dipole interactions, and hydrophobic effects. These interactions are finely tuned to achieve thermodynamically stable or metastable aggregates with desired functionalities.[132] The advantage of self-assembly lies in its ability to amplify intermolecular interactions and guide the spatial arrangement of molecules without the need for extensive manipulation. For instance, assembly systems with specific packing geometries can be realized by altering functional groups and the formation of hydrogen bonding.[133]

Moreover, external stimuli offer dynamic control over molecular aggregates, enabling reversible or irreversible transformations in their packing and properties. Upon the introduction of some external force, such as force, light, heat, electronic field, and others, dynamic optical properties can be achieved. Mechano-stimulation, including rubbing, stretching, or compressing, can induce changes in intermolecular interactions and alignments, disrupt crystalline packing, or alter molecular conformations. For example, the planarization of molecular conformations or the transition from crystalline to amorphous states under mechanical stress can lead to mechanochromic effects or mechanoluminescence phenomena, yielding applications in stress sensors.[134,135] Mechanical stretching or compression can induce molecular alignment along the direction of the applied force. These alignments improve the corresponding properties, such as charge carrier mobility, birefringence, or optical transparency. Shearing forces, such as those applied during solution processing or coating, can orient molecular assemblies in specific directions. This is particularly effective for achieving uniaxial alignment in liquid crystals or polymer films.[136,137] Furthermore, the physical rubbing of a substrate using a specific tool (such as a soft cloth or a rubber roller), creates microscopic grooves on the surface. These grooves serve as topographical cues that guide the orientation of the molecules that are subsequently deposited onto the surface, causing them to align in the direction of the rubbing motion. This alignment can be transferred to the liquid crystal molecules deposited on top of the rubbed surface, which further align due to molecular interactions. It is widely used to control molecular orientation in liquid crystal.[138]

As for the photo-stimulation, some mechanisms are involved. Photo-stimulation has been found to utilize light to trigger molecular changes through processes such as isomerization, photodimerization, and photo-induced charge transfer. These processes can alter molecular geometry, dipole moment, or electronic distribution, which in turn modulate how molecules interact and arrange. Photoisomerization involves a reversible change in molecular conformations upon light irradiation.[19] For example, azobenzene switches between *trans* and *cis* forms under UV and visible light, respectively. This conformational change can alter molecular alignment and packing density, influencing properties such as charge transport and refractive index.[139] Certain molecules, such as anthracenes and coumarins, undergo light-induced dimerization. This process can create covalent bonds between molecules, driving them into specific packing arrangements or disrupting preexisting assemblies.[140,141] Photochromic molecules, such as spiropyrans, exhibit a reversible transformation between two states with different electronic and structural properties.[142] This enables light-triggered changes in molecular packing, orientation, or aggregation behavior. Besides, light could influence molecular packing by enhancing π-π interactions or suppressing quenching effects through oxygen consumption.[143]

Heat and electric fields play crucial roles in the modulation of molecular aggregation. Thermal energy provides the necessary activation for molecular systems to overcome energy barriers and reorganize their packing. This process is often governed by thermodynamic principles. The effects of heating processes include the increase of molecular mobility, phase transitions, and annealing of crystalline or semi-crystalline structures. Electric fields interact with molecular dipoles, charges, or polarizabilities, inducing reorientation and alignment. This mechanism leverages the ability of electric fields to exert torque or directional forces on polarizable molecules, promoting anisotropic packing and alignment.[144] Molecules with permanent or induced dipoles can align parallel to an applied electric field, creating ordered domains. This effect is widely used in liquid crystal technologies.[145,146] Heat and electric fields offer complementary and versatile tools for adjusting molecular packing and orientation, which are beneficial for the achievement of the desired orientation in materials such as second-order non-linear optical and liquid crystal materials.[147,148] In these materials, thermal energy facilitates mobility and phase transitions, while electric fields provide directional alignment and polarization control.

Chemical stimuli could also change molecule packing, which affects molecular systems through a variety of mechanisms, including solvation, selective interactions, and chemical reactions. These interactions can alter the thermodynamic and kinetic pathways of molecular organization, enabling the precise tuning of packing density, phase behavior, and orientation. Selective solvents can promote specific interactions, guiding the self-assembly of molecules. These solvents can influence the solubility and molecular interactions of solutes, providing a thermodynamic driving force for specific rearrangement pathways.[149,150] The addition of specific ligands, ions, or templates can trigger supramolecular assembly driven by hydrogen bonding, coordination, or host-guest interactions.[151,152] Adjusting pH or ionic strength can influence charge distribution and electrostatic interactions in polyelectrolytes or ionic materials, enabling controlled orientation and packing.[153,154] By utilizing internal and external forces, these strategies continue to promote the development of next-generation devices, highlighting the critical intersection of molecular aggregate science in opto-electronics.

1.1.4 Overview of Topics Covered

Molecular aggregation plays a pivotal role in the functional properties and performance of opto-electronic materials. The proposed concept of "Molecular Uniting Set Identified Characteristic (MUSIC)" highlights the key role of molecular uniting in aggregated states. The exploration of molecular aggregates provides a profound understanding of their behaviors and potential in diverse applications, urging further research into their structure–property relationships. This section focuses on the key role of molecular aggregates in opto-electronic materials, including RTP, ML,[155–157] mechanoresponsive luminescence (MRL),[134,158,159] hole-transporting layer in perovskite solar cells, active layer in organic photovoltaics, OLED devices, and second-order non-linear optical materials. Their functionalities are closely related to the molecular aggregates,[24,45,16] including molecular packing and orientation, intermolecular interactions (hydrogen bond, π–π interactions, electrostatic interactions),[160] molecular configurations and conformations.[161] The diverse aggregation states of molecules such as crystals, co-crystals, thin films, nanoparticles, and fluids, exhibit distinct structural and functional characteristics, influencing their applicability in various advanced technologies.[162–164] The strategic design and control of these aggregated states are indispensable for achieving tailored opto-electronic properties (Figure 1.4).

Crystal or crystalline states represent highly ordered structures that exhibit reliable molecular packing and conformation for studying opto-electronic properties. Cocrystals consist of two or more components that form a unique crystalline structure. The precise packing modes and molecular conformations of both crystals and cocrystals can be characterized by single crystal/powder X-ray diffraction analysis, affording a convenient way to get insights into the possible relationship between

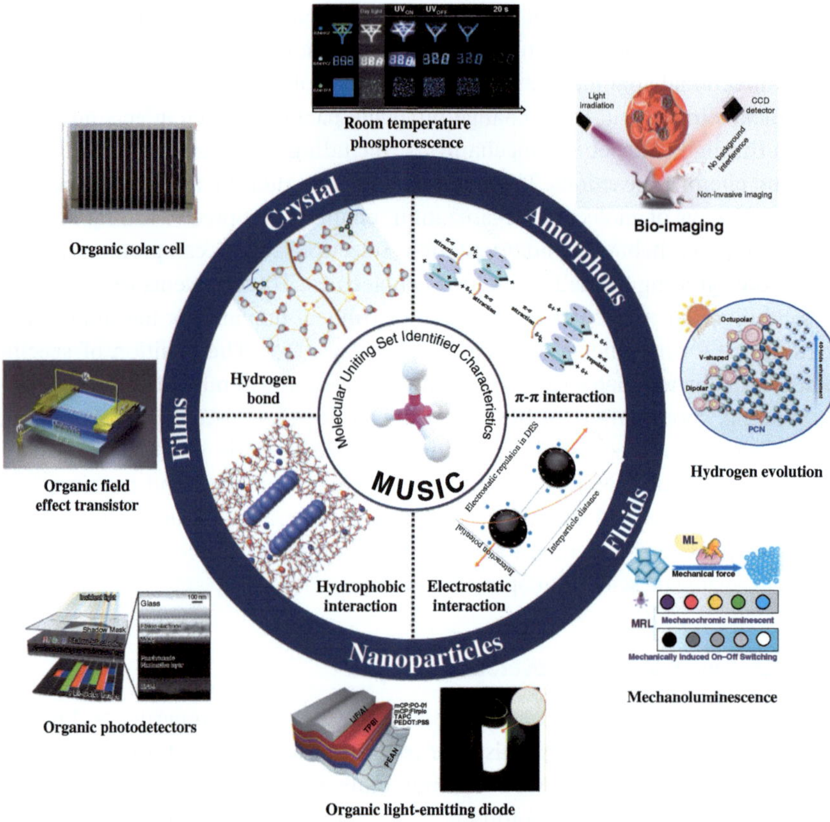

Figure 1.4 Schematic illustration of the possible aggregation states of molecules – crystals, films, nanoparticles, and fluids – to achieve applications in various fields. *Source*: Reproduced with permission from [165], Copyright 2023, American Chemical Society; [166], Copyright 2025, Royal Society of Chemistry; [167], Copyright 2007, American Chemical Society; [168], Copyright 2021, American Chemical Society; [169], Copyright 2020, American Chemical Society; [170], Copyright 2023, American Chemical Society; [171], Copyright 2024, American Chemical Society; [172], Copyright 2021, American Chemical Society; [173], Copyright 2022, Elsevier; [45], Copyright 2020, American Chemical Society; [174], Copyright 2014, American Chemical Society; [175], Copyright 2019, American Chemical Society.

molecular arrangement and opto-electronic properties.[158,176–180] For example, the herringbone packing mode with the ability to effectively suppress molecular slippage has been proven to be essential to ML crystals.[155,181] Controlled dimer formation in phenothiazine derivatives[182], anthracene derivatives and pyrene derivatives.[157,183] plays a key role in MRL materials. π–π coupling has been found to stabilize excited triplet states, enhancing RTP performance.[184–189] In other fields, such as the hole-transporting layer in perovskite solar cells, active layer in organic photovoltaics, and thermally activated delayed fluorescence (TADF) materials, molecular packing influences their charge transport and exciton behaviors, providing indirect evidence to explain the mechanism and guiding the material design.[176,190,191,30] However, it is not easy to get perfect crystals to fit the requirements of many opto-electronic devices, and most crystal-based organic materials face challenges in processability, flexibility, and reproducibility, limiting their application in practical devices.[192]

Alternatively, thin films with amorphous or partially crystalline states, ranging from fractions of a nanometer (monolayer) to several micrometers in thickness, provide diverse functional properties in opto-electronic applications due to their ease of processing and adaptability to device integration. These films can be fabricated through spin coating, vapor deposition, and solution casting, enabling controlled aggregation and tunable functionality. For example, in dye-doped polymer films,[193] the aggregation behaviors can be controlled by changing the concentrations of dyes and the temperatures in the solution casting process, and the emission from different triplet states can be achieved, enabling adjustable afterglow colors. In stimulus-responsive copolymer films,[194–196] rigid cross-linked polymer network and hydrogen bonding among polymer chains effectively suppress the motion of polymer chains, and prevent the quenching of triplet excitons from oxygen and moisture, enabling ultralong RTP emission.

In organic semiconductor layer of OFETs, molecular aggregation is equally vital. The introduction of additives promotes electrostatic interactions that facilitate dynamic nucleation and crystal growth, resulting in long-range ordered crystalline domains. These ordered domains enhance charge transport, leading to high charge-carrier mobility, which is critical for device performance.[169,197,198] For active layers in polymer light-emitting diodes,[199,200] hole-transporting layers in perovskite solar cells,[30,201,202] and active layers in hybrid solar cells,[191,203,204] aggregation and morphology significantly affect exciton dissociation and carrier transport properties, and change the performance of devices accordingly,[205–209] which can be achieved through the optimization in film preparation processes. Thin films offer a practical alternative to crystalline materials, balancing structural randomness with functional performance while addressing the challenges of scalability and flexibility.

A nanoparticle or ultrafine particle usually exhibits the dimension ranging from 1 to 100 nanometers (nm) in diameter. The term is sometimes used for larger particles up to 500 nm in diameter, or fibers and tubes that are smaller than 100 nm in only two directions. They exhibit unique opto-electronic properties dictated by their size, morphology, and aggregated states. These properties can be finely tuned through controlled fabrication processes, including the modulation of nucleation, crystal growth, intermolecular interactions, and the ratio of orthogonal solvent or adding different surfactants. In the field of photocatalytic hydrogen production, by changing the solvent ratios in the fabrication process of nanoparticles, the morphology can be optimized effectively due to the fine tuning of molecular packing in aggregated states, and better photocatalytic performance can be achieved.[210,211] For healthcare, nanoparticles are one of the most popular aggregate states to be investigated in bioimaging, labeling, diagnosis, and treatment. Traditional fluorescent dyes, such as fluorescein, rhodamine B, or cyanine dye,[212] can help to solve problems including labeling and observation as nanoparticles. Until now, to satisfy the requirement of high signal-to-noise ratio, low toxicity, single targeting, long afterglow nanoparticles are designed and fabricated through a top-down preparation method. The selected compounds include carbazole,[213] phenothiazine,[214,51] acridine,[215] xanthene,[216] triphenylamine,[217] phenylboronic derivatives,[218] and benzothiazole[219] derivatives. Requirements for fluorescent/phosphorescent materials in the applications in biomedical field keep increasing. For example, in the COVID-19 test, gold nanoparticles,[220,221] fluorescent nanoparticles (time-resolved fluorescence, fluorescent microspheres, and quantum dots),[222–224] magnetic nanoparticles, liposomes, rare earth nanomaterials, etc. have been developed to detect COVID-19 antigen due to the urgent need for effective detection.[225]

Furthermore, some nanoparticles demonstrate properties of stimulus-responsive due to the presence of specific groups. For instance, pH changes can promote the protonation or deprotonation, or the formation or breaking process of chemical bonds, while light irradiation can lead to isomerization or formation of chemical bonds due to the presence of some light-sensitive structures. Thus, environmental changes have a direct impact on intermolecular interactions in these nanoparticles, inducing morphological and functional transformations and enabling applications in diagnostics and responsive devices.

Fluids, including liquids and other nearly incompressible states of matter, represent a unique aggregate state with distinct physical and molecular characteristics. While fluids conform to the shape of their containers, they maintain a nearly constant volume regardless of external pressure, exhibiting fluidity, compressibility, and viscosity. These properties arise from the molecular dynamics in the liquid state, where molecules exhibit long-range disorder yet maintain short-range order. This semicrystalline behavior manifests as the local molecular organization that is transient and dynamic, distinguishing liquids from other aggregated states. Each molecule in liquid is surrounded by several adjacent molecules, forming regular geometric configurations. However, the regularity can only be maintained within several molecules, and the geometric configurations in different molecular clusters can be different. The geometric configurations can also fluctuate over time due to thermal motions. Such dynamic aggregation properties enable fluids to exhibit unique behaviors, making them invaluable for various opto-electronic applications, such as liquid crystal and RTP materials. Liquid crystal displays (LCDs), an early flat-screen technology, utilize the directional flow and alignment of liquid crystalline molecules to manipulate light under an electric field. This alignment controls light transmission, enabling image formation when combined with color filters and a backlight. Recent breakthroughs in fluid-based RTP materials, such as malic-acid-based systems,[163] have resolved challenges in achieving stable phosphorescence in fluid environments. These malic-acid-based materials demonstrate that supramolecular assembly in fluid state can mitigate these effects, opening pathways for in vivo applications and other advanced opto-electronic functionalities.

In conclusion, the versatility of molecular aggregates is reflected in their broad applicability across opto-electronic fields. For instance, nanoparticles in biological imaging and healthcare improve diagnostic accuracy and therapeutic outcomes. Thin films in photovoltaics and OLEDs enhance charge transport and exciton management. Crystalline materials provide insights into structure–property relationships, guiding the design of complex multifunctional devices. From organic solar cells to anticounterfeiting technologies, the aggregate states of molecules – whether nanoparticles, films, or crystals – play a decisive role in achieving desired performance. Atoms make up molecules through chemical bonds, and molecular aggregates are formed through weak intermolecular interactions. The fancy performances of these materials not only depend on the properties of the well-designed molecules with specific function groups or/and fixed energy level, but also on their molecular packing, orientation, and interactions. The MUSIC concept emphasizes the centrality of aggregation in molecular design, urging further exploration of these states to unlock new development in opto-electronics. By integrating advanced aggregation strategies with innovative molecular design, researchers can continue to push the boundaries of performance and application in this field.

References

1 Zhao, W., He, Z., and Tang, B.Z. (2015). Room-temperature phosphorescence from organic aggregates. *Nature Reviews Materials* 5:869–885.

2 Yu, X., Zhang, H., and Yu, J. (2021). Luminescence anti-counterfeiting: from elementary to advanced. *Aggregate* 2: 20–34.

3 Banal, J.L., Zhang, B., Jones, D. J. et al. (2017). Emissive molecular aggregates and energy migration in luminescent solar concentrators. *Accounts of Chemical Research* 50: 49–57.

4 Cornil, J., Verlaak, S., Martinelli, N. et al. (2013). Exploring the energy landscape of the charge transport levels in organic semiconductors at the molecular scale. *Accounts of Chemical Research* 46: 434–443.

5 Kim, J.H., Schembri, T., Bialas, D. et al. (2022). Slip-stacked J-aggregate materials for organic solar cells and photodetectors. *Advanced Materials* 34: 2104678.

6 Uoyama, H., Goushi, K., Shizu, K. et al. (2012). Highly efficient organic light-emitting diodes from delayed fluorescence. *Nature* 492: 234–238.

7 Liu, Y., Li, C., Ren, Z. et al. (2018). All-organic thermally activated delayed fluorescence materials for organic light-emitting diodes. *Nature Reviews Materials* 3: 18020.

8 Fung, M.K., Li, Y. Q., and Liao, L.S. (2016). Tandem organic light-emitting diodes. *Advanced Materials* 28: 10381–10408.

9 Cicoira, F. and Santato, C. (2007). Organic light emitting field effect transistors: Advances and perspectives. *Advanced Functional Materials* 17: 3421–3434.

10 Liu, H., Xu, J., Li, Y., and Li, Y. (2010). Aggregate nanostructures of organic molecular materials. *Accounts of Chemical Research* 43: 1496–1508.

11 Guo, Z., Park, S., Yoon, J., and Shin, I. (2014). Recent progress in the development of near-infrared fluorescent probes for bioimaging applications. *Chemical Society Reviews* 43: 16–29.

12 Xu, W., Wang, D., and Tang, B. Z. (2021). NIR-II AIEgens: A win-win integration towards bioapplications. *Angewandte Chemie International Edition* 60: 7476–7487.

13 Li, J., Wang, J., Li, H. et al. (2020). Supramolecular materials based on AIE luminogens (AIEgens): Construction and applications. *Chemical Society Reviews* 49: 1144–1172.

14 Tsutsui, T., Aminaka, E., Lin, C.P., and Kim, D.U. (1997). Extended molecular design concept of molecular materials for electroluminescence: Sublimed-dye films, molecularly doped polymers and polymers with chromophores. *Philosophical Transactions of the Royal Society A* 355: 801–814.

15 Sugiyama, K., Yoshimura, D., Miyamae, T. et al. (1998). Electronic structures of organic molecular materials for organic electroluminescent devices studied by ultraviolet photoemission spectroscopy. *Journal of Applied Physics* 83: 4928–4938.

16 Li, Q. and Li, Z. (2017). The strong light-emission materials in the aggregated state: What happens from a single molecule to the collective group. *Advanced Science* 4: 1600484.

17 Zhang, H., Zhao, Z., Turley, A.T. et al. (2020). Aggregate science: From structures to properties. *Advanced Materials* 32: 2001457.

18 Ma, S., Du, S., Pan, G. et al. (2021). Organic molecular aggregates: From aggregation structure to emission property. *Aggregate* 2: e96.

19 Gu, J., Li, Z., and Li, Q. (2023). From single molecule to molecular aggregate science. *Coordination Chemistry Reviews* 475: 214872.

20 Liao, Q., Li, Q., and Li, Z. (2023). The key role of molecular packing in luminescence property: From adjacent molecules to molecular aggregates. *Advanced Materials* 37: 2306617.

21 Xie, Y. and Li, Z. (2018). Triboluminescence: Recalling interest and new aspects. *Chem* 4: 943–971.

22 Ma, X., Wang, J., and Tian, H. (2019). Assembling-induced emission: An efficient approach for amorphous metal-free organic emitting materials with room-temperature phosphorescence. *Accounts of Chemical Research* 52: 738–748.

23 Zhang, J., He, B., Hu, Y. et al. (2021). Stimuli-responsive AIEgens. *Advanced Materials* 33: 2008071.

24 Yang, J., Fang, M., and Li., Z. (2021). Stimulus-responsive room temperature phosphorescence materials: Internal mechanism, design strategy, and potential application. *Accounts of Materials Research* 2: 644–654.

25 Kenry, C.C. and Liu, B. (2019). Enhancing the performance of pure organic room-temperature phosphorescent luminophores. *Nature Communications* 10: 2111.

26 Swick, S.M., Zhu, W., Matta, M. et al. (2018). Closely packed, low reorganization energy π-extended postfullerene acceptors for efficient polymer solar cells. *Proceedings of the National Academy of Sciences* (Vol. 115, p. E8341).

27 Aldrich, T.J., Matta, M., Zhu, W. et al. (2019). Fluorination effects on indacenodithienothiophene acceptor packing and electronic structure, end-group redistribution, and solar cell photovoltaic response. *Journal of the American Chemical Society* 141: 3274.

28 Lai, H. and He, F. (2020). Crystal engineering in organic photovoltaic acceptors: A 3D network approach. *Advanced Energy Materials* 10: 2002678.

29 Xu, F., Testoff, T.T., Wang, L., and Zhou, X. (2020). Cause, regulation and utilization of dye aggregation in dye-sensitized solar cells. *Molecules* 25: 4478.

30 Yao, Y., Cheng, C., Zhang, C. et al. (2022). Organic hole-transport layers for efficient, stable, and scalable inverted perovskite solar cells. *Advanced Materials* 34: 2203794.

31 Zhang, K.Y., Yu, Q., Wei, H. et al. (2018). Long-lived emissive probes for time-resolved photoluminescence bioimaging and biosensing. *Chemical Reviews* 118: 1770–1839.

32 Cheng, H., Li, Y., Tang, B.Z., and Yoon, J. (2020). Assembly strategies of organic-based imaging agents for fluorescence and photoacoustic bioimaging applications. *Chemical Society Reviews* 49: 21–31.

33 Celli, J. P., Spring, B.Q., Rizvi, I. et al. (2010). Imaging and photodynamic therapy: Mechanisms, monitoring, and optimization. *Chemical Reviews* 110: 2795–2838.

34 Gao, M. and Tang, B. Z. (2020). AIE-based cancer theranostics. *Coordination Chemistry Reviews* 402: 213076.

35 Stokes, G.G. (1853). On the change of refrangibility of light. *Royal Society* 143: 0261–0523.

36 Schmidt, G.C. (1921). Über lumineszenz von festen lösungen. *Annalen der Physik* 370:247–256; Oster, G. and Nishijima, Y. (1956). Fluorescence and internal rotation: Their dependence on viscosity of the medium. *Journal of the American Chemical Society* 78: 1581–1584.

37 Luo, J., Xie, Z., Lam, J.W.Y. et al. (2001). Aggregation-induced emission of 1-methyl-1, 2, 3, 4,5-pentaphenylsilole. *Chemical Communications* 21: 1740–1741.

38 Chen, J., Law, C.C.W., Lam, J.W.Y. et al. (2003). Synthesis, light emission, nanoaggregation, and restricted intramolecular rotation of 1,1-substituted 2, 3, 4, 5–tetraphenylsiloles. *Chemistry of Materials* 15: 1535–1546.

39 Yu, G., Yin, S., Liu, Y. et al. (2005). Structures, electronic states, photoluminescence, and carrier transport properties of 1,1-disubstituted 2, 3, 4, 5–tetraphenylsiloles. *Journal of the American Chemical Society* 127: 6335–6346.

40 Wang, H., Zhao, E., Lam, J.W.Y., and Tang, B.Z. (2015). AIE luminogens: Emission brightened by aggregation. *Materials Today* 18: 365–377.

41 Gong, J., Gong, W., Wu, B. et al. (2023). ASBase: The universal database for aggregate science. *Aggregate* 4: e263.

42 Sklar, A.L. (1937). The near ultraviolet absorption of substituted benzenes. *Journal of Chemical Phyics* 7: 984–993.

43 Schulman, E.M. and Parker, R.T. (1977). Room temperature phosphorescence of organic compounds. The effects of moisture, oxygen, and the nature of the support-phosphor interaction. *The Journal of Physical Chemistry* 81: 1932–1939.

44 Yuan, W.Z., Shen, X.Y., Zhao, H. et al. (2010). Crystallization-induced phosphorescence of pure organic luminogens at room temperature. *The Journal of Physical Chemistry* 114: 6090–6099.

45 Li, Q. and Li, Z. (2020). Molecular packing: Another key point for the performance of organic and polymeric optoelectronic materials. *Accounts of chemical research* 53: 962–973.

46 Yang, J., Zhen, X., Wang, B. et al. (2018). The influence of the molecular packing on the room temperature phosphorescence of purely organic luminogens. *Nature Communication* 9: 840.

47 Zhou, Q., Yang, C., and Zhao, Y. (2023). Dynamic organic room-temperature phosphorescent systems. *Chem* 14: 2446–2480.

48 Bacon, F. (1605). *The Advancement of Learning. Book IV, Chap. 3.* Oxford.

49 Hardy, G.E., Kaska, W.C., Chandra, B.P., and Zink, J.I. (1981). Triboluminescence-structure relationships in polymorphs of hexaphenylcarbodiphosphorane and anthranilic acid, molecular crystals, and salts. *Journal of the American Chemical Society* 103: 1074–1079.

50 Ubba, E., Tao, Y., Yang, Z. et al. (2018). Organic mechanoluminescence with aggregation-induced emission. *Chemistry-An Asian Journal* 13: 3106–3121.

51 Yang, J., Gao, X., Xie, Z. et al. (2017). Elucidating the excited state of mechanoluminescence in organic luminogens with room-temperature phosphorescence. *Angewandte Chemie International Edition* 56: 15299–15303

52 Xie, Z., Xu, Y., Zhang, X. et al. (2024). Isostructural doping for organic persistent mechanoluminescence. *Nature Communication* 15: 3668.

53 Pope, M., Kallmann, H.P., and Magnante, P. (1963). Electroluminescence in organic crystals. *The Journal of Physical Chemistry* 38: 2042–2043.

54 Yang, J., Huang, J., Li, Q., and Li, Z. (2016). Blue AIEgens: The approaches to control the intramolecular conjugation and the optimized performance of OLED devices. *Journal of Materials Chemistry C* 4: 2663–2684.

55 Han, P., Xia, E., Qin, A., and Tang, B.Z. (2022). Adjustable and smart AIEgens for nondoped blue and deep blue organic light-emitting diodes. *Coordination Chemistry Reviews* 473: 214843.

56 Endo, A., Sato, K., Yoshimura, K. et al. (2011). Efficient up-conversion of triplet excitons into a singlet state and its application for organic light emitting diodes. *Applied Physics Letters* 98: 083302.

57 Xu, S., Liu, T., Mu, Y. et al. (2015). An organic molecule with asymmetric structure exhibiting aggregation-induced emission, delayed fluorescence, and mechanoluminescence. *Angewandte Chemie International Edition* 54: 874–878.

58 Huan, J., Nie, H., Zeng, J. et al. (2017). Highly efficient nondoped OLEDs with negligible efficiency roll-off fabricated from aggregation-induced delayed fluorescence luminogens. *Angewandte Chemie International Edition* 56: 12971–12976.

59 Ju, H., Wang, K., Zhang, J. et al. (2017). 1,6- and 2,7-trans-β-Styryl substituted pyrenes exhibiting both emissive and semiconducting properties in the solid state. *Chemistry of Materials* 29: 3580–3588.

60 Hepp, A., Heil, H., Weise, W. et al. (2003). Light-emitting field-effect transistor based on a tetracene thin film. *Physical Review Letters* 91: 157406.

61 Fang, M., Yang, J., and Li, Z. (2022). Light emission of organic luminogens: Generation, mechanism and application. *Progress in Materials Science* 125: 100914.

62 Zhan, X., Zhang, J., Tang, S. et al. (2015). Pyrene fused perylene diimides: Synthesis, characterization and applications in organic filed-effect transistors and optical limiting with high performance. *Chemical Communications* 51: 7156–7159.

63 Franken, P.A., Hill, A.E., Peters, C.W., and Weinreich, G. (1961). Generation of optical harmonics. *Physics Review Letters* 7: 118–120.

64 Rentzepis, P.M. and Pao, Y. H. (1964). Later-induced optical second harmonic generation in organic crystals. *Applied. Physics Letters* 5: 156–158.

65 Davydov, B.L., Derkacheva, L.D., and Duina, V.V. (1970). Connection between charge transfer and laster second harmonic generation. *Soviet Journal of Experimental and Theoretical Physics Letters* 12: 16–18.

66 Robinson, B.H. and Dalton, L.R. (2000). Monte Carlo statistical mechanical simulations of the competition of intermolecular electrostatic and poling-field interactions in defining macroscopic electro-optic activity for organic chromophore/polymer materials. *Journal of Physics Chemistry A* 104: 4785–4795.

67 Tang, R., Zhou, S., Xiang, W. et al. (2015). New "X-type" second-order nonlinear optical (NLO) dendrimers: Fewer chromophore moieties and high NLO effects. *Journal of Material Chemistry C* 3: 4545–4552.

68 Tang, R., Zhou, S., Cheng, Z. et al. (2017). Janus second-order nonlinear optical dendrimers: Their controllable molecular topology and corresponding largely enhanced performance. *Chemical Science* 8: 340–347.

69 Li, Z., Li, Z., Di, C. et al. (2006). Structural control of the side-chain chromophores to achieve highly efficient nonlinear optical polyurethanes. *Macromolecules* 39: 6951–6961.

70 Li, Z., Li, Q., and Qin, J. (2011). Some new design strategies for second-order nonlinear optical polymers and dendrimers. *Polymer Chemistry* 2: 2723–2740.

71 Wu, W., Qin, J., and Li, Z. (2013). New design strategies for second-order nonlinear optical polymers and dendrimers. *Polymer* 54: 4351–4382.

72 Wu, W., Tang, R., Li, Q., and Li, Z. (2015). Functional hyperbranched polymers with advanced optical, electrical and magnetic properties. *Chemical Society Review* 44: 3997–4022.

73 Li, Z., Wu, W., Li, Q. et al. (2010). High generation second-order nonlinear optical dendrimers: Convenient synthesis from click chemistry, and the increasing trend of NLO effects. *Angewandte Chemie International Edition* 49: 2763–2767.

74 Singh, R., Giussani, E., Mro'z, M.M. et al. (2014). On the role of aggregation effects in the performance of perylene-diimide based solar cells. *Organic Electronics* 15: 1347–1361.

75 Yuan, J., Zhang, Y., Zhou, L. et al. (2019). Single-junction organic solar cell with over 15% efficiency using fused-ring acceptor with electron-deficient core. *Joule* 3: 1140–1151.

76 Zhang, H., Tian, C., Zhang, Z. et al. (2023). Concretized structural evolution supported assembly-controlled film-forming kinetics in slot-die coated organic photovoltaics. *Nature Communications* 14: 6312.

77 Mahadevan, S., Liu, T., Pratik, S.M. et al. (2024). Assessing intra- and inter-molecular charge transfer excitations in non-fullerene acceptors using electroabsorption spectroscopy. *Nature Communications* 15: 2393.

78 Jo, B., Park, H., Kamaraj, E. et al. (2021). Synergistic effect of excited state property and aggregation characteristic of organic semiconductor on efficient hole-transportation in perovskite device. *Advanced Functional Materials* 31: 2007180.

79 Zhang, D., Wang, X., Tian, T. et al. (2023). Multi-functional buried interface engineering derived from in-situ-formed 2D perovskites using π-conjugated liquid-crystalline molecule with aggregation-induced emission for efficient and stable NiOx-based inverted perovskite solar cell. *Chemical Engineering Journal* 469: 143789.

80 Yang, P., Zhang, Q., Zhang, Y. et al. (2023). Aggregation triggers red/near-infrared light hydrogen production of organic dyes with high efficiency. *ACS Catalysis* 13: 3723–3734.

81 He, T., Zhang, Y., Zhang, H. et al. (2023). Aggregation-induced structural symmetry breaking promotes charge separation for efficient photocatalytic hydrogen production. *ChemSusChem* 16: e202300500.

82 Yuan, X., Yang, K., Grazon, C. et al. (2024). Tuning the aggregates of thiophene-based trimers by methyl side-chain engineering for photocatalytic hydrogen evolution. *Angewandte Chemie International Edition* 63: e202315333.

83 Albrecht, M. (2007). Supramolecular chemistry-general principles and selected examples from anion recognition and metallosupramolecular chemistry. *Naturwissenschaften* 94: 951–956.

84 Konig, B. (1995). Molecular recognition. The principle and recent chemical examples. *Journal fur praktische Chemie Chemiker-Zeitung* 337: 339–346.

85 Goshe, A.J., Steele, I.M., Ceccarelli, C. et al. (2002). Supramolecular recognition: On the kinetic lability of thermodynamically stable host-guest association complexes. *Proceedings of the National Academy of Sciences of the United States of America* (Vol. 99, pp. 4823–4829).

86 Liu, X., Liao, Q., Yang, J. et al. (2023). Unveiling one-to-one correspondence between excited triplet states and determinate interactions by temperature-controllable blue-green-yellow afterglow. *Angewandte Chemie International Edition* 62: e202302792.

87 Lu, H., Li, D., Liu, W. et al. (2024). Designing A-D-A type fused-ring electron acceptors with a bulky 3D substituent at the central donor core to minimize non radiative losses and enhance organic solar cell efficiency. *Angewandte Chemie International Edition* 63: e202407007.

88 Karas, L.J., Wu, C.H., Das, R., and Wu, J.I.C. (2020). Hydrogen bond design principles. *Wiley Interdisciplinary Reviews-Computational Molecular Science* 10: e1477.

89 Sandorfy, C. (2006). Hydrogen bonding: How much anharmonicity? *Journal of Molecular Structure* 790: 50–54.

90 Liu, Y., Wang, L., Zhao, L. et al. (2024). Multiple hydrogen bonding driven supramolecular architectures and their biomedical applications. *Chemical Society Reviews* 53: 1592–1623.

91 Lutz, H.D. (2003). Structure and strength of hydrogen bonds in inorganic solids. *Journal of Molecular Structure* 646: 227–236.

92 Jansze, S.M., Felipe, A.M., Storey, J.M.D. et al. (2015). A twist-bend nematic phase driven by hydrogen bonding. *Angewandte Chemie International Edition* 54: 643–646.

93 Song, P. and Wang, H. (2020). High-performance polymeric materials through hydrogen bond cross-linking. *Advanced Materials* 32: 1901244.

94 Li, F., Deng, X., Shi, Z. et al. (2023). Hydrogen-bond-bridged intermediate for perovskite solar cells with enhanced efficiency and stability. *Nature Photonics* 17: 478–484.

95 Lin, L., Jones, T.W., Yang, T.C.J. et al. (2024). Hydrogen bonding in perovskite solar cells. *Matter* 7: 38–58.

96 Hwang, D. and Schlenker, C.W. (2021). Photochemistry of carbon nitrides and heptazine derivatives. *Chemical Communications* 57: 9330–9353.

97 Nalaoh, P., Sungworawongpana, N., Chasing, P. et al. (2021). A dimeric π-stacking of anthracene inducing efficiency enhancement in solid-state fluorescence and non-doped deep-blue triplet-triplet annihilation organic light emitting diodes. *Advanced Optical Materials* 9: 2100500.

98 Liu, C., Zhou, Y., Li, T. et al. (2023). Efficient nondoped deep-blue electrofluorescence benefiting from structural hindrance and regular C-H...π stacking. *Chemical Engineering Journal* 471: 144505.

99 Ma, L., Zhang, S., Hou, J. et al. (2021). Completely non-fused electron acceptor with 3D-interpenetrated crystalline structure enables efficient and stable organic solar cell. *Nature Communications* 12: 5093.

100 George, S.J. and Ajayaghosh, A. (2005). Self-assembled nanotapes of oligo(p-phenylene vinylene)s: Sol-gel controlled optical properties in fluorescent p-electronic gels. *Chemistry-A European Journal* 11: 3217–3227.

101 Hunter, C.A. and Sanders, J.K.M. (1990). The nature of π-π interactions. *Journal of the American Chemical Society* 112: 5525–5534.

102 Rashkin, M.J. and Waters, M.L. (2002). Unexpected substituent effects in offset π-π stacked interactions in water. *Journal of the American Chemical Society* 124: 1860.

103 Chen, Z., Lohr, A., Saha-Moller, C.R., and Würthner, F. (2009). Self-assembled π-stacks of functional dyes in solution: Structural and thermodynamic features. *Chemical Society Reviews* 38: 564–584. 2009

104 Kashida, J., Shoji, Y., Taka, H. et al. (2024). Peripheral fusion of carbon-based aromatic rings to B_4N_4-heteropentalene leading to close π-stacking in the solid state. *Chemistry-A European Journal* 2: e202402862.

105 Cho, H.J., Kim, S.W., Kim, S. et al. (2020). Suppressing π-π stacking interactions for enhanced solid-state emission of flat aromatic molecules via edge functionalization with picket-fence-type groups. *Journal of Materials Chemistry C* 8: 17289–17296.

106 Ma, L., Cui, Y., Zhang, J. et al. (2023). High-efficiency and mechanically robust all-polymer organic photovoltaic cells enabled by optimized fibril network morphology. *Advanced Materials* 35: 2208926.

107 Du, Z., Deng, S., Bei, Y. et al. (2014). Adsorption behavior and mechanism of perfluorinated compounds on various adsorbents-a review. *Journal of Hazardous Materials* 274: 443–454.

108 Liu, Z., Dai, X., Sun, Y., and Liu, Y. (2020). Organic supra molecular aggregates based on water-soluble cyclodextrins and calixarenes. *Aggregate* 1: 31–44.

109 Peng, Q., Ma, H., and Shuai, Z. (2021). Theory of long-lived room-temperature phosphorescence in organic aggregates. *Accounts of Chemical Research* 54: 940–949.

110 Xu, Y., Yao, H., Ma, L. et al. (2020). Tuning the hybridization of local exciton and charge-transfer states in highly efficient organic photovoltaic cells. *Angewandte Chemie International Edition* 59: 9004–9010.

111 Ma, L., Yao, H., Wang, J. et al. (2021). Impact of electrostatic interaction on bulk morphology in efficient donor-acceptor photovoltaic blends. *Angewandte Chemie International Edition* 60: 15988–15994.

112 Zhuang, W., Wang, Y., Cui, P. et al. (2019). Applications of π-π stacking interactions in the design of drug-delivery systems. *Journal of Controlled Release* 294: 311–326.

113 Yu, Y., Wang, C., Wei, Y. et al. (2019). Halogen-containing TPA-based luminogens: different molecular packing and different mechanoluminescence. *Advanced Optical Materials* 7: 1900505.

114 Wang, C., Yu, Y., Yuan, Y. et al. (2020). Heartbeat-sensing mechanoluminescent device based on a quantitative relationship between pressure and emissive intensity. *Matter* 2: 181–193.

115 Biedermann, F., Nau, W.M., and Schneider, H. (2014). The hydrophobic effect revisited-studies with supramolecular complexes imply high-energy water as a noncovalent driving force. *Angewandte Chemie International Edition* 53: 11158–11171.

116 Ma, X. and Tian, H. (2014). Stimuli-responsive supramolecular polymers in aqueous solution. *Accounts of Chemical Research* 47: 1971–1981.

117 Trouillas, P., García, J.C.S., Freitas, V.D. et al. (2016). Stabilizing and modulating color by copigmentation: insights from theory and experiment. *Chemical Reviews* 116: 4937–4982.

118 Jiang, H., Duan, L., Ren, X., and Gao, G. (2019). Hydrophobic association hydrogels with excellent mechanical and self-healing properties. *European Polymer Journal* 112: 660–669.

119 Liptrot, D.J. and Power, P.P. (2017). London dispersion forces in sterically crowded inorganic and organometallic molecules. *Nature Reviews Chemistry* 1: 0004.

120 Berland, K., Cooper, V.R., Lee, K. et al. (2015). Van der Waals forces in density functional theory: a review of the vdW-DF method. *Reports on Progress in Physics* 78: 066501.

121 Söderhjelm, P. and Ryde, U. (2009). How accurate can a force field become? A polarizable multipole model combined with fragment-wise quantum-mechanical calculations. *Journal Of Physical Chemistry A* 113: 617–627.

122 Wang, J., Wang, C., Gong, Y. et al. (2018). Bromine-substituted fluorene: Molecular structure, Br-Br interactions, room-temperature phosphorescence, and tricolor triboluminescence. *Angewandte Chemie International Edition* 57: 16821–16826.

123 Shi, H., Yao, W., Ye, W. et al. (2022). Ultralong organic phosphorescence: from material design to applications. *Accounts of Chemical Research* 55: 3445–3459.

124 Khasbaatar, A., Lee, Z.X.J., Campillo-Alvarado, G. et al. (2023). From solution to thin film: molecular assembly of π-conjugated systems and impacton (Opto) electronic properties. *Chemical Reviews* 123: 8395–8487.

125 Yoon, S., Chung, J.W., Gierschner, J. et al. (2010). Multistimuli two-color luminescence switching via different slip-stacking of highly fluorescent molecular sheets. *Journal of the American Chemical Society* 132: 13675–13683.

126 Yagai, S., Seki, T., Karatsu, T. et al. (2008). Transformation from H- to J-aggregated perylene bisimide dyes by complexation with cyanurates. *Angewandte Chemie International Edition* 47: 3367–3371.

127 Wang, J. and Zhan, X. (2021). Fused-ring electron acceptors for photovoltaics and beyond. *Accounts of Chemical Research* 54: 132–143.

128 Chang, Y., Zhang, J., Chen, Y. et al. (2021). Achieving efficient ternary organic solar cells using structurally similar non-fullerene acceptors with varying flanking side chains. *Advanced Energy Materials* 11: 2100079.

129 Yang, J., Gao, H., Wang, Y. et al. (2019). The odd-even effect of alkyl chain inorganic room temperature phosphorescence luminogens and the corresponding in vivo imaging. *Materials Chemistry Frontiers* 3: 1391–1397.

130 Xie, Y., Ge, Y., Peng, Q. et al. (2017). How the molecular packing affects the room temperature phosphorescence in pure organic compounds: Ingenious molecular design, detailed crystal analysis, and rational theoretical calculations. *Advanced Materials* 29: 1606829.

131 Wang, J., Xue, P., Jiang, Y. et al. (2022). The principles, design and applications of fused- ring electron acceptors. *Nature Reviews Chemistry* 6: 615.

132 Hunter, C.A. (2004). Quantifying intermolecular interactions: guidelines for the molecular recognition toolbox. *Angewandte Chemie International Edition* 43: 5310–5324.

133 Fang, M., Yang, J., Xiang, X. et al. (2018). Unexpected room-temperature phosphorescence from a non-aromatic, low molecular weight, pure organic molecule through the intermolecular hydrogen bond. *Materials Chemistry Frontiers* 2: 2124–2129.

134 Sagara, Y., Yamane, S., Mitani, M. et al. (2016). Mechanoresponsive luminescent molecular assemblies: An emerging class of materials. *Advanced Materials* 28: 1073–1095.

135 Yuan, Y., Yuan, W., and Chen, Y. (2016). Recent advances in mechanoluminescent polymers. *Science China Materials* 59: 507–520.

136 Ware, T.H., McConney, M.E., Wie, J.J. et al. (2015). Voxelated liquid crystal elastomers. *Science* 347: 6225.

137 Hou, W., Wang, J., and Lv, J. (2023). Bioinspired liquid crystalline spinning enables scalable fabrication of high-performing fibrous artificial muscles. *Advanced Materials* 35: 2211800.

138 Srivastava, A.K., Zhang, W., Schneider, J. et al. (2019). Luminescent down-conversion semiconductor quantum dots and aligned quantum rods for liquid crystal displays. *Advanced Science* 6: 1901345.

139 Kumar, K., Knie, C., Bleger, D. et al. (2016). A chaotic self-oscillating sunlight-driven polymer actuator. *Nature Communications* 7: 11975.

140 Liang, X., Guo, Z., Wei, H. et al. (2018). Selective photooxidation of sulfides mediated by singlet oxygen using visible-light-responsive coordination polymers. *Chemical Communication* 54: 13002–13005.

141 Zhang, X., Ma, X., Wang, K. et al. (2018). Recent advances in cyclodextrin-based light-responsive supramolecular systems. *Macromolecular Rapid Communications* 39: 1800142.

142 Xiao, X., Zheng, W., Zhao, Y., and Li, C. (2023). Visible light responsive spiropyran derivatives based on dynamic coordination bonds. *Chinese Chemical Letters* 34: 107457.

143 Gmelch, M., Thomas, H., Fries, F., and Reineke, S. (2019). Programmable transparent organic luminescent tags. *Science Advances* 5: e7310.

144 Bisoyi, H.K. and Li, Q. (2022). Liquid crystals: Versatile self-organized smart soft materials. *Chemical Reviews* 122: 4887–4926.

145 Meredith, G.R., VanDusen, J., and Williams, D. J. (1982). Optical and nonlinear optical characterization of molecularly doped thermotropic liquid crystalline polymers. *Macromolecules* 15: 1385–1389.

146 Liu, Z., Bisoyi, H.K., Huang, Y. et al. (2022). Thermo- and mechanochromic camouflage and self-healing in biomimetic soft actuators based on liquid crystal elastomers. *Angewandte Chemie International Edition* 61: e202115755.

147 Wu, J., Guo, Y., Qi, J. et al. (2023). Multi-stimuli responsive luminescence and domino phase transition of hybrid copper halides with nonlinear optical switching behavior. *Angewandte Chemie International Edition* 62: e202301937.

148 Huang, Y., Hsiang, E., Deng, M., and Wu, S. (2022). Mini-LED, micro-LED and OLED displays: present status and future perspectives. *Light: Science & Applications* 9: 105.

149 Mizoshita, N., Tani, T., and Inagaki, S. (2012). Isothermally reversible fluorescence switching of a mechanochromic perylene bisimide dye. *Advanced Materials* 24: 3350–3355.

150 Wang, W., Li, H., Zhang, J. et al. (2020). Design of highly efficient deep-blue organic afterglow through guest sensitization and matrices rigidification. *Nature Communications* 11: 4802.

151 Kakuta, T., Yamagishi, T., and Ogoshi, T. (2018). Stimuli-responsive supramolecular assemblies constructed from Pillar[n]arenes. *Accounts of Chemical Research* 51: 1656–1666.

152 Dong, S., Zheng, B., Wang, F., and Huang, F. (2014). Supramolecular polymers constructed from macrocycle-based host-guest molecular recognition motifs. *Accounts of Chemical Research* 47: 1982–1994.

153 Huang, L., Chen, B., Zhang, X. et al. (2018). Proton-activated “off-on” room-temperature phosphorescence from purely organic thioethers. *Angewandte Chemie International Edition* 57: 16046–16050.

154 Tian, Y., Gong, Y., Liao, Q. et al. (2020). Adjusting organic room-temperature phosphorescence with orderly stimulus responsive molecular motion in crystals. *Cell Reports Physical Science* 1: 100052.

155 Di, B.H. and Chen, Y.L. (2018). Recent progress in organic mechanoluminescent materials. *Chinese Chemical Letters* 29: 245–251.

156 Tu, L., Xie, Y., and Li, Z. (2022). Advances in pure organic mechanoluminescence materials. *The Journal of Physical Chemistry Letters* 13: 5605–5617.

157 Xie, Y. and Li, Z. (2020). The development of mechanoluminescence from organic compounds: Breakthrough and deep insight. *Materials Chemistry Frontiers* 4: 317–331.

158 Huang, L., Qian, C., and Ma, Z. (2020). Stimuli-responsive purely organic room-temperature phosphorescence materials. *Chemistry-A European Journal* 26: 11914–11930.

159 Xue, S., Qiu, X., Sun, Q., and Yang, W. (2016). Alkyl length effects on solid-state fluorescence and mechanochromic behavior of small organic luminophores. *Journal of Materials Chemistry C* 4: 1568–1578.

160 Liao, Q., Li, Q., and Li, Z. (2021). Substituent Effects in organic luminogens with room temperature phosphorescence. *ChemPhotoChem* 5: 694–701.

161 Huang, A., Li, Q., and Li, Z. (2022). Molecular uniting set identified characteristic (MUSIC) of organic optoelectronic material. *Chinese Journal of Chemistry* 40: 2356–2370.

162 Singh, M., Liu, K., Qu, S. et al. (2021). Recent advances of cocrystals with room temperature phosphorescence. *Advanced Optical Materials* 9: 2002197.

163 Sun, S., Wang, J., Ma, L. et al. (2021). A universal strategy for organic fluid phosphorescence materials. *Angewandte Chemie International Edition* 60: 18557–18560.

164 Liao, P., Huang, J., Yan, Y., and Tang, B.Z. (2021). Clusterization-triggered emission (CTE): one for all, all for one. *Materials Chemistry Frontiers* 5: 6693–6717.

165 Liu, J., Lv, J., Liu, Z. et al. (2023). Enhanced interfacial H-bond networks promote glycan-glycan recognition and interaction. *ACS Applied Materials & Interfaces* 15: 17592–17600.

166 Jing, L., Li, P., Li, Z. et al. (2025). Influence of π-π interactions on organic photocatalytic materials and their performance. *Chemical Society Reviews* 54: 2054–2090.

167 Zangi, R., Hagen, M., Berne B. J. (2007). Effect of ions on the hydrophobic interaction between two plates. *Journal of the American Chemical Society* 129: 4678–4686.

168 Sanchez-Fernandez, A., Jackson, A. J., Prévost, S. F. et al. (2021). Long-range electrostatic colloidal interactions and specific ion effects in deep eutectic solvents. *Journal of the American Chemical Society* 143: 14158–14168.

169 Chen, H., Zhang, W., Li M. et al. (2020). Interface engineering in organic field-effect transistors: Principles, applications, and perspectives. *Chemical Reviews* 120: 2879–2949.

170 Liao, Y.-J., Hsieh, Y.-C., Chen, J.-T. et al. (2023). Large-area nonfullerene organic photovoltaic modules with a high certified power conversion efficiency. *ACS Applied Materials & Interfaces* 15: 7911–7918.

171 Zhou, L., Mu, S., Ma, L. et al. (2024). Achieving ultralong room-temperature phosphorescence via single-bond locking planarization strategy. *ACS Materials Letters* 6: 5384–5391.

172 Liang, Y.-C., Cao, Q., Liu, K.-K. et al. (2021). Phosphorescent carbon-nanodots-assisted Förster resonant energy transfer for achieving red afterglow in an aqueous solution. *ACS Nano* 15: 16242–16254.

173 Liu, S., Lin, P., Wu, M. et al. (2022). Organic dyes with multi-branched structures for highly efficient photocatalytic hydrogen evolution under visible-light irradiation. *Applied Catalysis B: Environmental* 309: 121257.

174 Zhou, L., Xiang, H.-Y., Shen S. et al. (2014). High-performance flexible organic light-emitting diodes using embedded silver network transparent electrodes. *ACS Nano* 8: 12796–12805.

175 Yoon, S., Sim, K. M., Chung, D. S. (2019). Bifunctional etalon-electrode to realize high-performance color filter free image sensor. *ACS Nano* 13: 2127–2135.

176 Shao, W. and Kim, J. (2022). Metal-free organic phosphors toward fast and efficient room-temperature phosphorescence. *Accounts of Chemical Research* 55: 1573–1585.

177 Garain, S., Ansari, S.N., Kongasseri, A.A. et al. (2022). Room temperature charge-transfer phosphorescence from organic donor–acceptor co-crystals. *Chemical Science* 13: 10011–10019.

178 Azzali, A., d'Agostino, S., Capacci, M. et al. (2022). Assembling photoactive materials from polycyclic aromatic hydrocarbons (PAHs): Room temperature phosphorescence and excimer-emission in co-crystals with 1, 4-diiodotetrafluorobenzene. *CrystEngComm* 24: 5748–5756.

179 Yang, X.G., Qin, W.J., Zhang, J.R. et al. (2021). Room-temperature phosphorescent co-crystal showing direct white light and photo-electric conversion. *Frontiers in Chemistry* 9: 765374.

180 Xu, Z., He, Y., Shi, H., and An, Z. (2022). Room-temperature phosphorescence materials from crystalline to amorphous state. *SmartMat* 4: e1139.

181 Jiang, Y., Shi, Y., Hu, D. et al. (2022). Insight into isomeric effect on the photoluminescence and mechanoluminescence of cyanostilbene derivatives. *The Journal of Physical Chemistry Letters* 13: 7681–7688.

182 Liu, Z., Tian, Y., Yang, J. et al. (2022). Direct demonstration of triplet excimer in purely organic room temperature phosphorescence through rational molecular design. *Light: Science & Applications* 11: 142.

183 Gong, Y., He, S., Li, Y. et al. (2020). Partially controlling molecular packing to achieve off-on mechanochromism through ingenious molecular design. *Advanced Optical Materials* 8: 1902036.

184 Miao, Y., Liu, S., Ma, L. et al. (2021). Ultralong and color-tunable room-temperature phosphorescence based on commercial melamine for anticounterfeiting and information recognition. *Analytical Chemistry* 93: 4075–4083.

185 Li, S., Fu, L., Xiao, X. et al. (2021). Regulation of Thermally activated delayed fluorescence to room-temperature phosphorescent emission channels by controlling the excited-states dynamics via J- and H-aggregation. *Angewandte Chemie International Edition* 60: 18059–18064.

186 Lucenti, E., Forni, A., Botta, C. et al. (2019). Intrinsic and extrinsic heavy-atom effects on the multifaceted emissive behavior of cyclic triimidazole. *Chemistry-A European Journal* 25: 2452–2456.

187 Gao, J., Zhao, Y., You, X. et al. (2022). Theoretical search of a simple characteristic for long-lived organic room-temperature phosphorescence materials with H aggregation. *Journal of Materials Chemistry C* 10: 5425–5432.

188 Godumala, M., Kumar, A.V., and Chandrasekar, R. (2021). Room-temperature phosphorescent organic materials for optical waveguides. *Journal of Materials Chemistry C* 9: 14115–14132.

189 Pan, S., Chen, Z., Zheng, X. et al. (2018). Ultralong room-temperature phosphorescence from supramolecular behavior via intermolecular electronic coupling in pure organic crystals. *The Journal of Physical Chemistry Letters* 9: 3939–3945.

190 Nielsen, C.B., Holliday, S., Chen, H.Y. et al. (2015). Non-fullerene electron acceptors for use in organic solar cells. *Accounts of Chemical Research* 48: 2803–2812.

191 Müller-Buschbaum, P. (2014). The active layer morphology of organic solar cells probed with grazing incidence scattering techniques. *Advanced Materials* 26: 7692–7709.

192 Guo, J., Yang, C., and Zhao, Y. (2022). Long-lived organic room-temperature phosphorescence from amorphous polymer systems. *Accounts of Chemical Research* 55: 1160–1170.

193 Su, Y., Zhang, Y., Wang, Z. et al. (2020). Excitation-dependent long-life luminescent polymeric systems under ambient conditions. *Angewandte Chemie International Edition* 59: 9967–9971.

194 Mishra, P.K., Yoo, C.M., Hong, E., and Rhee, H.W. (2020). Photo-crosslinking: An emerging chemical tool for investigating molecular networks in live cells. *Chembiochem* 21: 924–932.

195 Gu, L., Wu, H., Ma, H. et al. (2020). Color-tunable ultralong organic room temperature phosphorescence from a multicomponent copolymer. *Nature Communications* 11: 944.

196 Gao, L., Zhang, Y., Chen, X. et al. (2021). Water-induced blue-green variable nonconventional ultralong room temperature phosphorescence from cross-linked copolymers via click chemistry. *Advanced Optical Materials* 9: 2101284.

197 Kwon, S., Yu, K., Kweon, K. et al. (2014). Template-mediated nano-crystallite networks in semiconducting polymers. *Nature Communications* 5: 4183.

198 Kwon, S., Kim, J., Kim, G. et al. (2015). Organic single-crystal semiconductor films on a millimeter domain scale. *Advanced Materials* 27: 6870–6877.

199 Singh, I. (2021). Luminance behaviour of PLEDs with integration of nanomaterials. *Displays* 69: 102046.

200 Low, J.Y., Merican, Z., Merican, A., and Hamza, M.F. (2019). Polymer light emitting diodes (PLEDs): An update review on current innovation and performance of material properties. *Materials Today: Proceedings* (Vol. 16, pp. 1909–1918).

201 Meng, F., Jia, P., Li, X. et al. (2022). Efficient perovskite solar cells with a cui-modified polymer hole-transport layer. *ACS Applied Energy Materials* 5: 11034–11041.

202 Ding, Y., Guo, Q., Geng, Y. et al. (2022). A low-cost hole transport layer enables $CsPbI_2Br$ single-junction and tandem perovskite solar cells with record efficiencies of 17.8% and 21.4%. *Nano Today* 46: 101586.

203 Ruderer, M.A., Guo, S., Meier, R. et al. (2011). Solvent-induced morphology in polymer-based systems for organic photovoltaics. *Advanced Functional Materials* 21: 3382–3391.

204 Du, X., Zeng, Q., Zhang, H., and Yang, B. (2017). Hybrid solar cells from aqueous polymers and colloidal nanocrystals. *Chinese Journal of Chemistry* 35: 551–561.

205 Li, Z. and Li, Z. (2017). Recent advances in second order nonlinear optical polymers with dendritic structure. *Acta Polymerica Sinica* 2: 155–177.

206 Liu, J., Ouyang, C., Huo, F. et al. (2020). Progress in the enhancement of electro-optic coefficients and orientation stability for organic second-order nonlinear optical materials. *Dyes and Pigments* 181: 108509.

207 Wei, M.K., Lin, C.W., Yang, C.C. et al. (2010). Emission characteristics of organic light-emitting diodes and organic thin-films with planar and corrugated structures. *International Journal of Molecular Sciences* 11: 1527–1545.

208 Yan, D.H. (2005). Morphology of thin-film of semi-crystalline organic semiconductors and influences on lifetime of OLED devices. *Journal of Molecular Science* 21: 21–25.

209 Lim, E.L., Yap, C.C., Teridi, M.A.M. et al. (2016). A review of recent plasmonic nanoparticles incorporated P3HT: PCBM organic thin film solar cells. *Organic Electronics* 36: 12–28.

210 Kosco, J., Bidwell, M., Cha, H. et al. (2020). Enhanced photocatalytic hydrogen evolution from organic semiconductor heterojunction nanoparticles. *Nature Materials* 19: 559–565.

211 Yang, H., Li, C., Liu, T. et al. (2023). Packing-induced selectivity switching in molecular nanoparticle photocatalysts for hydrogen and hydrogen peroxide production. *Nature Nanotechnology* 18: 307–315.

212 Li, H., Kim, H., Xu, F. et al. (2022). Activity-based NIR fluorescent probes based on the versatile hemicyanine scaffold: Design strategy, biomedical applications, and outlook. *Chemical Society Reviews* 51: 1836–1836.

213 Chen, C., Chi, Z., Chong, K.C. et al. (2020). Carbazole isomers induce ultralong organic phosphorescence. *Nature Materials* 20: 175–180.

214 Ren, J., Wang, Y., Tian, Y. et al. (2021). Force-induced turn-on persistent room-temperature phosphorescence in purely organic luminogen. *Angewandte Chemie International Edition* 60: 12335–12340.

215 Li, S., Xie, Y., Li, A. et al. (2021). Different molecular conformation and packing determining mechanochromism and room-temperature phosphorescence. *Science China Materials* 64: 2813–2823.

216 Liao, Q., Gao, Q., Wang, J. et al. (2020). 9, 9-Dimethylxanthene derivatives with room-temperature phosphorescence: Substituent effects and emissive properties. *Angewandte Chemie International Edition* 59: 9946–9951.

217 Wang, J., Chai, Z., Wang, J. et al. (2019). Mechanoluminescence or room-temperature phosphorescence: Molecular packing-dependent emission response. *Angewandte Chemie International Edition* 58: 17297–17302.

218 Chai, Z., Wang, C., Wang, J. et al. (2017). Abnormal room temperature phosphorescence of purely organic boron-containing compounds: The relationship between the emissive behavior and the molecular packing, and the potential related applications. *Chemical Science* 8: 8336–8344.

219 Fan, Y., Liu, S., Wu, M. et al. (2022). Mobile phone flashlight-excited red afterglow bioimaging. *Advanced Materials* 34: 2201280.

220 Karakuş, E., Erdemir, E., Demirbilek, N., and Liv, L. (2021). Colorimetric and electrochemical detection of SARS-CoV-2 spike antigen with a gold nanoparticle-based biosensor. *Analytica Chimica Acta* 1182: 338939.

221 Pramanik, A., Gao, Y., Patibandla, S. et al. (2021). The rapid diagnosis and effective inhibition of coronavirus using spike antibody attached gold nanoparticles. *Nanoscale Advances* 3: 1588–1596.

222 Wang, C., Cheng, X., Liu, L. et al. (2021). Ultrasensitive and simultaneous detection of two specific SARS-CoV-2 antigens in human specimens using direct/enrichment dual-mode fluorescence lateral flow immunoassay. *ACS Applied Materials & Interfaces* 13: 40342–40353.

223 Liu, J., Mao, J., Hou, M. et al. (2022). A rapid SARS-CoV-2 nucleocapsid protein profiling assay with high sensitivity comparable to nucleic acid detection. *Analytical Chemistry* 94: 14627–14634.

224 Hu, R., Liao, T., Ren, Y. et al. (2022). Sensitively detecting antigen of SARS-CoV-2 by NIR-II fluorescent nanoparticles. *Nano Research* 15: 7313–7319.

225 Porte, L., Legarraga, P., Iruretagoyena, M. et al. (2021). Evaluation of two fluorescence immunoassays for the rapid detection of SARS-CoV-2 antigen—new tool to detect infective COVID-19 patients. *PeerJ* 9: e10801.

2

The Molecular Engineering and Fabrication Processes for Molecular Aggregates

Yujie Yang, Boxi Wu, Wentao Yuan, Shiyue Tang, Panpan Qiao, Yifan Niu, Peidong Xie, Ruixing Wang, Yan Gao, Yuexin Li, Wanni Yao, Kai Wang, Qianqian Li, and Zhen Li**

Department of Chemistry, Wuhan University, Wuhan, China

*Corresponding authors: liqianqian@whu.edu.cn, lizhen@whu.edu.cn

2.1 Crystal Engineering

Compared to isolated organic molecules, organic molecules in the aggregated state exhibit the unique characteristics of extended assemblies with distinct molecular packing arrangements.[1] Organic crystals, as typical molecular aggregates, have been formed by various intermolecular interactions, including dispersion forces, electrostatic interactions, and hydrogen bonding. These crystals show great promise in a wide range of applications, such as organic field-effect transistors (OFETs),[2–5] organic light-emitting diodes (OLEDs),[6–9] nonlinear optical (NLO),[10–13] and organic solar cells (OSCs),[14–16] due to their structural defect-free, long-range ordered structures, excellent thermal and optical stability, and high carrier mobility. Moreover, organic crystals serve as ideal models for research due to their high purity and low defect density,[17] making them invaluable for investigating structure-performance relationships,[18] and intrinsic mechanisms, with the aid of theoretical calculations and simulations.[19–21] The growth of these crystals plays a crucial role in determining their properties, and the size of crystals significantly influences the determination of their crystal structure.

Crystal engineering, also the design of molecular solids, is the utilization of intermolecular interactions to assemble building blocks of neutral or ionic states into functional solid structures. These interactions can regulate both the nucleation and crystallization processes during crystal growth, and the regulation of crystal morphology and molecular arrangement can be achieved by combining the control of relevant thermodynamic and kinetic parameters. During the crystal growth process, certain defects are possibly formed by chemical impurities and structural defects caused by steps on the crystal surface, etc. Therefore, optimization of crystal growth methods and conditions is essential to culture high-quality crystals.

Crystal growth technology uses the physicochemical properties of a substance to control the phase transition process and obtain crystals with certain structure, size, shape, and property.[22] The history of people engaged in crystal growth can be traced back to ancient times. Throughout the long history, our ancestors have mastered the method of obtaining salt from seawater. Until the beginning of the twentieth century, modern science and technology were continuously used to control the crystal growth processes with evolution from skill to science. Especially since the 1950s, the development of semiconductor materials represented by monocrystalline silicon has driven the development of crystal growth theoretical research and technology.

The following conditions are required for the preparation of intact crystals.[23] First, the temperature of the culturing system should be controlled uniformly to prevent local overcooling or overheating, which may affect the nucleation and growth of the crystals. Second, the crystallization process should be as slow as possible to prevent the appearance of spontaneous nucleation, because once spontaneous nucleation occurs, many fine pins will be generated, which will hinder the growth of the crystals. Third, the cooling rate should be matched with the nucleation and growth rate of the crystals so that the crystals can grow uniformly. In recent years, there has been rapid development in the preparation methods for organic crystals. Crystal growth methods can be grouped into three main categories, based on the type of their parent phase, including solution phase growth,[24–26] vapor phase growth, and melt growth. With the change of control conditions, dozens of crystal growth techniques have been proposed.

2.1.1 Organic Crystals Grown from Solution

The growth of organic crystals from solution is a critical process in the development of opto-electronic materials, as the morphology and structural properties of these crystals strongly influence their performance in applications. Several techniques are commonly employed for crystallizing organic compounds, which generally rely on manipulating the concentration of solute and solvent conditions to induce crystallization, offering unique advantages depending on the desired crystal properties. The most widely used techniques are the solvent evaporation method, the slow cooling method, the vapor diffusion method, and the liquid diffusion method, with distinct characteristics that suit different material types and applications.

2.1.1.1 Solvent Evaporation Method

The solution evaporation method of crystal growth is one of the most widely used techniques for growing organic crystals, which has the longest history and the widest range of applications.[27] As shown in Figure 2.1a, the basic principle involves dissolving the solute in a suitable solvent to achieve a concentration close to saturation,

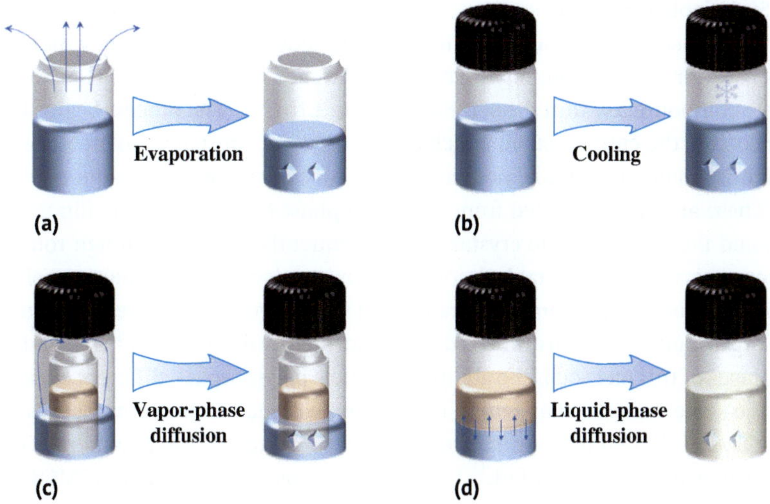

Figure 2.1 Schematic diagram of the principle of solution crystallization (a) Solvent evaporation method. Crystallization occurs as the solvent evaporates, causing the concentration of solute to increase. As the concentration reaches supersaturation, the solute begins to crystallize. (b) Slow cooling method. Crystallization takes place as the temperature decreases, resulting in a reduction in the solubility of the solute in the solution. As the solution cools, the solute reaches supersaturation and begins to crystallize. (c) Vapor diffusion method. Crystallization occurs when solvent 1 evaporates and diffuses into solvent 2. This diffusion leads to supersaturation of the solution, causing the solute to crystallize from the solution. (d) Liquid–liquid diffusion method. Crystallization happens when solvents 1 and 2 diffuse into each other. This diffusion decreases the solubility of the solution in solvent 2, leading to supersaturation and the subsequent crystallization of the solute.

but still less than its solubility. To initiate crystallization, various methods can be employed, such as gradually adding a non-solvent or slowly evaporating the good solvent. These methods gradually lead to saturation or even supersaturation of the solution, at which point the solute molecules begin to aggregate, nucleate, and crystallize.[28] Most organic compounds can be dissolved in common solvents such as methylene chloride, chloroform, and chlorobenzene, which allow the solute molecules to move freely in the solvent. When the concentration in the system exceeds the saturation concentration, the molecules gradually aggregate, dissolve, and diffuse, and these reversible processes allow the molecules to assemble into highly ordered aggregates. In general, relatively large single crystals can be obtained by using this method, which is widely used in organic opto-electronic materials.[27] In addition, some solvent molecules can be contained in the organic crystals, leading to changes in the crystal structure, which can affect their photophysical properties. Zhao, Tang, and Chi et al. demonstrated that solvent-containing phenothiazine derivative CP crystals, grown via the solvent evaporation method, exhibited enhanced phosphorescence.[29] The solvent molecules incorporated within the restricted molecular motions by intermolecular interactions improved the photophysical performance of organic crystalline materials.

2.1.1.2 Slow Cooling Method

Another widely used crystallization technique is the slow cooling method, which is primarily used for purification, relying on the difference in freezing points of various organic compounds within the solution (Figure 2.1b).[30,31] The process begins by gradually cooling the entire system, which causes different substances to precipitate in a sequential manner. Initially, non-crystalline substances precipitate out of the solution. These are then separated from the liquid phase by filtration. The filtrate is collected, and the desired solute crystals are subsequently obtained through rotary evaporation. This method is suitable for organic semiconductors with moderate solubility in organic solvents at room temperature but with large temperature variations, and large organic crystals, even mm-sized, can be formed by this method.[28] One of the key advantages of this method is the ability to produce polymorphs – different crystal forms of the same material that exhibit distinct optical and electronic properties.[24] Tian and Zhou et al. used the slow cooling method to crystallize two polymorphs of a tetraphenylene derivative APMOB.[32] These polymorphs, APMOB-1 and APMOB-2 exhibited different fluorescence properties due to variations in their molecular packing in the crystal lattice. APMOB-1 exhibited a weaker deep blue emission at 440 nm, while APMOB-2 gave a stronger cyan emission at 465 nm. These differences in fluorescent properties were due to their different molecular arrangement and conformation in the aggregated state, highlighting the importance of crystal growth methods in tuning the photophysical properties of opto-electronic materials.

2.1.1.3 Vapor Diffusion Method

The vapor diffusion method is employed to culture organic crystals of compounds that are highly soluble in certain organic solvents but exhibit poor solubility in others (Figure 2.1c).[28,33] The experiment setup consists of two glass containers: a larger container, such as a beaker with a lid, to hold the undesirable solvent, and a smaller container to hold the solution of the target compound dissolved in the good solvent. The smaller container is placed inside the larger one and sealed with a sealing film. The undesirable solvent, which has a low boiling point, is volatile and slowly diffuses from the large container into the smaller container, leading to supersaturation and subsequent crystallization of the target compound. This method is particularly advantageous for obtaining well-defined crystals and can also be used to obtain polymorphs with different molecular packing and photophysical properties. Zou and Ma et al. demonstrated the use of vapor diffusion to prepare two polymorphs, B-phase and G-phase, from a CN-DBS (cyano-substituted distyrylbenzene).[34] Quadrilateral-like crystal CN-DBS-B emitted blue fluorescence at 445 nm, and strip-like crystal CN-DBS-G exhibited green fluorescence at 503 nm. These distinct fluorescence characteristics were attributed to the molecular configuration in CN-DBS-G being almost planar, whereas the molecular configuration in 8-B was highly torsional. The molecules in CN-DBS-B were stacked side-to-side with a minimum contact distance of 3.00 Å. The molecules in 8-G were stacked face-to-face with a minimum contact distance of 3.48 Å.

2.1.1.4 Liquid–Liquid Diffusion Method

The liquid–liquid diffusion method involves two kinds of solvents with significantly different solubility for the target compound (Figure 2.1d).[28] The good and poor solvents must be miscible with each other, allowing for controlled diffusion between the two phases. Glass containers, such as vials or nucleation tubes, are typically employed for this method. The size of the container plays a critical role in crystallization, smaller containers have a reduced contact area, which in turn slows the diffusion rate, allowing for more gradual crystallization. The two-dimensional (2D) single crystals can be obtained by this method. Wang and Ye et al. demonstrated the liquid–liquid diffusion method by generating two 2D thin single crystals, yellow block crystal SADA-Y, and orange plate crystal SADA-O, using CH_2Cl_2/MeOH mixtures with ratios of 1:2 and 2:1, respectively.[35] The resultant SADA-Y exhibited a weak yellow emission at 565 nm, while SADA-O exhibited a stronger orange emission at 583 nm. Although both crystals of SADA-Y and SADA-O adopted similar end-to-end slip stacking modes in the aggregated state, their molecular conformations differed, contributing to the distinct emission characteristics.

2.1.2 Vapor Phase Growth Method

The vapor phase method is another crystal growth technique commonly used in the fabrication of high-performance opto-electronic devices (Figure 2.2).[28] The basic principle is that when solid molecules are heated to temperatures near their melting points, causing them to sublimate into the gas phase. In this high-energy gaseous state, the molecules move freely and reassemble themselves, eventually gathering and crystallizing into highly ordered crystals when they are cooled in a low-temperature region.[36,37] The experimental process involves placing the raw material in a high-temperature zone, where molecules with specific sublimation temperatures deposit or crystallize in different temperature regions during the heating process. This crystallization method is considered to be free from the induction of solvent molecules and can reflect the physicochemical properties of the molecules themselves.

2.1.3 Melt Growth

In addition to vapor phase growth, various melt-based crystal growth methods are also utilized in the development of opto-electronic materials. The melt method involves heating to exceed their melting point and subsequently cooling in a controlled manner to achieve directional crystallization,[38] which is pivotal for preparing large-sized and geometrically specific single crystals. Key techniques include the Bridgman–Stockbarger method,[39–41] Czochralski process (commonly referred to as crystal pulling),[42–44] subcooled melting,[45,46] zone melting,[47,48] and laser-heated pedestal growth.[49]

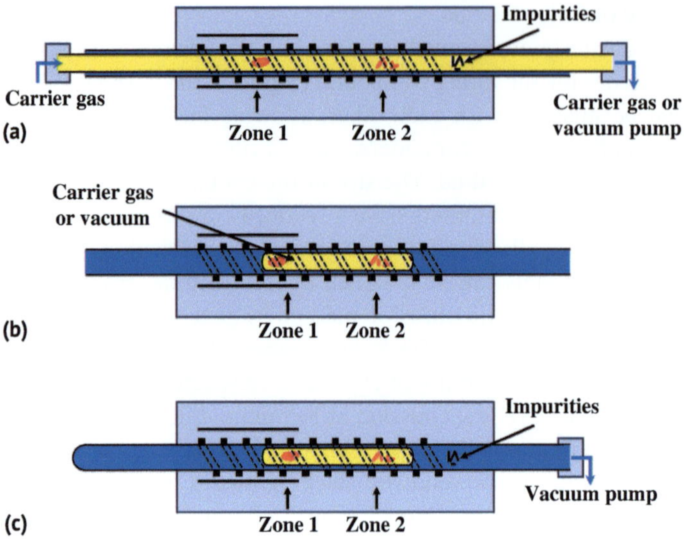

Figure 2.2 Schematic diagram of the vapor phase method for crystal growth. (a) Open system. The material is heated in Zone 1, where it sublimates into the vapor phase within the carrier gas stream. The molecular vapor travels downstream to Zone 2, where it crystallizes at a lower temperature, resulting in the formation of pure crystals that are separated from any remaining impurities. (b) Closed system. The material, including impurities and the target compounds, is heated inside a sealed glass ampoule. The sublimation and crystallization processes occur within this confined environment. (c) Semi-closed system. The material is sealed inside the glass ampoule, but a small hole is provided to allow the escape of impurities and partially crystallized material. Zone 1: Sublimation zone; zone 2: Crystal growth zone. *Source*: [28]/Cambridge University Press.

2.1.3.1 Bridgman–Stockbarger Method

The conventional Bridgman method, also known as the crucible descent method, is a melt growth method invented by Bridgman and improved by Stockbarger,[50] which has been widely used in inorganic functional materials (Figure 2.3a). In this method, the quartz ampoule filled with powdered material is sealed either in a vacuum or an inert gas and is subsequently passed through a temperature gradient. Nucleation is initiated at a certain temperature at the top of the ampoule, with the crystallization front progressively moving through the melted material. However, this method encounters limitations in impurity segregation if the solubility of the impurity is similar in both the melt and crystalline phases. Despite this, materials, such as polyphenylene derivatives (e.g. anthracene, naphthalene, perylene) and biphenyls, known for their high stability and good crystallinity, are effectively grown in large crystals using this method. It is especially advantageous for synthesizing large, high-quality crystals suitable for applications in NLO materials.

2.1.3.2 Subcooled Melt Method

The subcooled melt method involves heating the raw material in a crucible until it is completely melted, followed by the introduction of seed crystals (Figure 2.3b).[46]

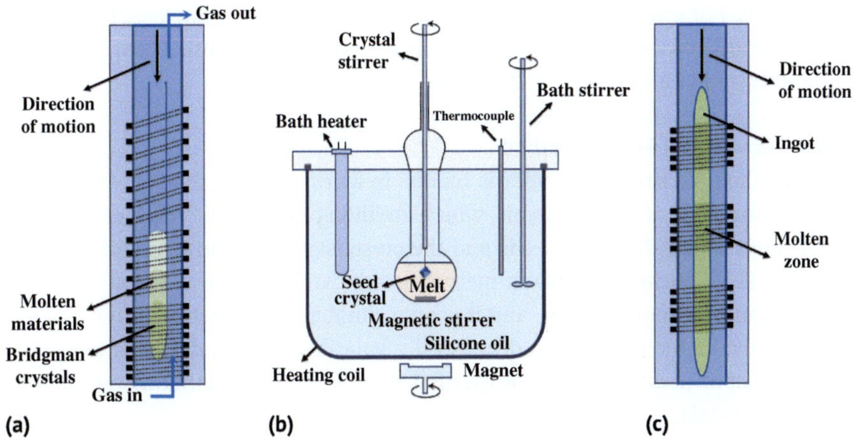

Figure 2.3 (a) Schematic diagram of the modified Bridgman-Stockbarger experimental apparatus. *Source*: [51]/with permission of Elsevier. (b) Apparatus for subcooled melt growth; Melt: supercooled melt. *Source*: [52]/with permission of Elsevier. (c) Schematic diagram of crystal growth using the zone melting method. *Source*: [28]/Cambridge University Press.

After stabilizing for a while, the temperature is lowered to a slightly lower temperature than the melting point, thus the crystals grow gradually in the subcooled melt in the sub-stable zone. In general, the subcooled melt method is particularly suitable for growing organic crystals with low melting points, as it operates at low temperatures with minimal temperature gradients, thereby avoiding the deterioration of crystals and reducing the thermal stresses inside the crystals. Different from the solvent, vapor and Bridgman–Stockbarger methods, the subcooled melt method allows precise control over the orientation and growth of crystals along preferred crystallographic planes. Moreover, the grown crystals and sealed crystallizers can be applied as nonlinear optics without the subsequent cutting and polishing.[45]

2.1.3.3 Zone-melting Recrystallization

Zone-melting recrystallization is based on the principle of moving a narrow melting zone continuously along a specimen in an inert atmosphere (Figure 2.3c).[47] This method leverages the differential solubility of impurities in the solid (unmelted portion) and molten phases to purify the substance. The distribution coefficient, which reflects the ratio of solute concentration in the solid phase to that in the liquid phase, is determined by the system's thermodynamic properties. As shown in Figure 2.3c, only a few narrow zones are melted, allowing the crystal–melt interfaces to migrate along with the temperature gradient. This method allows the repeated movement of the heating element and the repeated growth of the crystal, enhancing the purity of the resultant single crystal. The zone melting method is also frequently used for purification and is considered to be one of the effective methods for obtaining high-purity single crystals. This capability is particularly advantageous in the development of organic scintillator materials, which require extremely pure crystals. For example, zone-melting recrystallization has been used to produce high-purity anthracene single crystals, which are crucial for scintillator applications in opto-electronics and radiation detection.[48]

2.1.3.4 Czochralski Growth Method

In this method, an organic powder is melted in a crucible and then slowly dipped into a rotating seed of small crystals, which are then pulled out of the melt.[42,43] The crystals grow along the small seeds and the crystals grow in the same direction as the seeds. The Czochralski growth method is not carried out in a closed space, so the evaporation is more serious, and it is only suitable for crystals with a relatively low saturated vapor pressure and not easily volatilized. Organic melts generally have high vapor pressure and poor thermal stability, so this method is rarely used in organic crystals.

2.1.3.5 Laser Heating Base Method

The laser-heated base method involves directing a laser beam onto the surface of a substance to rapidly heat it to a molten state, thereby facilitating crystal growth within a controlled thermal environment.[49] This method allows for the independent manipulation of temperature gradients in both the solid and melt phases of the material, ensuring that the crystals grow in a stable temperature field and avoiding the effects of natural and forced convection. High-purity crystals can be obtained by using this method, as it avoids direct contact with the crucible, thus eliminating potential contamination. The main advantage of the laser-heated base method is its capability to produce long, rod-shaped organic crystals in any crystal orientation. This feature is especially valid for the production of second-order NLO crystals, as precisely aligned long crystals' phase-matched orientations can achieve highly efficient wavelength conversion.

Among the diverse strategies for crystal growth, the solution method stands out as the most prevalent for fabricating organic crystals. The factors, such as solvent types, concentrations, and the interaction forces between solvent and solute molecules are pivotal to the growth of crystals. Single crystals, which are highly sought after, serve as ideal models for the investigation of aggregated states. Scientists have explored numerous growth techniques to produce single crystals in high quantities. Solution methods are typically used for compounds that exhibit good solubility in organic solvents, offering a straightforward route to crystal formation. In contrast, physical vapor transport methods are suitable for organic semiconductors that can sublimate without decomposition, delivering high-quality organic crystals ideal for advanced electronic applications. The Bridgman growth method, tailored for organic semiconductors with low melting points that do not decompose, enables the production of sizeable single crystals. Vapor phase methods, which avoid solvent-related disruptions, are particularly effective in yielding high-quality crystals, though they require materials that exhibit robust thermal stability. These crystal growth methods can be employed complementarily depending on the specific requirements of the material and desired properties of the crystals.

2.2 Self-assembly System

Self-assembly refers to the spontaneous organization of molecules into ordered structures without external control, driven by intrinsic molecular interactions and synergistic effects.[53–55] This process involves transitioning from a disordered state to a more ordered aggregated state, where molecules or molecular segments form larger, more stable aggregates. The resultant aggregation structures are strongly influenced by the structural properties of the individual molecules and the governing conditions during the self-assembly process. Self-assembly has been prevalently applied to the fabrication of advanced and complex materials with specific sizes, structures, and functionalities.[56,57] This section presents the key factors of molecular self-assembly, including the various types of driving forces, formations of aggregation, and methods for controlling the process. It will be further analyzed to illustrate how these factors can be manipulated to achieve targeted materials through self-assembly in opto-electronic fields.[58,59]

2.2.1 Drivers of Molecular Self-assembly

By tailoring the driving forces and conditions that govern self-assembly, it is possible to control the formation of aggregates with specific structural characteristics and functionalities, thereby addressing a wide range of practical applications in opto-electronic fields.[60] These interactions are key to the development of advanced materials used in opto-electronic applications, such as organic semiconductors, light-emitting diodes (LEDs), solar cells, and sensors.[61,62] The major non-covalent interactions that drive molecular self-assembly, include hydrogen bonding, hydrophobic effect, electrostatic interactions, π–π interactions, and coordination interactions, each contributing to the stabilization, structure, and performance of opto-electronic materials (Figure 2.4).

2.2.1.1 Hydrogen Bonding Interactions

Hydrogen bonding is a form of association between a strongly electronegative atom (usually as N, O, or F) and a hydrogen atom attached to another relatively electronegative atom (Figure 2.4a). The interaction between the positive charge on the hydrogen and the negative charge on the electronegative atom creates a strong, directional bond.[63] Hydrogen bonding is highly versatile and can occur both intermolecularly and intramolecularly, making it an ideal force for guiding the formation of ordered aggregates. The bond strength, typically ranging from 9 to 40 kJ mol^{-1}, combined with its directionality, enables the formation of highly ordered aggregates with specific spatial arrangements.[64] In nature, hydrogen bonding is critical to the formation of complex structures, such as the double helix of DNA, where base pairs are stabilized by hydrogen bonding interactions. Similarly, in opto-electronic materials, hydrogen bonding plays a significant role in stabilizing molecular assemblies, controlling the packing of organic semiconductors, and facilitating the formation of nanostructures with predictable optical properties.

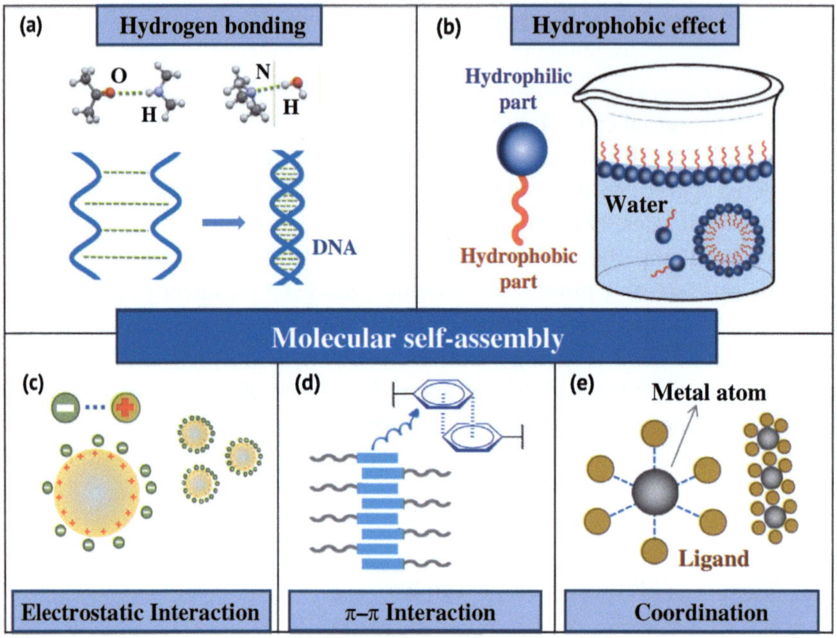

Figure 2.4 Schematic diagram of the driving forces of molecular self-assembly. (a) Hydrogen bonding interactions. (b) Hydrophobic effect. (c) Electrostatic interactions. (d) π–π interactions. (e) Coordination interactions.

For example, in OLEDs, intra- and intermolecular hydrogen-bonding interactions can restrict the rotation between different donor/acceptor moieties and inhibit the vibrational coupling of excited states, which could obtain high luminous efficiency and color purity. Also, the multiple H-bonds interactions could further enhance horizontal orientation in amorphous organic semiconductor films and significantly increase hole and electron mobilities, which is beneficial for efficiency stability with negligible roll-off.[65] Similarly, in OSCs, hydrogen bonding can help organize donor and acceptor molecules into well-ordered nanostructures, improving energy conversion efficiency.

2.2.1.2 Hydrophobic Effect

The hydrophobic effect is a crucial driving force in the self-assembly of nonpolar in aqueous environments (Figure 2.4b). When hydrophobic molecules interact with water, the system tends to minimize the exposure of hydrophobic groups to the solvent. This leads to the aggregation of hydrophobic molecules in a more regular shape, reducing the overall free energy and enhancing the thermodynamic and kinetic stability of the system.[66,67] A classic example is the spontaneous aggregation of oil into droplets in water or the formation of lipid bilayers in cell membranes, both of which are driven by the fundamental principles of the hydrophobic effect. It provides an essential mechanism for the self-assembly of amphiphilic colloidal particles, particularly in organic photonic crystals. The hydrophobic domains of these particles aggregate into well-ordered structures, which significantly extend the photonic band gap. This structural refinement enhances the selectivity of transmitted light wavelengths, enabling precise control over photonic properties.[68]

2.2.1.3 Electrostatic Interactions

Electrostatic interactions, both electrostatic attraction between oppositely charged species and electrostatic repulsion between like charges, are long-range, nonselective interactions, which play a pivotal role in driving the self-assembly of many functional systems (Figure 2.4c). These interactions are highly versatile and can be tuned by adjusting the charge density or the dielectric environment. When two such components, carrying opposite charges, come into proximity, they are naturally attracted to each other and begin to form aggregates.[69–71] The aggregation process continues until an equilibrium is reached, where the number of positive charges equals that of negative charges, resulting in electrically neutral aggregates. It is less constrained by the species, shapes, and types of components involved, offering a broader selection of materials for assembly. Furthermore, this method is relatively straightforward, yielding inherently stable aggregates.[72] For instance, in organic photovoltaics, electrostatic interactions between the donor and acceptor materials can enhance charge separation at the interface, leading to increased power conversion efficiency. Similarly, in organic field-effect transistors (OFETs), the electrostatic interactions between the semiconductor and dielectric layers influence the charge transport properties, affecting device performance.[73]

2.2.1.4 π–π Interaction

π–π interaction refers to the interactions between aromatic rings when they are oriented face-to-face (Figure 2.4d). This interaction is primarily driven by the overlap of π-electrons in the aromatic systems and is highly directional, often leading to the formation of extended, one-dimensional structures.[74,75] The strength of π–π interactions typically ranges from 1 to 50 kJ mol^{-1}. π–π interactions play a critical role in enhancing charge transport, improving luminescence efficiency, and facilitating the formation of highly ordered molecular films. The aggregates formed by π–π interactions have been used to design environmentally responsive self-assembly systems.[76] For example, π–π interactions are a key in the design of organic room temperature phosphorescence (RTP) materials, where it helps align the molecules in a way that promotes the stabilization of triplet excitons. In OSCs, π–π interactions in conjugated polymers lead to enhanced charge mobility and exciton diffusion, both of which are essential for optimizing the device performance.[77]

2.2.1.5 Coordination Interactions

Coordination interactions involve the centered metal ions to ligands, typically through the donation of electron pairs from the ligand to the metal center (Figure 2.4e). These interactions are highly selective and can be precisely controlled, making it an ideal tool for designing molecular architectures with specific properties.[78,79] Coordination self-assembly has been widely employed for the synthesis of dynamically regulated and chiral aggregates. For example, by using Fe and Pt ions as the core, enantiomers of molecule cages can be obtained through self-assembly, driven by chiral recognition and ligand binding, resulting in a high optical polarization purity.[80]

In most self-assembly processes, multiple non-covalent interactions act in concert, rather than in isolation, to direct the formation of aggregates. Combining different types of driving forces, such as hydrogen bonding interactions, hydrophobic effect, and π–π interactions can optimize the sizes and spatial structure in the aggregated state for achieving the targeted modulation of nanoparticles.[81,82]

2.2.2 Aggregation Types in Molecular Self-assembly

Molecular self-assembly is a fundamental process by which monomers spontaneously organize into structured aggregates driven by the specific interactions discussed above. The type of aggregate formed depends on the nature of these interactions, as well as the chemical properties of the monomers involved.[83] These aggregates can take various types, each exhibiting unique structural characteristics and dimensions, making them suitable for a wide range of applications in fields like imaging, drug delivery and photovoltaic material. The most common aggregation types are nanoparticles, membranes and micelles, each with distinct structures features and functions. Understanding these different aggregation types is crucial for tailoring materials for specific technological uses.

2.2.2.1 Nanoparticles

Nanoparticles are one of the most common types of self-assembled aggregates, due to their ability to minimize the overall free energy, which results in a spherical morphology (Figure 2.5a).[84] For example, Tang and Ding et al. prepared nanoparticles by encapsulating aggregation-induced emission (AIE) molecules within a triblock copolymer (polyethylene glycol 2000).[85,86] These nanoparticles were then ultrasonically dispersed in a water-oil miscible system, where hydrophobic interactions between the AIE molecules and the polymer drove the self-assembly, resulting in stable aggregates in aqueous solutions. Polymer encapsulation in nanoparticles by the polymers effectively inhibits molecular movement, maintaining the excellent luminescence performance of luminogens, which is essential for application in bioimaging.

2.2.2.2 Membranes

Membranes typically form when monomers are situated at the interface between two distinct phases, such as solid-liquid, gas-solid, or immiscible liquids, where the self-assembly process unfolds due to interactions with the interface, such as hydrophobicity, adsorption, or coordination bonding (Figure 2.5b).[87] These

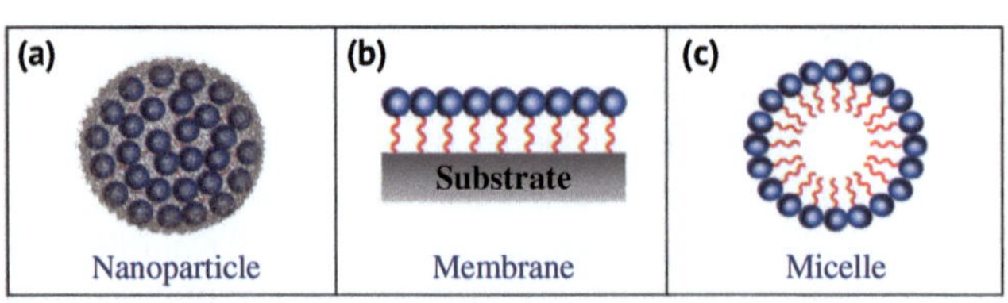

Figure 2.5 Schematic diagram of the aggregation types in molecular self-assembly. (a) Nanoparticles. (b) Membranes. (c) Micelles.

self-assembled membranes possess a high degree of stability and can cover large areas, making them suitable for applications in surface modification, catalysis, and sensing. For instance, Huang et al. reported the self-assembly of sulfobetaine-thiol molecules on the surface of gold particles.[88] The thiol groups formed coordination bonds with gold atoms, leading to a membrane-like structure with hydrophilic sulphonic acid groups oriented outward. This self-assembly process significantly enhanced the hydrophilicity of the gold surface and conferred stable antifouling properties.

2.2.2.3 Micelles

Micelles are spherical structures that form when amphiphilic molecules, those with both hydrophobic and hydrophilic regions aggregate in aqueous solutions. This configuration can minimize the free energy of the system, leading to the spontaneous formation of micelles when the concentration of amphiphilic molecules exceeds a certain point (Figure 2.5c).[89] The hydrophilic parts afford the spherical micelles solubility in water, while the hydrophobic parts generate an intrinsic hollow structure, providing an ideal location for encapsulation. Furthermore, as the constituent molecules continuously exchange with the surrounding solution, micelles can respond to environmental changes such as temperature and concentration, leading to variations in the size and shape of micelles. This dynamic nature is particularly advantageous for applications in drug delivery and diagnostic imaging. Hong and Zhang et al. utilized di-block copolymers as solvophilic chain-transfer agents that bind to tertiary amine groups, producing pH-responsive micelles with hollow structures.[90] The micelles can take sufficient payloads of rhodamine B, realizing luminescent signal detection. At pH = 4, these micelles continued swelling and then disaggregated gradually to form smaller aggregates. As pH increased, the volume of the micelles progressively diminished, accompanied by a notable reduction in the rate of payload release. In addition, the stability of the micelles could be enhanced by increasing the molecular weight of the copolymer, which also slowed down the contraction and release, further exploiting the flexibility of micelles' self-assembly.

The variety of aggregation types ranging from nanoparticles and membranes to micelles demonstrates the flexibility of molecular self-assembly in generating materials with specific functions and properties. As researchers move toward more complex and advanced material designs, it becomes increasingly important to understand how these aggregation types can be controlled and combined to achieve the desired performance in opto-electronic devices, biomedical applications, and beyond.

2.2.3 Control Methods of Molecular Self-assembly

The increasing demand for advanced self-assembling materials has surged in fields such as opto-electronics, nanotechnology, and biomaterials. This demand underscores the need for precise control over the mechanisms that govern the self-assembly process. By fine-tuning various parameters, such as the type of driving force, external stimuli, and solvent conditions, researchers can modulate the structure, size, and properties of the resultant aggregates.[91] This ability to control molecular assembly is crucial for the development of functional materials with tailored properties, particularly in the context of opto-electronic applications where precise control over morphology and aggregation directly influences performance.

The primary control methods for molecular self-assembly, provide insight into how different strategies can be used to direct the formation of aggregates and create materials with desired properties. These methods include stimuli-responsive self-assembly, solvent evaporation, template-guided self-assembly, and adsorption-based self-assembly. Each method plays a unique role in the design of opto-electronic materials, where the precise control of molecular organization is essential for optimizing their performance.

2.2.3.1 Stimuli-responsive Self-assembly

Stimuli-responsive self-assembly relies on the ability to trigger changes in the structure or properties of a material in response to external stimuli, such as light, temperature, pH, magnetic, or ultrasound.[92] The ability to dynamically alter aggregation behavior in response to environmental stimuli is a powerful tool for creating reconfigurable materials. Incorporating functional groups that are characteristically responsive to an environmental factor, such as azo groups that produce conformational transformation under light irradiation or long polymer chains that move sharply at high temperatures, enables the modulation of the rate, extent, and morphology of aggregation. For example, Thordarson and Beves et al. reported isomeric arylazopyrazole photoswitches by incorporating a carboxylic acid group, aromatic rings, and a pyrazole ring together, each of which contributes distinct driving forces such as hydrogen bonding, π–π stacking, and hydrophobic interactions (Figure 2.6a).[93] Upon light exposure at 360 nm, the azo group undergoes a reversible isomerization from the *E*-form to the Z-form, altering the molecule's solubility and allowing the gelation to undergo gradual shrinkage until precipitation occurs, and the loaded dyes are released. Furthermore, under illumination at 520 nm, Z-form conformations present reversible transformation to E-form, allowing the precipitate to reassemble into a gelation. This light-induced process enables precise control over the aggregation process, allowing for applications such as responsive drug delivery and adaptive photonic materials.

2.2.3.2 Solvent Evaporation Self-assembly

Solvent evaporation is one of the most commonly used methods for self-assembly, particularly for materials that begin in a disordered, dilute solution state. As the solvent evaporates, the concentration of monomers increases, which enhances the probability of interacting and forming aggregates through various driving mechanisms.[94] This process is particularly advantageous due to its simplicity and wide applicability. For instance, Mai et al. used solvent evaporation to co-assemble block copolymers with reactive precursors into a bi-continuous self-assembled material (Figure 2.6b).[95] The block copolymer and precursor molecules were primarily dissolved in the solvent to form a homogeneous solution, and the precursors could be

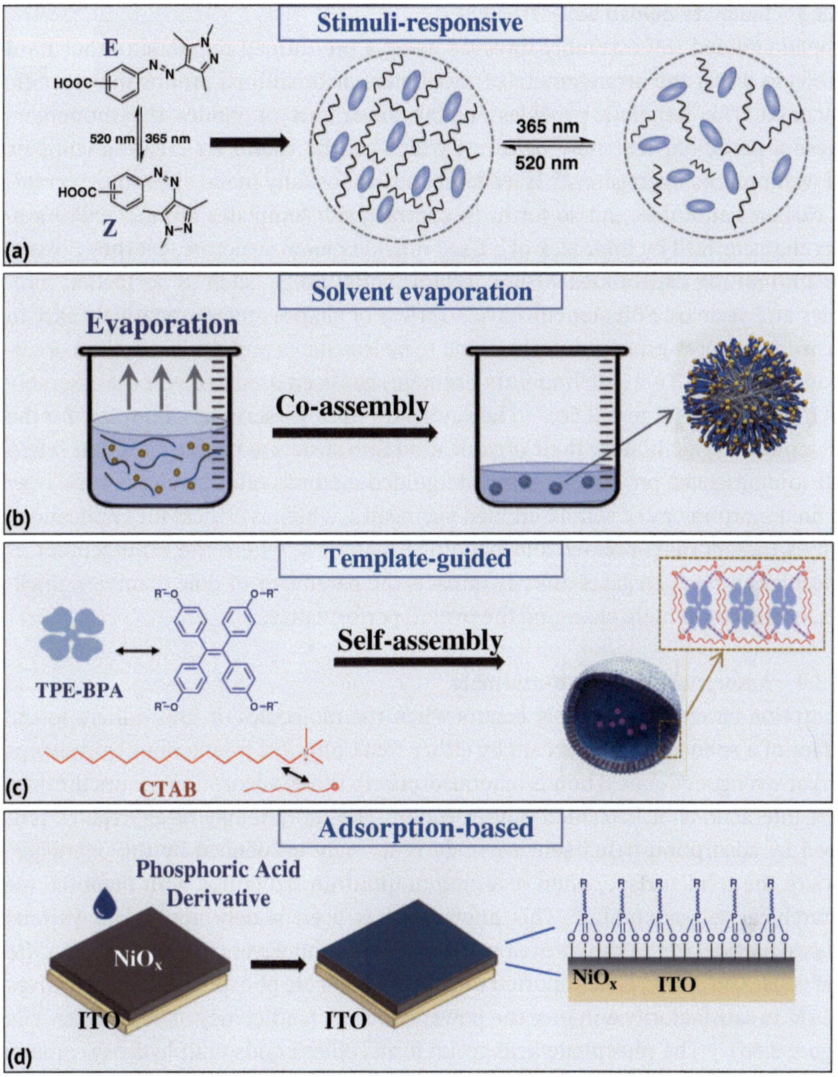

Figure 2.6 The control methods of molecular self-assembly. (a) Stimuli-responsive.[93] (b) Solvent evaporation.[95] (c) Template-guided.[97] (d) Adsorption-based.[99]

adsorbed onto the hydrophilic chain segments through intermolecular interactions such as hydrogen bonding or electrostatic interactions. As the solvent evaporates, the block copolymer and precursor molecules separate into distinct phases, and the precursor molecules cross-link to form a continuous network and then assemble into ordered aggregates. By carefully controlling the rate of solvent evaporation, it is possible to control the size and shape of the self-assembled aggregates, making this method highly versatile for the fabrication of multilayered and encapsulated nanoparticulate materials for catalysis and photovoltaics applications.

2.2.3.3 Template-guided Self-assembly

Template-guided self-assembly involves using a pre-formed template (either hard or soft) to direct the arrangement of monomers or functional groups into specific structures. The template provides spatial constraints or guides the monomers toward a particular assembly pathway, which can be useful for creating complex and well-ordered aggregates.[96] Hard templates are usually metal substrates, ceramics, carbon nanotubes and so forth. In contrast, soft templates are molecular systems characterized by their lack of a fixed organizational structure, yet they possess domain-limiting capabilities with a certain spatial range, such as surfactant molecules and vesicles. Soft stencils have a variety of shapes and are relatively easy to construct and post-process, but they tend to be less stable and exhibit limited orientational control. CTAB (cetrimonium bromide) has been used to direct the assembly of AIE molecules (Figure 2.6c).[97] The surfactant micelles serve as a template for the AIE molecules, facilitating their organization into structured aggregates that retain their luminescence properties. Template-guided methods offer precise control over the final morphology of self-assembled aggregates, which is critical for applications such as mesoporous, hollow, and nanotube materials, where the arrangement of molecules at the aggregates directly impacts the parameter of dole diameter, thickness, or size, ultimately changing the overall performance.

2.2.3.4 Adsorption-based Self-assembly

Adsorption-based self-assembly occurs when the molecules or ions adhere to the surface of a solid substrate, driven by either weak physical interactions (physisorption) or stronger chemical bonds (chemisorption). When adsorption occurs through weak interactions, it is termed physisorption. The morphology of aggregates produced by adsorption-based self-assembly is strongly influenced by the characteristics of the solid surface, often resulting in ultrathin structures with potential for hierarchical self-assembly.[98] This approach has been widely applied in various devices, particularly in the form of self-assembled monolayers (SAM) in perovskite solar cells. Zhu and Li et al. reported the use of carbazole phosphate acid derivatives as SAM to satisfactorily enhance the power conversion efficiency of perovskite cells (Figure 2.6d).[99] The phosphate acid group forms ether bonds with hydroxyl groups and coordination bonds with nickel atoms on the substrate, thus enabling the monomer to be stably adsorbed to the substrate. Moreover, the carbazole groups engage in π–π interactions, promoting an ordered molecular arrangement at the interface. The self-assembly membrane can effectively reduce the voltage loss of perovskite cells and further maintain a strong fixation effect. Adsorption-based methods are critical for the development of interfaces in opto-electronic devices, where precise control of molecular orientation and packing density is necessary to optimize device efficiency, particularly in applications like OSCs, OLEDs, and sensors.

Each of the self-assembly methods described above offers unique advantages, and in many cases, combining multiple strategies can provide enhanced control over the aggregation process. Through the action of multiple driving forces, diverse aggregation types and various control methods, molecular self-assembly could obtain aggregated materials with expected structures and properties. This has become an important synthetic method in various fields, especially for organic opto-electronic functional materials. As advancements in real-time monitoring of the aggregation process and precise characterization of the spatial arrangement of aggregates continue to evolve, molecular self-assembly will play an increasingly vital role in the production of advanced materials.

2.3 Gel System

A gel is comprised of a three-dimensional polymer network structure saturated with a medium, typically a liquid or a gas. This structure allows the gel to maintain a certain shape while being deformed under external forces. Gels are characterized by a high diffusion coefficient, functioning as solvents within a highly swollen medium.

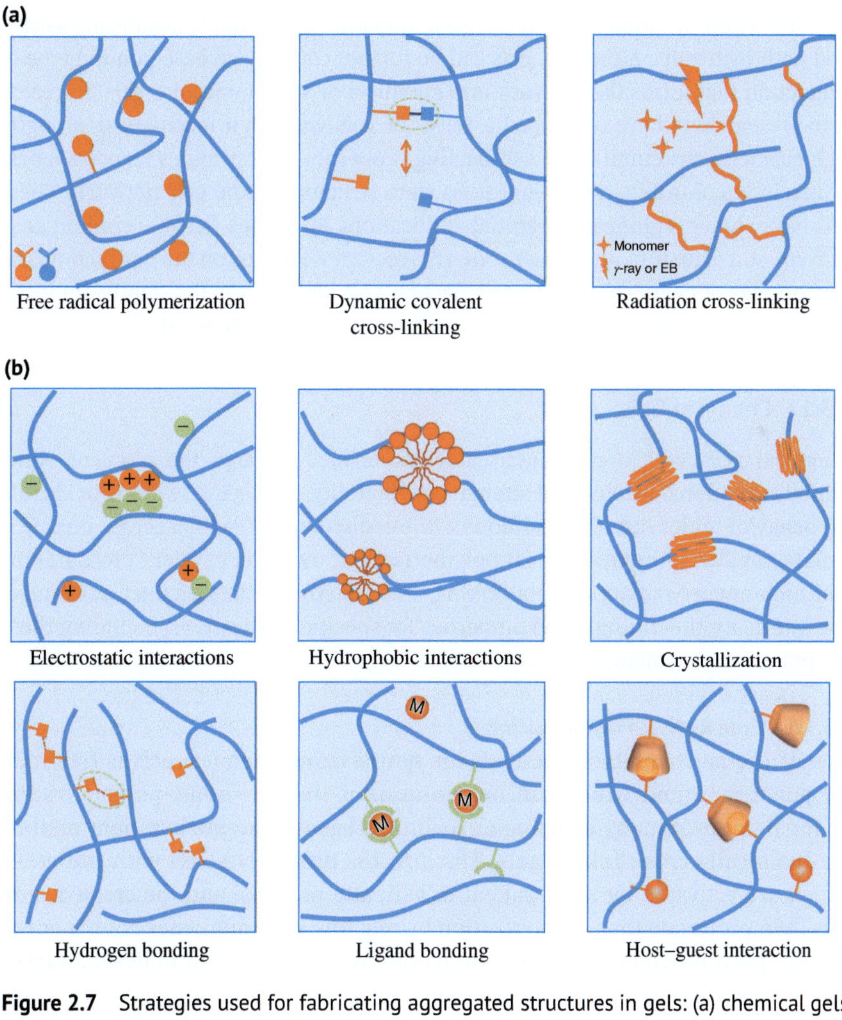

Figure 2.7 Strategies used for fabricating aggregated structures in gels: (a) chemical gels; (b) supramolecular gels.

The most important characteristic of the gel is swelling behavior, where the network dynamically adjusts to external stimuli by achieving thermodynamic equilibrium through the absorption and the expulsion of solvent. This equilibrium is influenced by various factors, including temperature, pH, and solvent type, making gels open systems that are in non-equilibrium and capable of exchanging energy, materials, and information with the environment. Furthermore, they can alter their shape, volume, and state in response to external environmental changes such as temperature, pH, and light exposure.

Based on their origin, gels can be classified into two main categories, including natural and synthetic ones (Figure 2.7). Natural gels are often derived from biological organisms and include materials like proteins, konjac, and agar. Synthetic gels, on the other hand, are produced through chemical synthesis with common examples including polyvinyl alcohol (PVA) and polyacrylic acid (PAA) gels. The majority of natural and synthetic gels utilize water as the medium, and these are typically referred to as hydrogels. In discussions about gels, the term "gel" is often specifically associated with hydrogels. Moreover, gels can be further categorized based on the type of interaction that forms the network into chemical or supramolecular gels. In recent years, researchers have developed a series of gels with high mechanical strength, high elasticity, structural color, self-healing properties, and stimulus responsiveness by precisely controlling the aggregation state structure of the gels network. These gels have shown significant potential applications in various fields, including drug delivery, soft actuators, and biosensors. This section will focus on the typical preparation strategy of chemical and supramolecular gels and analyze the significant influence of the aggregation structure on the properties of these materials.[100–102]

2.3.1 Chemical Gels

Chemical gels, with a permanent network formed through the covalent bond, exhibit exceptional mechanical strength and stability, displaying swelling or shrinking behavior under specific conditions without dissolving. Chemical cross-linking is mainly conducted by free radical polymerization, dynamic covalent cross-linking and high-energy radiation cross-linking. Each method offers distinct advantages and allows for the tuning of gel properties for specific applications, including those in opto-electronic materials.

2.3.1.1 Free Radical Polymerization

One of the most common methods for synthesizing chemical gels is free radical polymerization, which can be synthesized through homo-polymerization or copolymerization by utilizing cross-linked agents, the predominant method for chemically cross-linking gels. This process primarily starts with initiators, which are activated by light or heat to generate radicals, and the created radicals then propagate the polymerization by reacting with monomer double bonds to form polymer chains. The polymerization terminates when radicals undergo combination, disproportionation, or transfer reactions. Vinyl-containing compounds, such as acrylic acid and its derivatives, are commonly used monomers for free radical polymerization, while common initiators are azo diisobutyronitrile (AIBN) and benzoyl peroxide (BPO).[103–105]

This method provides a high degree of control by carefully regulating parameters such as the concentration of monomers and initiators, polymerization temperature, and the choice of suitable solvents; thus, the corresponding mechanical and adhesive properties can be fine-tuned. For example, Zhou and Wu et al. successfully fabricated self-adhesive gels with tunable mechanical and adhesive properties by conducting free radical polymerization in both aqueous and micellar phases.[106] The study revealed that, as the concentration of dopamine (DA) increased, the strength of the resulting gels decreased, while the elongation at break increased (Figure 2.8a). This phenomenon was attributed to the reaction of the phenolic hydroxyl groups present in DA with free radicals, which exerted an inhibitory effect on the free radical polymerization process. Conversely, an increase in the concentration of 2-methoxyethyl acrylate (MEA) likely enhances the formation of the poly(2-methoxyethyl acrylate) (PMEA) network, thereby elevating the modulus and tensile strength of the gels. The highly mechanical properties and adhesiveness of gels in aggregated states were effectively utilized in the fabrication of capacitive pressure sensors, demonstrating their capability to monitor a range of human movements and physiological activities.

2.3.1.2 Dynamic Covalent Cross-linking

In addition to free radical polymerization, dynamic covalent cross-linking provides another approach for synthesizing chemical gels. In this method, the polymer with active functional groups such as amino, carboxyl, and hydroxyl in the side chain or the end of the backbone, react with molecules that have aldehyde, hydroxyl, carboxyl, and epoxy groups, forming tightly aggregated networks through various reactions.[107,108] For example, the dynamic properties of acylhydrazone and imine bonds enable the self-healing capability of gels, as exemplified by the combination of hexanedihydrazide (ADH), water-soluble chitosan (CEC), and oxidized sodium alginate (OSA)-based.[109] Upon fracture, the damaged gels exhibit the capacity for spontaneous aggregation. The dynamic bonds can reform at the fracture interface, facilitating self-healing in a physiological environment with a remarkable healing efficiency of up to 95% (Figure 2.8b).

Moreover, other dynamic covalent reactions, such as esterification between boron ions and hydroxyl groups, and traditional condensation or addition reactions like the Diels-Alder and Michael additions, are also instrumental in constructing strong gels cross-linked networks that are used in flexible electronic devices.[110] For example, conductive hydrogels synthesized through dynamic covalent bonding have shown promise as materials for stretchable interconnects in wearable electronics or as components in organic field-effect transistors (OFETs), where mechanical flexibility combined with electrical conductivity is essential.

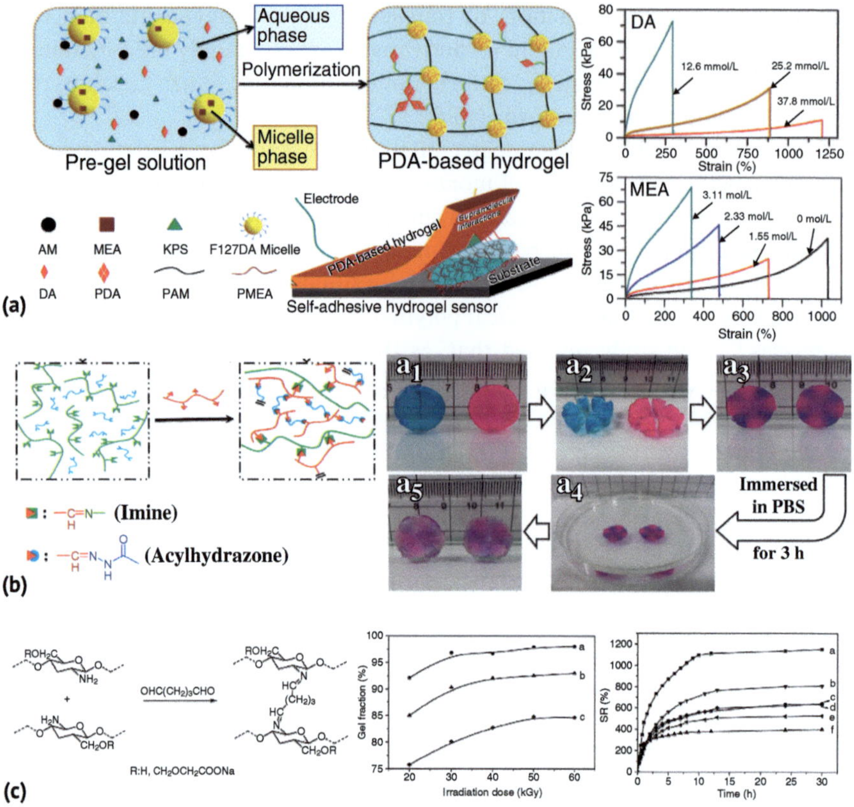

Figure 2.8 Effect of aggregation on mechanical properties, self-healing properties, and expansion rates of gels. (a) Stress–strain curves of PMEA-PAM-PDA gels prepared under different DA and MEA concentrations. *Source*: Reproduced from [106]/with permission of Elsevier. (b) Damaged gels aggregate spontaneously to achieve efficient self-healing. *Source*: [108]/with permission of John Wiley & Sons. (c) Relationship between irradiation dose and gelation, and swelling kinetics with various irradiation doses. *Source*: Reproduced from [113]/with permission of John Wiley & Sons.

2.3.1.3 Radiation Cross-linking

Another alternative gel synthesis method is radiation cross-linking, which involves irradiating polymers or their aqueous solutions with high-energy radiation, such as γ-ray and electron beams (EB), creating cross-linking structures without the need for additional chemicals.[111,112] This process breaks the C—H bonds in the polymer backbone uniformly, generating radicals that lead to polymerization and cross-linking. Yin and Chen et al. used poly(vinyl methyl ether) (PVME) and carboxymethyl chitosan (CMCS) as monomers, demonstrating how the variations in the irradiation dose of EB and glutaraldehyde (GA) concentration as a cross-linked agent affect the swelling behavior and cross-linked density of gels (Figure 2.8c).[113] Increased cross-linking density, which is characterized by enhanced aggregation of polymer chains within the network, resulted in a decreased swelling ratio of the gels. Conversely, a reduction in cross-linking density, indicative of diminished polymer chain aggregation, facilitated the penetration of water molecules into the network structure, thereby increasing the swelling ratio.

2.3.2 Supramolecular Gels

In contrast to chemical gels, supramolecular gels are assembled from small molecules or polymers through non-covalent interactions such as electrostatic interactions, hydrophobic effects, crystallization, hydrogen bonding, ligand binding, and host–guest interactions, resulting in a reversible supramolecular network. These non-covalent interactions endow supramolecular gels with dynamic reversibility, allowing them to transform into sols upon the disruption of cross-linking points. Consequently, supramolecular gels hold broad application potential in the realms of drug delivery, biosensing, controllable microvalves, and self-healing applications.

2.3.2.1 Electrostatic Interactions

One of the most widely studied mechanisms for supramolecular gel formation is electrostatic interaction. Ionic composite gels can be synthesized by mixing at least two polymer electrolyte solutions or polymer electrolyte solutions with oppositely charged multivalent ions under appropriate conditions. The resultant molecular aggregates are formed primarily through electrostatic interactions, the strength of which is closely related to the distribution of charges and the specific pairs of anions and cations involved.[114,115] The commonly used anionic electrolytes are alginate, sodium polystyrene sulfonate, and sodium polyacrylate, while the cationic electrolytes are polyelectrolytes, such as chitosan, polyacrylamide, or multivalent metal ions. The polymer chains in these gels are tightly intertwined, often resulting in excellent mechanical properties. For instance, the self-assembly of a triblock polyelectrolyte with an oppositely charged homopolyelectrolyte through electrostatic interactions forms polyelectrolyte complex (PEC) gels.[116] The aggregation state of the polymer chains and the formation of cross-linking points can be modulated by altering intrinsic and extrinsic factors such as the content of cross-linking agents, the irradiation dose, temperature, and pH values. These variables influence the microstructure of PEC gels, particularly the interdomain distance or mesh size, which affects the swelling behavior, ion transport, and mechanical stability of gels.

Recently, Wang, Yang, and Jiao et al. reported on an in situ multilayered gel polymer electrolyte (ISMGPE), where the potassium ions (K^+) exhibited electrostatic interactions with the electronegative functional groups within the ethoxylated trimethylolpropane triacrylate (ETPTA) polymer chains, facilitating the rapid migration of free K^+.[117] Compared to lithium ions (Li^+) and sodium ions (Na^+), K^+ possesses a larger ionic radius and lower charge density, which influences its migration behavior within the polymer electrolyte matrix (Figure 2.9a). The cross-linking aggregation structure of the ETPTA-based polymer electrolyte further enhances the migration of cations and restricts the migration of anions. By meticulously designing the gel polymer electrolyte with specific aggregated structures and chemical compositions, the ion migration rate was altered, achieving high ion transport efficiency

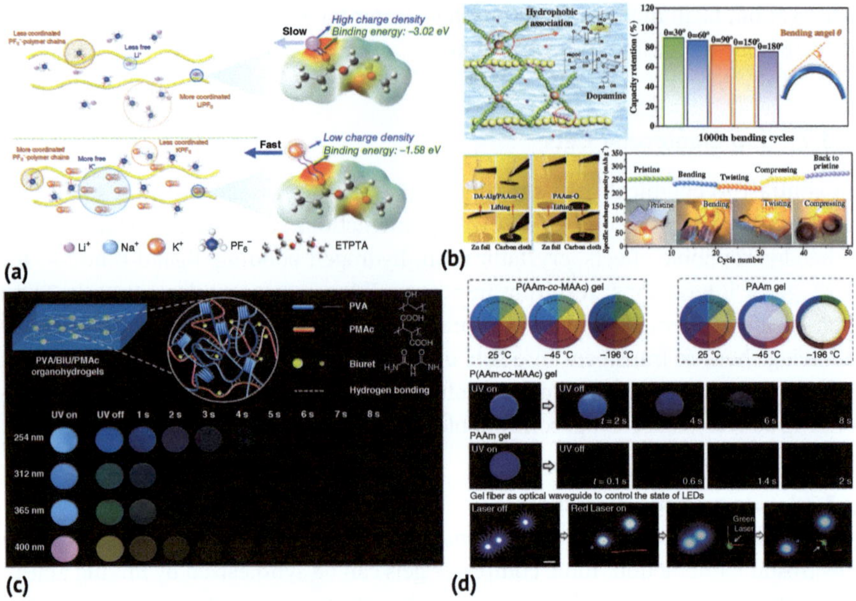

Figure 2.9 (a) Electrostatic interaction promotes rapid migration of K^+. *Source*: Adapted from [130]. (b) Schematic depictions of the DA-Alg/PAAm-O GPE, adhesiveness of PAAm-O, and DA-Alg/PAAm-O GPEs. Assembly and performance of Zn//MnO_2 batteries with DA-Alg/PAAm-O GPEs. *Source*: [119]/with permission of Elsevier. (c) The schematic illustration of the cross-linking mechanism in PVA/BIU/PMAc gels. *Source*: [123]/with permission of John Wiley & Sons. (d) Hydrogen-bonding cross-linked glassy gels have excellent freeze resistance and RTP properties. *Source*: [126]/with permission of John Wiley & Sons.

and electrochemical stability, thereby endowing the quasi-solid-state potassium-ion batteries (SSPIBs) with superior performance.

2.3.2.2 Hydrophobic Interactions

Gels can also be engineered through hydrophobic interactions when monomers or polymers with hydrophobic units, functional groups, or side chains are combined. These components tend to aggregate, forming micelle-like microdomains that serve as physical cross-linking points, thereby enhancing the network structure of gels.[118] The majority of hydrophobically associated gels are prepared through micellar copolymerization techniques with stable cross-linking networks. This structure configuration not only enhances the mechanical strength of the gel but also possibly endows it with unique responsiveness and self-healing capabilities.

In the context of opto-electronic materials, hydrophobic gels can play a pivotal role in the fabrication of flexible, stretchable devices. Wang and Cheng et al., drawing inspiration from the adhesive mechanism of mussels, have developed gel polymer electrolytes (GPEs) with self-healing and high adhesion properties, modulated by hydrophobic interactions, to enhance the durability and flexibility of flexible aqueous zinc-ion batteries (AZIBs).[119] The GPEs are composed of a polymer network of dopamine-grafted sodium alginate (DA-Alg) and a hydrophobic association network of poly(acrylamide-co-octadecyl methacrylate) (PAAm-O). The hydrophobic segments within the DA-Alg/PAAm-O GPEs reinforce the mechanical robustness and augment the healing capabilities, thereby extending the service life of flexible AZIBs and mitigating safety issues arising from device failure during mechanical deformations (Figure 2.9b).

2.3.2.3 Crystallization

Crystallization is another fundamental non-covalent interaction that contributes to supramolecular gel formation. The conformation of polymer chains in solution is usually a solvated random wire cluster. As the temperature decreases, the clusters increasingly order and twist into an organized structure to form microcrystals, which act as physical cross-linking points in the network. PVA is a typical polymer that forms microcrystals through repeated freeze-thaw ordering of molecular chains. The cross-linking method of polyvinyl alcohol with microcrystals relies heavily on the cyclically repeating freeze-thaw steps, where the solvent in the polymer solution freezes, forming ice crystals that segregate the polymer chains. The growing ice crystals divide the polyvinyl alcohol solution into two phases. The polymer chains are expelled from the growing ice crystals, leading to an increase in the local PVA concentration and the formation of a polymer-rich phase. The polymer-rich phase begins to entangle continuously under hydrogen bonding, eventually forming microcrystalline polymer chains. During repeated freezing and thawing, the molecular chains are ordered to form microcrystals, and these microcrystals are used as cross-linking points to form gels. The preparation of gels by crystalline cross-linking is relatively simple, but the selection of the appropriate molecular weight, concentration, and number of cycles is the most important factor to obtain the required morphology and properties of physically cross-linking gels.[120–122]

Recently, Wang et al. prepared an RTP gel based on non-aromatic polymers by subjecting PVA, biuret (BIU), and poly(maleic acid) (PMAc) to a freeze-thaw process.[123] During the freeze-thaw process, the increased crystallinity of the gel not only contributes to the enhancement of its mechanical properties and stability, but also facilitates the formation of nontraditional chromophores with specific optical characteristics. Moreover, the hydrogen bonds formed during the freeze-thaw process help to prevent nonconventional chromophores (NCCs) from contacting water molecules. Consequently, this gel exhibits significant excitation-dependent and metal cation-responsive phosphorescence under ambient conditions, with a maximum lifetime of up to 782.8 ms (Figure 2.9c).

2.3.2.4 Hydrogen Bonding

Hydrogen bonding is the most common non-covalent interaction in supramolecular gels and plays a critical role in the synthesis of gels due to their suitable strength, orientation, and affinity. Many functional groups, such as carboxylic acids, hydroxyl groups, amides, and ureidopyrimidinones can form hydrogen bonds at aggregated states. Although weaker than covalent bonds, reversible hydrogen bonds can impart unique properties to gels through the formation of energy dissipation mechanisms, such as self-healing, deformation memory, and shear thinning. The strength of hydrogen bonding is closely related to solvent polarity, and polar solvents like water can compete for binding sites and hinder the ability of hydrogen bonding between polymer chains and their acceptor unit. Therefore, hydrophobic spacer groups are usually introduced to shield the hydrogen bonding groups in gels which are cross-linked by hydrogen bonding networks.[124]

For example, the complexation of polyacrylic acid (PAA) with polyvinylpyrrolidone (PVPON) in an aqueous solution allows PVPON to interact with PAA through hydrogen bonding, forming transparent PAA-PVPON composites.[125] These composites, uniformly dispersed as nanoparticles within the PAA-PVPON complex, exhibited exceptionally high mechanical strength due to dense hydrogen bonding between the carbonyl groups of PVPON and the carboxyl groups of PAA. Additionally, the self-assembled hydrophobic domains within the nanoparticles contribute to the overall structural integrity of the gels. He and Wu et al. synthesized a supramolecular gel P(AAm-co-MAAc) by employing acrylamide (AAm) and methacrylic acid (MAAc).[126] This gel forms through dense entanglements and hydrogen-bonded cross-links, endowing it with intrinsic anti-freezing properties and superior optical and mechanical performance. The mobility of water molecules within the gel is reduced by the confining effects of the glassy matrix, which maintains the gel with transparency even at liquid nitrogen temperatures (−196 °C). Furthermore, the P(AAm-co-MAAc) gel exhibits blue fluorescence at room temperature. This fluorescence is sustained by the aggregation of carbonyl clusters within the gels, facilitated by internal hydrogen bonds and dipole-dipole interactions, resulting in a sustained blue afterglow lasting several seconds after exposure to ultraviolet irradiation (Figure 2.9d). In contrast, the PAAm gels with fewer hydrogen bonding interactions fail to manifest this phosphorescent effect.

2.3.2.5 Ligand Binding

The formation of coordination bonds between polymers with coordination groups on the side chains and multivalent metal ions is another effective supramolecular cross-linking method for preparing gels. Gels cross-linked by ligand bonding are thermally reversible and can be dynamically assembled/decomposed or joined/dissociated. This feature is particularly widely used in areas such as self-healing gels.[127,128] Inspired by the structure and properties of mussels, researchers have introduced catechol analogues-tannic acid (TA) into the gel structure formed by the coordination of HOPO 3-hydroxy-4-pyridinone with metal ions, thereby constructing a self-healing double-cross-linked gel.[129] By varying the amount of tannic acid incorporated, the stiffness of the gel can be tailored, which can be attributed to the increased cross-linking density of the gel due to the addition of tannic acid (TA), which influences the dynamic performance and the ability of the network to rearrange under deformation, thereby affecting the fracture behavior. Li and Li et al. fabricated robust luminescent nanocomposite gels possessing high extensibility, favorable mechanical strength, tunable luminescence, and efficient self-healing capabilities.[130] The dynamic reversible coordination of lanthanide metals and ligands, coupled with hydrogen bond interactions, endows the gels with exceptional

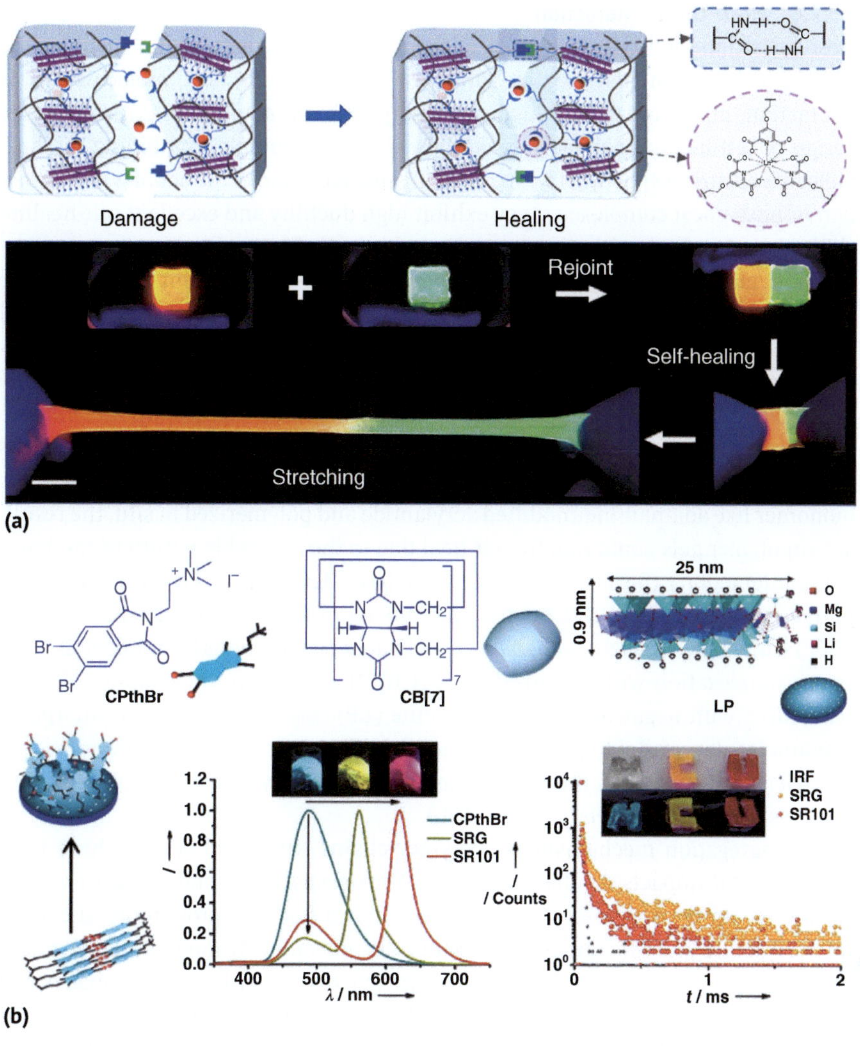

Figure 2.10 (a) Schematic illustration of the self-healing mechanism. Photographs of the self-healing process and the healed gels being stretched (scale bar, 1 cm). *Source*: Reproduced from [130]/with permission of American Chemical Society. (b) Molecular structures and schematic representation of CPthBr, CB[7], and LP. Normalized gated emission spectra, and delayed fluorescence lifetime decay plot of hybrid gels. *Source*: [134]/with permission of John Wiley & Sons.

self-healing ability at room temperature. Upon fracture, these non-covalent bonds can reform to a certain extent, thereby restoring the integrity and performance of the gel. Furthermore, by altering the ratio of lanthanide ions, such as Eu^{3+} and Tb^{3+}, the emission colors of the gel can be modulated, achieving a range of colors from red to green (Figure 2.10a).

2.3.2.6 Host–Guest Interaction

Host–guest interaction is a type of supramolecular interaction that involves the specific recognition and binding of guest molecules by host molecules. This interaction, characterized by high specificity and reversibility, is critical for the design of self-healing polymeric materials.[131,132] For example, combining the host molecule cucurbit[8] (CB[8]) with a vinyl-ligated guest molecule enables the formation of host–guest complexes. They exhibit high ductility and excellent self-healing ability in gels, obtained by in situ polymerization with acrylamide monomers.[133] The host/guest complex could act as a dynamic cross-linking point to enhance the mechanical strength of the polyacrylamide network and dissipate energy to external stresses by continuous opening/re-formation, thus offering the gels excellent tensile properties. At the same time, the gels had efficient repair properties due to the reversible dissociation and association of the host/guest complex at room temperature. Cyclodextrins (CDs), particularly β-cyclodextrin (β-CD), are frequently used as hosts to prepare self-healing gel materials. When β-CD is mixed with a hydrophobic monomer like adamantine-modified acrylamide and polymerized in situ, the resultant copolymer gels could rapidly self-heal due to the reversible nature of the host–guest interactions (β-CD-Ad). Researchers have developed a structurally simple phthalimide phosphor (CPthBr) that exhibits room-temperature phosphorescence (RTP) with quantum yield as high as 41.8% in aqueous solutions by employing host–guest complexation with cucurbit[7]uril (CB[7]) and electrostatic supramolecular scaffolding with negatively charged laponite (LP) clays.[134] This organic–inorganic supramolecular scaffolding anchors and rigidifies the phosphor while providing a barrier against oxygen, thereby enhancing the stability of the phosphor in both aqueous solutions and gels (Figure 2.10b).

The aggregation mechanisms of polymer chains within gel systems have multidimensional impacts on the properties of materials. Such impacts are not only manifested in the macroscopic physical attributes but also profoundly affect the microstructure and functional performance. By meticulously modulating the cross-linking patterns of polymer chains, the aggregated state within the gel matrix can be altered, thereby optimizing key characteristics such as opto-electronic properties, mechanical performance, and chemical stability. For instance, by adjusting the cross-linking density, the pore structure and network robustness can be controlled, which influences its light transmittance, energy dissipation capacity, and environmental adaptability in turn.

A deep understanding of the molecular mechanisms in the aggregated state within gel systems and their relationship to material performance is crucial for guiding the design and development of novel opto-electronic functional materials. Future research may focus on exploring innovative methods for regulating the aggregation mechanisms of gels. The development of stimulus-responsive polymers and intelligent nanoparticles represents a significant advancement in the organic opto-electronic field. These materials exhibit groundbreaking capabilities by enabling precise manipulation of their internal structures through external stimuli such as light, electricity, and magnetic fields. This dynamic control facilitates the modulation of functional properties, opening new frontiers in material performance and adaptability. Such investigations not only propel the evolution of gel materials toward higher performance but also possibly uncover new physicochemical phenomena, providing a theoretical foundation for the development of new types of intelligent materials and systems.

2.4 Cross-linking System

Cross-linking is the process of constructing three-dimensional network structures through chemical reactions or physical interactions, which is important in the development of advanced opto-electronic materials. By forming covalent or non-covalent bonds between or within polymer chains, cross-linking technology has profound implications across diverse fields, including biological coupling, medical materials, polymer science, and industrial production of chemical products, and the ability to precisely control cross-linking mechanisms and aggregated states enables researchers to tailor material properties, facilitating innovation in opto-electronics. Cross-linking can be realized by chemical cross-linking (light, heat, radicals), physical cross-linking, radiation cross-linking and other diversified methods, each offering unique advantages for specific applications.

2.4.1 Chemical Cross-linking

Chemical cross-linking relies on chemical reactions such as condensation, addition, or radical polymerization, etc., which can be triggered by heat, light, catalysts, or other chemical agents (Figure 2.11). The chemical cross-linking mechanism can significantly improve the mechanical strength, thermal stability and solvent resistance of materials, making it widely applied in the preparation of materials like rubber, plastic, and resin.

2.4.1.1 Triggered by Chemical Agents

Chemical cross-linking agents are widely used in molecular modification of nucleic acids, drugs, and solid surfaces (such as microspheres, nanoparticles, etc.) in

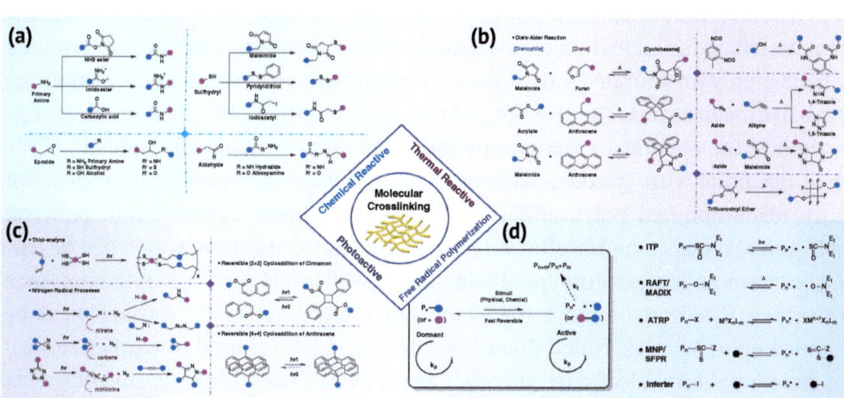

Figure 2.11 The types of chemical cross-linking and the characteristic reactive groups.

scientific research. Applications of cross-linking agents, with reactive groups such as amino groups, sulfhydryl groups, carboxyl groups, and hydroxyl groups, provide a powerful tool for the directional functionalization of macromolecules.

Cross-linkers like *N*-hydroxysuccinimide (NHS) esters, imidoesters, maleimides, and carbodiimides can be used for targeted conjugation. NHS esters and imidoester cross-linkers react with primary amines in alkaline conditions to form stable amide,[135,136] and amidine bonds,[137,138] respectively, while maleimides can form thioether linkages with sulfhydryl groups under neutral conditions.[139–141] Furthermore, epoxy groups create versatile chemical bonds with nucleophiles,[142–145] and aldehydes react with primary amines to form Schiff bases, which can be stabilized via reduction.[146–148] These chemical agents are extensively used in bioconjugation, protein modification, and the synthesis of biomedical materials. For example, the properties of spacer arms connecting two reactive groups, such as their length, composition, cleavability, water solubility, and cell membrane permeability, significantly influence the conjugation between fluorescent dye cross-linkers and biomolecules. These factors play a crucial role in the development of functional detection materials for life science research, as they directly impact the stability, specificity, and overall performance of the resulting conjugates.

2.4.1.2 Triggered by Heat

Thermal cross-linking can be triggered by heat, which utilizes temperature-sensitive reactions, offering precise control over the formation and fracture of bonds. The key reactions include Diels–Alder (D–A) reaction, isocyanate reaction, "Click Chemistry" reaction and [2+2] cycloaddition, etc. D–A reaction refers to the [4+2] cycloaddition between electron-rich conjugated dienes (such as furan or anthracene) and electron-deficient dienophiles (such as maleimides or acrylates).[149–152] The formation and fracture of covalent bonds can be controlled due to the temperature sensitivity of the thermoreversible D–A reaction. The introduction of dynamic covalent bonds into polymer materials has great significance in flexible sensing, electronic technology, and biomedicine. Reversible cross-linked polymers equipped with excellent self-healing capabilities, recyclable processing, and closed-loop recovery properties, are powerful tools for the development of sustainable opto-electronic functional materials. Isocyanates with strong reactivity determined by their highly unsaturated structures are easy to react with compounds containing active hydrogen atoms. The reactions of isocyanates with hydroxyl or amino groups are important chemical reactions in the polyurethane synthesis industry,[153–156] while reactions with epoxides have been mainly used for the related products of epoxy resin-modified polyurethane.[157,158] The 1, 3-dipole cycloaddition between azide and alkyne is a highly efficient "Click chemistry" reaction, which can be carried out at room temperature, producing only 1, 4-triazole products at the presence of Cu(I) catalyst. Furthermore, the isomers of 1, 4-triazole and 1, 5-triazole can be obtained upon heating. "Click Chemistry" has been widely used to synthesize various functional opto-electronic materials attributed to the enormous advantages of high selectivity and compatibility of functional groups, exceptional reactivity with minimal by-products, and straightforward reaction conditions.[159,160] Besides, the resultant five-member rings formed through the Click reaction can also act as isolation groups to reduce dipole-dipole interactions between chromophore moieties, which is beneficial to promoting the poling efficiency of organic second-order NLO materials under electric fields.[11,161–163] The [2+2] cycloaddition of trifluoroethylene groups at high temperatures form a four-membered ring system with good selectivity, which endows polymer films with high tenacity, transparency, and hydrophobicity due to the existence of fluoro-containing groups.[164–167] In one word, cross-linked polymers possessing prominent chemical and thermal stabilities are promising candidates for optical waveguides or optical device materials. However, high temperature also presents strict requirements for the thermal stability of substrates.

2.4.1.3 Triggered by Light

In contrast to the thermal process, photochemical cross-linking triggered by light offers unique advantages, including low-temperature reactions and spatial control through light irradiation. Several mature techniques for the photochemical cross-linking are presented here. Thiol–alkene/alkyne reaction is a typical Click chemical process, which has been extensively used in the formation of photopolymer networks due to its extremely fast kinetics and great efficiency.[168–171] The cross-linked system is produced by the reaction between tri- or tetrafunctional thiol crosslinkers and monomers or polymers containing alkene or alkyne. The opto-electronic performance of the resultant polymer network can be selectively controlled by adjusting the percentage of crosslinkers. Nitrogen radicals and diazirine intermediates enable protein cross-linking and labeling, with applications in hydrogels, fluorescent imaging, and microparticle synthesis.[172–176] These processes utilize photogenerated species for precise network formation. Photoinduced nitrilimine intramolecular cross-linking of linear precursor polymers serves as a versatile platform technology for developing novel imaging agents.[177–179] The varied tetrazole content of the lateral polymer chains allows precise control over the size, fluorescence, and reactive properties of the resultant functional nanomaterials. Systems such as coumarin, cinnamic acid, and anthracene undergo wavelength-specific cycloadditions, allowing for fine-tuning of network density and properties.[180–185] Substituent modification enables responsiveness to visible light, expanding their applicability. Most [2+2] photocycloaddition systems require strict parallel alignment of the double bonds and a distance of no bigger than 4.5 Å to gain the reaction activity.[186,187] Taking the spatiotemporal advantages of light, the hierarchical patterns of wrinkles are generated by programmed photomask exposure, and the morphology of which can be regulated by adjusting the monomer proportion and the film thickness.[188] Moreover, the photo-controlled reversible cross-linking allows the recording, erasure, and rewriting of the surface pattern, offering potential applications in intelligent luminescent coatings with anti-counterfeit and encapsulation of LED chips for displaying.[189,190]

2.4.1.4 Triggered by Radicals

Radical polymerization has long been a cornerstone technique in the synthesis of polymer networks, owing to its broad monomer compatibility and tolerance for diverse functional groups. Among these, conventional free radical polymerization (FRP) remains widely used due to its simplicity and effectiveness in producing cross-linked polymers. This process involves the copolymerization of a monovinyl monomer with a small proportion of divinyl monomer, which serves as a cross-linking agent to interconnect polymer chains, forming a three-dimensional network. However, FRP suffers from inherent limitations, such as a lack of control over chain growth, resulting in the formation of "dead" polymer chains incapable of undergoing post-functionalization or chain extension.[191] This restricts the adaptability of such networks for advanced applications, particularly in opto-electronic materials, where precision and tunability are critical.

To address these shortcomings, controlled radical polymerization (CRP) techniques, also known as reversible-deactivation radical polymerization (RDRP),[192] have emerged as a transformative advancement in polymer science. Techniques such as nitroxide-mediated polymerization (NMP),[193] atom transfer radical polymerization (ATRP),[194] and reversible addition–fragmentation chain transfer (RAFT) polymerization,[195] have enabled precise control over polymerization processes. These methods work well by reversibly deactivating propagating radial species and offering unparalleled control over molecular weight, which can achieve narrow molecular weight distributions, and the synthesis of complex architectures, such as block copolymers with unique topologies like graft and star polymers.[103] A significant advantage of RDRP lies in its ability to produce polymer network with dormant and reactivatable species, which can undergo post-synthetic transformations, granting immense flexibility in tailoring polymers for specific applications.[196] This capability has positioned ATRP and RAFT as pivotal techniques in preparing advanced polymer networks, particularly in designing opto-electronic materials, where precise molecular organization and customizable properties are essential.[197] By eliminating the dependence on the use of elevated temperature, photoactivated RDRP systems (photoRDRP) introduce spatial and temporal control over polymerization, enabling the precise fabrication of advanced polymer networks. This innovation has led to the development of novel concepts such as structurally tailored and engineered macromolecular (STEM) gels, living additive manufacturing, and photo expandable/transformable-polymer networks.[196] These systems provide unique advantages, including the ability to create dynamic materials with on-demand properties, self-healing capabilities, and enhanced mechanical performance, all of which are critical for high-performance opto-electronic applications.

2.4.2 Physical Cross-linking

In contrast to chemical cross-linking, physical cross-linking offers flexibility and responsiveness to external stimuli and relies on non-covalent interactions such as hydrogen bonding, van der Waals forces, or ionic interactions. This approach is usually reversible and depends on external factors such as temperature, pressure, or solvent conditions, making it a dynamic method for tailoring material properties in opto-electronic applications. Various mechanisms of physical cross-linking with illustrative examples are discussed here.

2.4.2.1 Physical Entanglement

Long-chain polymers, such as high molecular weight polyethylene, can become physically entangled, accompanying with the increased viscosity and improved mechanical properties.[198–200] This entanglement is a non-covalent yet robust means of cross-linking that supports enhanced durability in opto-electronic devices.

2.4.2.2 Hydrogen Bonding

Polymers like polyvinyl alcohol (PVA) can form hydrogen bonds at aggregated states, which contribute to reversible cross-linking.[201,202] This mechanism is particularly useful in materials requiring adaptive mechanical and optical properties.

2.4.2.3 Crystallization

Semi-crystalline polymers such as polyethylene contain crystalline regions that serve as physical cross-linking sites.[203,204] These regions enhance mechanical strength and thermal resistance.

2.4.2.4 Ionic Interactions

Polyelectrolytes, such as sodium polyacrylate, rely on ionic interactions for physical cross-linking.[205–207] These interactions improve structural integrity and provide unique ion-conducting properties beneficial in energy storage and conversion devices.

2.4.2.5 Self-assembly and Supramolecular Interactions

Non-covalent interactions, including π–π stacking and host–guest interactions, contribute to the physical cross-linking of supramolecular polymers.[208–211] These self-assembled systems enable dynamic and reversible properties, which are essential for smart opto-electronic materials.

2.4.3 Radiation Cross-linking

Radiation cross-linking involves high-energy electron beams or other ionizing radiation acting on polymer chains, where the original chemical bonds are broken to produce reactive radicals to form new cross-links. During this process, the broken polymer chains may either combine with other chains, or capture a hydrogen atom to form shorter chains. The latter usually leads to a reduction in molecular weight, while residual free radicals cause oxidative degradation, which undermines the long-term stability of the polymer network.[212] The cross-linking reaction and degradation reaction are simultaneously competitive processes for most polymers under radiation. The relative dominance of these two processes depends on several factors, including the radiation dose, dose rate and radiation environment, such as temperature and atmosphere, and the chemical structures of the polymers.[213] Radiation cross-linking not only improves the physical and mechanical properties of the polymer, such as heat resistance, stress cracking resistance, and electrical insulation, but also endows its new properties by modifying the secondary structure of the polymer.[214] For instance, high-performance engineering plastics, used in opto-electronic applications, can achieve superior mechanical and thermal properties through irradiation cross-linking. Cross-linked polyethylene can retain its structural integrity at high temperatures, improving its values in applications such as cables and pipes. In addition, unlike traditional chemical cross-linking methods, radiation cross-linking avoids the generation of harmful substances due to the absence of chemical cross-linking agents. As an efficient, environmentally friendly, and advanced modification method, radiation cross-linking has significant potential to expand the application scope of organic opto-electronic materials. Different crosslinking mechanisms are essential for adjusting polymer properties. Appropriate cross-linking methods should be chosen to adjust the mechanical, thermal, and chemical resistance of materials, making them indispensable for specific applications such as insulating layers in electronic components or optical lenses requiring extreme durability.

2.4.4 The Advantages of Cross-linking in the Regulation of Aggregated Structures

Conventional linear polymers often exhibit inferior mechanical properties, thermal stability, and solvent resistance. These shortcomings primarily originate from the weak interactions among polymer chains, which can be easily disrupted by stimuli such as heat and force. In contrast, cross-linked technology, through its diverse methods and precise control in aggregated states, creates a strong three-dimensional network by linking neighboring polymer chains through covalent bonds. By carefully controlling aggregation states, cross-linking not only enhances the mechanical, thermal and chemical properties of the polymers, but also enables the fine-tuning of optical and electronic behaviors, serving as a foundational technology in the design of opto-electronic materials.

In organic opto-electronic materials, the role of cross-linking extends beyond structural reinforcement. It enables precise molecular alignment, enhances stability under operational stresses, and provides the means to create multifunctional materials for applications such as LEDs, photovoltaic devices, and electro-optic modulators. Cross-linking systems have been pivotal in overcoming issues like orientation instability, mechanical degradation, and chemical resistance in these cutting-edge technologies.

2.4.4.1 Fixed Orientation and Structural Stability

One of the most critical contributions of cross-linking is its ability to stabilize molecular orientations within polymer networks, a property essential for opto-electronic applications requiring anisotropy. Covalent cross-linked networks improve the overall stability of polymers by strengthening intermolecular connections and influencing their aggregation structure. This feature is crucial for materials used in second-order nonlinear optics or liquid crystal polymers, where precise alignment is required for optimal performance. In contrast to linear polymers, which typically struggle to maintain molecular orientation due to chain mobility, cross-linked networks can effectively stabilize the orientation, realizing the material from isotropic to anisotropic crossing (Figure 2.12a). Liquid crystals (LCs), often considered the fourth state of matter due to their intrinsic anisotropy, combine the fluidic

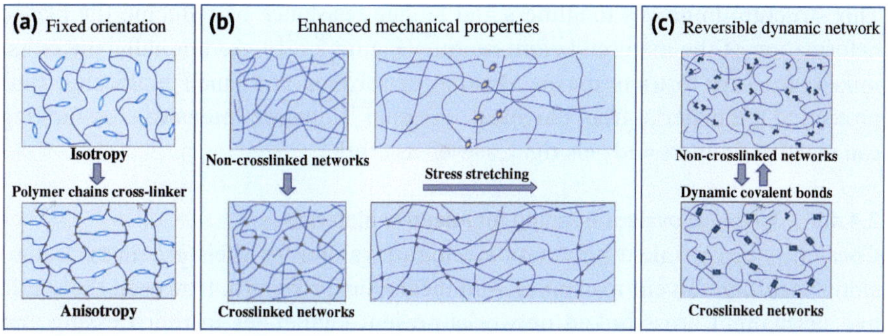

Figure 2.12 (a) Schematic diagram showing the function of covalently cross-linked network fixed orientation. (b) Schematic diagram showing the difference between linear and cross-linked polymers and the corresponding mechanical properties. (c) Schematic showing the reversible cycle of dynamic covalent bonding in polymers.

properties of liquids with the crystalline long-range ordering of solids. LCs orientation is usually achieved by mechanical stretching,[215–217] surface induction,[218–221] magnetoelectric induction,[222–224] and shear orientation.[225–228] Subsequent cross-linking solidifies this orientation, and the resultant macroscopically anisotropic polymer network exhibits a variety of anisotropic properties of material properties, such as optical properties, mechanical properties, dielectric properties, polarization behaviors, and thermal expansion coefficients.

Traditionally, single-domain liquid crystal elastomers (LCEs) are prepared by a one- or two-step cross-linked process. The process directly mixes liquid crystal monomers or liquid crystal polymer precursors with cross-linking agents to form a cross-linked network structure. The most prevalent one-step preparation involves the radical polymerization of (meth)acrylate-based liquid crystal monomers, sometimes referred to as reactive liquid crystal radicals.[217] In 1991, Finkelmann et al.[215] first proposed a two-step method for the preparation of single-domain LCEs based on polysiloxanes, employing a hydrosilylation reaction catalyzed by platinum, in which the Si—H bonds are reacted by addition to unsaturated functional groups such as vinyl monomers. The initial step forms a weakly cross-linked polymer with moderate mechanical strength by controlling the extent of the Si—H addition reaction. The second step involves stretching and orienting the weakly cross-linked polymer under mechanical force, completing the cross-linking reaction under sustained mechanical load to achieve orientation fixation and obtain a fully reacted LCE.

2.4.4.2 Enhanced Mechanical Properties

Building upon the structural stability provided by cross-linking, the mechanical properties of polymers are significantly enhanced. The formation of covalently cross-linked networks significantly enhances the tensile strength, toughness, modulus, and stiffness of the polymers, making the material more resistant to wear. These properties are particularly valuable in opto-electronic applications where materials are subjected to repetitive mechanical stress or extreme environmental conditions.

This structure improves toughness and fatigue resistance by reducing the plastic deformation of the material before rupture (Figure 2.12b). Additionally, the cross-linked structure restricts the penetration of solvents and small molecules, thus protecting the material from chemical corrosion. However, intensive cross-linking can make elastomers and gels rigid, as well as brittle.

2.4.4.3 Dynamic Covalent Bonding for Adaptive Materials

Covalently cross-linked polymers are highly valued for their excellent dimensional stability, solvent resistance, and mechanical strength. However, the stable and permanent cross-linked networks present challenges in reprocessing and recycling, leading to serious environmental pollution and waste of resources. Introducing dynamic covalent bonds into these networks to construct covalent adaptive networks (CANs), allows for the reversible exchange of bonds under specific stimuli, such as temperature, light, and pH (Figure 2.12c). This adaptability grants thermoset materials properties like plasticity, self-healing, recyclability, and shape memory, enhancing their functionality. The introduction of dynamic covalent bonds realizes the dynamic regulation of the aggregation structure.

Dynamic covalent bond-based polymer plastics have great potential advantages in degradability and recycling. Polyurethane (PU) plastics, commonly used daily, are often downcycled into lower-value products due to chemical inertness. In 2020, Ellison and Dichtel et al. developed a method to reprocess PU plastics by introducing dibutyltin dilaurate (DBTDL) to catalyze the dynamic carbamate exchange reaction, enabling recycling or even product upgrading.[229] Furthermore, plastic materials based on dynamic covalent bonding have also been applied to microscale fabrication. Zhao and Xie et al. demonstrated this strategy by combining solid-state plasticity with conventional treatments to fabricate complex, porous objects. Initially, porous polydimethylsiloxane (PDMS) was first made by using conventional methods.[230] Subsequently, using an alkali catalyst, 2D PDMS films can be permanently molded into 3D objects using thermally induced solid-state plasticity without sacrificing the original pores. In contrast to commercial PDMS and anionic PDMS, which are limited in their deformation capabilities, this dynamically shaped PDMS maintains its macroscopic shape and microstructure integrity. As the result, highly porous ceramics with complex shapes are obtained, which are difficult to be prepared by other means. With the utilization of these porous ceramics as templates, similar reshaping processes showcase the promising potential of dynamic chemistry in developing micro-sized materials.

In summary, the chemical bonding within covalently cross-linked networks not only significantly affects the aggregation structure of polymers, but also has an important impact on their mechanical, thermal and chemical properties. The functional endings of the cross-linkers and the number of reactive sites on the monomers or polymers are variable. Careful molecular design should be carried out to ensure the required cross-linking density and mechanical properties, and appropriate crosslinking mechanisms should be selected to meet the requirements of the final products.

Cross-linking technology can regulate the aggregated structures, which makes it an indispensable tool in the molecular design of advanced opto-electronic materials. The technology helps to stabilize molecular orientations by strengthening intermolecular connections, which is crucial for materials requiring precise alignment. The modulated intermolecular interactions after cross-linking can also suppress the quenching effect or facilitate special emission behaviors, which is critical in opto-electronic applications such as OLEDs, solar cells, and sensors. Furthermore, cross-linking enables the formation of ordered structures, such as microphase-separated domains or hierarchical networks, which are essential for achieving specific opto-electronic properties. By providing enhanced thermal and chemical resistance, cross-linking technology can directly influence the mechanical properties of materials and allow researchers to fine-tune rigidity, elasticity, and durability based on specific application needs. Advanced cross-linking strategies, including dynamic covalent bonds and reversible interactions, provide self-healing capabilities and enable recyclability. These features address issues of material degradation and environmental sustainability, offering significant advantages for long-term applications. Cross-linked polymers with enhanced mechanical and opto-electronic properties are ideal for flexible displays, sensors, and energy-harvesting devices. In the future, intelligent cross-linking technologies, such as photoresponsive or stimuli-responsive systems, can further enable the development of materials capable of adapting to environmental changes, broadening their utility in dynamic applications like wearable electronics and soft robotics. Cross-linking processes compatible with additive manufacturing and microfabrication techniques, such as 3D printing and microfluidics, will pave the way for next-generation opto-electronic devices. In addition, micro- and nanoscale cross-linked structures should enable the design of advanced drug delivery systems and bioimaging tools with controlled release and targeted functionality. By addressing challenges such as sustainability, recyclability, and advanced manufacturing compatibility, future developments in cross-linking will drive developments across numerous fields, solidifying its role in opto-electronic materials.

2.5 Host–Guest System

The concept of "host–guest chemistry" introduced in the 1970s by Charles J. Pedersen, Donald James Cram, and Jean-Marie Lehn et. al. describes the chemical properties of complexes formed by two or more components, such as molecules, ions, or atoms, and held together by non-covalent interactions.[231–233] The host–guest system is based on two fundamental principles: complementarity and preorganization. Complementarity refers to the structural and geometric compatibility between the host and guest, which is akin to fitting a key into a lock. Once the host is structurally predisposed to interact with the guest, the binding efficiency enhances in preorganization systems, thereby maximizing the stability of the complex.

As illustrated in Figure 2.13a, crown ethers exhibit size-specific binding to metal cations based on ring size, thus, larger crown ethers preferentially bind larger

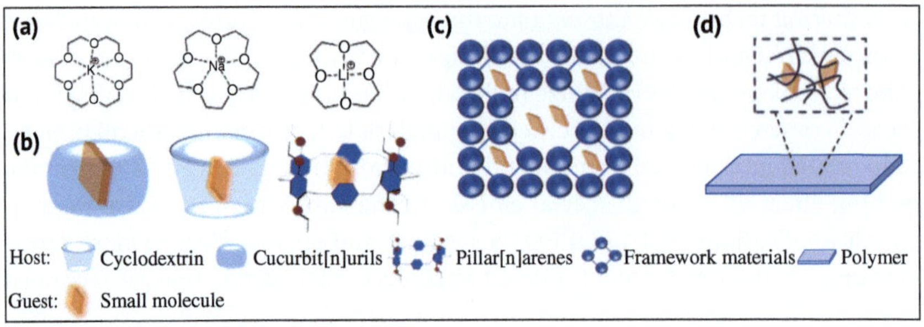

Figure 2.13 The different host–guest systems with (a) crown ethers, (b) Cyclodextrins, Cucurbit[n]urils, and Pillar[n]arenes, (c) frameworks, and (d) polymers as hosts.

cations, and smaller ones selectively interact with smaller cations. Hosts like CBs, CDs, and Pillar[n]arenes demonstrate their ability to encapsulate organic molecules (Figure 2.13b).[234–236] Recently, advancements have extended the scope of host–guest systems to include frameworks and polymers that incorporate physically blended components or small-molecule dopants (Figure 2.13c,d).[237–239] These developments showcase the host–guest systems based on the versatility of non-convent interactions with tailored properties, particularly in opto-electronic materials.

2.5.1 Non-covalent Interactions in the Host–Guest System

Non-covalent interactions are the cornerstone of host–guest systems, offering precise control over molecular aggregation and functional organization, making them invaluable in opto-electronic materials, including hydrogen bonding, hydrophobic interactions, electrostatic interactions, and π–π interactions (Figure 2.14).

2.5.1.1 Hydrogen Bonding

Hydrogen bonding is the common interaction in host–guest systems, also a special type of interaction that exists between a hydrogen atom and an atom with a relatively large electronegativity (such as nitrogen, oxygen, fluorine, etc.). For example, polyvinyl alcohol (PVA) with a large number of hydroxyl groups as a host matrix, can form hydrogen bonds with guest molecules containing atoms with high electronegativity. Through doping heterocyclic polynuclear aromatic chromophores into the PVA matrix by Zhao and Yang et. al.[239] the long-lived RTP emission with lifetime up to 2044.86 ms has been achieved, due to the abundant hydrogen bonding. Similarly, polymers such as polymethyl methacrylate (PMMA) and polylactic acid (PLA) with abundant carbonyl groups can also form hydrogen bonding with guest molecules containing active hydrogen atoms. By immobilizing phosphors onto β-cyclodextrin (β-CD), RTP emission can be detected for the suppressed nonradiative relaxation and quenching effect.[235,240]

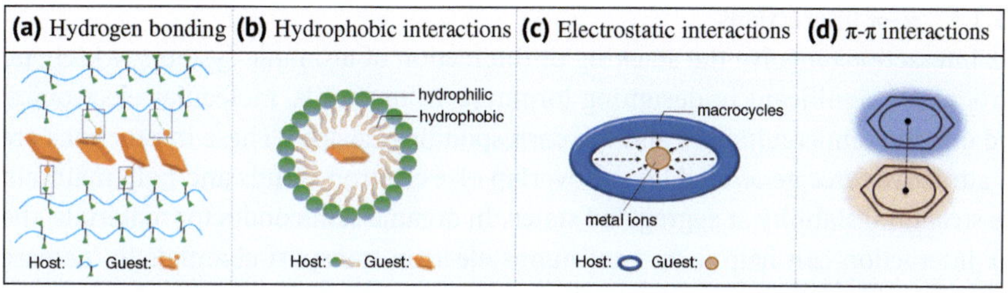

Figure 2.14 The types of non-covalent interactions in the host–guest system: (a) Hydrogen bonding. (b) Hydrophobic interactions. (c) Electrostatic interactions. (d) π–π interactions.

2.5.1.2 Hydrophobic Interactions

In contrast to the directional nature of hydrogen bonding, hydrophobic interactions are non-directional and tend to drive the aggregation of non-polar molecules in polar solvents. These interactions are fundamental in the formation of organized assemblies, such as micelles, which are widely used in opto-electronic applications. The hydrophobic effect drives the clustering of non-polar molecules in polar solvents, exemplified by the separation of oil and water. This phenomenon can be exploited in the design of materials with tunable electronic properties by controlling the aggregation state of molecules. For example, CBs and CDs, which possess an adjustable hydrophobic cavity, can selectively bind hydrophobic guest molecules, enhancing the stability and functionality of opto-electronic materials. This property is particularly useful in applications like molecular sensors and organic semiconductors, where the selective binding of hydrophobic molecules can enhance the overall performance.[234,241] Furthermore, hydrophobic interactions can also influence solubility and dispersibility, which are crucial for fabricating thin films in devices like OSCs. For example, the hydrophobic interactions between guest molecules and surfactants in polymer blends can improve the charge transport properties in organic photovoltaics, thus enhancing their efficiencies.

2.5.1.3 Electrostatic Interactions

Electrostatic interaction refers to the attractions or repulsions of charged particles or polar molecules (dipolar interactions), influencing the organization of host–guest systems. For example, Adachi and Tanaka et. al. utilized dipolar interactions between the host molecule 3,3′-di(9H-carbazol-9-yl)–1,1′-biphenyl (mCBP) and the guest molecule HDT-1 to form an efficient emission layer (EML) for OLEDs.[242] The dipolar interactions counteracted the large surface potential generated by the orientation of guest molecules, improving carrier transport and overall device performance. Hu et. al. constructed a dynamic cross-linking supramolecular binder by forming strong ion-dipole interactions between β-cyclodextrin polymer (β -CDp), choline glycerophosphate (Cg), and supramolecular crosslinking agent guest (2AD).[243] The system achieved a relatively high reversible specific capacity of 703 mAh g^{-1} at 1C, even after 800 cycles with an average CE over 97% due to strong ion-dipole interactions and fast Li_2S redox reaction.

2.5.1.4 π–π Interactions

π–π interactions involve the stacking or interaction of aromatic systems, which are particularly significant in designing luminescent materials, molecular electronics, and organic semiconductors and the corresponding devices. These interactions are an attractive force generated by the overlap of π electron clouds and help maintain the structural stability at aggregated states. In organic semiconductor materials, the π–π interaction can help form continuous electron transport channels to increase carrier mobility.[244]

Host–guest chemistry has been instrumental in advancing the design of functional materials for opto-electronic applications. By leveraging non-covalent interactions precisely, host–guest systems enable the assembly of molecular components into highly organized structures with tunable properties. This aggregation drives unique photophysical behaviors, such as prolonged phosphorescence lifetimes, efficient charge transport, and controlled light emission – advances critical for applications in organic photovoltaics, OLEDs, molecular sensors, and information storage. The incorporation of frameworks and polymers as host matrices further broadens their scope, offering unprecedented opportunities for the development of next-generation opto-electronic devices. The versatility of host–guest systems supported by their diverse interaction mechanisms, ensures their wide applications in cutting-edge material science and industrial innovation. By uniting molecular components into aggregated states with superior properties, host–guest chemistry exemplifies the power of supramolecular design in advancing opto-electronic technologies.

2.6 Conclusion

In summary, various molecular engineering and fabrication processes for molecular aggregates have been introduced, which can be applied in opto-electronic materials with different applications. With the combination of rational molecular design and external forces, the different types of molecular aggregates with optimized molecular packing and intermolecular interactions can promote the development of opto-electronic fields from aggregation levels.

References

1 Li, Q. and Li, Z. (2020). Molecular packing: Another key point for the performance of organic and polymeric optoelectronic materials. *Accounts of Chemical Research* 53: 962–973.

2 Di, C., Zhang, F., and Zhu, D. (2013). Multi-functional integration of organic field-effect transistors (OFETs): Advances and perspectives. *Advanced Materials* 25: 313–330.

3 Muccini, M. (2006). A bright future for organic field-effect transistors. *Nature Materials* 5: 605–613.

4 Torsi, L., Magliulo, M., Manoli, K., and Palazzo, G. (2013). Organic field-effect transistor sensors: A tutorial review. *Chemical Society Reviews* 42: 8612–8628.

5 Zaumseil, J. and Sirringhaus, H. (2007). Electron and ambipolar transport in organic field-effect transistors. *Chemical Reviews* 107: 1296–1323.

6 Yook, K.S. and Lee, J.Y. (2012). Organic materials for deep blue phosphorescent organic light-emitting diodes. *Advanced Materials* 24: 3169–3190.

7 Song, J., Lee, H., Jeong, E.G. et al. (2020). Organic light-emitting diodes: Pushing toward the limits and beyond. *Advanced Materials* 32: 1907539.

8 Uoyama, H., Goushi, K., Shizu, K. et al. (2012). Highly efficient organic light-emitting diodes from delayed fluorescence. *Nature* 492: 234–238.

9 Farinola, G.M. and Ragni, R. (2011). Electroluminescent materials for white organic light emitting diodes. *Chemical Society Reviews* 40: 3467–3482.

10 Wu, W., Tang, R., Li, Q., and Li, Z. (2015). Functional hyperbranched polymers with advanced optical, electrical and magnetic properties. *Chemical Society Reviews* 44: 3997–4022.

11 Li, Z.A., Wu, W., Li, Q. et al. (2010). High-generation second-order nonlinear optical (NLO) dendrimers: convenient synthesis by click chemistry and the increasing trend of NLO effects. *Angewandte Chemie International Edition* 49: 2763–2767.

12 Jiang, M.H. and Fang, Q. (1999). Organic and semiorganic nonlinear optical materials. *Advanced Materials* 11: 1147–1151.

13 Mondal, S., Tanari, P., Roy, S. et al. (2023). Autonomous self-healing organic crystals for nonlinear optics. *Nature Communications* 14: 6589.

14 Günes, S., Neugebauer, H., and Sariciftci, N.S. (2007). Conjugated polymer-based organic solar cells. *Chemical Reviews* 107: 1324–1338.

15 Cheng, P. and Zhan, X. (2016). Stability of organic solar cells: Challenges and strategies. *Chemical Society Reviews* 45: 2544–2582.

16 Duan, L. and Uddin, A. (2020). Progress in stability of organic solar cells. *Advanced Science* 7: 1903259.

17 Li, R., Hu, W., Liu, Y., and Zhu, D. (2010). Micro- and nanocrystals of organic semiconductors. *Accounts of Chemical Research* 43: 529–540.

18 Yang, X., Wang, L., Wang, C. et al. (2008). Influences of crystal structures and molecular sizes on the charge mobility of organic semiconductors: Oligothiophenes. *Chemistry of Materials* 20: 3205–3211.

19 Podzorov, V., Menard, E., Borissov, A. et al. (2004). Intrinsic charge transport on the surface of organic semiconductors. *Physical Review Letters* 93: 086602.

20 Sundar, V.C., Zaumseil, J., Podzorov, V. et al. (2004). Elastomeric transistor stamps: Reversible probing of charge transport in organic crystals. *Science* 303: 1644–1646.

21 Najafov, H., Lee, B., Zhou, Q. et al. (2010). Observation of long-range exciton diffusion in highly ordered organic semiconductors. *Nature Materials* 9: 938–943.

22 Hulliger, J. (1994). Chemistry and crystal growth. *Angewandte Chemie International Edition* 33: 143–162.

23 Klauk, H. (2010). Organic thin-film transistors. *Chemical Society Reviews* 39: 2643–2666.

24 Liang, C. (2024). Organic polymorphs based on AIE-active molecules: Preparation, characterization, and application. *Crystal Growth & Design* 24: 7322–7341.

25 Nancollas, G.H. (1979). The growth of crystals in solution. *Advances in Colloid and Interface Science* 10: 215–252.

26 Vekilov, P.G. (2007). What determines the rate of growth of crystals from solution? *Crystal Growth & Design* 7: 2796–2810.

27 Hu, R., Leung, N.L.C., and Tang, B.Z. (2014). AIE macromolecules: Syntheses, structures and functionalities. *Chemical Society Reviews* 43: 4494–4562.

28 Jiang, H. and Kloc, C. (2013). Single-crystal growth of organic semiconductors. *MRS Bulletin* 38: 28–33.

29 Li, W., Huang, Q., Mao, Z. et al. (2022). A dish-like molecular architecture for dynamic ultralong room-temperature phosphorescence through reversible guest accommodation. *Nature Communications* 13: 7423.

30 Zhu, Y., Haut, B., Halloin, V., and Delplancke-Ogletree, M.P. (2005). Investigation of crystallization kinetics of sodium bicarbonate in a continuous stirred tank crystallizer. *Journal of Crystal Growth* 282: 220–227.

31 Jiang, S., Zhang, Y., and Li, Z. (2019). A new industrial process of NaHCO3 and its crystallization kinetics by using the common ion effect of Na_2CO_3. *Chemical Engineering Journal* 360: 740–749.

32 Jiang, S., Wang, J., Qi, Q. et al. (2019). Organic polymorphs with fluorescence switching: Direct evidence for mechanical and thermal modulation of excited state transitions. *Chemical Communications* 55: 3749–3752.

33 Johannsen, I., Groth-Andersen, L., and Nielsen, K.F. (1981). Crystal growth of tetrathiafulvalece-tetracyanoquinodimethane (TTF-TCNQ) from oriented seeds. *Journal of Crystal Growth* 51: 627–628.

34 Xu, Y., Wang, K., Zhang, Y. et al. (2016). Fluorescence mutation and structural evolution of a π-conjugated molecular crystal during phase transition. *Journal of Materials Chemistry C* 4: 1257–1262.

35 Zhang, Z., Song, X., Wang, S. et al. (2016). Two-dimensional organic single crystals with scale regulated, phase-switchable, polymorphism-dependent, and amplified spontaneous emission properties. *The Journal of Physical Chemistry Letters* 7: 1697–1702.

36 Kloc, C., Simpkins, P.G., Siegrist, T., and Laudise, R.A. (1997). Physical vapor growth of centimeter-sized crystals of α-hexathiophene. *Journal of Crystal Growth* 182: 416–427.

37 Laudise, R.A., Kloc, C., Simpkins, P.G., and Siegrist, T. (1998). Physical vapor growth of organic semiconductors. *Journal of Crystal Growth* 187: 449–454.

38 Horowitz, A. and Horowitz, Y.S. (1986). In praise of the confined vertical growth of semiconductors. *Materials Research Bulletin* 21: 1123–1129.

39 Sherwood, J.N. and Thomson, S.J. (1960). Growth of single crystals of anthracene. *Journal of Scientific Instruments* 37: 242.

40 Ai, Q., Chen, P., Feng, Y., and Xu, Y. (2017). Enhanced crystalline perfection and fluorescence properties of p-terphenyl single crystals grown by the vertical

Bridgman technique with a novel modified growth vessel. *Journal of Applied Crystallography* 50: 278–282.

41 Suthan, T., Dhanaraj, P., Rajesh, N. et al. (2011). Growth and characterization of benzil single crystals using nanotranslation by the modified vertical Bridgman technique. *CrystEngComm* 13: 4018–4024.

42 Tomaszewski, P.E. (2002). Jan czochralski—father of the czochralski method. *Journal of crystal growth* 23: 1–4.

43 Shimizu, M., Tachibana, M., Inoue, K., and Kojima, K. (1997). Melt growth and defects in N-methylurea crystals. *Journal of Crystal Growth* 177: 135–139.

44 Arivanandhan, M., Sankaranarayanan, K., Sanjeeviraja, C. et al. (2005). Optical frequency doubling in microtube Czochralski (μT-CZ) grown benzophenone single crystals. *Journal of Crystal Growth* 281: 596–603.

45 Wang, W.L., Wang, M., and Huang., W.D. (2004). A novel growth method for organic nonlinear optical crystals. *Optical Materials* 27: 609–612.

46 Wang, W., Huang, W., Ma, Y., and Zhao, J. (2004). Oriented growth of benzophenone crystals from undercooled melts. *Journal of Crystal Growth* 270: 469–474.

47 Demianets, L., Emelchenko, G., Hesse, J. et al. (1980). Growth, properties and applications of narrow-gap semiconductors. *Organic Crystals, Germanates*, 145–219.

48 Hong, I.H., Jie Tan, K., Toh, M. et al. (2013). Impurities in zone-refining anthracene crystals. *Journal of Crystal Growth* 363: 61–68.

49 Yokoo, A., Tomaru, S., Yokohama, I. et al. (1995). A new growth method for long rod-like organic nonlinear optical crystals with phase-matched direction. *Journal of Crystal Growth* 156: 279–284.

50 Stockbarger, D.C. (1936). The production of large single crystals of lithium fluoride. *Review of Scientific Instruments* 7: 133–136.

51 Aggarwal, M.D., Wang, W.S., Choi, J. et al. (1996). Modified Bridgman-Stockbarger growth of a novel NLO organic crystal (2-methoxyphenyl)-methylene-propanedinitrile. *Journal of Crystal Growth* 166: 542–544.

52 Ryu, G. and Yoon, C.S. (1998). Seeded supercooled melt growth and polar morphology of organic nonlinear optical crystal, meta-nitroaniline (mNA). *Journal of Crystal Growth* 191: 190–198.

53 Whitesides, G.M. and Grzybowski, B. (2002). Self-assembly at all scales. *Science* 295: 2418–2421.

54 Chang, R., Yuan, C., Zhou, P. et al. (2024). Peptide self-assembly: From ordered to disordered. *Accounts of Chemical Research* 57: 289–301.

55 Whitesides, G.M. and Boncheva, M. (2002). Beyond molecules: Self-assembly of mesoscopic and macroscopic components. *Proceedings of the National Academy of Sciences* (Vol. 99, pp. 4769–4774).

56 Ariga, K., Nishikawa, M., Mori, T. et al. (2019). Self-assembly as a key player for materials nanoarchitectonics. *Science and Technology of Advanced Materials* 20: 51–95.

57 Mendes, A.C., Baran, E.T., Reis, R.L., and Azevedo, H.S. (2013). Self-assembly in nature: Using the principles of nature to create complex nanobiomaterials. *Wiley Interdisciplinary Reviews: Nanomedicine and Nanobiotechnology* 5: 582–612.

58 Kim, J., Lee, S., Kim, Y. et al. (2023). In situ self-assembly for cancer therapy and imaging. *Nature Reviews Materials* 8: 710–725.

59 Borah, R., Ag, K.R., Minja, A.C., and Verbruggen, S.W. (2023). A review on self-assembly of colloidal nanoparticles into clusters, patterns, and films: Emerging synthesis techniques and applications. *Small Methods* 7: e2201536.

60 Wang, C., Wang, D., Xu, H. et al. (2022). Research progress of regulation of driving forces in short peptide supramolecular self-assembly. *Chinese Journal of Applied Chemistry* 39: 1190–1201.

61 Liao, Q., Li, Q., and Li, Z. (2020). The key role of molecular packing in luminescence property: From adjacent molecules to molecular aggregates. *Advanced materials* 53: 962–973.

62 Huang, Z. and Ma, X. (2020). Tailoring tunable luminescence via supramolecular assembly strategies. *Cell Reports Physical Science* 1: 100167–100193.

63 Song, X., Wang, Y., Wang, C. et al. (2022). Design rules of hydrogen-bonded organic frameworks with high chemical and thermal stabilities. *Journal of the American Chemical Society* 144: 10663–10687.

64 Lin, R.B. and Chen, B. (2022). Hydrogen-bonded organic frameworks: Chemistry and functions. *Chem* 8: 2114–2135.

65 Liu, X., Li, J., Qiu, X., and Pan, Y. (2022). Organic electroluminescent materials possessing intra- and intermolecular hydrogen bond interactions: A mini-review. *Frontiers in Chemistry* 10: 954419.

66 Foster, W., Miyazawa, K., Fukuma, T. et al. (2020). Self-assembly of small molecules at hydrophobic interfaces using group effect. *Nanoscale* 12: 5452–5463.

67 Waghule, T., Saha, R.N., Alexander, A., and Singhvi, G. (2022). Tailoring the multi-functional properties of phospholipids for simple to complex self-assemblies. *Journal of Controlled Release* 349: 460–474.

68 Tian, Y., Zhu, Z., Li, Q. et al. (2021). Rapid visualized hydrophobic-force-driving self-assembly towards brilliant photonic crystals. *Chemical Engineering Journal* 420: 127582.

69 Zhang, Z. and Cui, H.F. (2010). Electrostatic self-assembly technology and its application in medicine field. *Food and Drug* 12: 351–354.

70 Mariani, G., Moldenhauer, D., Schweins, R., and Gröhn, F. (2016). Elucidating electrostatic self-assembly: Molecular parameters as key to thermodynamics and nanoparticle shape. *Journal of the American Chemical Society* 138: 1280–1293.

71 Bian, T., Gardin, A., Gemen, J. et al. (2021). Electrostatic co-assembly of nanoparticles with oppositely charged small molecules into static and dynamic superstructures. *Nature Chemistry* 13: 940–949.

72 Li, X., Chen, C., Niu, Q. et al. (2022). Self-assembly of nanoparticles at solid-liquid interface for electrochemical capacitors. *Rare Metals* 41: 3591–3611.

73 Chen, H., Zhang, W., Li, M. et al. (2020). Interface engineering in organic field-effect transistors: Principles, applications, and perspectives. *Chemical Reviews* 120: 2879–2949.

74 Han, L., Wang, M., Jia, X. et al. (2018). Uniform two-dimensional square assemblies from conjugated block copolymers driven by π-π interactions with controllable sizes. *Nature Communications* 9: 865.

75 Zhuang, W.R., Wang, Y., Cui, P.F. et al. (2019). Applications of π-π stacking interactions in the design of drug-delivery systems. *Journal of Controlled Release* 294: 311–326.

76 Li, G.L., Zhuo, Z., Wang, B. et al. (2021). Constructing π-stacked supramolecular cage based hierarchical self-assemblies via π-π stacking and hydrogen bonding. *Journal of the American Chemical Society* 143: 10920–10929.

77 Li, H., Shi, J., Deng, J. et al. (2020). Intermolecular π-π conjugation self-assembly to stabilize surface passivation of highly efficient perovskite solar cells. *Advanced Materials* 32: 1907396.

78 Pan, M., Wei, Z., Xu, Y., and Su, C.Y. (2017). Coordination assembly of metal-organic materials. *Progress in Chemistry* 29: 47–74.

79 Liang, L., Zhao, W., Yang, X.J., and Wu., B. (2022). Anion-coordination-driven assembly. *Accounts of Chemical Research* 55: 3218–3229.

80 Hou, Y.J., Wu, K., Wei, Z.W. et al. (2018). Design and enantioresolution of homochiral Fe(II)-Pd(II) coordination cages from stereolabile metalloligands: Stereochemical stability and enantioselective separation. *Journal of the American Chemical Society* 140: 18183–18191.

81 Zhao, Z., Lei, S., Zeng, M., and Huo, M. (2024). Recent progress in polymerization-induced self-assembly: From the perspective of driving forces. *Aggregate* 5: e418.

82 Sun, H., Liu, D., and Du, J. (2019). Nanobowls with controlled openings and interior holes driven by the synergy of hydrogen bonding and π-π interaction. *Chemical Science* 10: 657–664.

83 Mai, Y., An, Z., and Liu, S. (2022). Self-assembled materials and applications. *Macromolecular Rapid Communications* 43: 2200481.

84 Hou, C., Gao, L., Wang, Y., and Yan, L.T. (2022). Entropic control of nanoparticle self-assembly through confinement. *Nanoscale Horizons* 7: 1016–1028.

85 Zhao, Z., Zhang, H., Lam, J.W.Y., and Tang, B.Z. (2020). Aggregation-induced emission: new vistas at the aggregate level. *Angewandte Chemie International Edition* 59: 9888–9907.

86 Gu, X., Zhang, X., Ma, H. et al. (2018). Corannulene-incorporated AIE nanodots with highly suppressed nonradiative decay for boosted cancer phototheranostics in vivo. *Advanced Materials* 30: 1801065.

87 Zeng, P., Liu, Y., Hu, W., and Zhu, D. (1999). Film-forming techniques of molecular self-assembly. *Physics* 28: 713–719.

88 Wang, Y.S., Yau, S., Chau, L.K. et al. (2019). Functional biointerfaces based on mixed zwitterionic self-assembled monolayers for biosensing applications. *Langmuir* 35: 1652–1661.

89 Fan, H. (2008). Nanocrystal-micelle: Synthesis, self-assembly and application. *Chemical Communications* 12: 1383–1394.

90 Xu, X.F., Pan, C.Y., Zhang, W.J., and Hong, C.Y. (2019). Polymerization-induced self-assembly generating vesicles with adjustable pH-responsive release performance. *Macromolecules* 52: 1965–1975.

91 Li, L., Sun, R., and Zheng, R. (2021). Tunable morphology and functionality of multicomponent self-assembly: A review. *Materials & Design* 197: 109209.

92 Jiang, C.Q., Xu, G.H., and Gao, J.P. (2022). Stimuli-responsive macromolecular self-assembly. *Sustainability* 14: 11738.

93 Larik, F.A., Fillbrook, L.L., Nurttila, S.S. et al. (2021). Ultra-low molecular weight photoswitchable hydrogelators. *Angewandte Chemie International Edition* 60: 6764–6770.

94 Tinkler, J.D., Scacchi, A., Kothari, H.R. et al. (2021). Evaporation-driven self-assembly of binary and ternary colloidal polymer nanocomposites for abrasion resistant applications. *Journal of Colloid and Interface Science* 581: 729–740.

95 Xiang, L., Li, Q., Li, C. et al. (2023). Block Copolymer self-assembly directed synthesis of porous materials with ordered bicontinuous structures and their potential applications. *Advanced Materials* 35: e2207684.

96 Li, C., Rafique, S., and Zhan, Y. (2022). Synergy of Block Copolymers and perovskites: Template growth through self-assembly. *The Journal of Physical Chemistry Letters* 13: 11610–11621.

97 Li, J., Liu, K., Chen, H. et al. (2017). Functional built-in template directed siliceous fluorescent supramolecular vesicles as diagnostics. *ACS Applied Materials & Interfaces* 9: 21706–21714.

98 Schwartz, D.K. (2001). Mechanisms and kinetics of self-assembled monolayer formation. *Annual Review of Physical Chemistry* 52: 107–137.

99 Li, Z., Sun, X., Zheng, X. et al. (2023). Stabilized hole-selective layer for high-performance inverted p-i-n perovskite solar cells. *Science* 382: 284–289.

100 Zhao, X., Chen, X., Yuk, H. et al. (2021). Soft materials by design: Unconventional polymer networks give extreme properties. *Chemical Reviews* 121: 4309–4372.

101 Cheng, Q., Hao, A., and Xing, P. (2020). Stimulus-responsive luminescent hydrogels: Design and applications. *Advances in Colloid and Interface Science* 286: 102301.

102 Wang, J., Li, A., and Li, Z. (2022). The progress of room temperature phosphorescent gel. *Progress in Chemistry* 34: 487–498.

103 Ida, S. (2019). Structural design of vinyl polymer hydrogels utilizing precision radical polymerization. *Polymer Journal* 51: 803–812.

104 Hou, F., Xi, B., Wang, X. et al. (2019). Self-healing hydrogel with cross-linking induced thermo-response regulated light emission property. *Colloids and Surfaces B: Biointerfaces* 183: 110441.

105 Park, J., Park, J., Kim, Y. et al. (2020). Aromatic nonpolar organogels for efficient and stable perovskite green emitters. *Nature Communication* 11: 4638.

106 Li, S., Zhou, H., Li, Y. et al. (2022). Mussel-inspired self-adhesive hydrogels by conducting free radical polymerization in both aqueous phase and micelle phase and their applications in flexible sensors. *Journal of Colloid and Interface Science* 607: 431–439.

107 Qiao, L., Liu, C., Liu, C. et al. (2019). Self-healing alginate hydrogel based on dynamic acylhydrazone and multiple hydrogen bonds. *Journal of Materials Science* 54: 8814–8828.

108 Yu, G., Niu, C., Liu, J. et al. (2023). Preparation and properties of self-cross-linking hydrogels based on chitosan derivatives and oxidized sodium alginate. *ACS Omega* 8: 19752–19766.

109 Wei, Z., Yang, J.H., Liu, Z.Q. et al. (2015). Novel biocompatible polysaccharide-based self-healing hydrogel. *Advanced Functional Materials* 25: 1352–1359.

110 Chen, Y., Qian, W., Chen, R. et al. (2017). One-pot preparation of autonomously self-healable elastomeric hydrogel from boric acid and random copolymer bearing hydroxyl groups. *ACS Macro Letters* 6: 1129–1133.

111 More, A.P. and Chapekar, S. (2024). Irradiation assisted synthesis of hydrogel: A review. *Polymer Bulletin* 81: 5839–5908.

112 Dong, Z., Wang, Y., Wen, D. et al. (2022). Recent progress in environmental applications of functional adsorbent prepared by radiation techniques: A review. *Journal of Hazardous Materials* 424: 126887.

113 Yan, S., Yin, J., Yu, Y. et al. (2009). Thermo- and pH-sensitive poly(vinylmethylether)/carboxymethylchitosan hydrogels crosslinked using electron beam irradiation or using glutaraldehyde as a crosslinker. *Polymer International* 58: 1246–1251.

114 Yang, X., Li, N., Wang, B. et al. (2024). Mechanics-photophysics correlation in tough, stretchable and long-lived room temperature phosphorescence ionogels deciphered by dynamic mechanical analysis. *Angewandte Chemie International Edition* 64: e202419114.

115 Wu, Y., Mu, Y., Luo, Y. et al. (2022). Hofmeister effect and electrostatic interaction enhanced ionic conductive organohydrogels for electronic applications. *Advanced Functional Materials* 32: 2110859.

116 Senebandith, H., Li, D., and Srivastava, S. (2023). Advances, applications, and emerging opportunities in electrostatic hydrogels. *Langmuir* 39: 16965–16974.

117 Yang, H., Wang, W., Huang, Z. et al. (2024). Weak electrostatic force on K+ in gel polymer electrolyte realizes high ion transference number for quasi solid-state potassium ion batteries. *Advanced Materials* 36: 2401008.

118 Liu, X., He, X., Yang, B. et al. (2021). Dual physically cross-linked hydrogels incorporating hydrophobic interactions with promising repairability and ultrahigh elongation. *Advanced Functional Materials* 31: 2008187.

119 Wang, X., Yang, M., Ren, Z. et al. (2024). Mussel-inspired, hydrophobic association-regulated hydrogel electrolytes with super-adhesive and self-healing properties for durable and flexible zinc-ion batteries. *Energy Storage Materials* 70: 103523.

120 Lin, S., Liu, X., Liu, J. et al. (2019). Anti-fatigue-fracture hydrogels. *Science Advances* 5: eaau8528.

121 Deng, J., Liu, H., Liu, D. et al. (2024). Room-temperature phosphorescent tough hydrogels based on ionically crosslinked nonaromatic polymers. *Advanced Functional Materials* 34: 2308420.

122 Ju, H., Zhang, H., Hou, L.X. et al. (2023). Polymerization-induced crystallization of dopant molecules: An efficient strategy for room-temperature phosphorescence of hydrogels. *Journal of the American Chemical Society* 145: 3763–3773.

123 Deng, J., Liu, D., Liu, H. et al. (2024). Persistent room-temperature phosphorescent organohydrogels based on nonaromatic luminogens crosslinked by hydrogen bonds. *Advanced Functional Materials* 34: 2408821.

124 Han, Z., Wang, P., Lu, Y. et al. (2022). A versatile hydrogel network-repairing strategy achieved by the covalent-like hydrogen bond interaction. *Science Advances* 8: eabl5066.

125 An, N., Wang, X., Li, Y. et al. (2019). Healable and mechanically super-strong polymeric composites derived from hydrogen-bonded polymeric complexes. *Advanced Materials* 31: 1904882.

126 Hou, L.X., Ju, H., Hao, X.P. et al. (2023). Intrinsic anti-freezing and unique phosphorescence of glassy hydrogels with ultrahigh stiffness and toughness at low temperatures. *Advanced Materials* 35: 2300244.

127 Huang, Y., Xiao, L., Zhou, J. et al. (2021). Strong tough polyampholyte hydrogels via the synergistic effect of ionic and metal-ligand bonds. *Advanced Functional Materials* 31: 2103917.

128 Filippidi, E., Cristiani, T.R., Eisenbach, C.D. et al. (2017). Toughening elastomers using mussel-inspired iron-catechol complexes. *Science* 358: 502–505.

129 Andersen, A., Krogsgaard, M., and Birkedal, H. (2018). Mussel-inspired self-healing double-cross-linked hydrogels by controlled combination of metal coordination and covalent cross linking. *Biomacromolecules* 19: 1402–1409.

130 Li, B., Li, Z., and Li, H. (2022). Ultrastretchable luminescent nanocomposite hydrogel with self-healing behavior. *ACS Applied Polymer Materials* 4: 2329–2336.

131 Paatelainen, M., Lahikainen, M., Berdin, A. et al. (2023). Hydrogel lasers via supramolecular host-guest complexation. *Advanced Optical Materials* 11: 2300232.

132 Wang, Q., Qi, Z., Xu, H. et al. (2024). Transient cucurbit[7]uril-mediated host-guest complexes for time-dependent fluorescence and information-self-erasing hydrogel. *Chemical Communications* 60: 10342–10345.

133 Wang, J., Huang, Z., Ma, X., and Tian, H. (2020). Visible-light-excited room-temperature phosphorescence in water by cucurbit[8]uril-mediated supramolecular assembly. *Angewandte Chemie International Edition* 59: 9928–9933.

134 Garain, S., Garain, B.C., Eswaramoorthy, M. et al. (2021). Light-harvesting supramolecular phosphors: highly efficient room temperature phosphorescence in solution and hydrogels. *Angewandte Chemie International Edition* 60: 19720–19724.

135 Kratzer, U., Sommersdorf, C., Maier, S. et al. (2021). Tris(hydroxymethyl) aminomethane compatibility with *N*-hydroxysuccinimide ester chemistry: Biotinylation of peptides and proteins in TRIS buffer. *Bioconjugate Chemistry* 32: 1960–1965.

136 McGee, W.M., Mentinova, M., and McLuckey, S.A. (2012). Gas-phase conjugation to arginine residues in polypeptide ions via *N*-hydroxysuccinimide ester-based reagent ions. *Journal of the American Chemical Society* 134: 11412–11414.

137 Ma, X., Bussonniere, A., and Liu, Q. (2017). A facile sonochemical synthesis of shell-stabilized reactive microbubbles using surface-thiolated bovine serum albumin with the Traut's reagent. *Ultrasonics Sonochemistry* 36: 454–465.

138 Tripathi, S.K., Singh, V.P., Gupta, K.C., and Kumar, P. (2013). Hydrophobic and membrane permeable polyethylenimine nanoparticles efficiently deliver nucleic acids in vitro and in vivo. *Journal of Materials Chemistry B* 1: 2515–2524.

139 Auclair, J.R., Boggio, K.J., Petsko, G.A. et al. (2010). Strategies for stabilizing superoxide dismutase (SOD1), the protein destabilized in the most common form of familial amyotrophic lateral sclerosis. *Proceedings of the National Academy of Sciences* (Vol. 107, pp. 21394–21399).

140 Li, F., Chen, X., He, Y., and Peng, Z. (2024). Mucoadhesive thiolated hyaluronic acid/pluronic F127 nanogel formation via thiol–maleimide click reaction for intravesical drug delivery. *ACS Applied Bio Materials* 7: 1976–1989.

141 Magalhães, M.V., Débera, N., Pereira, R.F. et al. (2024). In situ crosslinkable multi-functional and cell-responsive alginate 3D matrix via thiol-maleimide click chemistry. *Carbohydrate Polymers* 337: 122144.

142 Guzmán, D., Ramis, X., Fernández-Francos, X., and Serra, A. (2014). New catalysts for diglycidyl ether of bisphenol A curing based on thiol–epoxy click reaction. *European Polymer Journal* 59: 377–386.

143 Konuray, A.O., Fernández-Francos, X., and Ramis, X. (2017). Analysis of the reaction mechanism of the thiol–epoxy addition initiated by nucleophilic tertiary amines. *Polymer Chemistry* 8: 5934–5947.

144 Niu, P., Li, C., Zhu, J. et al. (2024). Asynchronous ring opening of cyclic carbonate and glycidyl ether induced phase evolution towards heat-free and rapid-bonding superior epoxy adhesive. *Angewandte Chemie International Edition* 63: e202408840.

145 Wang, W., Li, Y., Zhang, H. et al. (2022). Double-interpenetrating-network lignin-based epoxy resin adhesives for resistance to extreme environment. *Biomacromolecules* 23: 779–788.

146 Calcio Gaudino, E., Acciardo, E., Tabasso, S. et al. (2020). Cross-linked cyclodextrins bimetallic nanocatalysts: Applications in microwave-assisted reductive aminations. *Molecules* 25: 410.

147 Kono, H., Noda, J., and Wakamori, H. (2022). Detailed structural characterization of oxidized sucrose and its application in the fully carbohydrate-based preparation of a hydrogel from carboxymethyl chitosan. *Molecules* 27: 6137.

148 Suo, H., Hussain, M., Wang, H. et al. (2021). Injectable and pH-sensitive hyaluronic acid-based hydrogels with on-demand release of antimicrobial peptides for infected wound healing. *Biomacromolecules* 22: 3049–3059.

149 Guillen Campos, J., Stricker, F., Clark, K.D. et al. (2023). Controlled diels–alder "click" strategy to access mechanically aligned main-chain liquid crystal networks. *Angewandte Chemie International Edition* 62: e202214339.

150 Kim, T.D., Luo, J., Ka, J.W. et al. (2006). Ultralarge and thermally stable electro-optic activities from Diels–Alder crosslinkable polymers containing binary chromophore systems. *Advanced Materials* 18: 3038–3042.

151 Shi, Z., Hau, S., Luo, J. et al. (2007). Highly efficient Diels–Alder crosslinkable electro-optic dendrimers for electric-field sensors. *Advanced Functional Materials* 17: 2557–2563.

152 Xu, H., Liu, F., Elder, D.L. et al. (2020). Ultrahigh electro-optic coefficients, high index of refraction, and long-term stability from Diels–Alder cross-linkable binary molecular glasses. *Chemistry of Materials* 32: 1408–1421.

153 Geng, S., Huang, Z., Chen, Y. et al. (2024). Polyurethane acrylate covalent adaptable networks based on the autocatalytic effect of palm oil-based diol. *ACS Applied Polymer Materials* 6: 14760–14771.

154 Jin, Y., Sun, G., Zhou, G. et al. (2024). Fabrication of hydroxyl modified hollow glass microsphere composite isocyanate-based polyimide foam and optimization strategy based on different bonding mechanisms. *Journal of Polymer Science* 63: 443–454.

155 Loo, S.S.L., Tan, K.Y., Zahid, N.I. et al. (2024). Integrating photothermal-responsive shape memory and self-healing polymers in 4D-printed thermally comfortable smart wearables. *ACS Applied Engineering Materials* 2: 2569–2582.

156 Zhang, C., Wang, C., Yang, J. et al. (2001). Electric poling and relaxation of thermoset polyurethane second-order Nonlinear optical materials: Role of cross-linking and monomer rigidity. *Macromolecules* 34: 235–243.

157 He, L., Attard, T. L., Zhou, H., and Brooks, A. (2019). Integrating energy transferability into the connection-detail of coastal bridges using reinforced interfacial epoxy-polyurea reaction matrix composite. *Composite Structures* 216: 89–103.

158 Jia, M., Hadjichristidis, N., Gnanou, Y., and Feng, X. (2021). Polyurethanes from direct organocatalytic copolymerization of p-tosyl isocyanate with epoxides. *Angewandte Chemie International Edition* 60: 1593–1598.

159 Qin, A., Lam, J. W. Y., and Tang, B. Z. (2010). Click polymerization. *Chemical Society Reviews* 39: 2522–2544.

160 Shi, Y., Cao, X., and Gao, H. (2016). The use of azide–alkyne click chemistry in recent syntheses and applications of polytriazole-based nanostructured polymers. *Nanoscale* 8: 4864–4881.

161 Li, Z.A., Yu, G., Wu, W. et al. (2009). Nonlinear optical dendrimers from click chemistry: Convenient synthesis, new function of the formed triazole rings, and enhanced NLO effects. *Macromolecules* 42: 3864–3868.

162 Wu, W., Xu, G., Li, C. et al. (2013). From nitro- to sulfonyl-based chromophores: Improvement of the comprehensive performance of nonlinear optical dendrimers. *Chemistry – A European Journal* 19: 6874–6888.

163 Wu, W., Ye, C., Yu, G. et al. (2012). New hyperbranched polytriazoles containing isolation chromophore moieties derived from AB_4 monomers through Click chemistry under copper(I) catalysis: Improved optical transparency and enhanced NLO effects. *Chemistry – A European Journal* 18: 4426–4434.

164 Cho, S.Y. and Allcock, H.R. (2007). Novel highly fluorinated perfluorocyclobutane-based phosphazene polymers for photonic applications. *Chemistry of Materials* 19: 6338–6344.

165 Ma, H., Chen, B., Sassa, T. et al. (2001). Highly efficient and thermally stable nonlinear optical dendrimer for electrooptics. *Journal of the American Chemical Society* 123: 986–987.

166 Qiu, F.Y., Zhang, M., Du, F.S., and Li., Z.C. (2017). Oxidation degradable aliphatic polycarbonates with pendent phenylboronic ester. *Macromolecules* 50: 23–34.

167 Wu, J., Iacono, S.T., McCandless, G.T. et al. (2015). Utilization of a Meldrum's acid towards functionalized fluoropolymers possessing dual reactivity for thermal crosslinking and post-polymerization modification. *Chemical Communications* 51: 9220–9222.

168 Blasco, E., Yameen, B., Quick, A.S. et al. (2015). Designing π-conjugated polymeric nano- and microstructures via light induced chemistry. *Macromolecules* 48: 8718–8728.

169 Cheng, Z., Tang, R., Wang, R. et al. (2018). Photo-crosslinkable second-order nonlinear optical polymer: Facile synthesis and enhanced NLO thermostability. *Polymer Chemistry* 9: 3522–3527.

170 Wang, R., Cheng, Z., Deng, X. et al. (2020). Photo-crosslinkable second order nonlinear AB2-type monomers: convenient synthesis and enhanced NLO thermostability. *Journal of Materials Chemistry C* 8: 6380–6387.

171 You, X., Wang, P., Tan, Y. et al. (2024). Photo-thermal double-crosslinked second-order nonlinear optical materials with high orientation stability. *Materials Today Chemistry* 37: 101971.

172 Bond, M.R., Zhang, H., Vu, P.D., and Kohler, J.J. (2009). Photocrosslinking of glycoconjugates using metabolically incorporated diazirine-containing sugars. *Nature Protocols* 4: 1044–1063.

173 Chou, C., Uprety, R., Davis, L. et al. (2011). Genetically encoding an aliphatic diazirine for protein photocrosslinking. *Chemical Science* 2: 480–483.

174 Gubbens, J., Ruijter, E., de Fays, L.E.V. et al. (2009). Photocrosslinking and click chemistry enable the specific detection of proteins interacting with phospholipids at the membrane interface. *Chemistry & Biology* 16: 3–14.

175 Png, R.Q., Chia, P.J., Tang, J.C. et al. (2010). High-performance polymer semiconducting heterostructure devices by nitrene-mediated photocrosslinking of alkyl side chains. *Nature Materials* 9: 152–158.

176 Yu, S.H., Boyce, M., Wands, A.M. et al. (2012). Metabolic labeling enables selective photocrosslinking of O-GlcNAc-modified proteins to their binding partners. *Proceedings of the National Academy of Sciences* (Vol. 109, pp. 4834–4839).

177 Täuber, K., Dani, A., and Yuan, J. (2017). Covalent cross-linking of porous poly(ionic liquid) membrane via a triazine network. *ACS Macro Letters* 6: 1–5.

178 Wang, C., Zieger, M.M., Schenzel, A. et al. (2017). Photoinduced tetrazole-based functionalization of off-stoichiometric clickable microparticles. *Advanced Functional Materials* 27: 1605317.

179 Willenbacher, J., Wuest, K.N.R., Mueller, J.O. et al. (2014). Photochemical design of functional fluorescent single-chain nanoparticles. *ACS Macro Letters* 3: 574–579.

180 Choi, W.J., Chung, J.S., Kim, J.j. et al. (2014). Healable properties of polymethacrylate derivatives having photo crosslinkable cinnamoyl side groups with surface hardness control. *Journal of Coatings Technology and Research* 11: 455–459.

181 Ma, Z.P., Song, X., Yang, B.Z. et al. (2023). Fabrication of PEG-anthracene/alginate double-network hydrogels and their application in photolithography. *Journal of Applied Polymer Science* 140: e54723.

182 Maddipatla, M.V.S.N., Wehrung, D., Tang, C. et al. (2013). Photoresponsive coumarin polyesters that exhibit cross-linking and chain scission properties. *Macromolecules* 46: 5133–5140.

183 Sang, Q., Zhu, B., Fan, J. et al. (2023). Waterproof, chemically durable, and closed-loop recyclable poly(vinyl alcohol) plastic enabled by dynamic covalent modification. *ACS Applied Polymer Materials* 5: 9403–9412.

184 Tian, Y., He, Y., Liu, P. et al. (2021). Mild and in situ photo-crosslinking of anthracene-functionalized poly (aryl ether ketone) for enhancing temporal stability of organic NLO materials. *Journal of Materials Science* 56: 5910–5923.

185 Zhang, X., Gärisch, F., Chen, Z. et al. (2022). Self-assembly and photoinduced fabrication of conductive nanographene wires on boron nitride. *Nature Communications* 13: 442.

186 Gnanaguru, K., Ramasubbu, N., Venkatesan, K., and Ramamurthy, V. (1985). A study on the photochemical dimerization of coumarins in the solid state. *The Journal of Organic Chemistry* 50: 2337–2346.

187 Schmidt, G.M.J. (1971). Photodimerization in the solid state. *Pure and Applied Chemistry* 27: 647–678.

188 Gao, X., Li, J., Li, T. et al. (2021). Photo-polymerization induced hierarchical pattern via self-wrinkling. *Advanced Functional Materials* 31: 2106754.

189 Connal, L.A., Vestberg, R., Hawker, C.J., and Qiao., G.G. (2008). Fabrication of reversibly crosslinkable, 3-dimensionally conformal polymeric microstructures. *Advanced Functional Materials* 18: 3315–3322.

190 Liu, Y., Tian, F., Hu, J. et al. (2023). Photochromic polyurethane coatings with cross-linked structure and self-healing behavior based on the FRET effect. *Macromolecular Rapid Communications* 44: 2300097.

191 Hamzehlou, S., Reyes, Y., and Leiza, J.R. (2016). Quantitative study on the homogeneity of networks synthesized by nitroxide-mediated radical copolymerization of styrene and divinylbenzene. *European Polymer Journal* 85: 244–255.

192 Corrigan, N., Jung, K., Moad, G. et al. (2020). Reversible-deactivation radical polymerization (Controlled/living radical polymerization): From discovery to materials design and applications. *Progress in Polymer Science* 111: 101311.

193 Nicolas, J., Guillaneuf, Y., Lefay, C. et al. (2013). Nitroxide-mediated polymerization. *Progress in Polymer Science* 38: 63–235.

194 Krys, P. and Matyjaszewski, K. (2017). Kinetics of atom transfer radical polymerization. *European Polymer Journal* 89: 482–523.

195 Moad, G. (2015). RAFT (Reversible addition–fragmentation chain transfer) crosslinking (co)polymerization of multi-olefinic monomers to form polymer networks. *Polymer International* 64: 15–24.

196 Bagheri, A., Fellows, C.M., and Boyer, C. (2021). Reversible deactivation radical polymerization: From polymer network synthesis to 3D printing. *Advanced Science* 8: 2003701.

197 Bainbridge, C.W.A., Wangsadijaya, A., Broderick, N., and Jin, J. (2022). Living polymer networks prepared by controlled radical polymerization techniques. *Polymer Chemistry* 13: 1484–1494.

198 Wang, Z., Liu, S., Zhu, C., and Xu, J. (2024). Physical-entanglements-supported polymeric form stable phase change materials with ultrahigh melting enthalpy. *Advanced Materials* 36: 2403889.

199 Yazdanbakhsh, A. and Motlagh, G.H. (2024). Simultaneous effects of temperature and backbone length on static and dynamic properties of high-density polyethylene-1-butene copolymer melt: Equilibrium molecular dynamics approach. *e-Polymers* 24: 20240072.

200 Ye, J., Chen, Y., Deng, R. et al. (2024). Robust tetra-armed poly (ethylene glycol)-based hydrogel as tissue bioadhesive for the efficient repair of meniscus tears. *MedComm* 5: e738.

201 Yang, X., Ling, X., Li, Y., and Wang, H. (2024). Simultaneous enhanced mechanical properties and foamability of biodegradable PBAT via physical cross-linking through reactive blending. *ACS Applied Polymer Materials* 6: 13969–13978.

202 Zheng, Y., Huang, H., Wang, Y. et al. (2021). Poly (vinyl alcohol) based gradient cross-linked and reprogrammable humidity-responsive actuators. *Sensors and Actuators B: Chemical* 34: 130735.

203 Gao, C., Wang, R., Feng, Y. et al. (2023). A novel conductive composite gel based on polyvinyl alcohol/polymerizable deep eutectic solvent with superior tear resistance, tensile properties and recyclability. *European Polymer Journal* 197: 112367.

204 Meng, L., Xi, J., Ye, W. et al. (2024). A novel poly(vinyl alcohol)/silica composite film with "spherical crystal interlocking structure" and its barrier and enhancement mechanism. *Surfaces and Interfaces* 46: 103964.

205 Han, Y., Li, Y., Liu, Y. et al. (2024). Carbon nanotube reinforced ionic liquid dual network conductive hydrogels: Leveraging the potential of biomacromolecule sodium alginate for flexible strain sensors. *International Journal of Biological Macromolecules* 282: 137123.

206 Shi, G., Zhang, J., Wang, X. et al. (2024). Ionic interactions induced anomalous segmental dynamics in polymer/salt complexes. *Macromolecules* 57: 11373–11385.

207 Yamada, S. and Honda, T. (2025). Supramolecular ionic gels for stretchable electronics and future directions. *ACS Materials Au* 5: 35–44.

208 Kwon, T.w., Jeong, Y.K., Deniz, E. et al. (2015). Dynamic cross-linking of polymeric binders based on host–guest interactions for silicon anodes in lithium ion batteries. *ACS Nano* 9: 11317–11324.

209 Xu, Y., Cui, M., Patsis, P.A. et al. (2019). Reversibly assembled electroconductive hydrogel via a host–guest interaction for 3D cell culture. *ACS Applied Materials & Interfaces* 11: 7715–7724.

210 Zhang, B., He, J., Shi, M. et al. (2020). Injectable self-healing supramolecular hydrogels with conductivity and photo-thermal antibacterial activity to enhance complete skin regeneration. *Chemical Engineering Journal* 400: 125994.

211 Zheng, C., Ju, M., Zhang, X. et al. (2024). A multi-stimulus responsive nanogel actuator with the π-π stacking interaction self-assembly and application for bioinspired smart skin with light, heat and humidity responses and superb UV resistance. *Chemical Engineering Journal* 499: 156092.

212 Wu, J., Dong, P., Wang, Z. et al. (2023). Improving wear resistance of ultrahigh molecular weight polyethylene: Cross-linking versus crystallinity. *Acta Polymerica Sinica* 54: 622–630.

213 Lei J. and Xu X. (1999). Recent progress in research of preparation and modification of polymeric materials by irradiation. *Modern plastics processing and applications* 3: 33–35.

214 Tian, T., Jiang, S., He, M. et al. (2021). Application progress of irradiation crosslinking in heat-resistant modification of polymer materials. *Sanghai Plastics* 49: 28–33.

215 Küpfer, J. and Finkelmann, H. (1991). Nematic liquid single crystal elastomers. *Die Makromolekulare Chemie, Rapid Communications* 12: 717–726.

216 Finkelmann, H., Kock, H.J., and Rehage, G. (1981). Investigations on liquid crystalline polysiloxanes 3. Liquid crystalline elastomers–a new type of liquid crystalline material. *Die Makromolekulare Chemie, Rapid Communications* 2: 317–322.

217 Zeng, H., Wani, O.M., Wasylczyk, P. et al. (2017). Self-regulating iris based on light-actuated liquid crystal elastomer. *Advanced Materials* 29: 1701814.

218 Weigert, F. (1921). Über einen neuen Effekt der Strahlung. *The Science of Nature* 9: 583–588.

219 Ware, T.H., McConney, M.E., Wie, J.J. et al. (2015). Voxelated liquid crystal elastomers. *Science* 347: 982–984.

220 Brannum, M.T., Steele, A.M., Venetos, M.C. et al. (2019). Light control with liquid crystalline elastomers. *Advanced Optical Materials* 7: 1801683.

221 Wang, Y., Dang, A., Zhang, Z. et al. (2020). Repeatable and reprogrammable shape morphing from photoresponsive gold nanorod/Liquid crystal elastomers. *Advanced Materials* 32: 2004270.

222 Liu, D., Bastiaansen, C.W.M., den Toonder, J.M.J., and Broer, D. J. (2012). Photo-switchable surface topologies in chiral nematic coatings. *Angewandte Chemie International Edition* 51: 892–896.

223 Yang, H., Buguin, A., Taulemesse, J.M. et al. (2009). Micron-sized main-chain liquid crystalline elastomer actuators with ultralarge amplitude contractions. *Journal of the American Chemical Society* 131: 15000–15004.

224 Buguin, A., Li, M.H., Silberzan, P. et al. (2006). Micro-actuators: When artificial muscles made of nematic liquid crystal elastomers meet soft lithography. *Journal of the American Chemical Society* 128: 1088–1089.

225 Lewis, J.A. (2006). Direct ink writing of 3D functional materials. *Advanced Functional Materials* 16: 2193–2204.

226 Kotikian, A., Truby, R.L., Boley, J.W. et al. (2018). 3D printing of liquid crystal elastomeric actuators with spatially programed nematic order. *Advanced Materials* 30: 1706164.

227 Saed, M.O., Ambulo, C.P., Kim, H. et al. (2019). Molecularly-engineered 4D-printed liquid crystal elastomer actuators. *Advanced Functional Materials* 29: 1806412.

228 Ambulo, C.P., Burroughs, J.J., Boothby, J.M. et al. (2017). Four-dimensional printing of liquid crystal elastomers. *ACS Applied Materials & Interfaces* 9: 37332–37339.

229 Sheppard, D.T., Jin, K., Hamachi, L.S. et al. (2020). Reprocessing postconsumer polyurethane foam using carbamate exchange catalysis and twin-screw extrusion. *ACS Central Science* 6: 921–927.

230 Zheng, N., Hou, J., Zhao, H. et al. (2019). Mechano-plastic pyrolysis of dynamic covalent polymer network toward hierarchical 3D ceramics. *Advanced Materials* 31: 1807326.

231 Pedersen, C.J. (1967). Cyclic polyethers and their complexes with metal salts. *Journal of the American Chemical Society* 89: 7017–7036.

232 Cram, D.J. and Cram, J.M. (1974). Host-Guest Chemistry: Complexes between organic compounds simulate the substrate selectivity of enzymes. *Science* 183: 803–809.

233 Lehn, J.M. (1978). Cryptates: Inclusion complexes of macropolycyclic receptor molecules. *Pure and Applied Chemistry* 50: 871–892.

234 Li, D., Feng, Z., Han, Y. et al. (2022). Time-resolved encryption via a Kinetics-tunable supramolecular photochromic system. *Advanced Science* 9: 2104790.

235 Li, D., Lu, F., Wang, J. et al. (2018). Amorphous metal-free room-temperature phosphorescent small molecules with multicolor photoluminescence via a host-guest and dual-emission strategy. *Journal of the American Chemical Society* 140: 1916–1923.

236 Ogoshi, T., Demachi, K., Kitajima, K., and Yamagishi, T. (2011). Monofunctionalized pillar[5]arenes: Synthesis and supramolecular structure. *Chemical Communications* 47: 7164–7166.

237 Liu, X., Qian, B., Zhang, D. et al. (2023). Recent progress in host-guest metal-organic frameworks: Construction and emergent properties. *Coordination Chemistry Reviews* 476: 214921.

238 Xu, X., Wu, X., Xu, K. et al. (2023). Pore partition in two-dimensional covalent organic frameworks. *Nature Communications* 14: 3360.

239 Zhang, Y., Su, Y., Wu, H. et al. (2021). Large-area, flexible, transparent, and long-lived polymer-based phosphorescence films. *Journal of the American Chemical Society* 143: 13675–13685.

240 Li, D., Liu, Z., Fang, M. et al. (2023). Ultralong Room-temperature phosphorescence with second-level lifetime in water based on cyclodextrin supramolecular assembly. *ACS Nano* 17: 12895–12902.

241 Watfa, N., Melgar, D., Haouas, M. et al. (2015). Hydrophobic effect as a driving force for host-guest chemistry of a multi-receptor Keplerate-type capsule. *Journal of the American Chemical Society* 137: 5845–5851.

242 Tanaka, M., Chan, C., Nakanotani, H., and Adachi, C. (2024). Simultaneous control of carrier transport and film polarization of emission layers aimed at high-performance OLEDs. *Nature Communications* 15, 5950.

243 Zhang, T., Mao, R., Jiang, W. et al. (2023). Dynamic cross-linking of zwitterionic polymer binder based on host-guest interactions for Li-S batteries with enhanced safety and electrochemical performance. *Nano Energy* 114: 108603.

244 Yu, G., Yang, Z., Fu, X. et al. (2018). Polyrotaxane-based supramolecular theranostics. *Nature Communications* 9: 766.

3

The Research Methodology of Molecular Aggregation Science

Jiajia Song[1], Aisen Li[1,2]*, Kun Yang[1]*, Weilong Che[1,3]*, and Zhen Li[1,4]**

[1] *Institute of Molecular Aggregation Science, Tianjin University, Tianjin, China*
[2] *School of Physics Science and Information Technology, Liaocheng University, Liaocheng, China*
[3] *Chemistry and Chemical Engineering, Yancheng Institute of Technology, Yancheng, China*
[4] *Department of Chemistry, Wuhan University, Wuhan, China*

*Corresponding author: jjsong@tju.edu.cn; liaisen@lcu.edu.cn; yangkuntj@tju.edu.cn; chewl867@ycit.edu.cn; lizhen@whu.edu.cn

3.1 The Theoretical Calculation

3.1.1 Introduction

Rational design of opto-electronic materials depends on a comprehensive understanding of how the material's properties are related to attributes such as composition, chemical bonding, and structural order. At the nanoscale, this understanding is often achieved through quantum chemistry calculations. These calculations not only explain or predict the energies, wave functions, charge density distribution, and various optical, wave and electron spectra at the electron and molecular levels, but also provide insight into how the predicted properties emerge from atomic-scale details. Quantum chemistry has been an essential tool for researching the structure–activity relationship of various multifunctional materials. Its role in chemical research has grown increasingly important with advancements in computational power and methods. This section begins with an introduction to the development of quantum chemistry, including various calculation methods and theories, and is followed by a summary of their applications. Then, we briefly review the software developed based on these theoretical methods. After that, we discuss the calculations used to characterize the materials' photoelectric properties. Finally, the related conclusion and outlook are presented.

3.1.2 Overview of Quantum Chemistry Calculation

Quantum mechanics is one of the greatest discoveries of the twentieth century, forming the foundation for the field of quantum chemistry. Quantum chemistry, a discipline grounded in quantum mechanics, has revolutionized our understanding of molecular systems. In 1998, Kohn and Pople were awarded the Nobel Prize in Chemistry for their groundbreaking contributions to the development of computational methods in quantum chemistry. The Nobel Committee's announcement highlighted the transformative impact of quantum chemical calculation: "Quantum chemical calculation have evolved into a tool utilized by a broad spectrum of chemists, ushering in a new era where experimental and theoretical work converge to explore the properties of molecular systems, making chemistry no longer a purely experimental science." The origins of quantum chemistry can be traced back to 1925 when Heisenberg formulated the matrix mechanics version of quantum mechanics. The following year, Schrödinger introduced the wave equation and demonstrated its equivalence to matrix mechanics in 1926. After that, the application of quantum mechanics to successfully study the chemical bond in hydrogen molecule in 1927 marked the formal beginning of quantum chemistry. Over the subsequent decades, quantum chemistry advanced rapidly with the development of various theories, such as valence bond theory, molecular orbital theory, and the frontier orbital theory, along with the refinement of the specific computational methods in quantum chemistry.[1–4] These advancements have led to significant improvements in both accuracy and efficiency of calculations, as well as the development of various computational programs.

Schrödinger equation is a fundamental equation that describes the motion of microscopic particles. It characterizes the behavior of a microscopic system in a stationary state, and can be expressed as follows:

$$\hat{H}\psi = \mathrm{E}\psi \tag{3.1}$$

The Hamiltonian operator, $\hat{H}$, is given by:

$$\hat{H} = \hat{T} + \hat{V} \tag{3.2}$$

Equation (3.1) is a typical eigen equation where $\hat{H}$, also called the Hamilton operator, represents the total energy of the system, comprising the kinetic energy operator $\hat{T}$ and the potential energy $\hat{V}$ (3.2). The wave function ψ describes the stationary state of the system and is a function of the position coordinates (x, y, z) and spin coordinates (s) of the particles within the system. The wave function (ψ) encapsulates all the microscopic properties of the system. The term "stationary state" refers to a situation where the probability density, $(|\psi|^2)$, does not change with time, a condition relevant to a variety of chemical phenomena, including the structure and reactivity of coal and other substances. The eigenvalue E represents the energy of the system in its stationary state and is obtained by solving the eigenvalue equation. This formulation applies to the many-body Hamiltonian of atomic, molecular or solid systems composed of M atoms and N electrons. The kinetic energy operator $\hat{T}$ is defined as:

$$\hat{T} = -\sum_I \frac{\hbar^2}{2M_I}\nabla_I{}^2 - \sum_i \frac{\hbar^2}{2m_i{}^2} \tag{3.3}$$

The potential energy operator $\hat{V}$ is given by:

$$\hat{V} = -\sum_{i}\sum_{I}\frac{Z_I e^2}{r_{ij}} + \sum_{i}\sum_{i<j}\frac{e^2}{r_{ij}} + \sum_{I}\sum_{I<J}\frac{Z_I Z_J e^2}{r_{IJ}} \tag{3.4}$$

In these equations, r is the distance between the two particles (in meters), M and m denote the mass of the nucleus and electron (in kilograms), respectively. Z is the atomic number, which corresponds to the number of charges in the nucleus, and e is the elementary charge $(1.60217733 \times 10^{-19}\ \mathrm{C})$. The first term of the kinetic energy operator $\hat{T}$ corresponds to the kinetic energy of the nucleus, while the second term corresponds to the kinetic energy of the electron. The potential energy operator $\hat{V}$ includes three terms: the first term describes the attraction between the electron and the nuclei, the second term accounts for the electron Coulomb repulsion between electrons, and the third term represents the Coulomb repulsion between nuclei. To solve the Schrödinger equation for multi-electron systems, it is necessary to consider the molecular wave function and introduce the nonrelativistic approximation[5] and Born–Oppenheimer approximation.[6] The nonrelativistic approximation neglects relativistic effects and defines the electron mass as: $m = m_0(9.1093897 \times 10^{-31})$Kg. The Born–Oppenheimer approximation decouples the electronic and nuclear degrees of freedom, effectively ignoring nuclear motion. Based on these approximations, the central task of quantum chemistry is to solve the Schrödinger equation for molecular systems. However, only in a few minimal cases can they be solved analytically with exact solutions under specified boundary conditions. It remains mathematically infeasible to obtain exact analytical solutions for Schrödinger's equation in general, even with the most advanced computers. This limitation has driven the development of methods that provide sufficiently accurate numerical solutions at a reasonable computational cost. The primary goal of quantum chemistry is to find ingenious methods for numerically approximating the wave function. In the following, we briefly overview some calculation methods, such as ab initio calculations based on an N-electron wave function, density functional theory (DFT) based on the electron density, and semiempirical molecular orbital methods that incorporate empirical fitting parameters.[7,8]

3.1.2.1 Ab Initio Calculation

Ab initio calculations are a class of computational methods that do not rely on empirical parameters or additional approximations beyond the fundamental physical constants: atomic number Z, Planck constant h, electron rest mass m_e, and electric quantity e. The foundation of solving multi-electronic systems was laid in 1928, when Hartree proposed the Hartree equation and the self-consistent field (SCF) method. In 1930, Fock refined the Hartree equation by adopting the wave function form of Slater's determinant, resulting in the Hartree–Fock (HF) equation. In 1951, Roothaan introduced the basis function to obtain the Hartree–Fock–Roothaan (HFR) equation, making practical calculation of the equation possible. The use of Gaussian basis functions in the HFR significantly simplifies multicenter integration and greatly reduces computational effort, thereby making the equation more tractable. Ab initio calculation involve solving the HFR equation using the SCF method to obtain molecular orbitals, energies, and other properties of multi-electron systems.

On the basis of nonrelativistic approximation and Born–Oppenheimer approximation, orbital approximation is introduced in ab initio calculation.[9] Orbital approximation assumes that each electron moves "independently" in the electrostatic field of the nuclei and the effective mean field of the other electrons, meaning that independent variables of the electron wave functions are also independent. Consequently, the wave function of an N-electron system can be written as the product of N single electron wave functions.

$$\psi(q_1, q_2, \cdots q_N) = \psi_1(q_1)\psi_2(q_2)\cdots \psi_N(q_N) \tag{3.5}$$

To describe the state of the multi-electron system, this function (3.5) is often expressed as a Slater determinant to satisfy the antisymmetric property of fermions:

$$\psi(q_1, q_2, \cdots q_N) = \frac{1}{\sqrt{N!}} \begin{vmatrix} \psi_1(q_1) & \psi_2(q_1) & \cdots & \psi_N(q_1) \\ \psi_1(q_2) & \psi_2(q_2) & \cdots & \psi_N(q_2) \\ \vdots & \vdots & \vdots & \vdots \\ \psi_1(q_N) & \psi_2(q_N) & \cdots & \psi_N(q_N) \end{vmatrix} \tag{3.6}$$

It is important to note that the HF method typically yields total energies and derived properties that can significantly deviate from experimental values because it does not account for electron correlation, except for the exchange interaction inherent in the antisymmetric requirement. This missing component, referred to as correlation, is addressed by post-HF methods, which incorporate electron correlation to the multi-electron wave function and allow for the construction of excited state wave functions. Examples of such methods include coupled cluster theory, configuration interaction (CI), and quantum Monte Carlo methods. While coupled cluster methods are generally limited to small systems due to their computational cost, they have been successfully applied to excited state calculations.[10–12] CI utilizes a variational wave function that is a linear combination of configuration state functions (CSFs) constructed from spin orbital Slater determinants. However, CI methods require significant CPU time and large memory, limiting their use to relatively small systems. Variational Monte Carlo, as presented by Zhao et al.,[13] applies a rigorous variational principle to both ground and excited state wave functions, and Quantum Monte Carlo has proven effective in performing large-scale calculations for extended systems like silicon crystal.[14]

Ab initio calculations offer high computational accuracy, but they are resource-intensive, especially for large systems. For example, in a system described by N basis functions, the computational cost for HF scales as N^4, for MP2 as N^5, and for more accurate methods like MP4 and the full configuration correlation method as N^7. In principle, the exact solution of the Schrödinger equation can be approached by infinitely enlarging the basis set and applying a complex correlation energy correction method. However, completing such calculations for large systems remains challenging. Additionally, as the atomic number increases, the electrons cluster closer to the nucleus, significantly increasing their speed and relativistic mass. The relativistic effect becomes more pronounced, leading to greater errors in the nonrelativistic approximation for heavy elements compared to light elements. Consequently, ab initio methods are typically more suitable for small systems or molecules with light elements.

3.1.2.2 Density Functional Theory

To address some limitations of the ab initio method, the DFT method was developed.[15] In 1964, Hohenbergy and Kohn introduced DFT, and in 1965, Kohn and Lu Jeu Sham derived the one-electron equation in DFT, known as the Kohn–Sham (KS) equation.[16] The KS method transforms the multiparticle problem into a single particle problem, making DFT more applicable in practice. This advancement accelerated the development of quantum chemistry. In DFT, the ground state energy of a system with a definite number of electrons is determined by the electron density function, which is a function of the position vector within the atom. Unlike ab initio methods that require the explicit N-electron wave function, DFT relies on the electronic density alone. If the molecule's ground-state electron density can be determined, it corresponds to the lowest possible energy state.

According to DFT, the wave function ψ, energy E, and other quantum mechanical properties are functions of the electron density n. The electron density n can be expressed as:

$$n = n(\bar{r}) = N\int d\bar{r}_1 \ldots \int d\overline{r_N}\psi^*(\bar{r}_1, \bar{r}_2, \ldots \bar{r}_N)\psi(\bar{r}_1, \bar{r}_2, \ldots \bar{r}_N) \tag{3.7}$$

The total energy of the system is given by:

$$E(n) = \psi^*(n)\hat{H}\psi(n) = \psi^*(n)\left(\hat{T} + \hat{U} + \hat{V}\right)\psi(n) \tag{3.8}$$

This expression involves only three degrees of freedom, allowing for the analysis of larger systems. The total energy is approximated and simplified into three terms: kinetic energy, external potential energy, and interaction energy. The kinetic energy, assuming noninteracting electrons in a single-electron orbital, is given by:

$$T(n) \approx T_s(n) = -\frac{\hbar^2}{2m}\sum_{i=1}^{N}\int d^3r\phi_i^*(\bar{r})\nabla^2\phi_i(\bar{r}) \tag{3.9}$$

Where ϕ_i represents the single-electron orbital of electron i, and $T(n)$ is the sum of the kinetic energies of these noninteracting single-electron orbitals. Based on the Born–Oppenheimer approximation, the external potential energy is expressed as:

$$V(n) = \int V(\bar{r})n(\bar{r})d\bar{r} \tag{3.10}$$

For electron-electron interactions, the Thomas–Fermi model provides an intuitive approximation.

$$U(n) = T_s(n) + U_H(n) + V(n) + E_{XC}(n) \tag{3.11}$$

The KS equation introduces the KS orbitals through a series of approximations, facilitating the calculation of kinetic energy and the electron interaction energy of the system with high accuracy:

$$E = \hat{T} + \int \rho(r)\hat{V}dr + \hat{U} = \hat{T}_s + \hat{V} + \hat{U} + \left(\hat{T} - \hat{T}_s\right) + \left(\hat{U} - \hat{U}_H\right) \tag{3.12}$$

$$\hat{T}_s = \int \rho(r)\hat{V}dr + \hat{U}_H + E_{XC} \tag{3.13}$$

After applying a variational principle to the energy functional, the KS equation is obtained:

$$\left[-\frac{\hbar}{2m}\nabla^2 + \hat{V} + \hat{V}_H + \hat{V}_{XC}\right]\varphi_i = \varepsilon_i \varphi_i \tag{3.14}$$

This equation is solved using the SCF method, and the total energy of the system is obtained by constructing the electron density and Hamiltonian operator. However, since the exchange-correlation E_{XC} in the above equation is unknown, handling this term is the key challenge in DFT. To date, several commonly used methods have been developed to address this, including the local density approximation (LDA) function, generalized gradient approximation (GGA) function, and hybrid functionals. LDA is the simplest approximation, based on the assumption that the charge density of the system, though not homogeneous, can be locally approximated as similar to a homogeneous electron gas.[16] An improvement over LDA is provided by semi-local GGA functionals,[17] which depend on both density at a point and its gradient. Hybrid functionals combine the GGA exchange energy functional with the HF exchange integral,[18,19] mixing them at a constant ratio.

Compared to ab initio calculation, the advantage of DFT is that it uses electron density ρ rather than wave function ψ as the basic variable. For an n-electron system, the wave function involves 3N variables, while the electron density is a function of just three variables. The accuracy of DFT calculation is less sensitive to the basis set size. However, DFT is constrained by the physical model and the artificially created exchange-related functionals, limiting its accuracy to the equivalent level of MP2 with the same basis set. Nonetheless, DFT greatly simplifies the calculations, shortens computation time, and offers a practical approach to studying the chemical properties of large systems. It is particularly effective for calculating the properties of organic macromolecules.

3.1.2.3 Semiempirical Molecular Orbital Methods

In ab initio calculation for larger molecular systems, the number of basis functions within the same basis set increases with the system size, leading to a number of integrals that scales proportional to N^4 (where N is the number of electrons). This scaling significantly increases the computational time required for multicenter integrals. Semiempirical calculation methods use various approximations to solve the HF equation, or directly apply fitted empirical parameters to approximate the SCF molecular orbital equation.[20] Based on these approximations, three types of semiempirical methods have been developed: (i) the one-electron approximation, where the equivalent Hamilton operator is chosen without considering the electron-electron interactions, such as in the extended Hückel molecular orbital (EHMO) method[21]; (ii) the X_α method, which calculates exchange potential energy using a statistical average model[22]; (iii) methods based on zero differential overlap (ZDO) approximation, including CNDO/2,[23–25] INDO,[26] improved MINDO,[27] AM1.[28] Below, we briefly introduce the first two semiempirical approaches.

In the 1930s, Hückel successfully applied molecular orbitals constructed from linear combinations of atomic orbitals to organic conjugated molecules, using several approximations.[29] His Hückel molecular orbitals theory enabled the treatment of conjugated molecular systems much larger than H_2, allowing the exploration of the stability and electronic structure of conjugated molecules, and predicted regarding hydrogenation or cyclization of alkenes. Many of Hückel qualitative results were satisfactory, including the 4N+2 rule and frontier orbital theory, which have become significant in practical applications. The advent of computers in the 1950s provided quantum chemistry with powerful tools. In 1963, Hoffman extended Hückel's work by proposing the extended Hückel molecular orbitals (EHMO), which successfully predicted relative conformational energies and favorable reaction pathways.[21] The EHMO method gained widespread use in the 1960s and 1970s. Since the Fock matrix data are mainly derived from experiments (e.g. ionization potential), adjusting the parameter k appropriately can yield fairly good qualitative results. However, the main disadvantage of EHMO is that it entirely neglects electron-electron interactions, lacks the two-electron integral term, and cannot provide excitation energies. Despite these limitations, the EHMO method remains an effective tool in cases where single-electron effects dominate chemical problems.

The X_α method, proposed by Slater in 1951, begins with the Schrödinger equation, expressing the wave function as a single Slater determinant. The corresponding HFR equation is derived using the variational method: $hU_K + V_cU_K + V_{X_\alpha}U_K = \varepsilon_K U_K$. In this equation, the noncommutative part of kinetic energy (h) and Coulomb action energy (V_C) can be calculated rigorously using quantum chemistry methods. However, calculating the commutative part of the Coulomb interaction energy is challenging, so the statistical average method of the Thomas–Fermi model is used: $V_{X_\alpha} = -3\alpha\left(\frac{3}{8\pi}\rho\right)^{\frac{1}{3}}$. Because the proportional coefficient in this expression contains α, it is known as the X_α method. Johnson further developed and implemented the program. This method is efficient and yields good results, particularly for a non-conjugated system with some group symmetry. However, it only provides the mean energy of multiples states and is less accurate for equilibrium geometries involving lone pairs. Generally, the X_α method is used for calculating atomic clusters and complexes and is an important tool in the study of macromolecules, systems containing heavy atoms, catalysts, catalytic reactions, materials science, solid state physics, and surface chemistry.

Semiempirical calculation methods have played an irreplaceable role in the development of quantum chemical calculations. With the rapid advancement of computer technology, the speed of calculations has increased, the computer capacity has expanded. Consequently, high-precision methods such as ab initio calculations have become increasingly popular, while semiempirical methods have gradually become less relevant.

In summary, different calculation methods in quantum chemistry calculations can be used to study the same system. However, the method that best explains and illustrates the experimental facts is highly recommended. For different systems, various computational methods may explain different aspects, even when applied to the same systems. The most appropriate method is not necessarily the one with the highest accuracy but should be chosen based on the specific requirements of the system, including calculation accuracy, computation time, and available resources. This consideration is crucial in practical applications.

3.1.3 Calculation Software

Quantum chemistry software is designed to automate the complex computational process involved in quantum chemistry, making it more accessible, improving computational efficiency, and ensuring broad applicability. Most quantum chemistry programs are written in Fortran (either Fortran 77 or Fortran 90) and typically consist of tens of thousands of lines of code. These programs are tailored to different research needs, with some specialized for finite-scale systems (such as molecules and clusters) and others for infinite periodic systems (such as crystals, solid surfaces, and chain polymers). Examples of software for finite-scale systems include Gaussian, ADF, Dalton, Games, MOPAC, EHMO, while Crystal, NNEW3, Vienna ab initio simulation package (VASP), Wien, etc. are commonly used for infinite periodic repetitive systems. Among them, Gaussian and VASP are the two most widely used computational software. The following sections provide a brief introduction to these two programs.

3.1.3.1 Gaussian

The introduction of Gaussian basis functions in the HFR equation significantly simplifies the problem of multicenter integration, thereby greatly reducing the computational effort required. Building on the Gaussian basis function, Pople designed the well-known commercial quantum chemistry program Gaussian 70, which has since evolved into Gaussian 23. Gaussian is now most widely used quantum chemistry software for both semiempirical and ab initio calculations. It allows researchers to study molecular energy and structure, transition state, chemical bonds, reaction energy, molecular orbital, dipole moment and multipole moments, proton charge and potentials, vibration frequencies, infrared and Raman spectra, NMR, polarizability and hyperpolarizability, thermodynamic properties, reaction path, and more. Excited states are crucial in understanding a wide range of phenomenon, including photosynthesis, photodecomposition, photochemical energy generation, light perception in animals, fluorescence, bioluminescence. Gaussian can calculate UV–visible spectra, model processes and reactions on excited state potential energy surfaces and predict vibronic absorption and emission spectra. Calculations can be performed in both the gas phase and in solution. With its analytic TD-DFT frequencies, Gaussian allows efficiently optimization of excited state transition structures, IRC calculations, and prediction of vibronic spectra. Analytic EOM-CCSD gradients enable the optimization of structures requiring high accuracy excited state treatments. Additionally, CASSCF calculations can be performed to model structures where multireference effects are significant, supporting active spaces of up to 16 orbitals.

3.1.3.2 VASP

VASP, the abbreviation of Virginia Academy of School Psychologists, is a software package developed by Hafner's group at the university of Vienna for electronic structure calculations and quantum mechanical-molecular dynamics simulations. It is one of the most widely used commercial software in material simulation and computational materials science research. VASP approximates the electronic state and energies of the systems by solving the Schrodinger equation. It is capable of solving KS equation within the framework of DFT, including hybrid functional calculation. Additionally, VASP can solved the Roothaan equation using HF approximation and supports methods such as Green's function (GW quasi-particle approximation, ACFDT-RPA) and perturbation theory (second-order Møler–Plesset). VASP employs a plane wave base set, with electron and ion interactions described by norm-conserving pseudopotentials (NCPP), ultra-soft pseudopotentials (USPP), or projector augmented wave (PAW) method. The software uses an efficient matrix diagonalization technique to solve the electronic ground state and Broyden and Pulay density mixing schemes are implemented to accelerate the convergence of self-consistent cycles. VASP can automatically determine the symmetry of any configuration, and Monkhorst–Pack special points are conveniently set based on symmetry, allowing for efficient calculations of bulk materials and symmetric clusters. For Brillouin region integration, VASP uses tetrahedral method with Blöchl corrections or a smearing technique to achieve faster K-point convergence. VASP applies periodic boundary conditions (for a supercellular model) to simulate various systems, including atoms, molecules, clusters, nanowires (or tubes), films, crystals, quasicrystals, and amorphous materials, and surfaces. It can calculate a wide range of material properties, including structural parameters (bond lengths, bond angles, lattice constants, and atomic positions), material configurations, equations of state, mechanical properties (bulk modulus, elastic constants), electronic structure (energy levels, charge density distributions, band structures, density of states (DOS), and electron localization functions), optical properties, and the magnetic properties.

3.1.4 Calculation of the Properties of Semiconductor Opto-electronic Materials

3.1.4.1 Structure Optimization

The determination of global or local minimum points is central to configuration optimization, which primarily depends on the energy gradient, or the first derivative of energy with respect to atomic coordinates. Configuration optimization is fundamental to chemical calculations, as any property is performed on an optimized molecular structure. Each molecular structure corresponds to a wave function, from which an energy value can be derived. Under natural conditions, system tends to exist in their lowest energy form, representing the most stable configuration. However, the initial molecular structure provided in the input file-whether hand-painted or from the crystal structure- often contains inaccuracies. These inaccuracies arise from the limitations of manual drawing (such as representing a three-dimensional structure in two dimensions) and changes in the specific chemical environment (e.g. optimizing in the gas phase by default, which differs significantly from the crystal environment). Therefore, structural optimization must be performed first to ensure accurate property calculations.

3.1.4.2 Band Structure and Density of States

According to quantum mechanics, the energy levels of electrons in a single atom are discrete. However, in a crystal, the interaction among multiple atoms splits these discrete level into N levels, resulting in what appears to be a continuous distribution of energy levels, known as a band. The band structure of a material is crucial in determining its properties, particularly its electronic and optical characteristics. By analyzing the band structure, one can determine whether a material is a metal, semiconductor or insulator. In intrinsic semiconductors, the band structure can also reveal whether the material has a direct or indirect bandgap. Band width plays a significant role in band analysis. The wider band, characterized by larger fluctuations in the band diagram, indicates a smaller the effective mass of the electrons, greater nonlocality, and stronger delocalization of the atomic orbitals that comprise the band. Conversely, a narrower band suggests that the eigenstates associated with that band are mainly composed of atomic orbitals localized at specific lattice points, and resulting in highly localized electrons with relatively large effective mass. In doped semiconductors, a new, narrower band often appear within the bandgap, commonly referred to as a doping state, the acceptor state, or the donor state. The DOS provides a visual representation of the band structure and is widely used in research due to its intuitive nature. Many aspect of DOS analysis correspond directly to the band structure analysis, making it a valuable tool in understanding material properties.

3.1.4.3 Thermodynamic Stability and Chemical Formation Energy

The first-principles calculation method provides the total energy (or cohesive energy), denoted as $E^{\varnothing}$, at the temperature $T = 0$ K. Consequently, the formation energy can be readily calculated using the following relations:

$$\Delta_f E = E^{\varnothing} - \sum x_i E_i^0 \tag{3.15}$$

Where E_i^0 represents the total energy (or cohesive energy) of i in its ground state at a temperature of $T = 0$ K. In certain cases, the enthalpy of formation is equivalent to the formation energy.

3.1.4.4 Carrier Effective Mass

The effective mass approximation is widely used in various calculations of semiconductor electronic properties, including shallow impurity states, transport phenomena, optical properties, and more. In this context, we assume that the perturbation potential $H_1(r)$ varies with the lattice constant a. This assumption allows for the simplification of the effective mass equation within the single-band approximation. The Schrodinger equation for one-electron Hamiltonian can be expressed as:

$$\left[H_0 + H_1(r)\right]\psi(r) = \varepsilon \cdot \psi(r) \tag{3.16}$$

Since the Wannier function is constructed from the Fourier transform of Bloch functions, it possesses both orthogonality and completeness. As a result, any function can be expanded using the Wannier function, allowing the eigenfunction $\psi(r)$ in Equation (3.16) to be expressed as:

$$\psi(r) = \sum_{n}\sum_{j} F_n(R_j)\, w_n(r - R_j) \tag{3.17}$$

Subsequently, since we focus on a specific band, we can omit the band index n. In the absence of perturbation, the Schrodinger equation can be represented using the Bloch function $b_{kn}(r)$ as follows:

$$H_0\, b_k(r) = \varepsilon_0(k)\, b_k(r) \tag{3.18}$$

Use the following expression:

$$\psi(r) = \sum_{j'} F(R_{j'})w(r - R_{j'}) \tag{3.19}$$

and substituting it into Equation (3.16), we multiply both side by $w^*(r - R_{j'})$ and integrate over the crystal volume, yielding:

$$\int \sum_{j'} w^*(r - R_j)\, H_0 F(R_{j'})w(r - R_{j'})\, d^3r + \int \sum_{j'} w^*(r - R_j)\, H_1 F(R_{j'})$$
$$w(r - R_{j'})\, d^3r = \int \varepsilon \sum_{j'} w^*(r - R_j)F(R_{j'})w(r - R_{j'})\, d^3r \tag{3.20}$$

When we define the following relationships:

$$(H_0)_{jj'} = \int w^*(r - R_j)\, H_0\, w(r - R_{j'})\, d^3r \tag{3.21}$$

$$(H_1)_{jj'} = \int w^*(r - R_j)\, H_0\, w(r - R_{j'})\, d^3r \tag{3.22}$$

Equation (3.20) can be written as:

$$\sum_{j'} (H_0)_{jj'} \cdot F(R_{j'}) + \sum_{j'} (H_1)_{jj'} \cdot F(R_{j'}) = \varepsilon \cdot F(R_j) \tag{3.23}$$

Given that the perturbation potential is assumed to vary very slowly, the following approximation is feasible. According to the characteristics of the Wannier function, $w^*(r - R_j)$ and $w(r - R_{j'})$ located at R_j and $R_{j'}$, respectively. Therefore, only the overlap between $w^*(r - R_j)$ and $w(r - R_{j'})$ contributes to the integral formula (3.22). In a more simplified approximation, only from the contribution from $R_j = R_{j'}$ is considered. Under this approximation, the summation over j yields the following result:

$$\sum_{j'} (H_1)_{jj'} \cong H_1(R_j) \int w^*(r - R_j)w(r - R_j)\, d^3r = H_1(R_j) \tag{3.24}$$

From Equation (3.22), we observe that $(H_0)_{jj'}$ is a function of $(R_j - R_{j'})$. This can be demonstrated using the Wannier function. Since H_0 is independent of and translation vector R, substituting r for $r - R_j$ yield:

$$(H_0)_{jj'} = \int w^*(r - R_j)\, H_0\, w(r - R_{j'} + R_j)\, d^3r \equiv h_0(R_j - R_{j'}) \tag{3.25}$$

According to the above results, Equation (3.23) can be rewritten in the following form:

$$\sum_{j'} h_0(R_{j'})(R_j - R_{j'})F(R_j - R_{j'}) + H_1(R_j)F(R_j) = \varepsilon \cdot F(R_j) \tag{3.26}$$

Based on Equation (3.16), H_0 can be expressed as follows:

$$H_0 = \frac{p^2}{2m} + V(r) \tag{3.27}$$

From this, the following relationship can be obtained:

$$\varepsilon_0(k) = \int b_k^*(r) H_0 b_k(r) d^3r \tag{3.28}$$

This expression can be further expanded in terms of the Wannier function using the Bloch function:

$$\begin{aligned}\varepsilon_0(k) &= \frac{1}{N}\sum_j \sum_{j'} \exp\left[-ik \cdot (R_j - R_{j'})\right](H_0)_{jj'} \\ &= \frac{1}{N}\sum_j \sum_{j'} \exp\left[-ik \cdot (R_j - R_{j'})\right] h_0(R_j - R_{j'})\end{aligned} \tag{3.29}$$

It should be pointed out that R_j and $R_{j'}$ represent the same lattice configuration. Therefore, for $j' = 1,2,3,\ldots$, N, the value of the sum remains constant, leading to the following formula:

$$\varepsilon_0(k) = \frac{1}{N}\sum_j \exp(-ik \cdot R_j)\, h_0(R_j) = \sum_j \exp(-ik \cdot R_j)\, h_0(R_j)\, h_0 \tag{3.30}$$

This result can be interpreted as follows: $\varepsilon_0(k)$ is a periodic function of k in reciprocal space, with a period defined by the lattice vector R_j. Consequently, $\varepsilon_0(k)$ is influenced by the lattice vector R_j. Additionally, the Fourier expansion of the coefficient $h_0(R_j)$ is given by the following formula:

$$h_0(R_j) = \frac{1}{N}\sum_k \varepsilon_0(k)\exp(ik \cdot R_j) \tag{3.31}$$

The function $F(r - R_j)$ can be expanded in a Taylor series at position r. For a three-dimensional crystal, this expansion yields:

$$\begin{aligned}F(r - R_j) &= F(r) - R_j \cdot \nabla F(r) + \frac{1}{2!}(R_j \cdot \nabla)\left[R_j \cdot \nabla F(r)\right] - \cdots \\ &= \exp(-R_j \cdot \nabla)F(r)\end{aligned} \tag{3.32}$$

Thus, the summation can be written as:

$$\sum_{j'} h_0(R_{j'})F(r - R_{j'}) = \sum_{j'} h_0(R_{j'})\exp(-R_{j'} \cdot \nabla)F(r) \tag{3.33}$$

Multiplying both sides of Equation (3.30) by $F(r)$ gives:

$$\varepsilon_0(k)F(r) = \sum_{j'} h_0(R_{j'})\exp(-ik \cdot R_{j'})F(r) \tag{3.34}$$

By comparing Equations (3.33) and (3.34), it can be observed that replacing k with $-i\nabla$ in Equation (3.34), results in the right side of Equation (3.33) being exactly the same. Therefore, the following relationship is obtained:

$$\sum_{j'} h_0\left(R_{j'}\right)F\left(r - R_{j'}\right) = \varepsilon(-i\nabla)F(r) \tag{3.35}$$

Using this result and substituting r for R_j, Equation (3.26) can be rewritten as:

$$\sum_{j'} h_0\left(R_{j'}\right)F\left(r - R_{j'}\right) + H_1\left(R_j\right)F(r) = \varepsilon \cdot F(r) \tag{3.36}$$

Alternatively, it can be rewritten in the following form:

$$\left[\varepsilon_0(-i\nabla) + H_1(r)\right]F(r) = \varepsilon \cdot F(r) \tag{3.37}$$

This is known as the effective mass equation, and the corresponding approximation is referred to as the effective mass approximation. It is important to note that using Equation (3.37), where $F(r)$ is an envelope function, the Fourier coefficient $F\left(R_j\right)$ of the Wannier function-expand wave function $\psi(r)$ is given by $F\left(R_j\right)$. This substitution is achieved by using R_j instead of the envelope function $F(r)$. Envelope function $F(r)$ is a slowly varying function of r. In other words, the coefficient $F\left(R_j\right)$ changes very slowly over the lattice spacing $\left(R_{j+1} - R_j\right)$.

$\varepsilon_0(k)$ represents the electron energy as a function of wave vector in the absence of a perturbing potential H_1, thus defining the electronic band structure. For simplicity, we can assume that the band is characterized by an isotropic effective mass $m*$, resulting in a spherical parabola dispersion relation.

$$\varepsilon_0(k) = \frac{\hbar^2 k^2}{2m*} \tag{3.38}$$

Therefore, the effective mass equation is given by the following formula:

$$\left[\frac{\hbar^2}{2m*}\nabla^2 + H_1(r)\right]F(r) = \varepsilon \cdot F(r) \tag{3.39}$$

By comparing equations (3.16) and (3.39), we observe the following key difference: in Equation (3.16), the periodic potential $V(r)$ is absent, and the free electron mass m was replaced by the effective mass $m*$. The effective mass equation is crucial for calculating various properties of semiconductors, making the determination of the effective mass very important.

3.1.4.5 Optical Absorption Spectrum

The imaginary part $\epsilon_2(E)$ of the dielectric constant is directly related to the light absorption in the material.[30] The linear expression of the energy loss due to the absorption of electromagnetic waves in materials is given by:

$$\frac{\partial u_{eff}}{\partial t}(\text{absorption}) = -2\omega_0 |E|^2 Im(\epsilon(\omega_0)) = 2\omega_0 |E|^2 \epsilon_2 \tag{3.40}$$

Where u_{eff} is the effective electromagnetic energy density, E is the time-dependent electric field vector of the incident light, and ω_0 is the frequency of the incident light.

To determine the absorption in crystal semiconductors, the single-electron approximation method can be used. The imaginary part $\epsilon_2(E)$ of the electric constant for crystal semiconductors is expressed as:

$$\epsilon_2(E) = \left(\frac{2\pi e\hbar}{m}\right)^2 \left(\frac{2}{VE^2}\right) \sum_{0,u} |P_{0,}u|^2 \delta(E_0 - E_u - E) \tag{3.41}$$

Where E is the energy of the incident light, E_0 is the energy of the occupied state, E_u is the energy of the unoccupied state, and V is the volume of the sample irradiated by incident light. Considering only the single-particle electronic state, the sum is normalized on the total volume V, with the unoccupied state located in the conduction band and the occupied state in the valence band. The transition matrix element $|P_{v,c}|^2$ is defined as:

$$P_{v,c} = \left\langle \psi_v(r) \middle| \left(\frac{\tau \cdot P}{\hbar}\right) \middle| \psi_c(r) \right\rangle \tag{3.42}$$

Where $\psi_v(r)$ is the wave function of carriers in the valence band and $\psi_c(r)$ is the wave function of the conduction band, τ is the polarization vector of the light wave, and P is the momentum operator. The momentum operator can be expressed in position space using its commutation relation:

$$P = \frac{im}{h}[H, R] \tag{3.43}$$

This expression allows the transition matrix element $P_{v,c}$ in Equation (3.42) to be rewritten as:

$$P_{v,c} = -\frac{imE}{h^2} \langle \psi_v(r) | \tau \cdot P | \psi_c(r) \rangle = -\frac{imE}{h^2} R_{v,c} \tag{3.44}$$

Using the transition operator represented in position space, Equation (3.41) can be rewritten as:

$$\epsilon_2(E) = \left(\frac{2\pi e}{\hbar}\right)^2 \left(\frac{2}{V}\right) \sum_{v,c} |R_{v,c}|^2 \delta(E_c - E_v - E) \tag{3.45}$$

This provides a general expression for the transition between the valence and conduction bands in crystalline semiconductors. However, for amorphous semiconductors, additional factors must be considered.

3.1.4.6 Exciton and Binding Energy of Exciton

Excitons can be classified based on the strength of the Coulomb attraction between electrons and holes, which in influenced by the properties of the solid. There are two main types of excitons: Frenkel excitons and Wannier–Mott excitons. In Frenkel excitons, the Coulomb attraction between electrons and holes is strong due to the small dielectric constant of material, resulting in an exciton binding energy in the range of 0.1–1.0 eV. In contrast, Wannier–Mott excitons exhibit a much weaker Coulomb interaction. For example, in semiconductors, the large dielectric constant significantly reduces the Coulomb interaction, leading to a larger separation between electrons and holes, and thus, a large radius of Wannier–Mott excitons. The binding energy of the Wannier–Mott excitons is much smaller than that of a hydrogen atom, typically around 0.01 eV. The expression for exciton binding energy E_{ex} is given by:

$$E_{ex} = -\frac{13.6\mu/m_e}{\left(\frac{\varepsilon}{\varepsilon_0}\right)^2 n^2} \tag{3.46}$$

Where μ is the reduced effective mass of the electron–hole pair, also referred to as the effective mass of the exciton, and is expressed as follows:

$$\mu = \frac{m_e^* m_h^*}{m_e^* + m_h^*} \tag{3.47}$$

Here, m_e^* and m_h^* are the effective masses of electrons and holes, respectively, while m_e is the free electron mass. The term $\varepsilon/\varepsilon_0$ represents the relative dielectric constant of the material, with ε is the dielectric constant of the material and ε_0 the dielectric constant of the free space. The quantum number n is a positive integer ($n = 1,2,3,\ldots,\infty$); when $n = 1$, it represents the ground state, while $n = \infty$ corresponds to zero exciton binding energy $E_{ex} = 0$, indicating that the exciton is dissociated or unbound, with the electrons and holes free. Additionally, the exciton binding energy E_{ex} is inversely proportional to the square of the dielectric constant and directly proportional to the effective mass.

The energy level distribution properties significantly influence carrier transport. For two isolated semiconductor materials C_1 and C_2, the energy at the conduction band minimum (CBM) and the valence band maximum (VBM) can be expressed as $E_{CBM}^{C_1}$, $E_{CBM}^{C_2}$ and $E_{VBM}^{C_1}$, $E_{VBM}^{C_2}$, respectively. To compare the band edge positions of these two semiconductors, it is necessary to align their energy band structures relative to the vacuum energy levels. The vacuum energy levels of two semiconductor materials C_1 and C_2 are denoted as $E_{vac}^{C_1}$, $E_{vac}^{C_2}$, respectively. The band edge positions after energy level alignment E_{la-CBM} and E_{la-VBM}, can be obtained using the following formula.

$$E_{la-CBM} = E_{CBM} - E_{vac} \tag{3.48}$$

$$E_{la-VBM} = E_{VBM} - E_{vac} \tag{3.49}$$

3.1.5 Conclusion and Outlook

The rational design of opto-electronic materials hinges on a thorough understanding of the relationships between their properties, composition, chemical bonding, and structural order. This understanding is typically achieved through quantum chemical calculations at the nanometer scale. These calculations not only explain or predict energy, wave functions, electron and molecular-level charge density distributions, and various spectra but also elucidate how these properties arise from atomic-scale details. As a result, quantum chemistry has become an indispensable tool for studying structure-activity relationships in multifunctional materials. With advancements in computational capabilities and methods, its role in chemical research is becoming increasingly prominent.

Among quantum chemical computation methods, ab initio approaches, such as the HF method, offer high precision but require significant computational resources, especially for large systems. To address this limitation, DFT was developed. DFT simplifies the computational process by describing systems in terms of electron density rather than N-electron wave functions. While the accuracy of DFT depends on the choice of exchange-correlation functional, its computational efficiency makes it highly suitable for studying large systems, such as organic macromolecules. Semiempirical molecular orbital methods further reduce computational complexity by simplifying HF equations or using empirical parameters. Although these methods sacrifice some precision, they are effective for macromolecules and complex systems. In terms of computational software, Gaussian and VASP are two widely used quantum chemical computation programs. Gaussian is versatile in calculating molecular properties such as energy, structure, and reaction pathways, while VASP excels in simulating materials, providing accurate calculations of structural parameters, electronic structures, optical properties, and magnetic properties.

Overall, quantum chemical computation methods offer powerful tools for probing the microscopic properties of materials. Different methods complement each other to provide the most accurate explanations of experimental results. As computational resources and methods continue to improve, both accuracy and efficiency of calculations will increase. Quantum chemical computations will remain central to material design, offering deeper insights into complex material systems. These tools not only uncover the fundamental physicochemical properties of materials but also guide experimental design and material preparation, advancing the development of opto-electronic technologies. Specifically, precise calculations of properties such as structural optimization, band structure, DOS, carrier effective mass, and optical absorption spectra will support the development and performance optimization of novel opto-electronic materials.

3.2 General Methods

At present, the characterization methods for materials at the solid state mainly include X-ray single-crystal diffractometer, nuclear magnetic resonance spectrum (NMR), transmission electron microscope (TEM), atom force microscope (AFM), the X-ray diffraction (XRD), 2D grazing incidence wide-angle X-ray scattering (GIWAXS) etc. Therein, X-ray single-crystal diffractometer is one of the most powerful tools, which can provide the information of the arrangement of atoms in single-crystal structure.

In a single crystal, the atom particles are regularly and periodically arranged in three-dimensional space, and the whole crystal is composed of the same spatial lattice in three-dimensional direction, while the spatial arrangement of the particles in the entire crystal is long-range ordered. Generally, the single-crystal structure and packing state of the molecules can be obtained by the single-crystal X-ray diffraction (SCXRD) method. In principle, XRD uses the diffraction phenomenon of X-rays in crystals to obtain the characteristics of X-ray signals after diffraction, and to obtain diffraction patterns after processing. From the spectral information, it is possible to determine the molecular structure, stacking state and to distinguish whether there are dislocations and lattice defects in the crystal. The single-crystal structures of light-emitting molecules help understand the properties of molecules in the aggregated state. By analyzing the single-crystal structure of the molecule and the packing state of the molecule, the opto-electronic properties and mechanism of the aggregated state of the molecule can be deeply investigated.

Knowing the single-crystal structure of a luminescent material is of great help in understanding the opto-electronic properties of the molecule in the aggregated state, because it is possible to more intuitively observe the specific stacking state of the luminescent molecule in the crystalline state. So that the more in-depth luminescent property analysis caused by different molecular relationships between in single-molecule state and the aggregated state can be achieved. Traditional fluorescent compounds are mostly with a conjugated planar structure, which is helpful for high luminous efficiency in the single-molecule state. But after transforming into a crystalline state, the luminous efficiency of the compound will generally decrease due to the intermolecular π–π stacking. In the practical application of organic light-emitting materials, it will inevitably be transformed into a crystalline state. So, the scientific researchers focus on inhibiting crystallization in order to develop high-efficiency light-emitting materials.[31–33] By introducing the molecular design concept of rigid helical and crosslinking structure, luminescent compounds with high glass transition temperatures are synthesized and processed into amorphous solids, thereby improving their luminous efficiency.[34,35]

For example, Tang's group found in a molecule that the five-membered silole ring and the six peripheral benzene rings pass through the single crystal of the aggregation-induced emission (AIE) compound (hexaphenylsilole) in an early systematic study.[36] Connected by a single bond, these benzene rings can rotate freely through the single bond, making the molecular conformation flexible. Through the crystal structure analysis, it was found that the highly distorted molecular configuration caused by the steric hindrance between adjacent molecules, makes it have almost no π–π stacking interactions in the crystalline state, but a large number of intermolecular interactions. These noncovalent interactions stabilize the crystal arrangement and limit the intramolecular rotation of the benzene ring, which is the main reason for the high luminescence efficiency of the crystal. It can be seen that, after obtaining the single-crystal structure from SCXRD, the packing mode of the molecules and the interaction between the organic molecules can be observed intuitively. This provides a guarantee for understanding the relationship between the luminescence property in crystalline state and the microscopic interaction between the molecules. Therefore, the single-crystal structure analysis provided by SCXRD provides direct evidence for the scientific researchers to study the opto-electronic properties from single molecules to molecular aggregates.

There are various methods for the preparation of single crystals. According to the physical and chemical properties of the crystalline substances, there are mainly melting and solution growth methods.

Growing crystals from the melting method is the earliest and most widely studied technique. During the growth process, the crystalline substance is heated to above the melting point, then cooled under a certain temperature decreasing rate, thus, the solid–liquid interface is slowly moved through different ways to gradually solidify into crystals. And the growth rate of melting method is mostly higher than the solution growth one.

The substance can also be crystallized in a solvent and makes the solution in a supersaturated state by appropriately controlling the conditions so that the crystal grows in the supersaturated solution. The advantage of the solution growth method is that the crystal can be grown at a temperature lower than the melting point, and it is easy to yield large crystals with good uniformity at room temperature. At the same time, the crystal growth process can be visually observed, providing convenient conditions for studying the crystal growth morphology and kinetics. The key factor for growing crystals in room temperature environment is to control the supersaturation of the solution. Only when the crystals grow in a stable supersaturated solution can the quality of the crystals be ensured. During the solution growth process, the solubility of the crystalline substance can be influenced by the temperature range of single-crystal growth. If the solubility of the crystalline substance is high, the slow cooling method can be adopted, and if the solubility is not so good, the constant temperature volatilization method would be more preferred. Besides, if the crystalline substance is hardly soluble in the solvent, the slow cooling method can be used for high-temperature dissolution. The slow cooling process is to dissolve the substance to be crystallized in the solution, and cool down according to a certain procedure during the crystal growth process to maintain a suitable degree of supersaturation and to promote the growth of the crystal. The constant temperature volatilization process is to dissolve the substance to be crystallized in the solvent to form a supersaturated solution, and the solution slowly volatilizes under constant temperature conditions, thereby promoting the growth of the crystal. The high-temperature dissolution and slow cooling process usually uses a kind of solvent with a high boiling point to dissolve the sample at high temperature, and then slowly cools down to room temperature in a certain procedure, so that crystals are precipitated in the supersaturated solution.

Besides, there is also a kind of method for slow volatilization at low temperature to culture a single crystal. The substance to be crystallized could be dissolved at room temperature and then placed in a refrigerator to make it slowly crystallize. This method is suitable for substances that are difficult to crystallize. In addition, there is also a kind of mixed solvent diffusion method, a volatile good solvent could be used through this method to dissolve the substance to be crystallized, and a poor solvent would be carefully and slowly added to keep the solution in a layered state, allowing it to diffuse slowly to promote the growth of crystals. The hydrothermal or solvothermal method is another kind of method for crystal growth, which refers to the crystallization in a supersaturated solution at high temperature and pressure. This method mainly relies on the high-pressure reaction kettle, and the crystalline substance and a suitable and appropriate amount of solvent are placed in the reaction kettle; the solvent used therein depends on the substance to be crystallized. During the heating process in the reaction kettle, strong convection will occur due to the temperature difference between the upper and lower parts of the kettle, so that the saturated solution at the bottom will form a supersaturated solution in the upper part, and crystals will be formed around the kettle wall. The distribution of the supersaturated solution in the kettle depends on the intensity of convection in the kettle, and it is continuously circulated to promote the growth of crystals in the kettle.

In the early stage, Tang's group found that the hexaphenylsilole (HPS) solution could obtain a piece of bright sky-blue film after slow volatilization.[37] Through scanning electron microscope, TEM, and other characterizations, it was found that the HPS film was transformed from solvent atmosphere into crystalline state with the phenomenon of luminescence enhancement during the transition process, and this phenomenon was named as crystallization-induced emission enhancement (CIEE). At present, although a variety of CIEE molecules have been reported, the mechanism is not clear enough.

In order to deeply understand the CIEE phenomenon, single-crystal structure analysis plays a crucial role. Dong's group[38] found that the amorphous film of compound DOPOD could convert into a crystalline film under solvent atmosphere or heating conditions, and its fluorescence intensity was significantly enhanced (Figure 3.1a–d). In order to explore the reasons for the different luminescent properties in these two states, the authors found, through single-crystal analysis, that there is interaction in crystalline state, which can fix the molecular conformation and

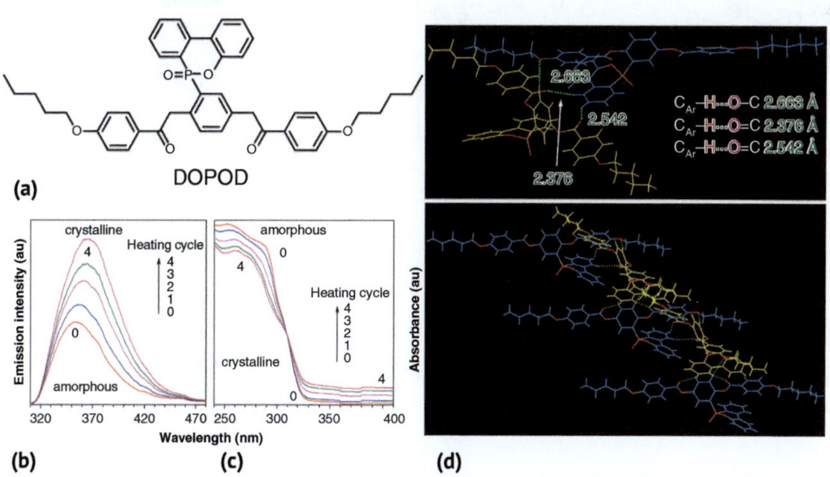

Figure 3.1 (a) Chemical structure of DOPOD. (b) Effect of heating on DOPOD emission spectrum and (c) light absorption by sample film. (d) DOPOD crystal packing and intermolecular interactions. *Source*: [38]/with permission of American Chemical Society.

suppress the nonradiation caused by molecular motion transition, resulting in the enhancement of luminescence. The luminescence mechanism was well revealed by the single-crystal structure analysis from SCXRD. In the research of CIEE system, the development of high-contrast CIEE materials has also attracted much attention of researchers, and when designing and improving high-contrast CIEE materials, single-crystal structure analysis is essential.

Ling and his partners designed and synthesized a series of diphenylmaleimide derivatives (DM1~DM4),[39] which hardly emit light in thin film state, but has strong luminescence in crystalline state with quantum efficiencies ranging from 28 to 80%. As shown in the structure of DM4 (Figure 3.2a–d), a large number of weak interactions between molecules fixed the molecular conformation and restricted the molecular motion, which was the main reason for the strong fluorescence.

Single-crystal structure analysis is crucial for understanding the photophysical properties of light-emitting molecules from single-molecule to aggregated states. This is also the case for understanding the photophysical properties of compounds with room temperature phosphorescence (RTP) properties.[40] The phenomenon, that some kinds of RTP compounds hardly emit phosphorescence in the amorphous state, but emits highly efficient phosphorescence in the crystalline state, is called crystallization induced phosphorescence (CIP). Many studies have been carried out on the exploration of CIP materials, but there are still many basic problems to be explored in this system, especially the luminescence mechanism. In 2013, Yuan's group discovered the CIP phenomenon of benzophenone derivatives, in which, BZL showed superior RTP performance. According to the crystal structure analysis, it was found that there were a lot of hydrogen bonds in the BZL crystal,

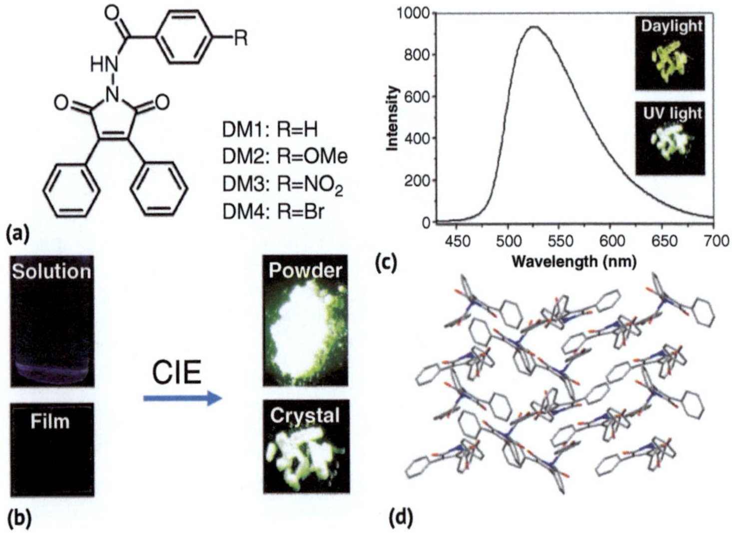

Figure 3.2 (a) Chemical structures of DM1~DM4. (b) Photographs of solution, film, powder, and crystal states of DM4 under UV light. (c) Emission spectra of DM4 crystals and their emission spectra under fluorescent and UV light Photograph. (d) Crystal packing of DM4. *Source*: [39]/with permission of Royal Society of Chemistry.

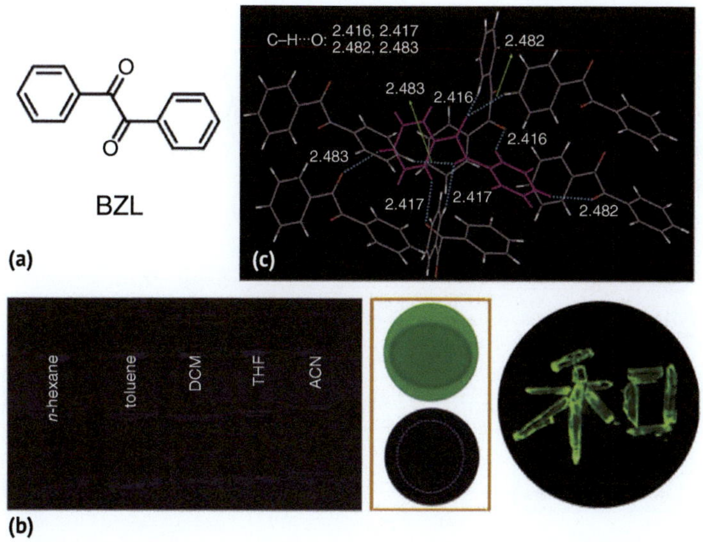

Figure 3.3 (a) Chemical structure of BZL. (b) Photographs of BZL on different anaerobic solvents (365 nm), TCL plates (upper: 254 nm, lower: 365 nm) and BZL crystals. (c) Crystal stacking and molecular interaction of BZ. *Source*: [41]/with permission of Springer Nature.

and six adjacent molecules formed 3D space network (Figure 3.3a–c).[41] This three-dimensional spatial network can lock the molecular conformation and restrict the molecular motion, contributing to its strong phosphorescence at room temperature.[41] In addition, through the analysis of the crystal structure, it can also be found that the lattice can prevent the luminescence quenching caused by the exposure of the luminophore to quenchers such as water and oxygen.

Besides, the Li group developed seven 10-phenyl-10H-phenothiazine-5,5-dioxide-based derivatives, revealed their different RTP properties and underlying mechanism.[42] The SCXRD results demonstrated that strong π–π interactions in solid state can promote the persistent RTP, and the phosphors with acceptor substituents (CS-Br, CS-Cl, and CS-F) show much enhanced π–π interactions, rather than CS-H without substituent and those (CS-CH3O and CS-CH_3) with donor substituents, in good accordance to the decreasing tendency in their RTP lifetimes for crystals (Figure 3.4). Based on the crystal data and this phenomenon, the concept of Molecular Uniting Set Identified Characteristic (MUSIC) was proposed.

In summary, crystal structure analysis has always played an important role in the rapid development of RTP materials, providing evidence to support exploring the relevant mechanism of the RTP phenomenon, and also providing a basis for designing and synthesizing high-efficiency and long-life RTP materials. NMR and TEM are also powerful tools to study the methodology of molecular aggregation science. For example, the Liu group prepared a new type of purely organic light-harvesting phosphorescence energy transfer (PET) supramolecular system. The host–guest binding mode of CB and CD-PY was investigated by ^{1}H NMR experiments,[43] and the results provided evidence that two CD-PY molecules were encapsulated into the

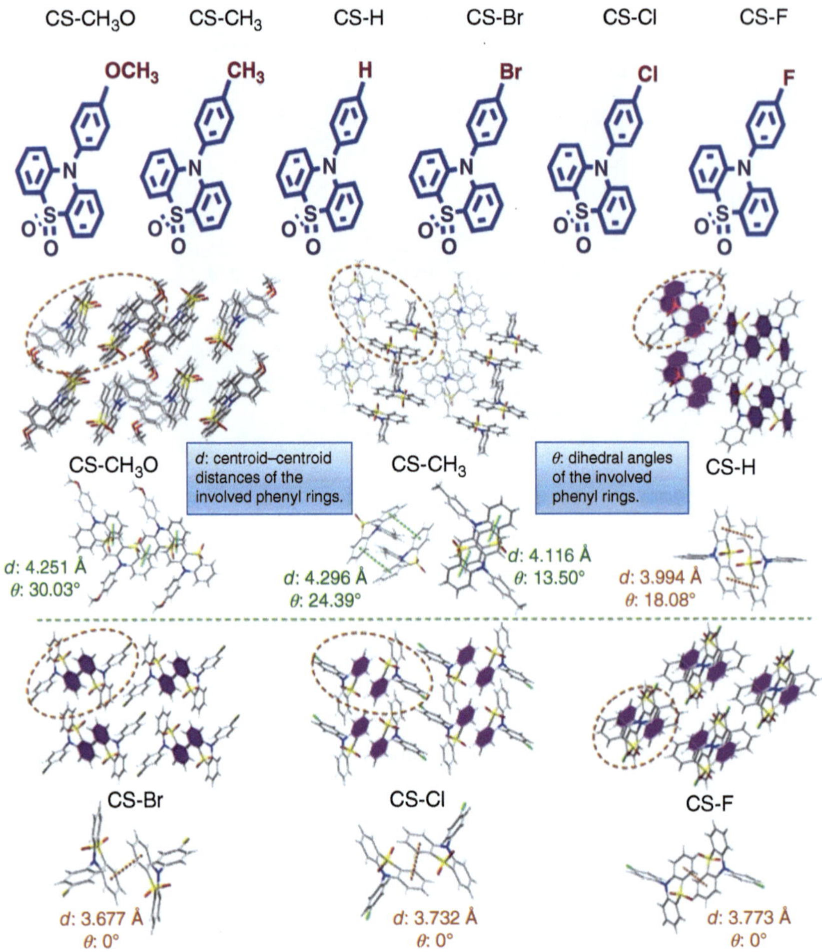

Figure 3.4 Chemical structure and single-crystal structures of seven kinds of purely organic RTP luminogens. *Source*: [42]/Springer Nature/CC BY 4.0.

cavity of CB[8] with a head-to-tail orientation to form a 2:1 complex. In addition, TEM was employed to investigate the morphological details and surface charge of final nanoparticles (Figure 3.5).[43]

3.3 Other Methods (Pressure, Light, and Temperature)

In recent years, the stimuli-responsive materials have raised tremendous attention because of their potential applications in sensors, safety devices, imaging and optoelectronic systems.[44,45] As well known, most of these materials displayed versatile optical responsive behaviors to different external stimuli with color or luminescence changes, and as the result, it is useful for translating macroscopic natural phenomena such like pressure, light, temperature and chemical substances into the changes

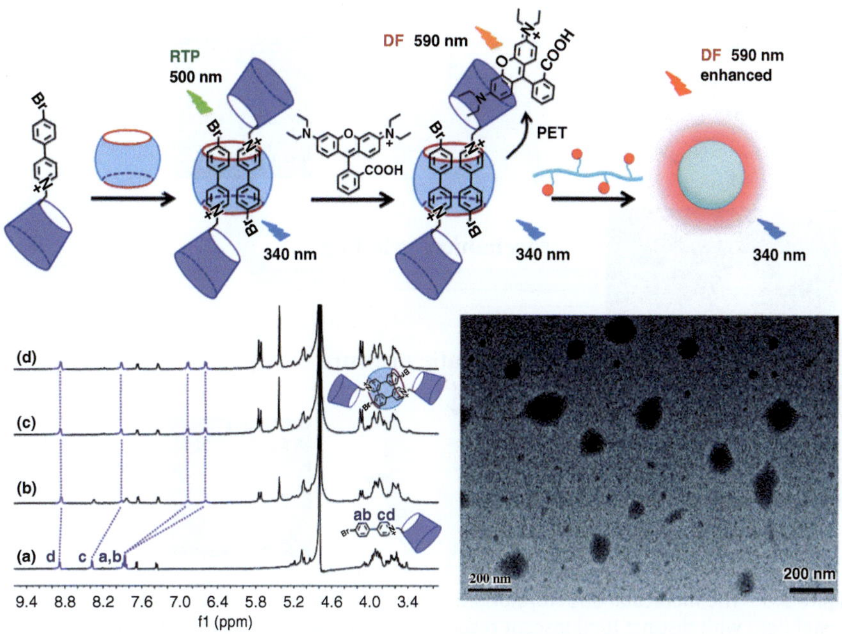

Figure 3.5 Construction of the supramolecular assembly for a purely organic light-harvesting PET system and the ^1H NMR titration spectra and TEM image of CD-PY. *Source*: [43]/with permission of Royal Society of Chemistry.

of photophysical properties by the modulation of the molecular conformation or aggregated stacking structure.[46]

3.3.1 Pressure

Mechanochromic luminescent molecules have been widely investigated due to their great potential in various fields, such as memory devices and sensors. In these reports,[47,48] pressure as an emerging energy input method, can directly act on intramolecular conformations and aggregated structures, and it can be disturbed by mechanical forces, such as shearing, grinding, tension, or hydrostatic pressure, to undergo a drastic conversion of photophysical properties. This can be further referred to as that the crystals exhibit a significant emission color change with a metastable form upon mechanical grinding, and the colors of the ground sample can revert to the original state when the fractured crystals are kept in dark or dissolved in some organic solvents to induce the occurrence of recrystallization behavior. In 1957, Alexander Schönberg found that some compounds can change color upon ground, and proposed the concept of piezochromic behavior for the first time.[49] So far, there are two common forms of force stimulation widely used in the investigations of mechanochromic luminescent materials including mechanical grinding and hydrostatic pressure. As shown in Figure 3.6, tetrathiazolylthiophene crystal showed distinct luminescent responses to these stresses.[50] The different mechanisms of a blueshift by grinding crystals and of a redshift under hydrostatic pressure

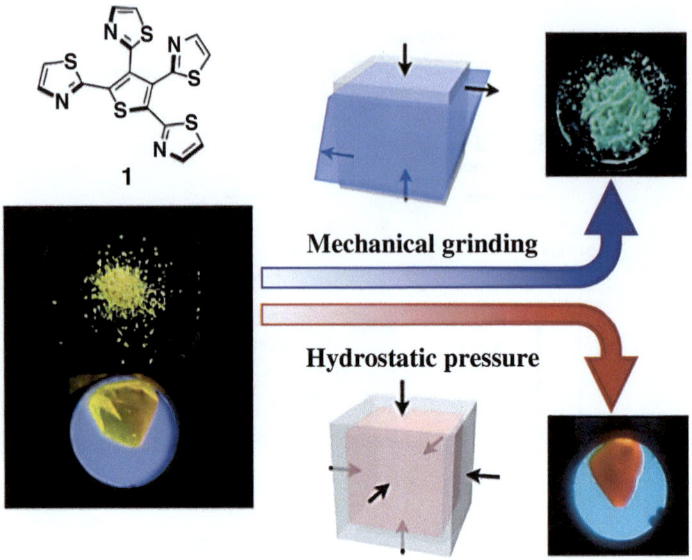

Figure 3.6 The molecular structure of tetrathiazolylthiophene and its yellow-emissive crystal (left) with distinct luminescent responses to mechanical grinding (green-emissive powder) and hydrostatic pressure (an orange-emissive crystal). *Source*: [50]/with permission of American Chemical Society.

are fully investigated, to reveal that the anisotropic and isotropic modes of mechanical loading suppress and enhance the excimer formation, respectively, in the 3D hydrogen-bond network. The difference between grinding and hydrostatic pressure will further promote and enrich the researches on the piezochromic mechanism of mechanochromic luminescent materials.

3.3.1.1 Mechanical Grinding

Generally, mechanical grinding, as a simple and direct method, is an effective means to investigate mechanochromic materials at the initial stage. Several organic materials have been reported to show mechanoresponsive luminescence behavior,[51] which include derivatives of cyanoethylene,[52] triphenylamine,[53] phenothiazine,[54] tetraphenylethylene,[55] phenanthroimidazole,[56] distyrylanthracene.[57] and many more. And the mechanochromic phenomenon mainly has been proved to depend on the changes of molecular packing and the related intermolecular interactions such as π–π interactions, hydrogen bonding, and dipole–dipole interactions caused by phase transitions.[58] For example, the Li group reported a molecular design strategy by introducing the isolation groups of tertiary butyl to regulate the molecular packing of pyrene derivatives, and they further found that compound 4-(2,7-di-tert-butylpyren-4-yl)benzaldehyde (PPCHO) showed an off-on mechanochromism with high performance (contrast ratio reaching 400 compared with the normal figure of <100) (Figure 3.7a), which is coupled with quenched emissions in monomer-like packing and bright emissions in dimer-like packing due to the transition from crystal to amorphous phase under the anisotropic pressure, which exhibited potential applications for pressure sensing and encryptions.[59]

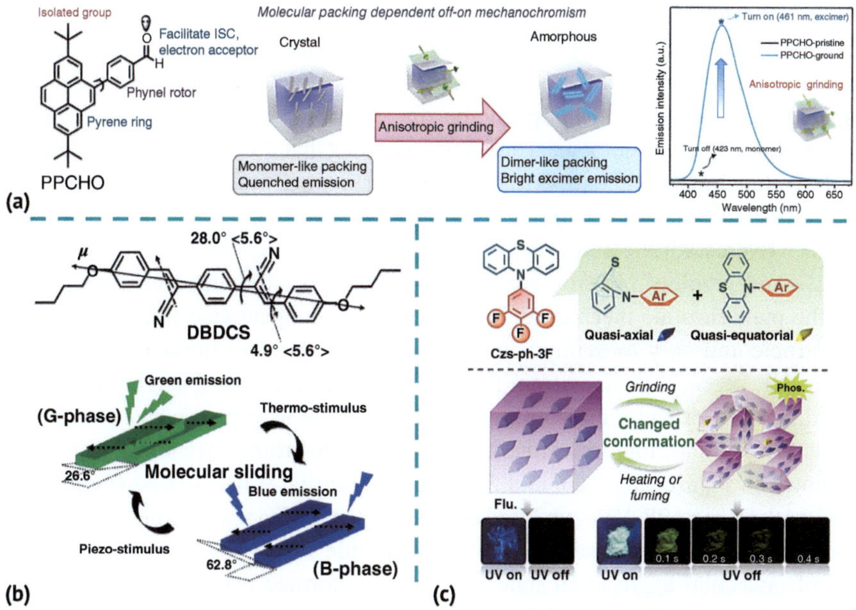

Figure 3.7 (a) The design strategy of molecular packing dependent off-on mechanochromic behaviors. *Source*: [59]/with permission of John Wiley & Sons. (b) Molecular Structure of DBDCS, and the illustration of two different modes of slip-stacking in DBDCS crystals, dictated by different ways of antiparallel/head-to-tail coupling of local dipoles. *Source*: [60]/with permission of American Chemical Society. (c) The molecular structure and two possible conformations of CzS-ph-3F, and a proposed diagram for the stimulus-responsive RTP effect based on conformational change along with the photographs of Czs-ph-3F taken before and after turning off the 365 nm UV irradiation under ambient conditions. *Source*: [61]/with permission of John Wiley & Sons.

However, in most conditions, it is difficult to study the changes of aggregated stacking structure and intermolecular interactions based on the amorphous state obtained after grinding, which limits the exhaustive study of the piezochromic mechanism. For some material systems, the piezochromic behavior mainly depends on the crystal-to-crystal phase transitions under grinding, which provides the possibility to analyze the piezochromic mechanism by comparing the corresponding single crystals before and after grinding. For example, Park et al. have synthesized a new AIEE-active compound, DBDCS and obtained two types of single crystals with blue and green emission.[60] Interestingly, when grinding the blue powder in a mortar or rubbing it with a spatula, it was observed that the fluorescence color changed immediately to green along with the changes of stacking modes, which exactly matches the emission spectra of the B- and G-phases of the single crystal. In all, it could be identified that two distinctively different phases- "green" (G) and "blue" (B) can be reversibly switched, from B to G by pressure and from G to B by thermal annealing through a specific phase transition induced by two-directional (2D) shear-sliding capability of the molecular sheets, as shown in Figure 3.7b.

Additionally, as described in Figure 3.7c, the Li group reported an abnormal stimulus-responsive room-temperature phosphorescence (RTP) effect,[61] in which the purely organic luminogen of Czs-ph-3F showed turn-on persistent phosphorescence under grinding derived from the transition of molecular conformation from quasi-axial to quasi-equatorial of the phenothiazine group. And it was found that the quasi-equatorial conformation of the phenothiazine derivatives was more likely to afford RTP emission than the quasi-axial one, demonstrating that the regulation of molecular conformation by external force is a good approach to modify the material properties.

3.3.1.2 Hydrostatic Pressure

Precise and gradient pressure control is welcome for studying and understanding the piezochromic mechanism. The shearing force generated from grinding is anisotropic and easy to manipulate, but uncontrollable and limited. In contrast, hydrostatic pressure attained from the diamond anvil cell (DAC) can be controlled accurately and provides an isotropic pressure as high as GPa, which becomes an impactful technique for further investigating the piezochromic mechanisms. In 2012, as shown in Figure 3.8a, Tian and Zou et al. found that 9,10-bis((E)-2-(pyrid-2-yl)vinyl)anthracene (BP2VA) exhibited spectacular and obvious luminescence characteristics with the increased external pressure due to the pressure-driven formation of molecular aggregation state,[62] which transformed from J-type aggregation (green emission without π–π interaction) to H-type aggregation (orange emission with weak π–π interaction), and further to the much tighter face-to-face stack like dimers (red emission, with strong π–π interaction), demonstrating that the mechanochromic behavior originated from the changes in their aggregation state under external pressure.

Generally, the gradually increasing external pressure would compress the molecules in an organic crystal to decrease its luminescence efficiency due to the formation of new excited state species with lower energy and lower emission between adjacent molecules by the enhanced interactions, including possibly enhanced π–π aggregation, hydrogen bond, excimer, etc. In 2017, a unique rehybridization-induced emission enhancement (RIEE) phenomenon under high pressure was found in 4-(2-(4′-(diphenylamino)-[1,1′-biphenyl]-4-yl)-1Hphenanthro[9,10-d]imidazol-1-yl)benzonitrile (TBPMCN) by Yang group,[63] which can be attributed to the change of excited state property from a CT-dominated state to a hybridized locally excited and charge-transfer (HLCT) state, as the result of the rehybridization of nitrogen atom driven by the repulsive force generated from the C—H···N weak hydrogen bonds during the compression process from 2 to 5 GPa (Figure 3.8b). When the external pressure was further raised, larger intermolecular action would be caused resting in normal piezochromic phenomenon: further redshift and decreasing fluorescence intensity.

Moreover, high pressure can effectively destroy and break some chemical bonds, thereby prompting the formation of new isomers or compounds. As shown in Figure 3.8c, Zhang and Li et al. reported that the tautomerization of a benzo[1,3]oxazine OX-1 in the crystal state can be quantitatively regulated by hydrostatic pressure,[64] which exhibited remarkable and continuous optical modulation covering nearly the entire visible light range (from ~430 to ~700 nm) due to the cleavage of C—O bond under high pressure. And this tautomerization process

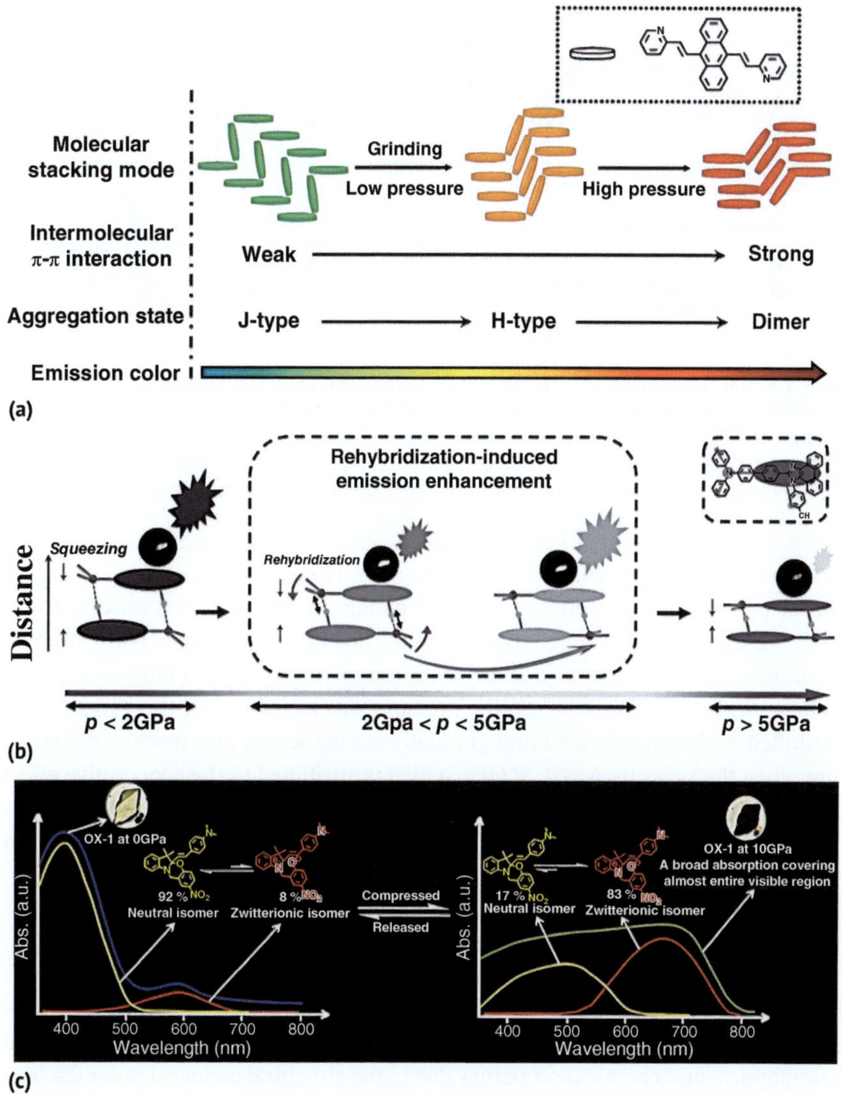

Figure 3.8 (a) Stacking modes and corresponding emission colors for the various molecular aggregation states in BP2VA powder. *Source*: [62]/with permission of John Wiley & Sons. (b) Assumed mechanism diagram of RHIEE in TBPMCN crystal. *Source*: [63]/with permission of John Wiley & Sons. (c) Illustration of optical modulations of OX-1 under high pressure. *Source*: [64]/with permission of American Chemical Society.

is strictly pressure-dependent reversibility. Also, as a comparison of the two control molecules, OX-2 and IA-1, this work revealed that (i) the low tautomerization energy barrier is a preliminary requirement to control the tautomerization by regulating hydrostatic pressure; (ii) difference in the compressibility of the two tautomeric species is the inherent cause of pressure-tunable tautomerization; and (iii) such a large-scale optical modulation would be resulted from molecular structural tautomerization rather than the configuration changes.

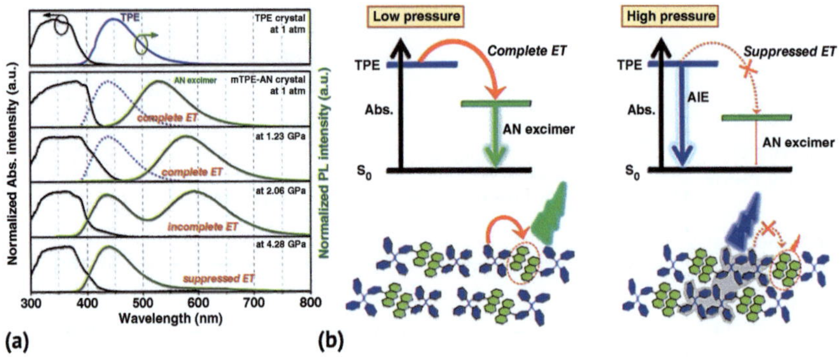

Figure 3.9 (a) Normalized absorption and emission spectra of TPE crystals at 1 atm and mTPE-AN crystals at 1 atm ~4.28 GPa. Dotted curve depicts the quenched emission spectrum of TPE to estimate the spectral overlap for ET process from TPE to AN dimer. (b) Schematic diagram of pressure-induced emission blueshift and enhancement in mTPE-AN crystals. *Source*: [65]/with permission of American Chemical Society.

As is well known, most organic piezochromic materials exhibited redshifted and quenched emission with increased pressure. However, an abnormal phenomenon of pressure-induced blueshifted and enhanced emission was observed in a 9-(3-(1,2,2-triphenylvinyl)-phenyl)anthracene (mTPE-AN) crystal with discrete π–π anthracene (AN) dimers stacking with tetraphenylethylene (TPE) as spacer.[65] As shown in Figure 3.9a, b, when the pressure was raised higher than 1.23 GPa, a blueshifted emission appeared and gradually strengthened, and reaches the maximum when the pressure was 4.28 GPa, which is attributed to the cooperative effect between the AIE of TPE units and energy-transfer (ET) suppression from TPE to an AN excimer. The high pressure impelled the changes of stacking mode and in turn achieved the modulation of luminescence path.

3.3.2 Light

Photochromic molecules can produce reversible photoisomerization under the stimulation of light with different wavelengths, accompanied by significant changes in photo-physical and chemical properties due to the structural change. Under the irradiation of another wavelength of light or by heat, it can return to its original form, which is a reversible chemical change. So far, many categories of typical photochromic materials have been investigated, such as azobenzene, spiropyran, fulgimide, dithiopheneethylene, which respond to light stimulation with remarkable color changes by cis-trans isomerization, bond heterosis, photocyclization, or bond homolysis (Figure 3.10a).[66] These materials have been widely used in the fields of optical data storage, anti-counterfeiting encryption, molecular logic gates, and so on. For example, in 2014, Park et al. reported two photo-switchable memory media by replacing one of the end tolyl groups of 1-cyano-trans-1,2-bis-(4'-methylbiphenyl)-ethylene (CN-MBE) with the photochromic 1,2-bisthienylethene (BTE) units (Figure 3.10b).[67] Compounds 1a and 2a with open-ring BTE units in THF solution and colloidal suspension were converted into 1b and 2b with closed-ring BTE units when irradiated with UV light accompanied by the decreased photoluminescence (PL) intensities, which was easily

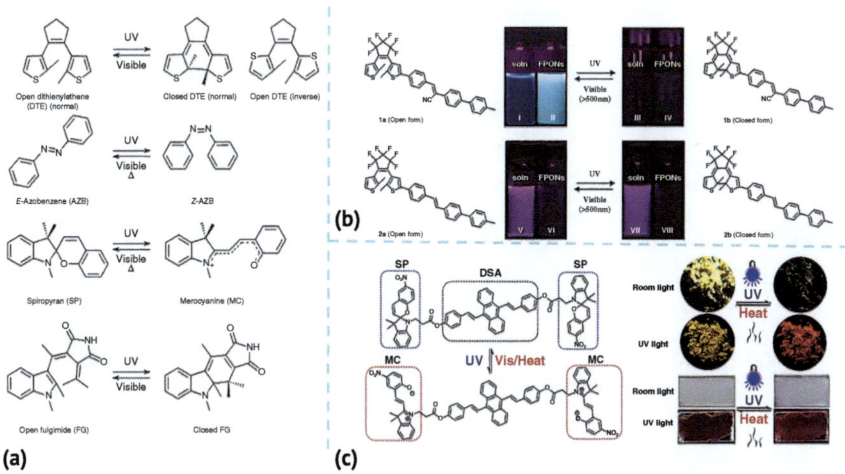

Figure 3.10 (a) Main families of organic photochromic compounds and the corresponding isomerization reactions. *Source*: [66]/with permission of Royal Society of Chemistry. (b) Molecular structure and the fluorescence images of its THF solution and the colloidal suspension of fluorescent photochromic organic nanoparticles before and after irradiation. *Source*: [67]/with permission of John Wiley & Sons. (c) Reversible structural isomerization between DSA-2SP and DSA-2MC under different stimuli, and corresponding visible and fluorescent pictures of DSA-2SP powders and films. *Source*: [68]/with permission of American Chemical Society.

monitored by the appearance of new absorption bands in the visible regions. Xu group discovered that the solid powder of a spiropyran-functionalized distyrylanthracene derivative DSA-2SP exhibited efficient photo-switching color changes from yellow to dark purple at room temperature after UV exposure (Figure 3.10c).[68] Meanwhile, a new UV–vis absorption band centered at 590 nm and an emission band peaked at 670 nm appeared, corroborating the photo-switching from unconjugated spiropyran (SP) form to conjugated merocyanine (MC) moiety. These materials with excellent photochromic property are highly applicable as both anticounterfeiting inks and super-resolution imaging agents.

Additionally, under certain conditions, light can be an efficient tool to modulate the photo-physical properties, such as photo-activated phosphorescence. Li group, for the first time, reported the interesting photoinduced RTP effect of CS-PhCF_3 crystal, which showed a persistent RTP emission with the lifetime of 299 ms after UV irradiation, while nearly no visible RTP was observed at the initial state (Figure 3.11a, b).[42] The main origin for the promoted RTP effect is attributed to the enhanced intermolecular π–π interaction by the π–π distance changing from 3.999 to 3.991 Å after UV irradiation, which would achieve the application of double anticounterfeiting with it. Then, their group also developed a new photoinduced RTP system based on the oxygen consumption mechanism under UV irradiation (Figure 3.11c, d),[69] where phenothiazine derivatives acted as phosphorescence emitter (guest) and poly(methyl methacrylate) (PMMA) as rigid matrix (host). Initially, nearly no RTP emission could be detected since the existence of trace oxygen molecules could interact with the triplet excitons in phenothiazine derivatives, and after UV irradiation with 30 s, the oxygen molecules in these systems would be consumed, leading to the turn-on RTP effect with

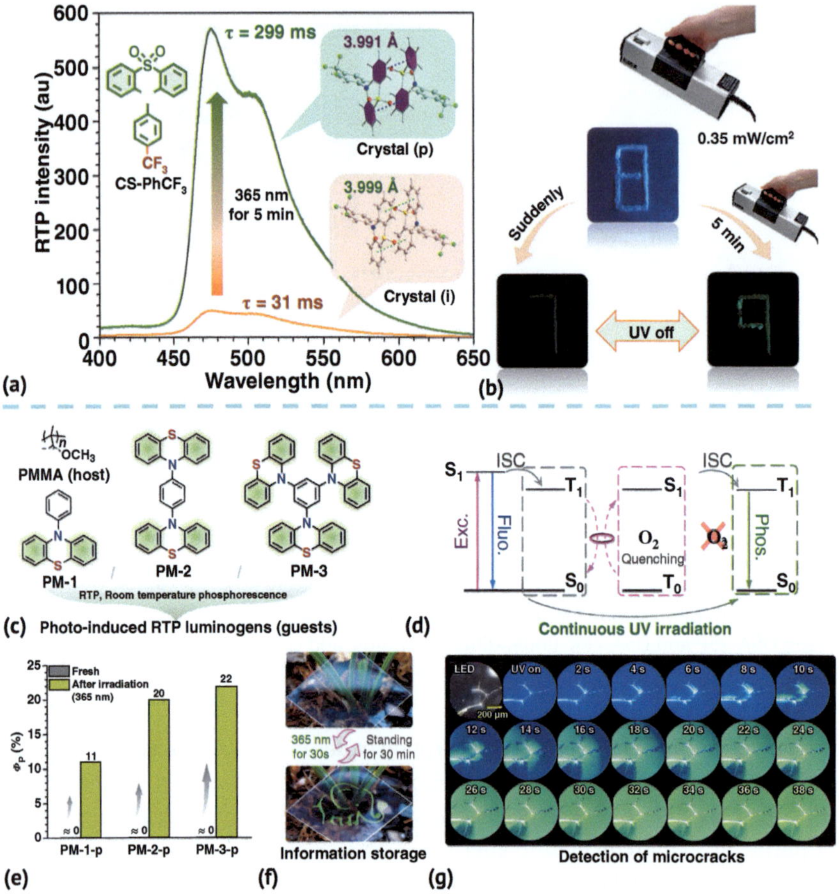

Figure 3.11 (a) The photoinduced RTP effect and corresponding internal mechanism of CS-PhCF$_3$ crystal. (b) Application of double anticounterfeiting based on the photoinduced RTP effect. *Source*: [69]/Springer Nature/CC BY 4.0. (c) Molecular structures of the photo-responsive RTP system. (d) Proposed internal mechanism for the photoinduced RTP effect. (e) The comparison of RTP efficiency for PM-1-p, PM-2-p, and PM-3-p before and after photo activation. (f) The application of information storage on the basis of the photoinduced RTP effect of PM-3-p. (g) The application of microcrack detection based on the oxygen sensitivity of PM-3-p. *Source*: [70]/with permission of John Wiley & Sons.

phosphorescence efficiency up to 22% (Figure 3.11e, f) just as those luminogens reported in literatures,[70–72] which is of great significance to the crack detection of special materials based on the oxygen sensitivity (Figure 3.11g).

3.3.3 Temperature

Thermochromism is defined as "the property of substances to change color due to a change in temperature." For the system of organic materials, heating alone could cause the changes in luminescence and color of a chromophore, as the result of the molecular rearrangement by the broken covalent bonds, the changes of molecular

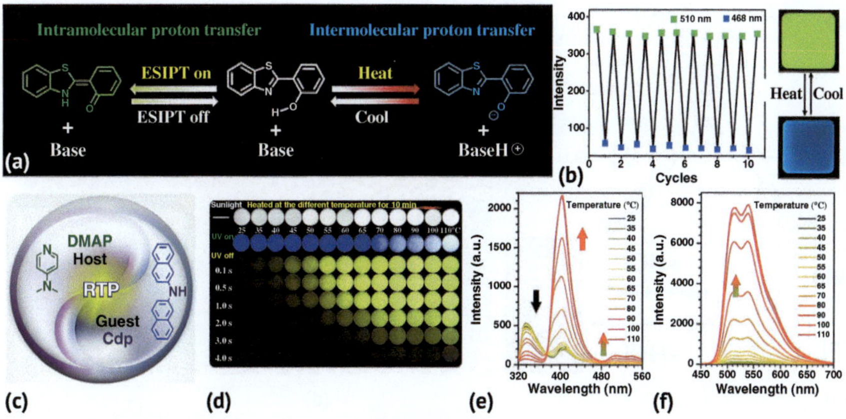

Figure 3.12 (a) Interconversion of intra- and intermolecular proton transfer processes by external thermal stimuli. (b) Switching the cycles of fluorescence at 510 and 468 nm for the blend of HBT-IM (1 wt% HBT) by alternately heating at 100 °C and cooling to room temperature. *Source*: [73]/The Royal Society of Chemistry. (c) The chemical structure of DMAP + Cdp; (d) Photographs of mixed samples (DMAP + Cdp with mass ratio of 100:1) after heating at different temperatures. (e) PL spectra of mixed samples after heating at different temperatures. (f) RTP spectra of mixed samples (DMAP and Cdp) after heating at different temperatures. *Source*: [74]/with permission of Elsevier.

conformation and/or stereoisomerism in molecules. Schiff bases exhibit remarkable thermochromic behavior, the reversible switching of luminescence of which can be achieved by controlling the proton transfer process between the enol-keto tautomeric forms and strongly depending on its crystal structure. As shown in Figure 3.12a, b, upon heating, the intermolecular proton transfer between the acid and base takes place to give a different emission between blue and green emission, and it is possible to reverse the proton transfer reaction when the heat source was removed.[73] The thermochromic phenomenon could be observed in doped system. Li and Tang et al. constructed a stimulus-responsive RTP system based on Förster resonance energy transfer (FRET), and the resultant material showed turn-on persistent RTP emission upon heating stimulus due to the sensitivity of the FRET system to distance (Figure 3.12c–f).[74]

3.3.4 Chemical Substances

Besides the stimulus mentioned above, the chemical substances are also an efficient stimulation to adjust the molecular packing and the corresponding photo-physical properties.[75] It is universal for the design of this stimulus-responsive fluorescence materials to introduce the pyridine group to modulate the molecular structure, owing to the nitrogen atom with lone pair electrons that could interact with acid and cause the changes of photo-physical properties. Taking CPHATN as example, as shown in Figure 3.13a, a porous hydrogen-bonded organic framework (HOF) was constructed, which is responsive to acid.[76] Remarkably, it showed reversible vapor acid-induced color changes from yellow to reddish-brown with an ON/OFF of its

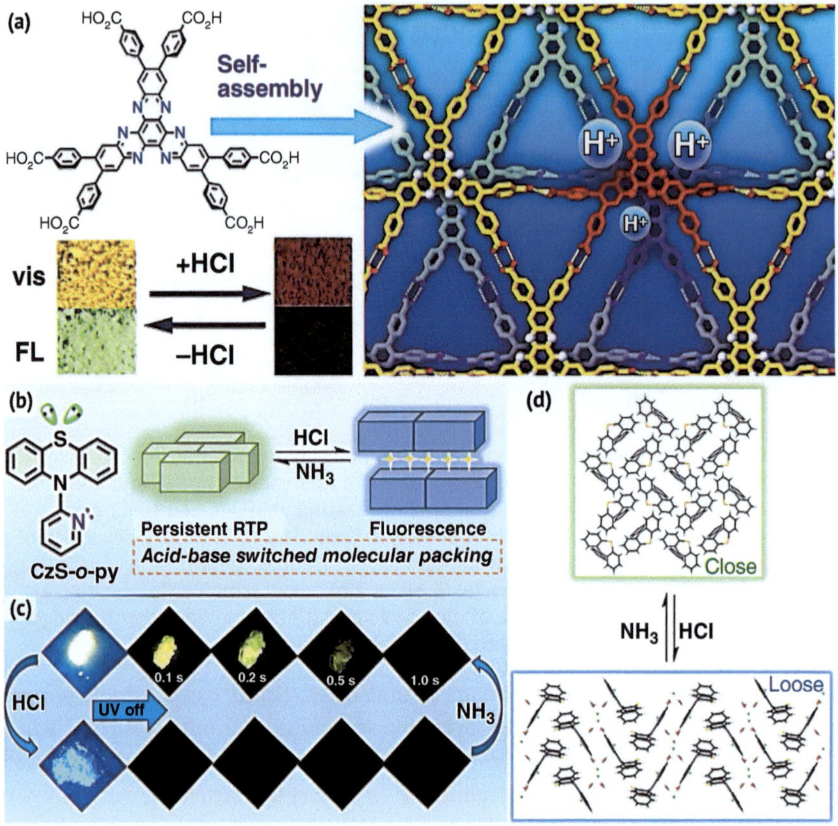

Figure 3.13 (a) Porous layered assembly of H-Hex Nets based on HATN derivative (CPHATN) with acid-responsive photo-physical properties. *Source*: [76]/with permission of American Chemical Society. (b and c) Schematic diagram of acid-base switched molecular packing and the acid-base switched RTP effect of CzS-o-py. (d) Crystal structures of CzS-o-py under the stimuli of HCl and NH_3. *Source*: [77]/Elsevier/CC BY-NC-ND 4.0.

emission because of the protonation of pyradyl nitrogen atoms embedded in its π-conjugated core. Moreover, Li group reported that CzS-*o*-py with pyridine group could achieve the reversible stimulus-responsive RTP effect by fuming with HCl and NH_3 or heating stimulus (Figure 3.13b, c).[77] According to the analysis of single-crystal structure (Figure 3.13d), the main origin for the changed RTP effect under HCl stimulation should be from the degree of the looseness of molecular packing.

As is well known, solvents are significant in crystal growth by liquid phase methods, which are highly related to morphology, size, and polymorph of the crystals to regulate the photo-physical properties effectively. Therefore, the bulk phase properties of the crystals can be adjusted by selecting different solvents during the crystal growth process. For example, an o-carborane-based anthracene was synthesized (Figure 3.14a), and its single crystals were cultured from the $CHCl_3$, CH_2Cl_2, and C_6H_6 solutions with different solvent molecules incorporated (Figure 3.14b), in which the anthracene ring is highly distorted by the formation of a π-stacked dimer between the anthracene units.[78] Moreover, solvent can also

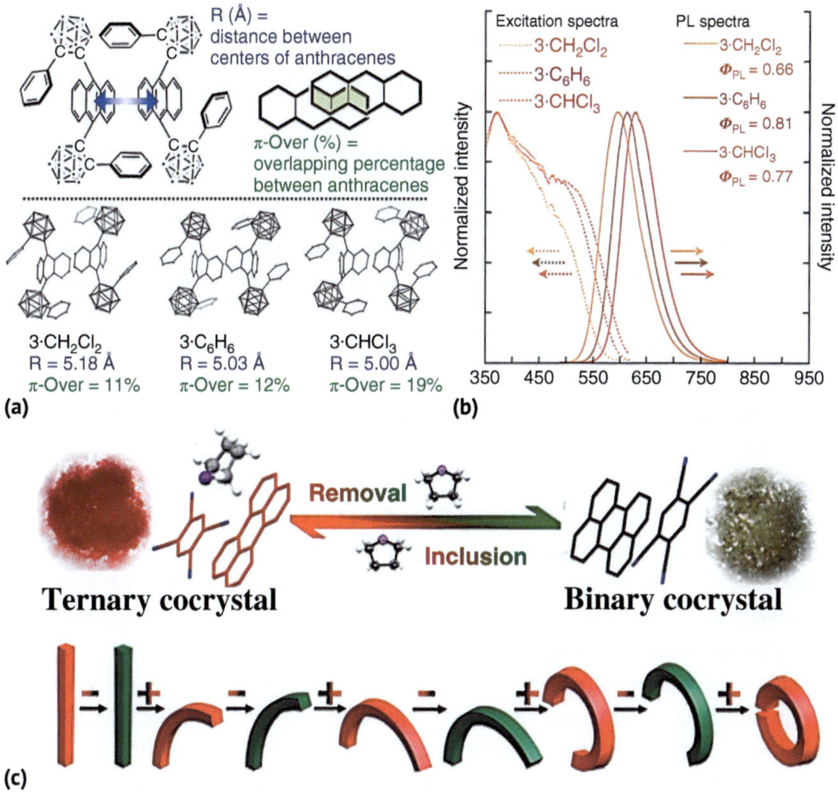

Figure 3.14 (a) Structure of π-stacked dimer and ORTEP drawings of $3{\cdot}CH_2Cl_2$, $3{\cdot}C_6H_6$, and $3{\cdot}CHCl_3$. (b) Excitation and PL spectra of $3{\cdot}CH_2Cl_2$, $3{\cdot}C_6H_6$, and $3{\cdot}CHCl_3$ crystals. *Source*: [78]/with permission of John Wiley & Sons. (c) Reversible structural and color changes between ternary PT2TC and binary PTC, and the corresponding schematic diagram of mechanical bending of a cocrystal rod ("+" represents the inclusion of THF, "−" displays the removal of THF). *Source*: [79]/with permission of American Chemical Society.

activate other property with the color changes. As shown in Figure 3.14c, Hu et al. found that a new ternary solvated (perylene-TCNB)·2THF cocrystal,[79] which can transform into binary perylene-TCNB cocrystal reversibly by successive desorption or absorption of THF solvent along with the macroscopic mechanical bending and color conversion from green to red when repeated stimulation with THF solvent, clearly demonstrating that solvent induced mechanical bending is driven by structural change.

References

1 Feng, J. and Guo, Y. (1987). *Basic Principles of Quantum Chemistry*. Higher Education Press: China.

2 Lin, M. (2005). *A Concise Course in Quantum Chemistry*. Chemical Industry Press: China.

3 Xu, G. and Li, L. (1980). *Quantum Chemistry: Basic Principles and Ab Initio Methods. Volume 1*. Science Press: China.
4 Chen, G. and Huang, Y. (2008). *Quantum Chemistry*. East China University of Science and Technology Press: China.
5 Landau, L.D. and Lifshitz, E.M. (2013). *Quantum Mechanics: Non-relativistic Theory*. Elsevier: Amsterdam.
6 Born, M. and Heisenberg, W. (1985). Zur quantentheorie der molekeln. In: *Original Scientific Papers Wissenschaftliche Originalarbeiten*. Springer: Germany.
7 Martin, R.M. (2020). *Electronic Structure: Basic Theory and Practical Methods*. Cambridge University Press: London.
8 Cramer, C.J. (2013). *Essentials of Computational Chemistry: Theories and Models*. John Wiley & Sons: USA.
9 Chen N., Gao, P., Le Z. et al. (2002). *Fundamentals of Quantum Chemistry Theory*. Harbin Institute of Technology Press: China.
10 Jeziorski, B. and Monkhorst, H.J. (1981). Coupled-cluster method for multideterminantal reference states. *Phys. Rev. A*. 24: 1668.
11 Lindgren, I. and Mukherjee, D. (1987). On the connectivity criteria in the open-shell coupled-cluster theory for general model spaces. *Phys. Rep*. 151: 93–127.
12 Sneskov, K. and Christiansen, O. (2012). Excited state coupled cluster methods. *Wiley Interdiscip. Rev. Comput. Mol. Sci*. 2: 566–584.
13 Zhao, L. and Neuscamman, E. (2019). Variational excitations in real solids: optical gaps and insights into many-body perturbation theory. *Phys. Rev. Lett*. 123: 036402.
14 Williamson, A., Hood, R.Q., Needs, R., and Rajagopal, G. (1998). Diffusion quantum Monte Carlo calculations of the excited states of silicon. *Phys. Rev. B*. 57: 12140.
15 Hohenberg, P. and Kohn, W. (1964). Inhomogeneous electron gas. *Phys. Rev*. 136: B864.
16 Kohn, W. and Sham, L.J. (1965). Self-consistent equations including exchange and correlation effects. *Phys. Rev*. 140: A1133.
17 Perdew, J.P. (1986). Density-functional approximation for the correlation energy of the inhomogeneous electron gas. *Phys. Rev. B*. 33: 8822.
18 Lee, C., Yang, W., and Parr, R.G. (1988). Development of the Colle-Salvetti correlation-energy formula into a functional of the electron density. *Phys. Rev. B*. 37: 785.
19 Becke, A. (1993). Density-functional thermochemistry. III. the role of exact exchange. *J. Chem. Phys*. 98: 5648–5652.
20 McIver Jr, J.W. and Komornicki, A. (1971). Rapid geometry optimization for semi-empirical molecular orbital methods. *Chem. Phys. Lett*. 10: 303–306.
21 Calzaferri, G., Forss, L., and Kamber, I. (1989). Molecular geometries by the extended Hueckel molecular orbital (EHMO) method. *J. Phys. Chem*. 93: 5366–5371.
22 Sambe, H. and Felton, R.H. (1975). A new computational approach to Slater's SCF–X α equation. *J. Chem. Phys*. 62: 1122–1126.
23 Pople, J.A. and Beveridge, D.L. (1970). *Molecular Orbital Theory*. McGraw-Hill, New York.

24 Pople, J.A., Santry, D.P., and Segal, G.A. (1965). Approximate self-consistent molecular orbital theory. I. Invariant procedures. *J. Chem. Phys.* 43: S129–S135.

25 Pople, J.A. and Segal, G.A. (1966). Approximate self-consistent molecular orbital theory. III. CNDO results for AB2 and AB3 systems. *J. Chem. Phys.* 44: 3289–3296.

26 Pople, J., Beveridge, D., and Dobosh, P. (1967). Approximate self-consistent molecular-orbital theory. V. Intermediate neglect of differential overlap. *J. Chem. Phys.* 47: 2026–2033.

27 Bingham. R.C., Dewar, M.J., and Lo, D.H. (1975). Ground states of molecules. XXV. MINDO/3. Improved version of the MINDO semiempirical SCF-MO method. *J. Am. Chem. Soc.* 97: 1285–1293.

28 Stewart, J.J. (2004). Optimization of parameters for semiempirical methods IV: Extension of MNDO, AM1, and PM3 to more main group elements. *J. Mol. Model.* 10: 155–164.

29 Yates, K. (2012). *Hückel Molecular Orbital Theory*. Elsevier: Amsterdam.

30 Hilfiker, J.N. and Tiwald, T. (2018). Dielectric function modeling. In: *Spectroscopic Ellipsometry for Photovoltaics*. Springer: Germany.

31 Joswick, M.D., Campbell, I.H., Barashkov, N.N., and Ferraris, J.P. (1996). Systematic investigation of the effects of organic film structure on light emitting diode performance. *J. Appl. Phys.* 80: 2883–2890.

32 Liu, S.J., He, F., Wang, H. et al. (2008). Cruciform DPVBI: Synthesis, morphology, optical and electroluminescent properties. *J. Mater. Chem.* 18: 4802–4807.

33 Yin, S.W., Shuai, Z., and Wang, Y.L. (2003). A quantitative structure-property relationship study of the glass transition temperature of OLED materials. *J. Chem. Inf. Comput. Sci.* 43: 970–977.

34 Poriel, C., Liang, J.J., Rault-Berthelot, J. et al. (2007). Dispirofluorence-indenofluorene derivatives as new building blocks for blue organic electroluminescent devices and electroactive polymers. *Chem. Eur. J.* 13: 10055–10069.

35 Wong, K.T., Chien, Y.Y., Chen, R.T. et al. (2002). Ter(9,9-diarylfluorene)s: highly efficient blue emitter with promising electrochemical and thermal stability. *J. Am. Chem. Soc.*, 124: 11576–11577.

36 Chen, J., Law, C.C.W., Lam, J.W.Y. et al. (2003). Synthesis, light emission, nanoaggregation, and restricted intramolecular rotation of 1,1-substituted 2, 3, 4, 5-tetraphenylsiloles. *Chem. Mater.* 15: 1535–1546.

37 Dong, Y.Q., Lam, J.W.Y., Li, Z. et al. (2005). Vapochromism of hexaphenylsilole. *J. Inorg. Organomet. Polym. Mater.* 15: 287–291.

38 Qian, L.J., Tong, B., Shen, J.B. et al. (2009). Crystallization-induced emission enhancement in a phosphorus-containing heterocyclic luminogen. *J. Phys. Chem. B.* 113: 9098–9103.

39 Zheng, R., Mei, X.F., Lin, Z.H. et al. (2015). Strong CIE activity, multi-stimuli-responsive fluorescence and data storage application of new diphenyl maleimide derivatives. *J. Mater. Chem. C.* 3: 10242–10248.

40 Narushima, K., Kiyota, Y., Mori, T. et al. (2019). Suppressed triplet exciton diffusion due to small orbital overlap as a key design factor for ultralong-lived room-temperature phosphorescence in molecular crystals. *Adv. Mater.* 31: 1807268.

41 Gong, Y.Y., Tan, Y.Q., Li, H. et al. (2013). Crystallization-induced phosphorescence of benzils at room temperature. *Sci. China Chem.* 56: 1183–1186.

42 Yang, J., Zhen, X., Wang, B. et al. (2018). The influence of the molecular packing on the room temperature phosphorescence of purely organic luminogens. *Nat. Commun.* 9: 840.

43 Shen, F.F., Chen, Y., Dai, X.Y. et al. (2021). Purely organic light-harvesting phosphorescence energy transfer by β-cyclodextrin pseudorotaxane for mitochondria targeted imaging. *Chem. Sci.* 12: 1851–1857.

44 Chen, W., Pan, Y., Chen, J. et al. (2018). Stimuli-responsive organic chromic materials with near-infrared emission. *Chin. Chem. Lett.* 29: 1429–1435.

45 Sun, L.J., Yang, F.X., Zhang, X.T., and Hu, W.P. (2020). Stimuli-responsive behaviors of organic charge transfer cocrystals: recent advances and perspectives. *Mater. Chem. Front.* 4: 715–728.

46 Yang, J., Fang, M., and Li, Z. (2021). Stimulus-responsive room temperature phosphorescence materials: internal mechanism, design strategy, and potential application. *Acc. Mater. Res.* 2: 644–654.

47 Li, A.S., Xu, S.P., Bi, C.J. et al. (2021). Piezochromic mechanism of organic crystals under hydrostatic pressure. *Mater. Chem. Front.* 5: 2588–2606.

48 Chi, Z.G., Zhang, X.Q., Xu, B.J. et al. (2012). Recent advances in organic mechanofluorochromic materials. *Chem. Soc. Rev.* 41: 3878–3896

49 Schönberg, A., Fateen, A.E.K., and Sammour, A.E.M.A. (1957). Organic sulfur compounds. XXXIV. synthesis of ethylenes and ethylene sulfides by action of diazoalkanes on thioketones. *J. Am. Chem. Soc.* 79: 6020–6023.

50 Nagura, K., Saito, S., Yusa, H. et al. (2013). Distinct responses to mechanical grinding and hydrostatic pressure in luminescent chromism of tetrathiazolylthiophene. *J. Am. Chem. Soc.* 135: 10322–10325.

51 Khan, F., Ekbote, A., Singh, G. et al. (2022). Mechanochromic luminogens with hypsochromically shifted emission switching property: Recent advances and perspective. *J. Mater. Chem. C.* 10: 5024–5064.

52 Zhao, J., Chi, Z.H., Zhang, Y. et al. (2018). Recent progress in the mechanoflurochromism of cyanoethylene derivatives with aggregation-induced emission. *J. Mater. Chem. C.* 6: 6327–6353.

53 Zhang, Y.J., Sun, J.W., Zhuang, G.L. et al. (2014). Heating and mechanical force-induced luminescence on-off switching of arylamine derivatives with highly distorted structures. *J. Mater. Chem. C.* 2: 195–200.

54 Xu, B., Mu, Y., Mao, Z. et al. (2016). Achieving remarkable mechanochromism and white-light emission with thermally activated delayed fluorescence through the molecular heredity principle. *Chem. Sci.* 7: 2201–2206.

55 Yang, Z.Y., Chi, Z.H., Mao, Z. et al. (2018). Recent advances in mechano-responsive luminescence of tetraphenylethylene derivatives with aggregation-induced emission properties. *Mater. Chem. Front.* 2: 861–890.

56 Ekbote, A., Han, S.H., Jadhav, T. et al. (2018). Stimuli responsive AIE active positional isomers of phenanthroimidazole as non-doped emitters in OLEDs. *J. Mater. Chem. C.* 6: 2077–2087.

57 Zhao, J., Chi, Z.H., Yang, Z.Y. et al. (2018). Recent progress in the mechanofluorochromism of distyrylanthracene derivatives with aggregation-induced emission. *Mater. Chem. Front.* 2: 1595–1608.

58 Wang, C. and Li, Z. (2017). Molecular conformation and packing: their critical roles in the emission performance of mechanochromic fluorescence materials. *Mater. Chem. Front.* 1: 2174–2194.

59 Gong, Y.B., He, S.Y., Li, Y.G. et al. (2020). Partially controlling molecular packing to achieve off-on mechanochromism through ingenious molecular design. *Adv. Opt. Mater.* 6: 1902036.

60 Yoon, S.J., Chung, J.W., Gierschner, J. et al. (2010). Multistimuli two-color luminescence switching via different slip-stacking of highly fluorescent molecular sheets, *J. Am. Chem. Soc.* 132: 13675–13683.

61 Ren, J., Wang, Y., Tian, Y. et al. (2021). Force-induced turn-on persistent room-temperature phosphorescence in purely organic luminogen. *Angew. Chem. Int. Ed.* 60: 12335–12340.

62 Dong, Y., Xu, B., Zhang, J. et al. (2012). Piezochromic luminescence based on the molecular aggregation of 9,10-bis((E)-2-(pyrid-2-yl)vinyl)anthracene. *Angew. Chem. Int. Ed.* 51: 10782–10785.

63 Zhang, S., Dai, Y., Luo, S. et al. (2017). Rehybridization of nitrogen atom induced photoluminescence enhancement under pressure stimulation. *Adv. Funct. Mater.* 27: 1602276.

64 Wang, Y., Tan, X., Zhang, Y.M. et al. (2015). Dynamic behavior of molecular switches in crystal under pressure and its reflection on tactile sensing. *J. Am. Chem. Soc.* 137: 931–939.

65 Liu, H., Gu, Y., Dai, Y. et al. (2020). Pressure-induced blue-shifted and enhanced emission: a cooperative effect between aggregation-induced emission and energy-transfer suppression. *J. Am. Chem. Soc.* 142: 1153–1158.

66 Fihey, A., Perrier, A., Browne, W.R., and Jacquemin, D. (2015). Multiphotochromic molecular systems. *Chem. Soc. Rev.* 44: 3719–3759.

67 Lim, S.J., An, B.K., Jung, S.D. et al. (2004). Photoswitchable organic nanoparticles and a polymer film employing multifunctional molecules with enhanced fluorescence emission and bistable photochromism. *Angew. Chem. Int. Ed.* 43: 6346–6350.

68 Qi, Q., Li, C., Liu, X. et al. (2017). Solid-state photoinduced luminescence switch for advanced anticounterfeiting and super-resolution imaging applications. *J. Am. Chem. Soc.* 139: 16036–16039.

69 Wang, Y., Yang, J., Fang, M. et al. (2021). New phenothiazine derivatives that exhibit photoinduced room-temperature phosphorescence. *Adv. Funct. Mater.* 31: 2101719.

70 Dou, X.Y., Wang, X., Xie, X.L. et al. (2024). Advances in polymer-based organic room-temperature phosphorescence materials. *Adv. Funct. Mater.* 34: 2314069.

71 Ding, B.B., Ma, X., and Tian, H. (2023). Recent advances of pure organic room temperature phosphorescence based on functional polymers. *Acc. Mater. Res.* 4: 827–838.

72 Li, D., Yang, Y., Yang, J. et al. (2022). Completely aqueous processable stimulus responsive organic room temperature phosphorescence materials with tunable afterglow color. *Nat. Commun.* 13: 347.

73 Qin, T., Du, J., Wang, C. et al. (2019). Thermally controlling the intra- and intermolecular proton transfer reaction: a distinct gateway to luminescent switching. *J. Mater. Chem. C.* 7: 9149–9153.

74 Wang, Y., Yang, J., Fang, M. et al. (2020). Förster resonance energy transfer: an efficient way to develop stimulus-responsive room-temperature phosphorescence materials and their applications. *Matter.* 3: 449–463.

75 Li, E., Jie, K., Liu, M. et al. (2020). Vapochromic crystals: understanding vapochromism from the perspective of crystal engineering. *Chem. Soc. Rev.* 49: 1517–1544.

76 Hisaki, I., Suzuki, Y., Gomez, E. et al. (2019). Acid responsive hydrogen-bonded organic frameworks. *J. Am. Chem. Soc.* 141: 2111–2121.

77 Tian, Y., Gong, Y., Liao, Q. et al. (2020). Adjusting organic room-temperature phosphorescence with orderly stimulus-responsive molecular motion in crystals. *Cell Rep. Phys. Sci.* 1: 100052.

78 Naito, H., Morisaki, Y., and Chujo, Y. (2015). o-Carborane-based anthracene: a variety of emission behaviors. *Angew. Chem. Int. Ed.* 54: 5084–5087.

79 Sun, Y., Lei, Y., Dong, H. et al. (2018). Solvatomechanical bending of organic charge transfer cocrystal. *J. Am. Chem. Soc.* 140: 6186–6189.

4

Photoluminescence of Molecular Aggregates

Jie Yang[1*]*, Manman Fang*[1]*, and Zhen Li*[1,2*]

[1] *Institute of Molecular Aggregation Science, Tianjin University, Tianjin, China*
[2] *Department of Chemistry, Wuhan University, Wuhan, China*

*Corresponding authors: jieyang2018@tju.edu.cn; lizhen@whu.edu.cn

4.1 Introduction

Photoluminescence (PL) is light emission from any form of matter after the absorption of photons.[1] The first documentation of PL could be traced back to the year 1565, Nicolás Monardes from Spain observed blue emission from the plant extracts used for the treatment of kidney and urinary diseases.[2] Possibly due to the limited knowledge at that time, this luminescent phenomenon received limited attention, and thus only sporadic reports of luminescence were presented for the next 300 years.[3] For instance, V. Cacariolo reported a famous phosphor of Bolognian stone in 1602, which could give red emission after being calcined; in 1834, D. Brewster found a complementary shade of red from the side when a strong beam of sunlight passed through the green chlorophyll solution; and Herschel saw the emissive behavior of quinine solution in 1845. In 1852, George Stokes realized that it was a process of light emission and began to use the term "fluorescence" to describe it.[4] Also, he carefully studied the relationship between emission intensity and solution concentration of luminogens, further proposing the possibility to use fluorescence for quantitative analysis. Henceforth, the research about photoluminescence has gained increasing attention with rapid development.[5] Particularly, the proposed Jablonski diagram in 1933 provided a basic theoretical framework for this research field and contributed a lot to its advancement.[6]

Through constant improvement, the Jablonski diagram has been able to describe almost all of the excited processes in photoluminescence.[1] As demonstrated in Figure 4.1, there are mainly two kinds of states in the diagram, namely singlet state and triplet one. Generally, the ground state for organics is in the singlet (S_0) state according to the Pauli-Prinzip. Thus, when the molecules in the S_0 state absorbed photons, they turned into singlet excitons, which then decay as fluorescence. On the other hand, for some

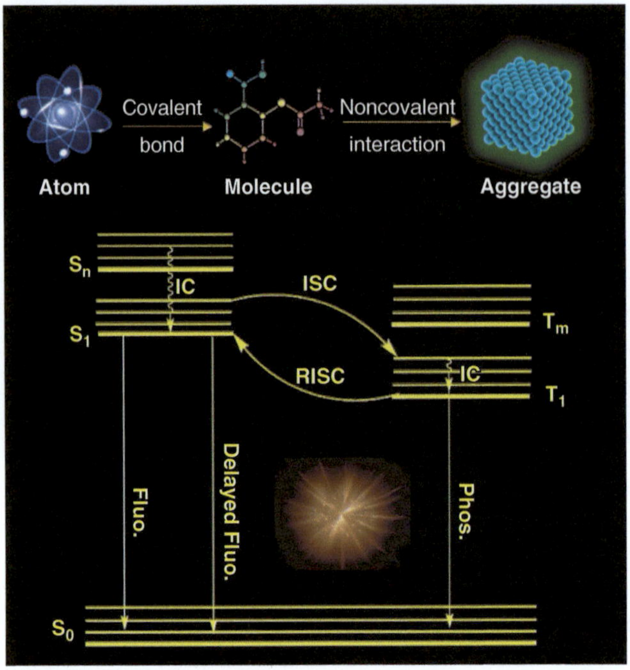

Figure 4.1 The composition of organic luminescent materials and corresponding internal mechanism (Fluo.: fluorescence; Phos.: phosphorescence; ISC: intersystem crossing; RISC: reverse intersystem crossing).

organic compounds with considerable spin-orbit coupling (SOC) between the excited singlet states and adjacent triplet states, the corresponding singlet excitons can be transformed into triplet ones through intersystem crossing (ISC), and then the phosphorescence emission becomes possible. Furthermore, if the formed triplet excitons return to a singlet through reverse ISC, it will lead to the delayed fluorescence. Thus, to adjust luminescent behaviors, the change of energy level and corresponding excited process in the Jablonski diagram should be the fundamental reason. To achieve this point, the tailoring of molecular structure has been widely utilized, because it directly determines the energy levels of organic luminogens. Therefore, most luminescent behaviors were understood based on the theory of molecular science until now.

Regardless, some emissive phenomena could still not. Particularly, it was found that many luminogens showed distinct properties in different states or aggregates, for example, excimer/exciplex with obviously red-shifted emission from solution to aggregation state, and polymorphisms with different PL behaviors, etc.[7–10] In the twenty-first century, scientists pay much attention to the effect of intermolecular interactions, as most luminescent materials are actually used in solid or aggregated state for practical applications.[11–15] It is found that the effect of aggregate widely exists in the luminescent behaviors, including both fluorescence and phosphorescence. That is, besides molecular structure, the molecular packing as well as intermolecular interaction can also fundamentally determine the resultant photophysical property (see Figure 4.1). In this chapter, we mainly discuss typical behaviors of luminescence in the solid state, particularly focusing on the molecular packing-property relationship.

4.2 Fluorescence

4.2.1 Aggregation-induced Emission

Aggregation-induced emission, abbreviated as AIE, first coined in 2001 by Chinese scientists Ben Zhong Tang and coworkers, indicates some luminogens are non-emissive in the solution state, while in aggregation or solid state, they show much enhanced emission.[16,17] This has been thought to be an abnormal PL behavior, since many conventional luminogens, especially those with planar molecular conformation, were subjected to the aggregation-caused quenching (ACQ) effect, just referring to Birks's classic book on *Photophysics of Aromatic Molecules* in 1970.[18] Thus, once the concept of AIE was proposed, it quickly attracted scientists' attention, as this could solve ACQ problem fundamentally, thus obtaining efficient luminescent materials in solid state.

As shown in Figure 4.2, some silole derivatives, such as 1-methyl-1,2,3,4,5-pentaphenylsilole (MPPS) and hexaphenylsilole (HPS), are non-emissive in tetrahydrofuran solution.[19] With the addition of a poor solvent of water, they tend to form aggregates, resulting in the enhanced green emission. The exploration of internal mechanisms finds that HPS presents a twisted molecular conformation, which can

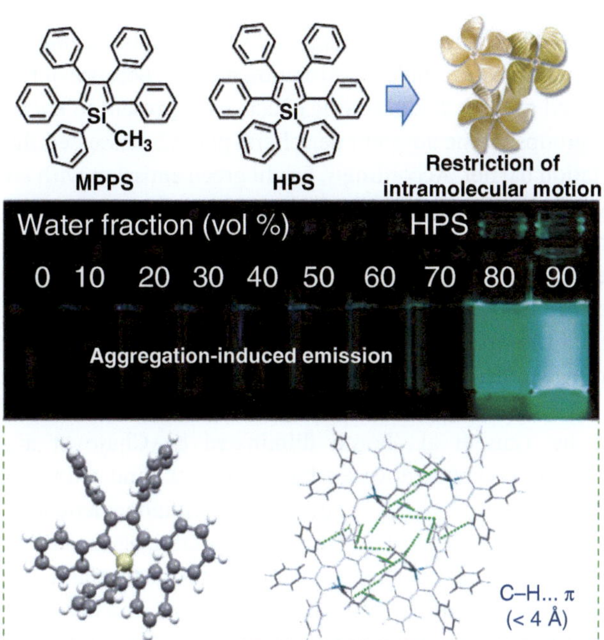

Figure 4.2 Molecular structures, PL behavior, and single crystal structure of silole derivatives. *Source*: [19]/with permission of John Wiley & Sons.

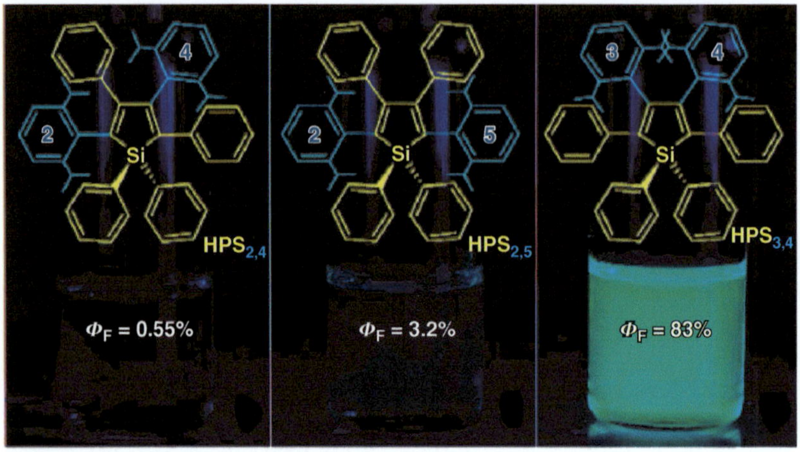

Figure 4.3 Molecular structures of three HPS derivatives and their PL behaviors in dilute solution. *Source*: [20]/with permission of American Chemical Society.

lead to the free intramolecular rotation of peripheral phenyls in a solution state. Therefore, the excitons will mainly decay through nonradiative transition. In the aggregation state, the formation of efficient intermolecular interactions can lead to the restriction of intramolecular motion (i.e., rotation), thus suppressing non-radiative decay and leading to bright emission.

To verify this mechanism, Li, Tang, and coworkers designed and synthesized three HPS derivatives, namely $HPS_{2,4}$, $HPS_{2,5}$, and $HPS_{3,4}$ (see Figure 4.3).[20] For these compounds, isopropyl (iPr) groups were attached to the peripheral phenyl rings of HPS to internally hinder the intramolecular rotation process. Because of the different positions of isopropyl groups, distinct steric effects were present in them. Especially, $HPS_{3,4}$ with iPr groups on the adjacent peripheral phenyls possessed the highest intramolecular rotation barrier. Accordingly, bright green emission with an efficiency of up to 83% could be achieved for its acetone solution. As for $HPS_{2,4}$ and $HPS_{2,5}$ solutions, the corresponding fluorescence efficiencies were only 0.55% and 3.2%, respectively, for the relatively free intramolecular rotation to facilitate non-radiation. Based on these results, the main origin of the AIE effect could be well certified to the restriction of intramolecular rotation (RIR).

According to the similar mechanism of RIR, a series of other AIE cores were successfully developed, such as tetraphenylethylene by Tang and Dong et al., 9,10-bis(2-phenylethenyl)anthracene by Tian et al., boron diiminated by Chujo et al., polyphenylbenzene by Li et al., polyphenylpyrrole by Dong et al., and tetraphenylpyrazine by Qin et al. (see Figure 4.4).[21–26] Specifically, tetraphenylethylene (TPE) was demonstrated to be one of the most excellent AIE building blocks, for its simple synthesis, easy modification, and outstanding AIE effect. Furthermore, its AIE effect could be easily inherited by the corresponding derivatives, and thus TPE group has been frequently utilized to construct multifunctional materials with AIE effect.[27] For instance, Chi and coworkers decorated the TPE core with polar groups to successfully yield AIE luminogens with mechanoluminescence (ML) effect,

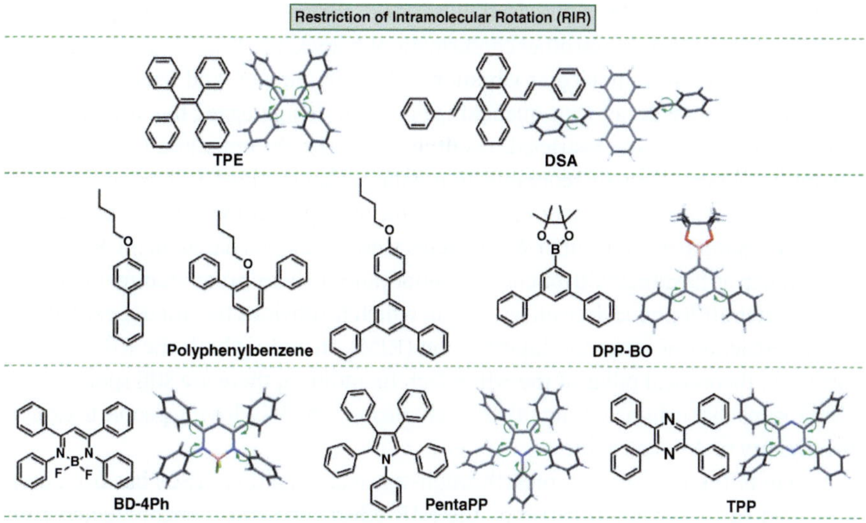

Figure 4.4 AIE luminogens based on the restriction of intramolecular rotation (RIR) mechanism.

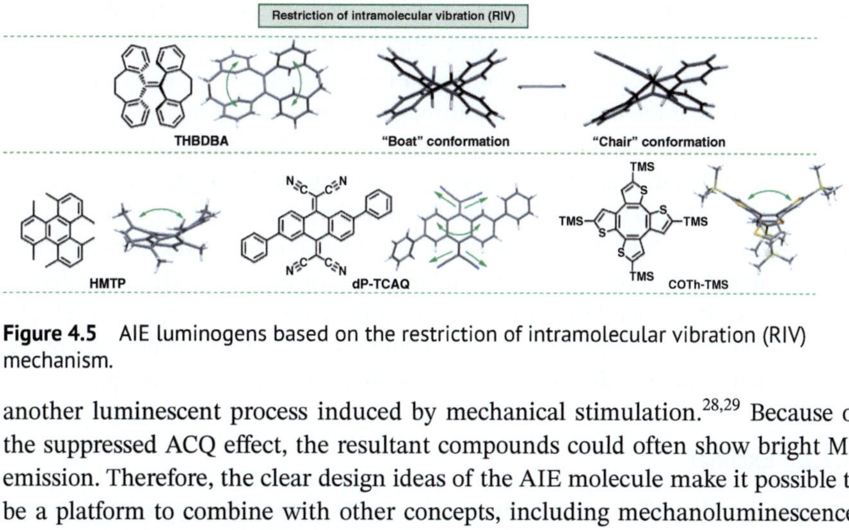

Figure 4.5 AIE luminogens based on the restriction of intramolecular vibration (RIV) mechanism.

another luminescent process induced by mechanical stimulation.[28,29] Because of the suppressed ACQ effect, the resultant compounds could often show bright ML emission. Therefore, the clear design ideas of the AIE molecule make it possible to be a platform to combine with other concepts, including mechanoluminescence, and then promote its further development.

Besides rotation, the vibration is also an efficient way to consume energy and accelerate non-radiative transition of excitons. Accordingly, the AIE effects of some luminogens have been found to be derived from the mechanism of restriction of intramolecular vibration (RIV) (see Figure 4.5). For instance, 10,10′,11,11′-tetrahydro-5,5-bidibenzo[a,d][7]annulenylidene (THBDBA), an analogue of TPE without freely rotatable phenyl rings, also shows a typical AIE effect, in which the fluorescence efficiency can increase from < 0.1% in solution to 23% in aggregate.[30] The single crystal structure of THBDBA indicates that each set of phenyl rings locked by the ethyl chain adopts a "boat" conformation for the stabilization of the neighboring intermolecular interactions. If THBDBA is dissolved in solution, theoretically it can adopt a "chair" conformation, as well as any other conformations between its "chair" and "boat" form. At this time, the intramolecular vibration will easily happen and lead to the dominated nonradiative decay, thus quenching emission. In aggregate, the intramolecular vibration can be largely restricted, resulting in the bright emission. Corresponding theoretical calculations of reorganization energy also certified this point, in which the RIV, rather than RIR, should be the main mechanism for the AIE behavior of THBDBA. Similarly, some other AIE systems were developed based on the RIV mechanism, such as hexamethyltriphenylene, tetracyano-9,10-anthraquinodimethane, and cyclooctatetrathiophene derivatives, etc., in which no obvious free rotator existed.[31–33] Thus, restriction of intramolecular motion (RIM), including both the RIR and RIV, forms the theoretical pillar of the AIE effect. In addition, there are still sporadic AIE luminogens that originate from other mechanisms, which is driving people to explore the mystery of AIE constantly.

As one of the typical examples, through-space conjugation (TSC) has been demonstrated to be the main cause of the AIE effect for some nonconjugated compounds.[34] TSC is a unique conjugate mode occurring in adjacent space, rather than covalent linkage. Because of its existence, some organic molecules without big aromatic conjugated structures could also give efficient emission in the visible region. As shown in Figure 4.6a, compound *s*-TPE with isolated phenyls linked by a non-conjugated ethyl group presents highly bright blue emission ($\Phi_F = 69\%$) in solid state, which even exceeds that of conjugated TPE ($\Phi_F = 44\%$).[35] Theoretical calculations indicate that the intramolecular through-space conjugation between the "isolated" phenyl rings should be mainly responsible for this abnormal phenomenon. On the other hand, a series of non-conjugated triphenylmethane (TPM) derivatives were reported to show noticeable blue emission in the solid state for the intra/intermolecular TSC effect (Figure 4.6b).[36] Also, it was found that the introduction of electron-donating groups into TPM could increase corresponding electronic density and stabilize TSC, then leading to the red-shifted emission and increased PL efficiency. Thus, the PL property of the non-conjugated AIE luminogens based on the TSC mechanism could also be adjusted through rational molecular design.

4.2.2 Excimer

Excimer is a class of dimeric excited-state species, which are formed by associating an excited-state molecule and a ground-state one.[37] As early as 1954, Förster and Kasper found that the PL behavior of pyrene solution was heavily related to its concentration for the formation of excimer.[38] As shown in Figure 4.7a, for the pyrene solution with low concentration (0.01–0.1 mmol/L), its PL behavior is independent of the changed solution concentration, in which just one emission band with fine structure from the monomer can be detected in the purple region (350–450 nm). Once further increasing the solution concentration, an additional emission band at 470 nm appears and gradually enhances, while its absorption shows no obvious change. This suggests that the intermolecular interaction occurs in the excited state, rather than the ground one. In 1960, Stevens and Hutton termed it

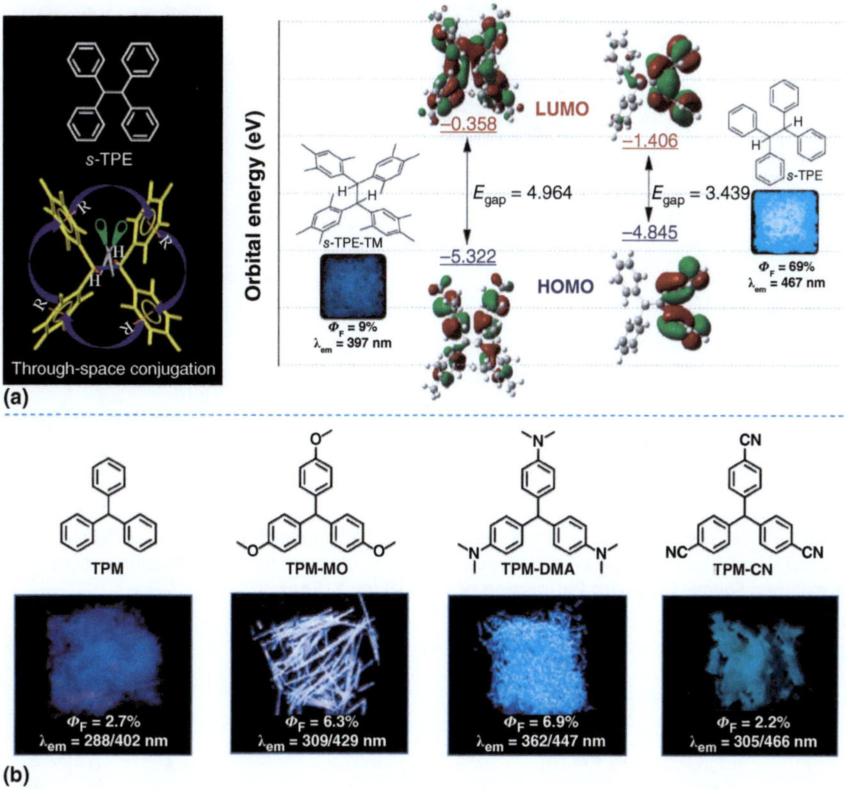

Figure 4.6 (a) The AIE effect of non-conjugated *s*-TPE based on intramolecular through-space conjugation (TSC) effect. *Source*: [35]/with permission of American Chemical Society; (b) the PL behaviors of non-conjugated triphenylmethane (TPM) derivatives with intra/intermolecular TSC effect. *Source*: [36]/with permission of American Chemical Society.

as "excimer," indicating it was a behavior in an excited state. For the formation of the pyrene excimer, the strong intermolecular π–π interaction is considered to be mainly responsible for it. Just as shown in its single crystal structure, the intermolecular π–π distance is as short as 3.528 Å within the molecular dimer, which can be much beneficial for the excited state molecule to interact with the adjacent one in the ground state, thus resulting in the excimer emission.

The formation process and emission property of excimer can be well understood as shown in Figure 4.7b.[39] In the ground state, the approach of two adjacent molecules will increase the repulsive energy quickly, leading to the sharply raised potential energy curve from flat to steep. As for the excited state, the molecule in the excited state approaches the one in the ground state, which will lead to the appearance of attraction, then causing the lowered potential energy curve in the excimer. Therefore, a minimum point for the potential energy can be obtained when the excimer reaches an equilibrium state. Since the energy of the excimer, decaying from the equilibrium position to the ground state, is much lower than that of the monomer, a large redshift emission will be presented. Besides, because of the

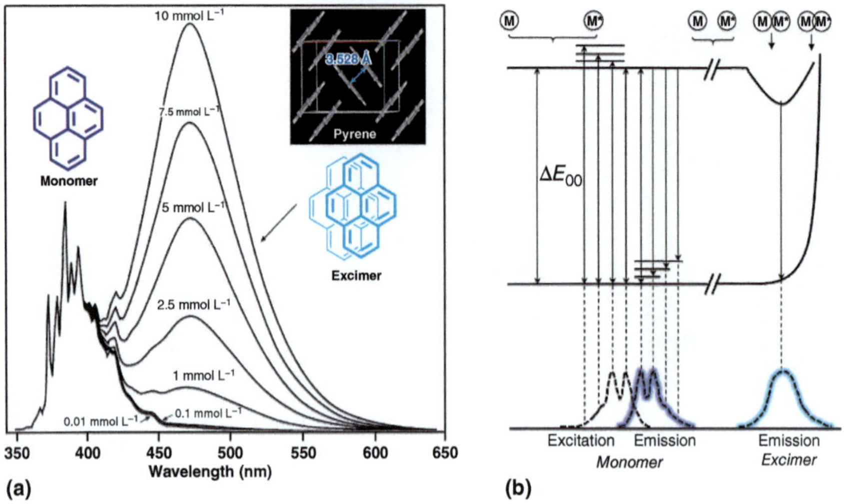

Figure 4.7 (a) The PL behavior of pyrene in hexahydrotoluene solution with different concentrations and its single crystal structure; (b) the diagram for the formation process of the excimer. *Source*: [38]/with Permission of John Wiley & Sons.

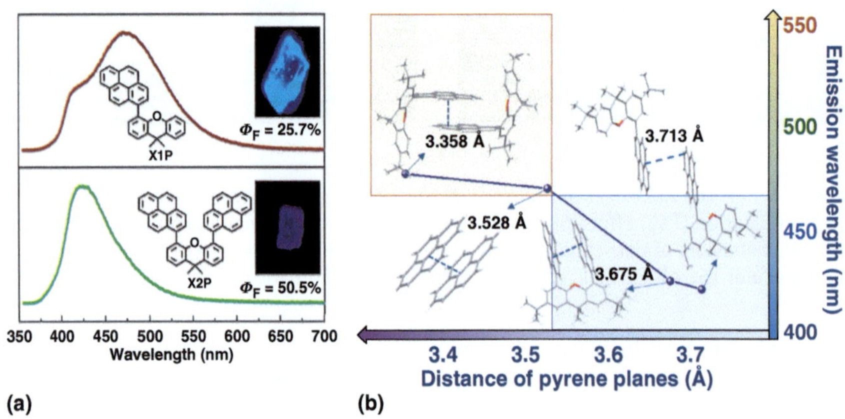

Figure 4.8 (a) Molecular structures and PL behaviors of X1P and X2P crystals. *Source*: [43]/Chinese Chemical Society; (b) the relationship between the π–π distance of pyrene dimer and the corresponding emission wavelength. *Source*: [43]/with permission of Chinese Chemical Society.

steep potential energy curve in the ground state and deep potential energy in the excited state, the thermal vibration will lead to the severe fluctuation of vibration energy level, which thus causes a broad emission band without any fine structure in the excimer. On the other hand, the emission from excimer often presents a much longer lifetime and greater temperature/viscosity sensitivity than the monomer one for the dynamic interaction in the excited state.[40–42]

In order to study the relationship of dimeric π–π distance and excimer emission, Li and coworkers designed and synthesized a series of pyrene derivatives (see Figure 4.8).[43] It is interesting to find that their emission peaks could well

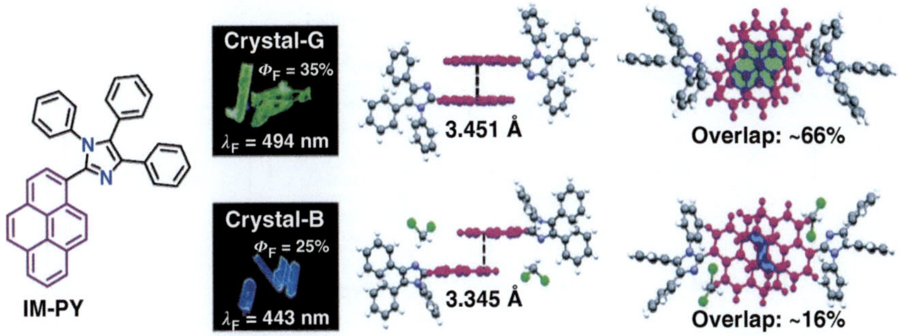

Figure 4.9 The polymorphisms of IM-PY and corresponding PL behaviors. *Source*: [44]/with permission of Royal Society of Chemistry.

correspond to the classes of intra/intermolecular pyrene dimers in the crystal state. For the X1P crystal, two kinds of pyrene dimers exist, of which the closest one with π–π distance of 3.358 Å corresponds to the emission peak at 475 nm, while the shoulder peak at 420 nm should be from the loose one with a distance of 3.713 Å. In a pyrene crystal with an emission peak at 468 nm, the π–π distance within the molecular dimer is 3.528 Å. Besides, the relatively longer π–π distance of 3.675 Å in the X2P crystal leads to a blue-shifted emission at 424 nm. These results clearly demonstrate that the distances within the pyrene dimer are heavily related to their formation of excimers and corresponding emission wavelengths, namely the shorter π–π distance, the more red-shifted emission.

Besides, Yang and coworker successfully demonstrated the important role of π–π overlap in the formation of excimer.[44] As shown in Figure 4.9, they cultured two polymorphisms with distinct PL emission properties for compound IM-PY. It was strange to find that although crystal-B presented a shorter π–π distance (3.345 Å) within the pyrene dimer than that (3.451 Å) of crystal-G, its PL emission was much blue-shifted. Detailed analyses of the single crystal structure revealed that the π–π overlap of the pyrene dimer was as large as 66% in crystal-G, while just 16% in crystal-B. Thus, it could be verified that large π–π overlap is also much important for the formation of an excimer, in addition to short π–π distance. Besides, it was interesting that crystal-G with obvious excimer emission gave much higher PL efficiency (35%) than crystal-B (25%). This indicates that the formation of excimer doesn't always lead to the quenched emission. Actually, a fair amount of excimers have been found to show highly bright PL emission, especially for those based on discrete dimers in the crystal state, including pyrene as mentioned above.[45–47] Therefore, through rational molecular design and crystal engineering, the construction of a particular excimer can also act as an efficient way to develop bright luminescent materials.

4.2.3 J/H/X Aggregate

For organic π-conjugated compounds, different packing modes can lead to distinct interactions between molecular transition dipoles in aggregate, which then affect the resultant photophysical properties. As early as 1963, Kasha and coworkers

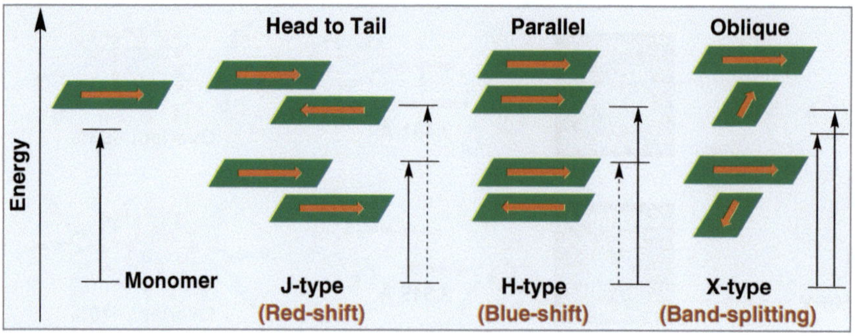

Figure 4.10 Excited state energy splitting diagram for molecular aggregates with J-aggregate, H-aggregate, and X-aggregate.

proposed the molecular exciton theory to describe the corresponding behaviors (see Figure 4.10).[48,49] Basically, the interaction between molecular transition dipoles can cause the excited state energy splitting in aggregate, while different types of aggregates will bring out different allowable transitions. For instance, in H-aggregates with parallel transition dipoles, the transition from the ground state to a higher exciton state is allowed, resulting in the blue-shifted absorption. When it turns to J-aggregate with head-to-tail transition dipoles, it is reversed. As for X-aggregate, the interaction of molecular transition dipoles is much weaker; thus, no obvious blue or red-shifted absorption can be observed. Besides, these three kinds of aggregates have also been found to show a significant influence on the corresponding PL behaviors.

To investigate the relationship between the molecular aggregate and luminescent property, Wang and coworkers prepared two kinds of perylene microcrystals by simply controlling the solution concentration.[50] As shown in Figure 4.11, strong intermolecular π–π interactions existed for both microcrystals, which drove the formation of perylene dimers. Among them, α phase microcrystals presented typical H-aggregation with a slip angle of 89.29° for perylene dimer, while that of β phase was 36.73° for the formation of J-aggregate. Correspondingly, the UV–vis absorption of β phase was red-shifted in comparison with α one. Further on, these two polymorphisms were found to exhibit much different emission properties, due to their different interactions of molecular transition dipoles. On one hand, the J-type microcrystals with small intermolecular π–π overlap gave a much blue-shifted emission (525 nm) than H-type ones (585 nm). On the other hand, the fluorescence decay rate was much larger in J-aggregate with a transition allowed to a lower exciton state, which then led to its higher PL efficiency (52%) and shorter emission lifetime (0.63 ns) in comparison with H-aggregate (35% and 19.61 ns). Thus, the changed molecular stacking modes can indeed modulate the luminescent properties of these organic aggregates.

In the formation of H/J-aggregate with cofacial transition dipoles, intermolecular π–π interaction often acted as the main driving force. But it was not always turned out that way. Taking compound AD-TPE as an example, it shows a twisted molecular conformation for the introduction of the tetraphenylethylene group, a

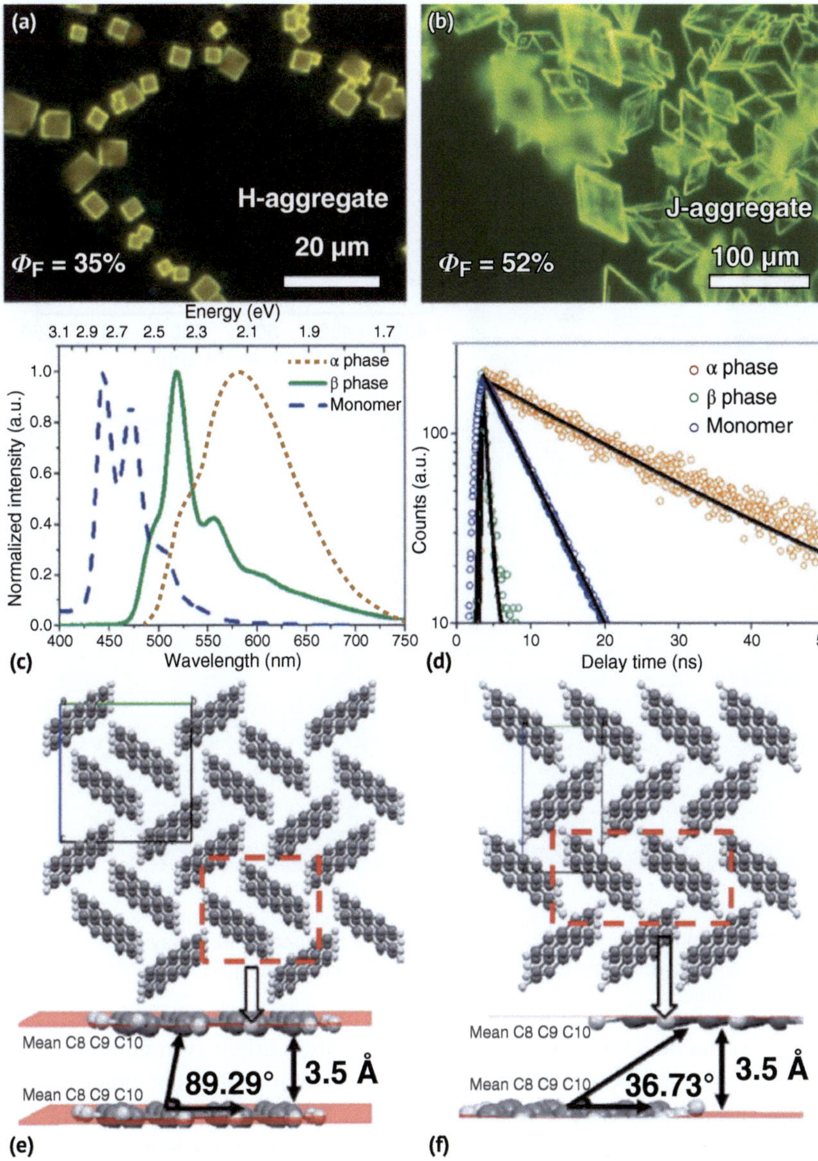

Figure 4.11 (a) The fluorescence microscope image of the α phase perylene microcrystals; (b) the fluorescence microscope image of the β phase perylene microcrystals; (c) the fluorescence spectra of the perylene solution and the α/β phase perylene microcrystals; (d) the fluorescence decay curves of the perylene solution and the α/β phase perylene microcrystals; (e) single crystal structure of the α phase perylene microcrystals; (f) single crystal structure of the β phase perylene microcrystals. *Source*: [50]/with permission of John Wiley & Sons.

typical AIE building block (see Figure 4.12).[51] Therefore, the intermolecular π–π interaction could be well restricted in its crystal. However, it was strange to find that AD-TPE pristine powders gave very weak emission ($\Phi_F = 1\%$) under 365 nm

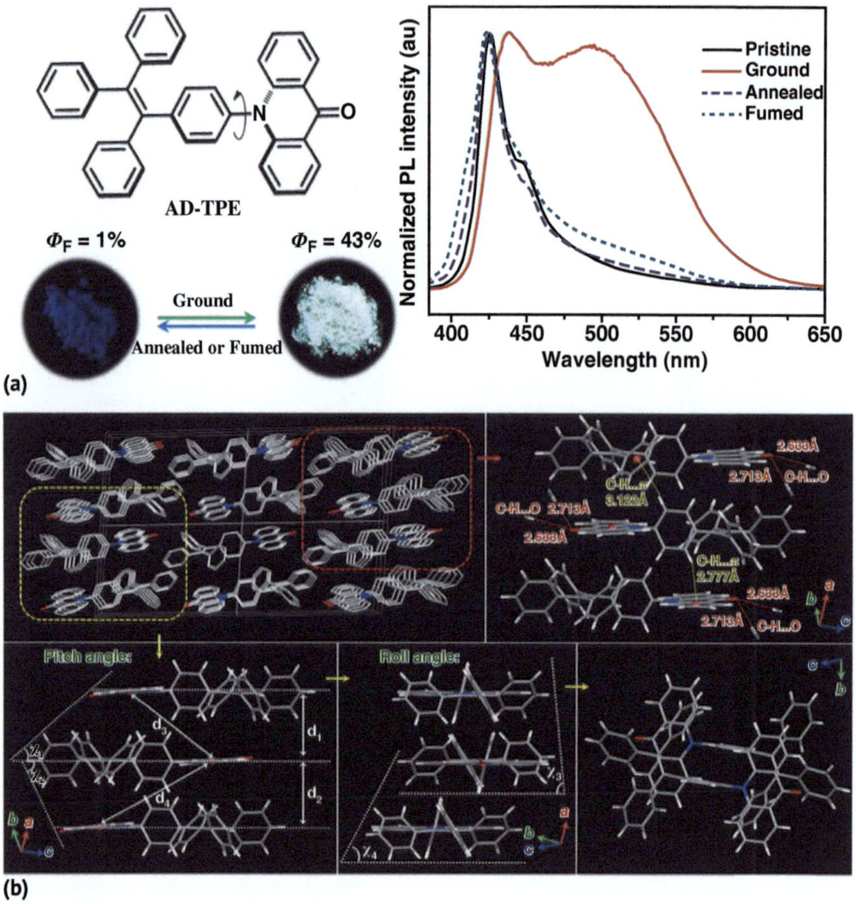

Figure 4.12 (a) Molecular structure of AD-TPE and corresponding PL behaviors under different treatments; (b) single crystal structure of AD-TPE. *Source*: [51]/with permission of John Wiley & Sons.

UV light irradiation. Only after being ground heavily, a bright cyan emission with a quantum efficiency of 43% could be achieved. Careful analyses of single crystal structure showed that typical H-aggregation has been formed, regardless of the absence of intermolecular π–π interaction. Instead, efficient dipole-dipole interactions and intermolecular hydrogen bond interactions (C—H···O and C—H···π etc.) could be observed in its crystal. These together brought about the specific dimers of AD-TPE with the antiparallel and side-by-side alignment in the crystal. At last, the formation of H-aggregation caused the quenched PL emission of AD-TPE pristine powders. In the ground state, destroyed H-aggregation and planarization of the molecular conformation jointly contributed to the turn-on fluorescence.

The X-one with cross-dipole stacking, compared to the H/J-aggregate, presents much weaker interactions between molecular transition dipoles, which often leads to the bluer emission and higher PL efficiency.[52,53] As shown in Figure 4.13, two polymorphisms of BDFVA with different morphologies were prepared through a physical vapor transport method inside a quartz tube.[54] For them, the ribbon-like crystal

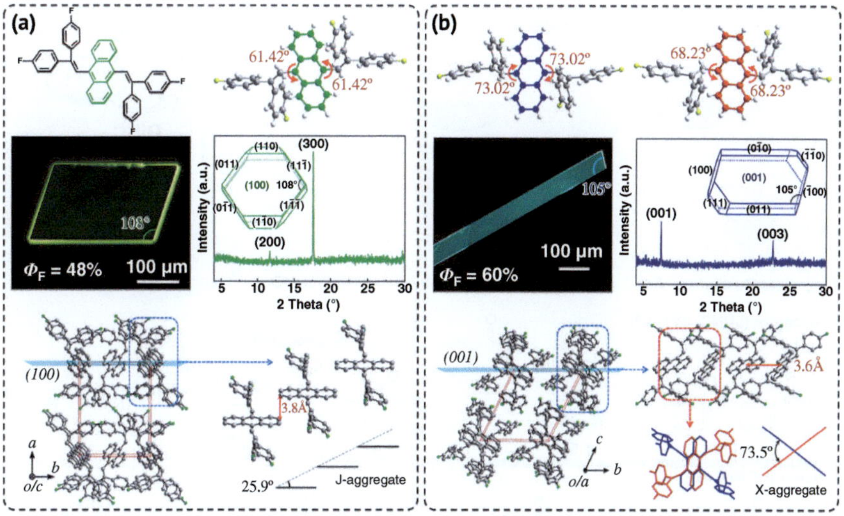

Figure 4.13 (a) The crystal of compound BDFVA in J-aggregate and corresponding PL property; (b) the crystal of compound BDFVA in X-aggregate and corresponding PL property. *Source*: [54]/with permission of American Chemical Society.

showed sky-blue emission of 498 nm and high PL efficiency of 60%, while the plate-like one was in green, peaked at 518 nm with lower efficiency of 48%. Further measurements of their single crystal structures found that a J-aggregate has been formed in the plate-like crystal, in which a pitch angle of 25.9° was observed along the anthracene planes. As for the ribbon-like one, the molecular dimers adopted cross-dipole stacking with a molecular rotation angle of 73.5°, showing a typical characteristic of X-aggregate. Correspondingly, the X-aggregates with higher PL efficiency were found to exhibit better performance in the applications of optical waveguide and light amplification compared to the J-aggregates. Therefore, the great potential of X-aggregate to achieve high PL efficiency could be well certified, which will be much beneficial for the practical applications of the luminescent materials in the crystal state.

4.3 Thermally Activated Delayed Fluorescence (TADF)

4.3.1 Generation of Delayed Fluorescence

For normal fluorescence, it comes from the direct decay of singlet excitons, which often results in a short lifetime on the nanosecond scale. If the singlet excitons undergo intersystem crossing (ISC) and reverse intersystem crossing (RISC) processes between singlet and triplet states, the delayed fluorescence can occur with a corresponding lifetime prolonged to microseconds or even milliseconds. As the delayed fluorescence is usually thermally activated for the endothermic RISC process, it was termed as TADF.[55] In recent years, TADF has attracted significant attention for its great advantages in the applications of bioimaging and OLEDs, etc.[56–58] For a typical TADF luminogen, a small energy difference between the S_1 and T_1 states ($\Delta E_{ST} < 0.30\,eV$) is required for the realization of RISC with thermal

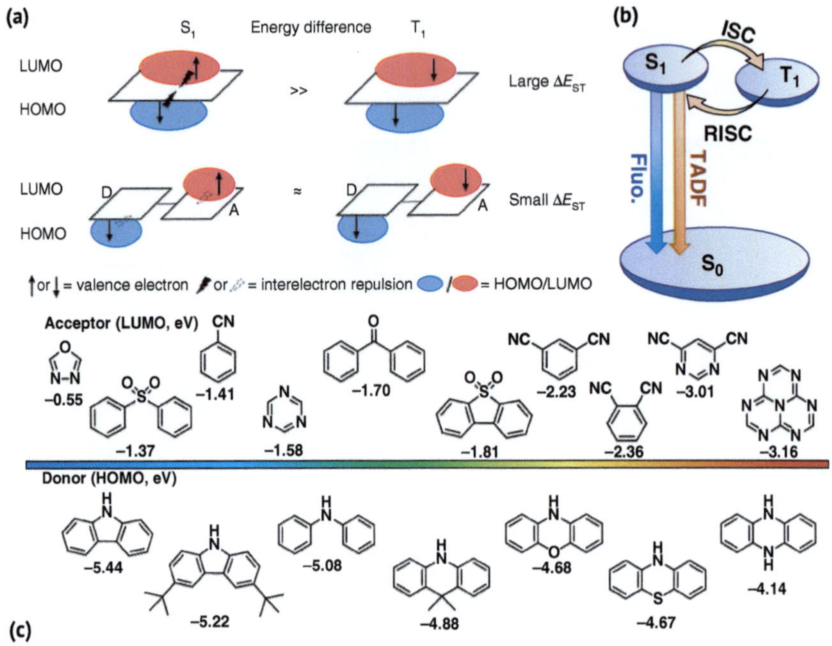

Figure 4.14 (a) Mechanism of fine-tuned ΔE_{ST} through HOMO–LUMO engineering; (b) diagram of thermally activated delayed fluorescence (TADF); (c) chemical structures of representative acceptor and donor groups for TADF molecules. *Source*: [60]/with permission of John Wiley & Sons.

activation.[59] Thus, the design of TADF molecules is actually the regulating of the corresponding ΔE_{ST}.

As shown in Figure 4.14a, the ΔE_{ST} of luminogens can be solely ascribed to the opposite spins of the two valence electrons in S_1, since this will induce more electron repulsion effect, then raising the total energy of the molecule.[60] Therefore, to reduce ΔE_{ST}, it is essential to alleviate the electron repulsion effect in the S_1 state. Since the effective separation of HOMO and LUMO can mitigate the repulsion effect between electrons, a spatially twisted D-A structure will be helpful (see Figure 4.14b). Accordingly, a series of electron donor (D) and acceptor (A) groups have been successfully used in the construction of TADF molecules (see Figure 4.14c). Partially for the extended π-conjugation of the donor or acceptor group, most of the TADF luminogens still presented π–π stacking in the aggregated state, which often led to the ACQ effect, thus hindering their practical applications.

4.3.2 Aggregation-induced Delayed Fluorescence

TADF luminogens were considered to be promising emitting materials in OLEDs. However, suffering from the ACQ effect, they were frequently doped into host materials to construct OLEDs. With it, the high OLED performance could be achieved, while some side problems were also brought, such as complicated process, inferior stability for the potential phase separation, and high cost, etc. Until 2015, Chi and coworkers successfully merged AIE and TADF into one luminogen to obtain the abnormal

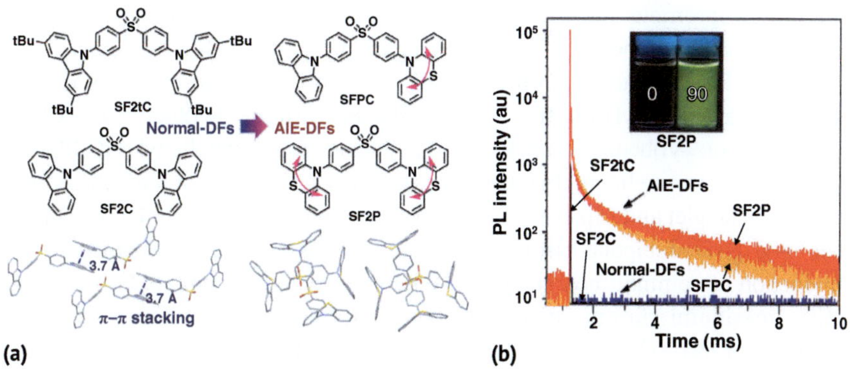

Figure 4.15 (a) Molecular structures and packing modes for normal delayed fluorescence molecules (SF2tC, SF2C) and aggregation-induced delayed fluorescence (AIDF) molecules (SFPC, SF2P). *Source*: Adapted from[61]; (b) the PL decay curves for SFPC, SF2P, SF2C, and SF2tC solids. Inset: the fluorescent pictures for SF2P in solution and aggregate. *Source*: [61]/with permission of John Wiley & Sons.

aggregation-induced delayed fluorescence (AIDF) effect.[61] As shown in Figure 4.15, because of the suppressed exciton quenching effect upon aggregation, high PL efficiencies up to 93.3% and obviously delayed components in PL decay curves could be both obtained for AIDF luminogens of SFPC and SF2P in the solid state, while the normal TADF emitters of SF2tC and SF2C not. These made AIDF materials show great advantage in nondoped OLEDs. Single crystal structures showed that strong π–π interaction existed in the SF2C crystal, while the intermolecular π–π stacking could be largely restricted for the introduction of the twisted phenothiazine group in the SF2P crystal. Also, two adjacent SF2P molecules presented X-type aggregation in their crystal, which could largely restrict the quenching of excitons. Thus, the molecular packing should also be of great importance in the realization of the AIDF effect.

Then, the applications of AIDF materials in OLEDs were explored, and great process has been made by Zhao et al.[62] For normal delayed fluorescence materials, they were often doped into a host to restrict the possible ACQ effect in OLEDs, while drastic efficiency roll-off at high voltages still frequently occurred. As for AIDF emitters, things were different (see Figure 4.16). Taking AIDF luminogen of DBT-BZ-DMAC as an example, the donor (DMAC) and acceptor (BZ) moieties were linked in a nearly perpendicular pattern, thus no close π–π stacking was presented for BZ-DMAC fragments.[63] Since the BZ-DMAC fragment, with an electronic push–pull effect, was mainly responsible for light emission, a high PL efficiency of 80.2% and a large delayed component of 64.5% were achieved in DBT-BZ-DMAC film (see Figure 4.16). On the other hand, the planar DBT moieties tended to form close π–π stacking with an interplane distance of 3.512 Å, which was favorable for charge transport. Thus, when DBT-BZ-DMAC acted as an emitter in a nondoped OLED, a maximum EQE of 14.2% with negligible efficiency roll-off was achieved. Besides, although comparable EQE values (13.0–17.9%) could be obtained in doped OLEDs based on DBT-BZ-DMAC, they showed severe efficiency roll-off up to 39%. Thus, the great advantages of AIDF emitters in the application of nondoped OLEDs could be well demonstrated.

To further promote the OLED performance, Zhao and coworkers designed and synthesized a series of TADF luminogens by grafting 4-(phenoxazin-10-yl) benzoyl (BP-PXZ), a moiety with AIDF effect, to common host materials including 1,4-di(carbazol-9-yl)benzene (DCB), 4,4'-bis(carbazol-9-yl)biphenyl (CBP), 1,3-bis(carbazol-9-yl)benzene (*m*CP), and 3,3'-di(carbazol-9-yl)biphenyl (*m*CBP) (see Figure 4.17).[64] On one hand, their neat films could give strong fluorescence and fully harvest both singlet and triplet excitons with suppressed exciton annihilation. On the other hand, good ambipolar carrier transport ability was successfully obtained for the introduction of common host materials. When they acted as emitters in nondoped OLEDs, maximum current efficiencies, power efficiencies, and external quantum efficiencies of 69.0–72.9 cd A^{-1}, 75.0–81.8 lm W^{-1}, and 21.4–22.6% were presented. Noticeably, excellent electroluminescence (EL) efficiencies could still be maintained

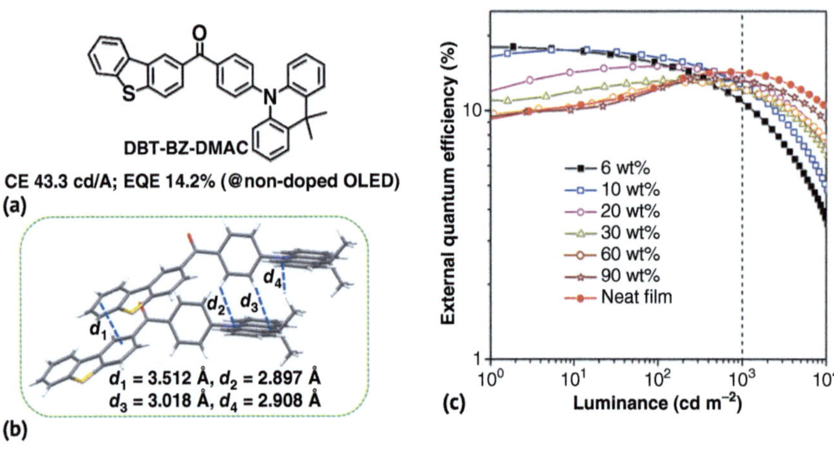

Figure 4.16 (a) Molecular structure and OLED performance of DBT-BZ-DMAC; (b) single crystal structure of DBT-BZ-DMAC; (c) external quantum efficiencies of OLEDs based on DBT-BZ-DMAC when it was doped in CBP with different concentrations. *Source*: [62]/with permission of John Wiley & Sons.

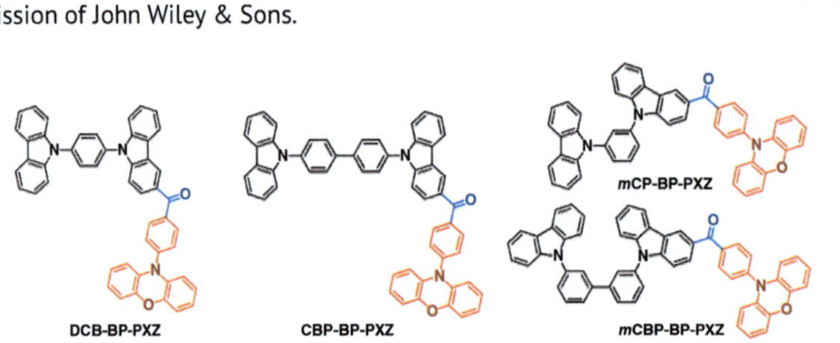

Device	V_{on} [V]	Maximum values				Values at 5000 cd m^{-2}					
Nondoped OLEDs		η_C [cd A^{-1}]	η_P [lm W^{-1}]	η_{ext} [%]	L [cd m^{-2}]	V [V]	η_C [cd A^{-1}]	η_P [lm W^{-1}]	η_{ext} [%]	RO [%]	λ_{EL} [nm]
DCB-BP-PXZ	2.5	72.9	81.8	22.6	95 577	4.0	64.6	60.8	20.1	11.4	548
CBP-BP-PXZ	2.5	69.0	75.0	21.4	98 089	4.2	62.0	46.4	19.2	10.1	546
*m*CP-BP-PXZ	2.5	72.3	79.0	22.1	10 0126	4.4	64.5	46.0	19.8	10.8	542
*m*CBP-BP-PXZ	2.5	70.4	76.5	21.8	96 815	4.2	63.4	47.5	19.6	9.9	542

Figure 4.17 Molecular structures and OLED performances of DCB-BP-PXZ, CBP-BP-PXZ, *m*CP-BP-PXZ, and *m*CBP-BP-PXZ.

at a high brightness of 5000 cd m^{-2} with η_C and η_{ext} of 62.0–64.6 cd A^{-1} and 19.2–20.1%, respectively, showing great potential in the application of solid-state lighting.

4.3.3 Through-space Charge Transfer-based TADF

In the construction of TADF luminogens, the minimized ΔE_{ST} to undergo ISC and RISC is the key point. To achieve it, the direct covalent connection of donor (D) and acceptor (A) has been frequently utilized. Actually, besides the covalent bond, the noncovalent interactions between donor and acceptor could also lead to a similar result, providing the spatially separated frontier molecular orbitals in resultant compounds. For instance, Swager and coworkers successfully developed three TADF luminogens based on through-space charge transfer (TSCT), namely XPT, XCT, and XtBuCT (see Figure 4.18).[65] For these compounds, a donor and an acceptor were bridged through the 4 and 5 positions of a 9,9-dimethylxanthene

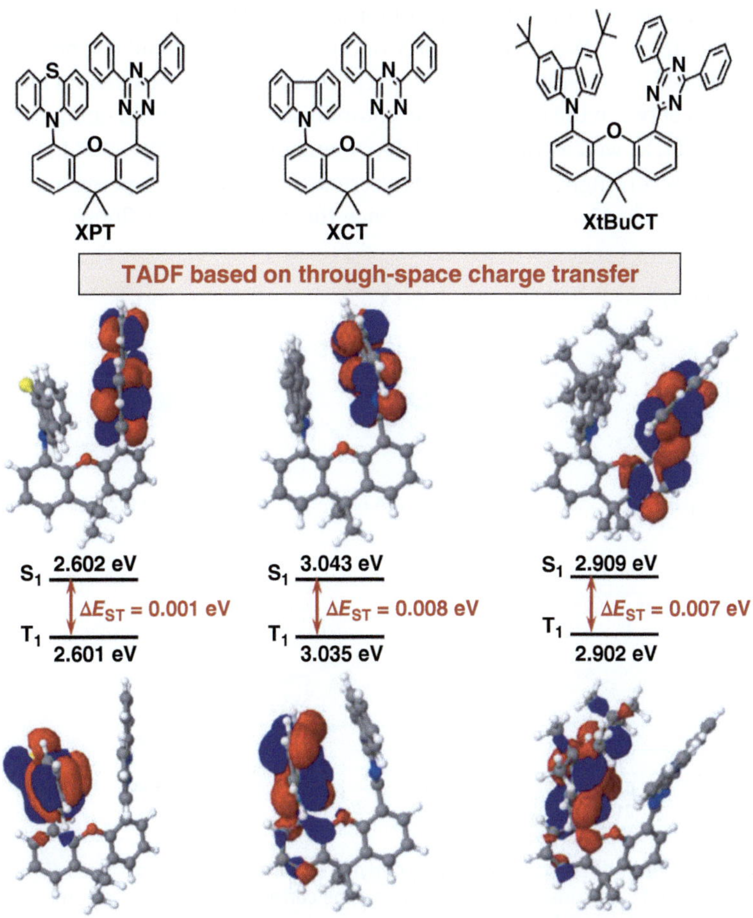

Figure 4.18 The molecular structures, calculated HOMO/LUMO orbital distributions, and energy levels of XPT, XCT, and XtBuCT. *Source*: [65]/with permission of American Chemical Society.

scaffold, which generated a relatively fixed interchromophore spacing. Their crystal structures confirmed the cofacial arrangement of the donor and acceptor groups at distances of 3.3-3.5 Å, which were short enough to occur through-space charge transfer. Because of the through-space interaction between donor and acceptor groups, the spatially separated frontier molecular orbitals were present for these compounds, which then led to the minimized ΔE_{ST} and an obvious TADF effect. Subsequently, based on the similar mechanism, a series of other TADF luminogens were successfully designed by replacing 9,9-dimethylxanthene with other bridges, such as the nonconjugated poly(cis, exo-2,3-diarylnorbornene) backbone reported by Wang, the asymmetric spiro-backbone reported by Jiang, and carbazole by Yang etc.[66–68]

To study the influence of intramolecular D-A electronic coupling on TSCT-type TADF, Liao and coworkers built three kinds of D/A pairs with 10-phenyl-9,10-dihydroacridine as donor (D) and 2,4,6-triphenyl-1,3,5-triazine as acceptor (A), namely, fixed D/A, flexible D/A, and distant D/A (see Figure 4.19).[69] When they were linked by a rigid bridge of fluorene to form a fixed D-A pair, quasi-cofacial arrangements with short D-A distances of 3.16, 3.06, and 2.83 Å could be obtained for compounds DM-B, DM-Bm, and DM-G, respectively. With these efficient electronic couplings between the donor and acceptor units, the minimized energy gap between S_1 and T_1, as well as the efficient TADF emission ($\Phi_F = 88 - 96\%$), were both obtained. Accordingly, their OLEDs presented excellent EL performance with η_{EQE}s up to 18.5–27.4%, respectively. Once the rigid linker of fluorene was

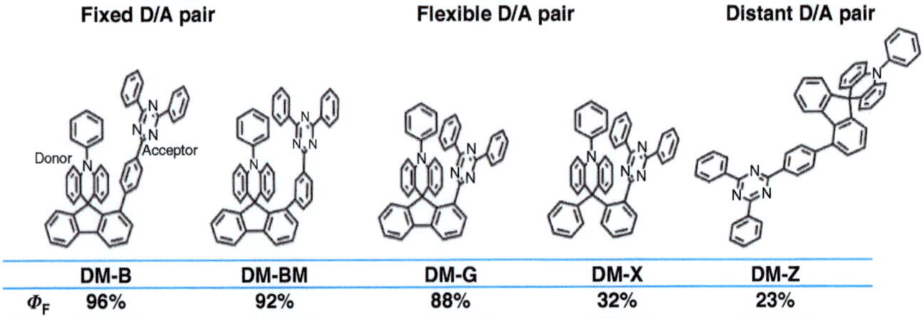

	DM-B	DM-BM	DM-G	DM-X	DM-Z
Φ_F	96%	92%	88%	32%	23%
ΔE_{ST}	0.17 eV	−0.08 eV	−0.11 eV	0.03 eV	0.54 eV
η_{EQE}	27.4%	21.7%	18.5%	4.3%	3.2%

Figure 4.19 Three kinds of D/A pairs with 10-phenyl-9,10-dihydroacridine as donor (D) and 2,4,6-triphenyl-1,3,5-triazine as acceptor (A), and the corresponding influence on fluorescence quantum yield (Φ_F), energy gap between S_1 and T_1 (ΔE_{ST}), and external quantum efficiency (η_{EQE}) in OLEDs. *Source*: [69]/with permission of Springer Nature.

replaced by flexible diphenylmethane to yield compound DM-X, a strongly distorted configuration was observed, which was not beneficial for the D-A coupling. Thus, its PL efficiency was largely decreased to 32% for the intensive nonradiative motion. Also, an inferior EL efficiency of 4.3% was obtained for the OLED based on DM-X, regardless of its TADF effect. As for DM-Z, no through-space charge transfer could be realized for the distant D/A pair, resulting in the disappearance of TADF effect and a low EL efficiency of 3.2%. Thus, the change of through-space electronic interaction opens up another way to regulate the opto-electronic property of organic luminogens.

4.3.4 Exciplex-based TADF

Further on, the direct physical mixture of donor and acceptor molecules could also trigger the formation of an exciplex with spatially separated frontier molecular orbitals, then resulting in the TADF effect.[56] In 2012, Adachi and coworkers successfully realized the typical TADF emission by utilizing the high RISC efficiency of the exciplex state between donor and acceptor molecules (see Figure 4.20).[70] In their work, *m*-MTDATA acted as a donor and *t*-Bu-PBD as an acceptor. It was interesting to find that their mixture showed an obviously red-shifted emission than the neat

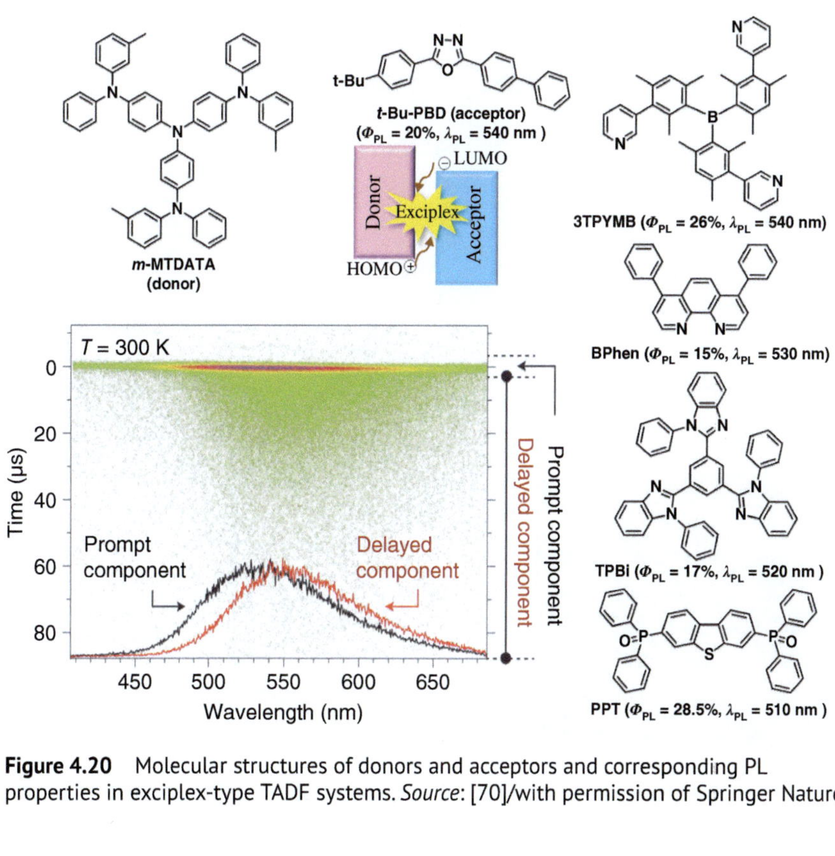

Figure 4.20 Molecular structures of donors and acceptors and corresponding PL properties in exciplex-type TADF systems. *Source*: [70]/with permission of Springer Nature.

m-MTDATA and *t*-Bu-PBD films. Besides, the corresponding ΔE_{ST} was just 50 meV for the well-separated HOMO and LUMO orbitals, and thus typical TADF emission was detected. Further on, some other acceptors were introduced, such as 3TPYMB, BPhen, TPBi, and PPT, etc.[71,72] When they were mixed with the *m*-MTDATA donor, the exciplex-typed TADF with tunable emission colors and quantum yields could be realized, verifying the wide applicability of this method.

To further clarify the internal mechanism of exciplex-type TADF, especially for the interaction mode of donor and acceptor, Hu and coworkers designed a CT cocrystal complex with TSB as donor and TCNB as acceptor (see Figure 4.21).[73] This cocrystal showed a new, featureless broad peak at 580 nm for the formation of exciplex, while TCNB and TSB both gave blue emission with wavelengths shorter than 400 nm. Further measurements of the temperature-dependent and time-resolved PL spectra indicated that this newly formed emission peak in the cocrystal was TADF active. To clarify the influence of intermolecular donor-acceptor interactions on the resultant TADF effect, the single crystal structure of this cocrystal was analyzed carefully. On one hand, the mean interplanar donor-acceptor (D-A) distance was measured to be 3.37 Å, where each benzene ring of the TSB was almost overlapped by a TCNB molecule along the a-axis, forming the strong π–π interaction. With it, the intermolecular

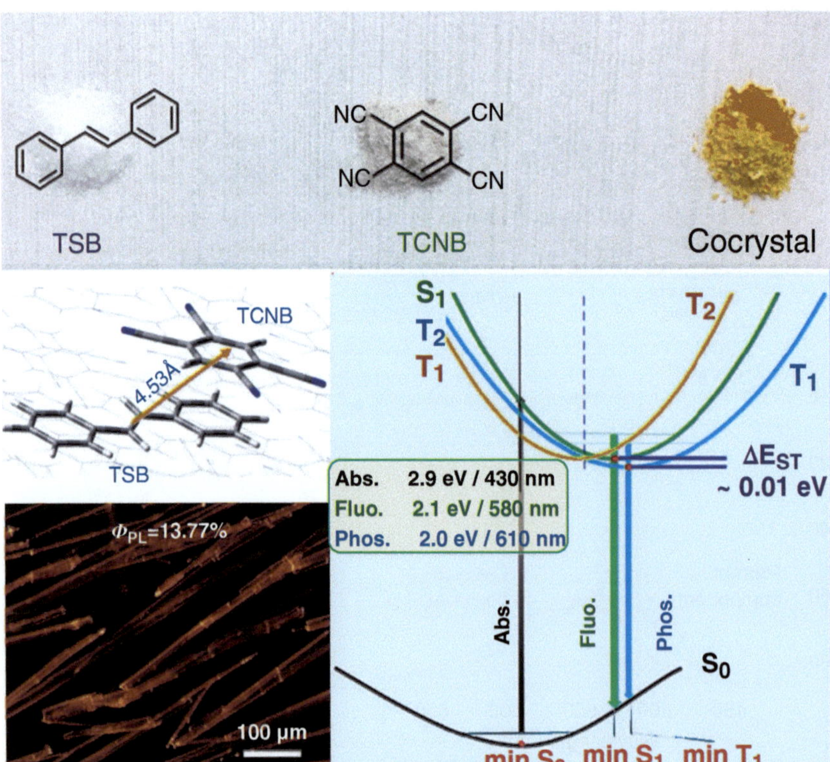

Figure 4.21 TADF material based on TSB/TCNB cocrystal. *Source:* [73]/with permission of John Wiley & Sons.

charge transfer could be achieved, thus leading to the minimized ΔE_{ST}(~0.01 eV). On the other hand, efficient intermolecular C—H···N interactions (2.773–3.313 Å) connected the adjacent TSB and TCNB molecules together in the same layer, forming a rigid three-dimensional network in the cocrystal. These efficient intermolecular interactions could largely restrict nonradiative relaxations and O_2 quenching, thus stabilizing the triplet excitons. Therefore, an obvious TADF effect could be easily detected for this cocrystal in the ambient conditions, well demonstrating the significant effect of intermolecular interaction on the TADF effect.

4.4 Phosphorescence

4.4.1 Room-temperature Phosphorescence

According to the Jablonski diagram, two kinds of excited states exist in the PL process, namely singlet (S) and triplet (T) states, while the ground state is usually in singlet (S_0). Therefore, the singlet excitons are easy to be formed and decay as fluorescence, while phosphorescence decaying from the T_1 state is hard for purely organic luminogens for their weak spin-orbit coupling (SOC) between singlet and triplet states. To enhance the corresponding SOC, a series of molecular design strategies have been proposed, including introducing heavy atoms (Br, I, etc.) or heteroatoms (N, O, S, P, etc.) with lone pair electrons.[74,75] However, although phosphorescence could appear at low temperature (i.e., 77 K) for these molecules, it would be quenched at room temperature for the dominate nonradiative transition ($k_{P,nr}$) of triplet excitons. Therefore, the organic room-temperature phosphorescence (RTP) was hard to achieve, especially for that with ultralong emission lifetime. In 2010, Ben Zhong Tang and coworkers reported the phenomenon of "crystallization-induced phosphorescence (CIP)" in some benzophenone derivatives, and scientists began to realize the significant influence of intermolecular interactions in the RTP effect.[76] Accordingly, more and more RTP luminogens were developed, including those with ultralong lifetimes.

A halogen bond is a kind of noncovalent interaction between halogen atoms (Lewis acids) and neutral or negatively charged Lewis bases, which has been widely used in many fields such as molecular recognition, chiral resolution, crystal engineering, and supramolecular assembly.[77] Also, because of the heavy atom effect from bromine and iodine, halogen bonds could enhance SOC between singlet and triplet states, thus contributing to the production of RTP emission. Accordingly, Kim and coworkers studied the PL property of 2,5-dihexyloxy-4-bromobenzaldehyde (Br6A) (see Figure 4.22a).[78] When it was dissolved in chloroform solution, just weak fluorescence at 420 nm could be observed. However, in the crystal state, green RTP at 500 nm appeared, with the phosphorescent efficiency of 2.9% and a lifetime of 5.4 ms. Single-crystal structure revealed an efficient halogen bond of C=O···Br existed in it with a distance of 2.86 Å and an angle of 126°. This might lead to the delocalized excited-state electrons from the carbonyl oxygen to the outer orbitals of the bromine atom, thus promoting ISC transition between singlet and triplet states. Also, this short bond could restrict the

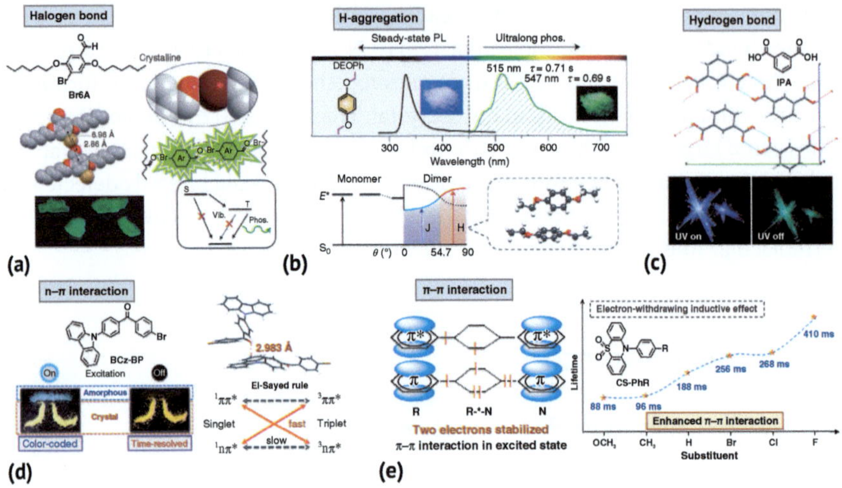

Figure 4.22 (a) Room-temperature phosphorescence based on halogen bond interaction. *Source*: [79]/with permission of Springer Nature; (b) room-temperature phosphorescence based on H-aggregation. *Source*: [80]/with permission of Springer Nature; (c) room-temperature phosphorescence based on hydrogen bond interaction. *Source*: [81]/with permission of The Royal Society of Chemistry; (d) room-temperature phosphorescence based on n-π interaction. *Source*: [83]/with permission of John Wiley & Sons; (e) room-temperature phosphorescence based on π–π interaction. *Source*: [84]/with permission of Springer Nature.

molecular motion, thus minimizing the nonradiative transition of triplet excitons. At last, the RTP effect was successfully achieved in the Br6A crystal.

For the RTP system based on halogen bonds, although the enhanced SOC between singlet and triplet states could promote RTP emission, the corresponding lifetimes were usually short for the fast phosphorescence decay rate. In 2015, Huang, Liu, Chen, and coworkers proposed the new strategy of H-aggregation to stabilize triplet excitons and prolong RTP lifetime (see Figure 4.22b).[79] The packing of the molecules by H-aggregation is known to increase the lifetime of singlet excitons through slowing the radiative decay rate. When it turned to RTP emission from triplet excitons, a similar result could be obtained. As shown in Figure 4.22b, the DEOPh crystal presented an ultralong green RTP emission ($\tau_{RTP} = 0.71$ s) for the formation of H-aggregation. On one hand, the molecular dimer with H-aggregation showed a much enhanced ISC transition than the monomer, due to an increased number of ISC channels. On the other hand, a lower T_1* state would be formed in H-aggregation, which functioned as an energy trap to suppress radiative and nonradiative decay rates in favor of long-lived excited states and ultralong phosphorescence. This work successfully ignited the enthusiasm for the study of persistent RTP effect, especially for the relationship between molecular packing and RTP lifetime.

Later, Tang and coworkers utilized strong intermolecular hydrogen bond interactions to restrict nonradiative motion and successfully realized the ultralong RTP emission (see Figure 4.22c).[80] Taking compound IPA as an example, it is almost non-emissive in solution; however, much enhanced blue emission with a quantum efficiency of 15.3% could be achieved in a crystal. Moreover, a long green phosphorescence (τ_{RTP} = 290 ms) lasting for several seconds was detected for IPA crystals after stopping the 254 nm UV irradiation. The analyses of its single crystal structure well certified the existence of strong hydrogen bonds (C=O···H—O 1.623–1.703 Å) for the introduction of the carboxy group. Also, partial edge-to-edge π–π interactions (3.396 Å) were presented in its crystal, which then led to the formation of rigid 3D networks. With these, the nonradiative transition of triplet excitons could be well restricted, thus contributing to the resultant persistent RTP emission. In 2017, Li and coworkers successfully achieved the ultralong phosphorescence emission with a lifetime up to 2.24 s in phenylboronic acid derivatives, further verifying the significant promotion of hydrogen bonds to persistent RTP effects.[81]

Besides, intermolecular *n*-π interaction could also contribute to the generation of efficient persistent RTP emission (see Figure 4.22d). According to El-Sayed's rule, the transition between $n\pi^*$ and $\pi\pi^*$ states is allowed, while the spin-orbit coupling between different $\pi\pi^*$ states is rather weak. Therefore, the introduction of heteroatoms with lone-pair electrons could often lead to efficient phosphorescence with a short lifetime, while the pure hydrocarbon compounds tended to show weak but long-lasting phosphorescence. To make a balance of it, Chi and coworkers designed and synthesized two kinds of twisted organic compounds containing a carbonyl or sulfonyl group as the *n* unit and a carbazolyl (Cz) group as the π unit (see Figure 4.22d).[82] In crystal, the electrostatic interactions between *n* and π units would make them approach each other to form close intermolecular stacking, thus causing the intermolecular electronic coupling of units and forming different excited-state configurations (i.e., $n\pi^*$ and $\pi\pi^*$). With it, the hybrid excited-state configurations with balanced $n\pi^*$ and $\pi\pi^*$ transitions could be achieved, which then resulted in the bright and persistent RTP emission for them.

In 2018, Li and coworkers found that efficient intermolecular π–π interaction could also promote the persistent RTP effect (see Figure 4.22e).[83,84] According to classical perturbation theory, the interaction between two π-systems will lead to the redistribution of electrons. Particularly, the electrons in an excited state would redistribute in a lower energy state, thus leading to the stabilized excitons. Accordingly, they successfully realized a persistent RTP effect in a series of phenothiazine 5,5-dioxide derivatives. More interestingly, it was found that the intermolecular π–π interaction and corresponding RTP lifetime could be adjusted to a large extent through rational molecular design. For instance, when the substituent of phenothiazine 5,5-dioxide changed from —OCH_3/—CH_3 (electron donor) to —H to —Br/—Cl/—F (electron acceptor), their π–π interactions were enhanced gradually within molecular dimers. Theoretical calculations suggested that the electron-withdrawing substituents could decrease the π-electron density of the substituted π-system and relieve the π–π repulsion between the two involved rings, thus leading to the closer π–π distance. Correspondingly, their RTP lifetimes were also prolonged from 88 to 188 ms, then to 410 ms, for the stabilized triplet excitons under π–π interaction. Based on the obtained exciting results, they proposed the concept of Molecular Uniting Set Identified Characteristic (MUSIC),[83] to emphasize the key role of the aggregated structure of materials, especially molecular packing and intermolecular interactions, on the material performance.

As presented above, the persistent RTP effect has been found to be heavily related to the intermolecular interactions in solids. Particularly, molecular dimers, the most basic intermolecular interaction form, have been explored frequently, such as dimers with H-aggregation, dimers with hydrogen bonding, dimers with intermolecular n–π/π–π interaction, etc. Then what about the exact role of molecular dimer in the RTP effect? As excimer was easy to form based on the molecular dimer with efficient intermolecular interaction, if the RTP also from a triplet excimer? To be exact, an ideal model was built by Li et al. (see Figure 4.23).[85] In their work, eight luminogens (2PtzO-nC, n = 3–10) were designed and synthesized accordingly, in which two RTP cores of phenothiazine 5,5-dioxide groups were linked by nonconjugated alkyl chains with different lengths to adjust the molecular packing in the crystal state. With this kind of molecular structure, one compound could contain both the monomer- and dimer-based RTP emissions. By comparing with the low-temperature phosphorescence of corresponding solutions, these two kinds of RTP bands would be distinguished easily, thus clearly clarifying the role of the molecular dimer in RTP emission. The study of their RTP behaviors and corresponding single crystal structures found that the crystals with strong π–π stacking at both bilateral dimers present pure triplet excimer emission, while those with strong π–π stacking just at the unilateral dimer or weak π–π stacking at both bilateral dimers gave two RTP bands from the monomer and triplet excimer. Specifically, dimers with strong π–π stacking $\left(\theta < 20.66°;\ d < 3.86\ \text{Å}\right)$ led to pure triplet excimer emission, while those with weak $\pi - \pi$ stacking $\left(27.02° < \theta < 40.64°;\ 3.84\ \text{Å} < d < 4.41\ \text{Å}\right)$ resulted in dual RTP emissions of both monomer and triplet excimer.

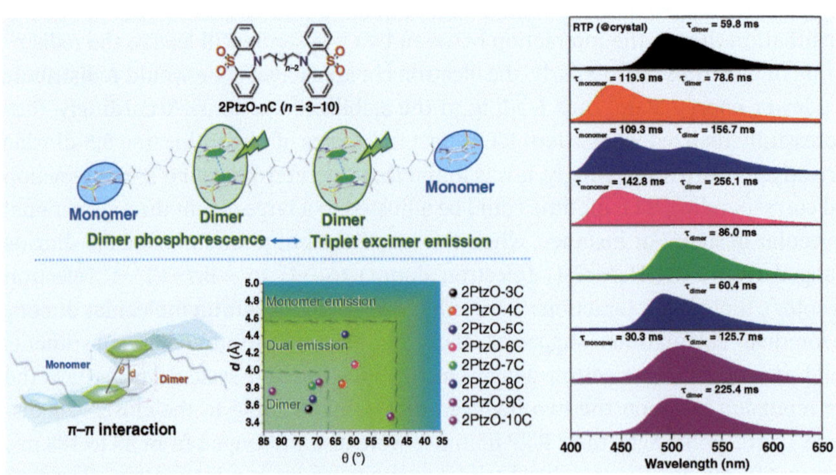

Figure 4.23 Different room temperature phosphorescence (RTP) emissions from monomer and dimer based on different molecular packing (π–π interaction) of phenothiazine-5,5-dioxide derivatives. *Source*: [86]/with permission of Springer Nature.

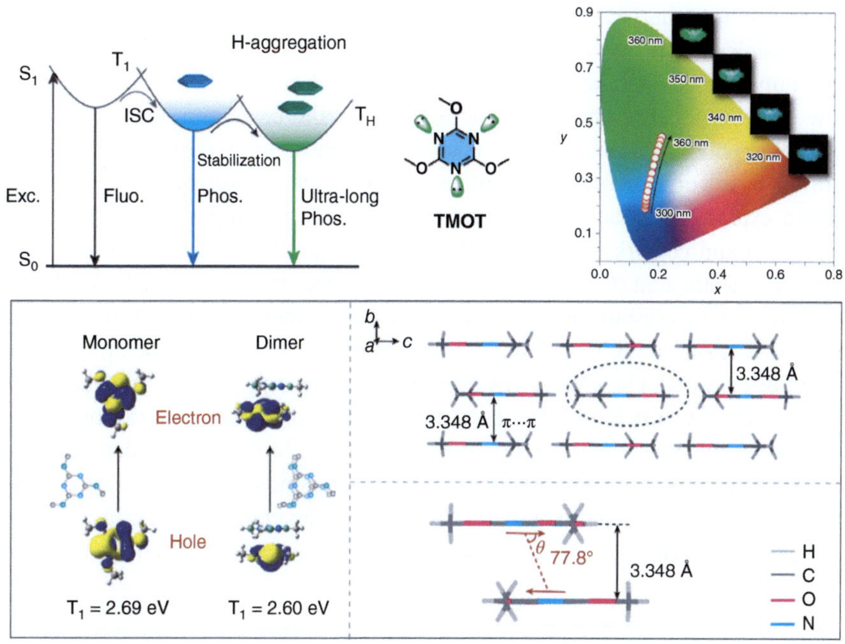

Figure 4.24 Color-tunable persistent RTP emission from a single-component molecular crystal and corresponding mechanism. *Source*: [87]/with permission of Springer Nature.

Thus, the direct demonstration of triplet excimer in purely organic room temperature phosphorescence was achieved.

Besides, through utilizing the different energy level features of monomer and dimer, Huang and coworkers successfully realized the color-tunable persistent RTP emission from a single-component molecular crystal.[86] As shown in Figure 4.24, the RTP luminogen of TMOT presented a planar conformation, which made it easy to form H-aggregation within the molecular dimer. The analyses of its single crystal structure also certified that the intermolecular π–π distance within the TMOT dimer was as short as 3.348 Å, and the corresponding slip angle was 77.8°. Thus, two kinds of stably excited triplet states with different energy levels could be formed, namely the higher one based on the TMOT monomer and the lower one of the TMOT dimer. Correspondingly, color-tunable persistent RTP emission could be observed for TMOT crystal, in which the long-lived phosphorescence tuned from sky-blue to green color as the excitation wavelength changed from 320 to 365 nm. Also, long RTP lifetimes of 0.58 and 0.75 s following excitation at 320 and 365 nm could be obtained for the stabilization of H-aggregation to triplet excitons. Again, the significant influence of aggregation structure on the energy level and corresponding luminescent behavior was demonstrated.

4.4.2 Nonaromatic Room-temperature Phosphorescence

In addition to the examples of organic aromatic systems in Section 4.4.1, the strong intermolecular interactions could also lead to the formation of a lower triplet energy level in nonaromatic compounds, which then leading to the RTP emission in the visible region. As early as 1937, A. C. Giese and P. A. Leighton found that some nonaromatic cellulosic materials, such as wood, leaves, and flowers, could show phosphorescence emission.[87] In 2016, Yuan and coworkers reported a non-conjugated polyacrylonitrile polymer with green phosphorescence.[88] These works were all based on polymers without exactly accurate structures, which made it hard to explore the relationship between molecular packing and RTP property. In 2018, Li and coworkers reported the first example of persistent RTP emission from cyanoacetic acid (CAA), a nonaromatic organic small molecule.[89] As shown in Figure 4.25, its crystal exhibited blue fluorescence under 365 nm UV irradiation. After the removal of the UV lamp, long-lasting green phosphorescence with a lifetime of 862 ms could be clearly observed by the naked eye. Careful analyses of its single crystal structure found that its green RTP emission was mainly ascribed to the strong intermolecular hydrogen bonding. In crystal, a layered molecular structure has been formed through self-assembly of hydrogen bonds, which could extend the conjugation of the electron cloud and thus narrow the energy gap; besides, strong hydrogen bonds $\left(d_{O-O} = 2.6337\ \text{Å}\right)$ extended over its whole crystal to form the

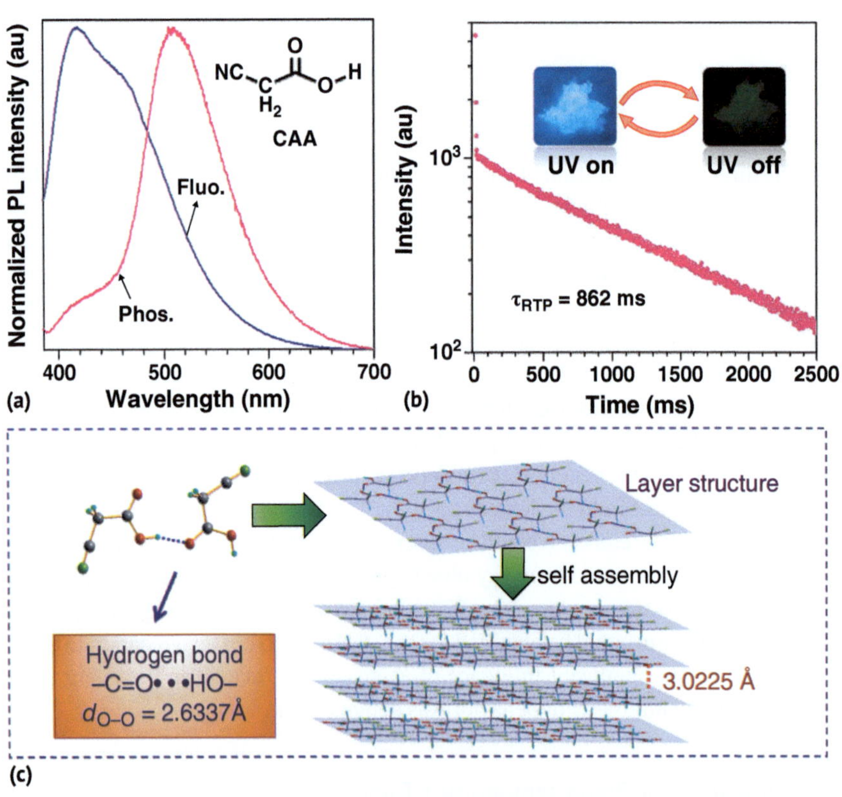

Figure 4.25 (a) The fluorescence and phosphorescence spectra of cyanoacetic acid (CAA) crystal at room temperature; (b) the RTP decay curve of CAA crystal; (c) the single crystal structure of CAA. *Source*: [90]/with permission of The Royal Society of Chemistry.

polymer-like structure through noncovalent interactions, then largely increasing the molecular rigidity and restricting the nonradiative motion. At last, persistent green RTP emission appeared from cyanoacetic acid in solids, showing the typical "Molecular Uniting Set Identified Characteristic (MUSIC)" effect.

Inspired by the great promotion of hydrogen bonding to persistent RTP emission, Yuan and coworkers studied the possible RTP effect of nonaromatic small molecules of HA and its derivatives, in which multiple hydrogen bonds were easy to form for the existence of N/O atoms with strong electronegativity (see Figure 4.26).[90] Impressively, efficient and tunable persistent RTP emission was achieved for HA crystal with phosphorescence efficiency and lifetime up to 21.8% and 1.74 s. For its crystal, the RTP colors could change from sky-blue to yellowish-green upon different excitation wavelengths. The analyses of single crystal structure certified the existence of multiple hydrogen bonds in the HA crystal. Furthermore, a notable supramolecular network was formed for the assembly effect of hydrogen bonds, in which adjacent molecules almost extended in the same plane. Besides, short distances among C=O groups were presented in different layers, which empowered strong π–π interactions and generated highly extended through-space delocalization. Then, the extended plane together with effective π–π interactions jointly contributed to the vast electronic communication, leading to the formation of diverse emissive clusters with

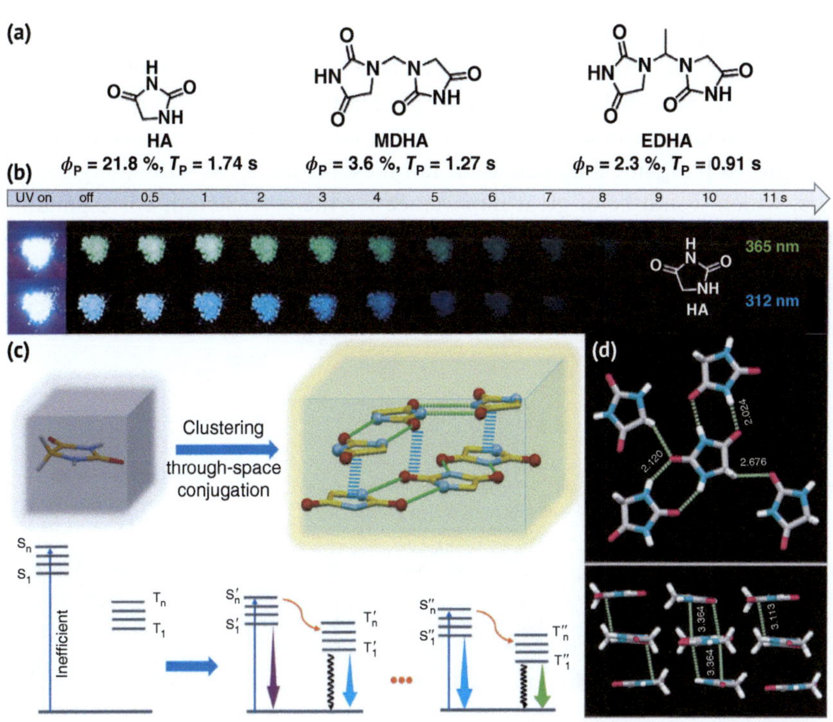

Figure 4.26 (a) Nonaromatic RTP luminogens of HA, MDHA, and EDHA; (b) the excitation-dependent RTP behavior of HA; (c) proposed RTP mechanism for HA and its derivatives; (d) crystal structure of compound HA. *Source*: [91]/The Royal Society of Chemistry.

excitation-dependent RTP emissions. Based on these examples, the significant influence of molecular packing on nonaromatic RTP could be further certified.

4.4.3 Long Persistent Luminescence

For common organic RTP behaviors, the emissions could just last for several seconds, which were still far from inorganic emitters. In previous literature, photo-induced ionized state, or charge-separated state, has been reported to be an efficient way to achieve long persistent emission (hours) at very low temperature (i.e., 20 K).[91] Inspired by it, Adachi and Kabe constructed a host-guest doping system with long persistent luminescence (LPL) at room temperature in 2017 (see Figure 4.27).[92] In their system, TMB, which could generate stable radical cations upon photo-irradiation, acted as a guest and electron donor, while PPT, with a high triplet level and rigid aggregation structure as a host and electron acceptor. As the radical cations were unstable in the presence of water or oxygen, the preparation and characterization of this doping system have both been carried out in a glove box. Upon photo-irradiation, the afterglow lasting for more than 30 min could be realized at 300 K, and the corresponding luminescence mechanism should be subdivided into five steps. During photo-excitation, charge-transfer states were formed between TMB and PPT (see Figure 4.27a, i and ii). Then, the generated PPT radical anions diffused by the hopping of charges among the PPT molecules, thus forming stable charge-separated states (see Figure 4.27a, iii). At last, the PPT radical anions and TMB radical cations gradually recombined to generate long persistent exciplex emission (see Figure 4.27a, iv, v). This work opened up a new way to develop organic long-persistent luminescent materials with afterglow for hours, which would be of great importance for the extension of corresponding applications.

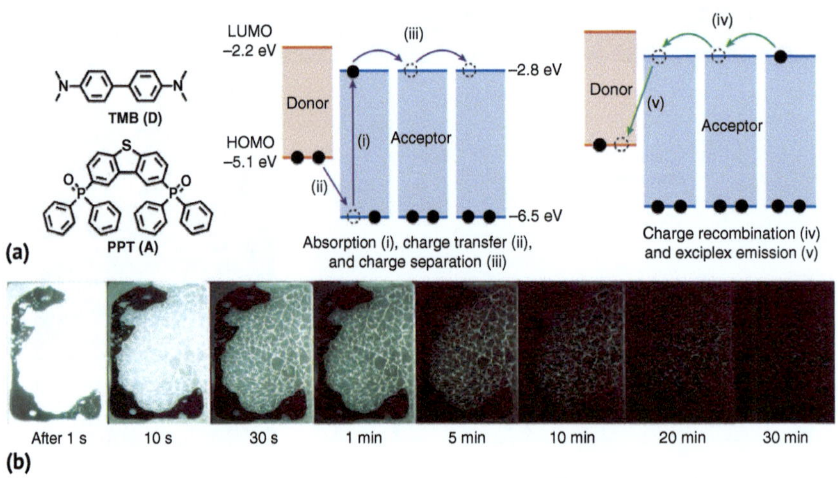

Figure 4.27 (a) Organic long persistent luminescence (LPL) system and corresponding mechanism; (b) LPL photographs of 1 mol% TMB/PPT film in the glove box. *Source:* [93]/with permission of Springer Nature.

To increase the oxygen stability and obtain the LPL under an atmospheric environment, Tang and coworkers introduced an RTP host of TPP-3C2B as an electron acceptor and a guest of *N,N*-dimethylaniline (DMA) as an electron donor to construct a doping system with photo-induced charge-transfer and charge-separation characteristics (see Figure 4.28a).[93] With RTP crystal as host, two additional advantages over the non-RTP one could be expected: (1) abundant triplet excitons can be produced by the RTP host for its strong SOC between excited singlet and triplet states; (2) the rigid host environment can effectively restrict molecular thermal motion and oxygen diffusion, thus promoting the stability of triplet excitons. Indeed, after 365 nm UV irradiation, long-lived charge-separated states could be formed in the crystal of this doping system, resulting in an afterglow that could last for 7 h at ambient conditions. To further simplify the structure of the long persistent luminescence system, they designed and synthesized a new phosphonium salt of DAP-Br by combining an electron-donating *N,N*-dimethylaniline moiety with the electron-accepting arylphosphonium into a single molecule.[94] As shown in Figure 4.28b, after DAP-Br crystals were excited by 365 nm UV light for 30 s, a long sky-blue afterglow lasting up to 12 min could be clearly observed under ambient conditions. Thus LPL was successfully realized in an organic single-component crystal for the first time. Nevertheless, differences still existed in the internal mechanisms between single-component and binary LPL systems. In particular, the generation of photo-induced charge separation in this single-component system was due to the two-photon ionization, while that of the above-mentioned binary LPL system originated from charge transfer exciton dissociation. Besides, the DAP-Br dimers with intermolecular π–π interaction were found to play an important role in stabilizing the formation of the photoinduced radical. Therefore, the adjustment of molecular packing should also be taken into consideration in the design of organic LPL systems.

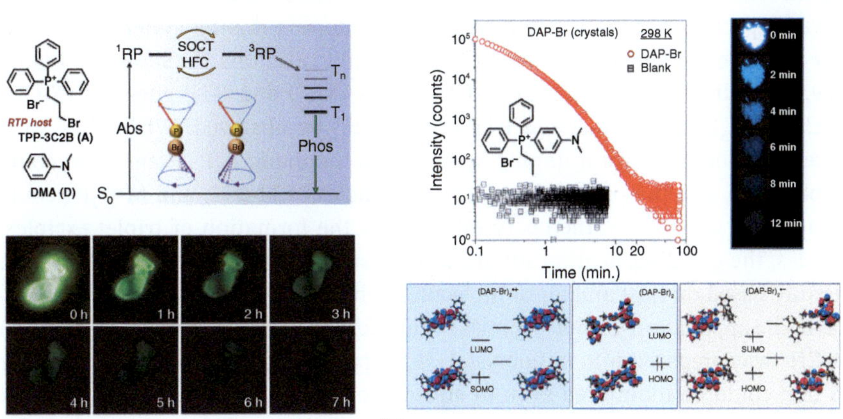

(a) (b)

Figure 4.28 (a) Long persistent luminescent system based on the mixture of TPP-3C2B/DMA. *Source*: [94]/with permission of John Wiley & Sons; (b) long persistent luminescent system based on the single-component crystal of DAP-Br. *Source*: [95]/with permission of American Chemical Society.

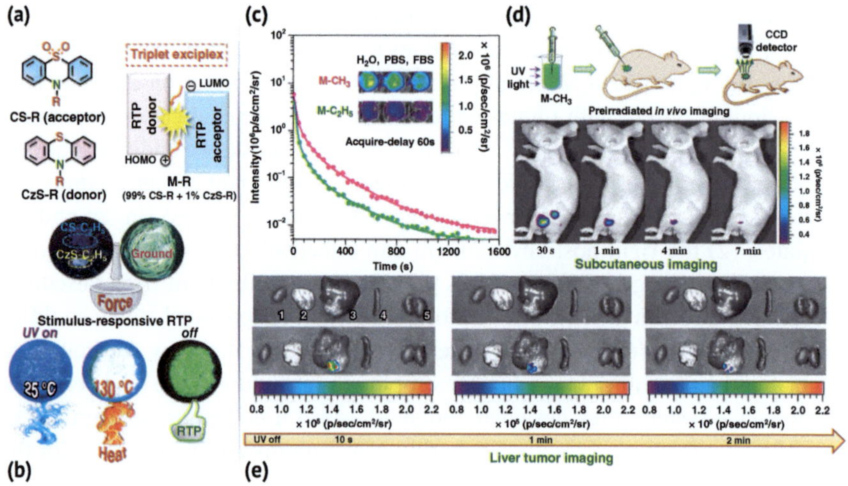

Figure 4.29 (a) Triplet exciplex system based on RTP donors and RTP acceptors; (b) heating/grinding-responsive RTP effect based on CS-C_2H_5 and CzS-C_2H_5 mixture; (c) time-resolved phosphorescence decays of M-CH_3 and M-C_2H_5 nanocrystals; (d) preirradiated *in vivo* subcutaneous imaging based on M-CH_3 nanocrystals; (e) time-resolved phosphorescence imaging results of different organs (1—heart, 2—lung, 3—liver, 4—spleen and 5—kidney) (upper: intravenous injection of saline; lower: intravenous injection of M-CH_3 nanocrystals). *Source*: [96]/with permission of John Wiley & Sons.

Further on, Li, Yang and coworkers developed a new binary LPL system with both high efficiency and long lifetime (see Figure 4.29).[95] In their system, two phenothiazine derivatives (CzS-CH_3 and CzS-C_2H_5) were selected as guests, and their corresponding dioxide derivatives (CS-CH_3 and CS-C_2H_5) acted as hosts. Notably, these hosts and guests could both show the RTP effect in a single-component state. It was expected that the integration of them into one doping system could further promote the generation and stability of triplet excitons. As shown in Figure 4.29, for individual RTP hosts (CS-CH_3 or CS-C_2H_5) and RTP guests (CzS-CH_3 or CzS-C_2H_5), just a moderate RTP effect could be observed. When they were mixed together to form a doping system, much enhanced RTP emission could be obtained, and the phosphorescence efficiencies of M-CH_3 and M-C_2H_5 could be up to 20 and 43% in the cocrystal state for the formation of triplet exciplex. Besides, the grinding/heating stimuli could also activate the RTP emission of the mixture based on host and guest molecules, which made them show great potential in the application of thermal anti-counterfeiting printing. Furthermore, when M-CH_3 was prepared into nanocrystals, long persistent luminescence lasting for 25 min could be monitored even in aqueous environments, indicating the great stability of triplet excitons in it. Based on the ultralong afterglow emission, the applications of time-resolved bioimaging were carried out. Notably, the preirradiated *in vivo* subcutaneous imaging and labeling of the liver tumor were achieved for the first time, in which the signal-to-background ratio (SBR) could be as high as 310 and 158, respectively. These SBR values were considerably higher than those of the reported fluorescence imaging, including NIR-II fluorescence imaging, demonstrating the advantage of time-resolved phosphorescent imaging with negligible background interference.[96,97]

4.5 Conclusion

In this chapter, we reviewed the typical luminescent behaviors in the solid state. It is clearly shown that the influences of molecular aggregates have gone through the whole photoluminescence process, including fluorescence, delayed fluorescence, and phosphorescence. Preliminarily, the relationships between aggregation structures and luminescent behaviors of organic solids have been understood. Therefore, to further verify these relationships and promote the corresponding PL properties, the adjustment of the aggregate structure, including intermolecular interactions and packing mode, etc., would be of great importance. During this process, the rational design of molecular structures can still be an efficient way.

As different emissive properties favor different packing modes, the corresponding molecular design strategy should also be changed. For instance, the AIE effect was mainly realized in the systems without π–π interaction; thus, the molecules with twisted molecular conformation would be much beneficial. However, the intermolecular π–π stacking would be helpful in some cases, such as excimer fluorescence and even exciplex-type RTP. At this time, the molecules with planar conformation were better. Further on, the inductive effect of electron-withdrawing substituents could lower the electron cloud density of π system at the surface, thus decreasing the intermolecular π–π repulsion and promoting the close π–π stacking. As for intermolecular hydrogen bonding, it's a form that could restrict intramolecular motion and resultant nonradiative transition, thus contributing to most of the emissive behaviors. Thus, the introduction of heteroatoms (N, O, F, etc.) with large electronegativity should be helpful.

Further on, to construct the single molecules with through-space electronic interaction is also very important, as they present high similarity with the molecular aggregate in interaction mode. On one hand, the more controlled electronic interaction and resultant emissive property could be achieved in this kind of single-molecular luminogens. On the other hand, the information gained from them can help us understand the internal mechanism of PL behaviors in aggregate. Therefore, the knowledge of molecular science can still be utilized in the research of PL property for molecular aggregates. In turn, the development of molecular aggregate science will promote the innovation of knowledge in molecular science.

References

1 Fang, M., Yang, J., and Li, Z. (2022). Light emission of organic luminogens: Generation, mechanism and application. *Prog. Mater. Sci.* 125: 100914.

2 Brewster, D. (1832). On the colours of natural bodies. *T. R. Soc. Edin.* 12: 538–545.

3 Valeur, B. and Berberan-Santos, M.N. (2011). A brief history of fluorescence and phosphorescence before the emergence of quantum theory. *J. Chem. Edu.* 88: 731–738.

4 Stokes, G.G. (1852). On the change of refrangibility of light. *Philos. Trans. R. Soc.* 142: 463–562.

5 Lakowicz, J.R. (2006). *Principles of Fluorescence Spectroscopy*. Springer: New York.

6 Jablonski, A. (1933). Efficiency of anti-stokes fluorescence in dyes. *Nature* 131: 839–840.

7 Ma, S., Du, S., Pan, G. et al. (2021). Organic molecular aggregates: From aggregation structure to emission property. *Aggregate* 2: e96.

8 Yang, J., Chi, Z., Zhu, W. et al. (2019). Aggregation-induced emission: A coming-of-age ceremony at the age of eighteen. *Sci. China Chem.* 62: 1090–1098.

9 Wang, C. and Li, Z. (2017). Molecular conformation and packing: Their critical roles in the emission performance of mechanochromic fluorescence materials. *Mater. Chem. Front.* 1: 2174–2194.

10 Dimitriev, O.P., Piryatinski, Y.P., and Slominskii, Y.L. (2018). Excimer emission in J-aggregates. *J. Phys. Chem. Lett.* 9: 2138–2143.

11 Xu, R., Wang, K., Chen, G., and Yan, W. (2019). Condensed-matter chemistry: From materials to living organisms. *Natl. Sci. Rev.* 6: 191–194.

12 Li, Q. and Li, Z. (2020). Molecular packing: Another key point for the performance of organic and polymeric optoelectronic materials. *Acc. Chem. Res.* 53: 962–973.

13 Zhang, H., Zhao, Z., Turley, A. et al. (2020). Aggregate science: From structures to properties. *Adv. Mater.* 32: 2001457.

14 Gu, L., Wang, X., Singh, M. et al. (2020). Organic room-temperature phosphorescent materials: From static to dynamic. *J. Phys. Chem. Lett.* 11: 6191–6200.

15 Sarma, M. and Wong, K. (2018). Exciplex: An intermolecular charge-transfer approach for TADF. *ACS Appl. Mater. Interfaces* 10: 19279–19304.

16 Luo, J., Xie, Z., Lam, J.W.Y. et al. (2001). Aggregation-induced emission of 1-methyl-1,2,3,4,5-pentaphenylsilole. *Chem, Commun.* 18: 1740–1741.

17 Mei, J., Leung, N.L.C., Kwok, R.T.K. et al. (2015). Aggregation-induced emission: Together we shine, united we soar! *Chem. Rev.* 115: 11718–940.

18 Birks, J.B. (1970). *Photophysics of Aromatic Molecules*. Wiley: London.

19 Mei, J., Hong, Y., Lam, J.W.Y. et al. (2014). Aggregation-induced emission: The whole is more brilliant than the parts. *Adv. Mater.* 26: 5429–5479.

20 Li, Z., Dong, Y., Mi, B. et al. (2005). Structural control of the photoluminescence of silole regioisomers and their utility as sensitive regiodiscriminating chemosensors and efficient electroluminescent materials. *J. Phys. Chem. B* 109: 10061–10066.

21 Dong, Y., Lam, J., Qin, A. et al. (2007). Aggregation-induced emissions of tetraphenylethene derivatives and their utilities as chemical vapor sensors and in organic light-emitting diodes. *Appl. Phys. Lett.* 91: 011111.

22 Feng, X., Tong, B., Shen, J. et al. (2010). Aggregation-induced emission enhancement of aryl-substituted pyrrole derivatives. *J. Phys. Chem. B* 114: 16731–16736.

23 Zeng, Q., Li, Z., Dong, Y. et al. (2007). Fluorescence enhancements of benzene-cored luminophors by restricted intramolecular rotations: AIE and AIEE effects. *Chem. Commun.* 1: 70–72.

24 He, J., Xu, B., Chen, F. et al. (2009). Aggregation-induced emission in the crystals of 9, 10-distyrylanthracene derivatives: the essential role of restricted intramolecular torsion. *J. Phys. Chem. C* 113: 9892–9899.

25 Chen, M., Li, L., Nie, H. et al. (2015). Tetraphenylpyrazine-based AIEgens: Facile preparation and tunable light emission. *Chem. Sci.* 6: 1932–1937.

26 Yoshii, R., Nagai, A., Tanaka, K., and Chujo, Y. (2013). Highly emissive boron ketoiminate derivatives as a new class of aggregation-induced emission fluorophores. *Chem. Eur. J.* 19: 4506–4512.

27 Li, Z.Z., Huo, Y.P., Yang, X.H., and Ji, S.M. (2016). Progress on research and application of tetraphenylethene derivatives. *Chin. J. Org. Chem.* 36: 2317–2332.

28 Xu, B., He, J., Mu, Y. et al. (2015). Very bright mechanoluminescence and remarkable mechanochromism using a tetraphenylethene derivative with aggregation-induced emission. *Chem. Sci.* 6: 3236–3241.

29 Xu, B., Li, W., He, J. et al. (2016). Achieving very bright mechanoluminescence from purely organic luminophores with aggregation-induced emission by crystal design. *Chem. Sci.* 7: 5307–5312.

30 Leung, N.L.C., Xie, N., Yuan, W. et al. (2014). Restriction of intramolecular motions: the general mechanism behind aggregation-induced emission. *Chem. Eur. J.* 20: 15349–15353.

31 Levell, J.W., Ruseckas, A., Henry, J.B. et al. (2010). Fluorescence enhancement by symmetry breaking in a twisted triphenylene derivative. *J. Phys. Chem. A* 114: 13291–13295.

32 Liu, J., Meng, Q., Zhang, X. et al. (2013). Aggregation-induced emission enhancement based on 11,11,12,12,-tetracyano-9,10-anthraquinodimethane. *Chem. Commun.* 9: 1199–1201.

33 Zhao, Z., Zheng, X., Du, L. et al. (2019). Non-aromatic annulene-based aggregation-induced emission system via aromaticity reversal process. *Nat. Commun.* 10: 2952.

34 Zhang, H. and Tang, B.Z. (2021). Through-space interactions in clusteroluminescence. *JACS Au* 1: 1805–1814.

35 Zhang, H., Zheng, X., Xie, N. et al. (2017). Why do simple molecules with "Isolated" phenyl rings emit visible light? *J. Am. Chem. Soc.* 139: 16264–16272.

36 Zhang, J., Hu, L., Zhang, K. et al. (2021). How to manipulate through-space conjugation and clusteroluminescence of simple AIEgens with isolated phenyl rings. *J. Am. Chem. Soc.* 143: 9565–9574.

37 Birks, J.B. (1975). Excimers. *Rep. Prog. Phys.* 38: 903–974.

38 Förster, T. (1969). Excimers. *Angew. Chem. Int. Ed.* 8: 333–343.

39 Yoshihara, K., Kasuya, T., Inoue, A., and Nagakura, S. (1971). Time resolved spectra of pyrene excimer and pyrene-dimethylaniline exciplex. *Chem. Phys. Lett.* 9: 469–472.

40 Chen, Y., Tang, K., Chen, Y. et al. (2016). Insight into the mechanism and outcoupling enhancement of excimer-associated white light generation. *Chem. Sci.* 7: 3556–3563.

41 Pownall, H.J. and Smith, L.C. (1973). Viscosity of the hydrocarbon region of micelles. measurement by excimer fluorescence. *J. Am. Chem. Soc.* 95: 3136–3140.

42 Liu, H., Yao, L., Li, B. et al. (2016). Excimer-induced high-efficiency fluorescence due to pairwise anthracene stacking in a crystal with long lifetime. *Chem. Commun.* 52: 7356–7359.

43 Wang, J., Dang, Q., Gong, Y. et al. (2021). Precise regulation of distance between associated pyrene units and control of emission energy and kinetics in solid state. *CCS Chem.* 3: 274–286.

44 Ge, Y., Wen, Y., Liu, H. et al. (2020). A key stacking factor for the effective formation of pyrene excimer in crystals: degree of π–π overlap. *J. Mater. Chem. C* 8: 11830–11838.

45 Liu, H., Cong, D., Li, B. et al. (2017). Discrete dimeric anthracene stackings in solids with enhanced excimer fluorescence. *Cryst. Growth Des.* 17: 2945–2949.

46 Zhao, Z., Chen, S., Lam, J.W.Y. et al. (2011). Pyrene-substituted ethenes: aggregation-enhanced excimer emission and highly efficient electroluminescence. *J. Mater. Chem.* 21: 7210–7216.

47 Shen, Y., Liu, H., Zhang, S. et al. (2017). Discrete face-to-face stacking of anthracene inducing high-efficiency excimer fluorescence in solids via a thermally activated phase transition. *J. Mater. Chem. C* 5: 10061–10067.

48 Kasha, M., Rawls, H.R., and Bayoumi, M.A. (1965). The exciton model in molecular spectroscopy. *Pure Appl. Chem.* 11: 371–392.

49 Kasha, M. (1963). Energy transfer mechanisms and the molecular exciton model for molecular aggregates. *Radiat. Res.* 20: 55–71.

50 Li, Z., Liao, L., and Wang, X. (2018). Controllable synthesis of organic microcrystals with tunable emission color and morphology based on molecular packing mode. *Small* 14: 1702952.

51 Qi, Q., Qian, J., Tan, X. et al. (2015). Remarkable turn-on and color-tuned piezochromic luminescence: Mechanically switching intramolecular charge transfer in molecular crystals. *Adv. Funct. Mater.* 25: 4005–4010.

52 Xie, Z., Yang, B., Li, F. et al. (2005). Cross dipole stacking in the crystal of distyrylbenzene derivative: The approach toward high solid-state luminescence efficiency. *J. Am. Chem. Soc.* 127: 14152–14153.

53 Cornil, J., Santos, D., Crispin, X. et al. (1998). Influence of interchain interactions on the absorption and luminescence of conjugated oligomers and polymers: a quantum-chemical characterization. *J. Am. Chem. Soc.* 120: 1289–1299.

54 Ma, S., Liu, Y., Zhang, J. et al. (2020). Polymorphism-dependent enhanced emission in molecular aggregates: J-aggregate versus X-aggregate. *J. Phys. Chem. Lett.* 11: 10504–10510.

55 Uoyama, H., Goushi, K., Shizu, K. et al. (2012). Highly efficient organic light-emitting diodes from delayed fluorescence. *Nature* 492: 234–238.

56 Zhang, M., Zheng, C., Lin, H., and Tao, S. (2021). Thermally activated delayed fluorescence exciplex emitters for high-performance organic light-emitting diodes. *Mater. Horiz.* 8: 401–425.

57 Ni, F., Li, N., Zhan, L., and Yang, C. (2020). Organic thermally activated delayed fluorescence materials for time-resolved luminescence imaging and sensing. *Adv. Opt. Mater.* 8: 1902187.

58 Bryden, M.A. and Zysman-Colman, E. (2021). Organic thermally activated delayed fluorescence (TADF) compounds used in photocatalysis. *Chem. Soc. Rev.* 50: 7587–7680.

59 Chen, F., Zhao, L., Wang, X. et al. (2021). Novel boron- and sulfur-doped polycyclic aromatic hydrocarbon as multiple resonance emitter for ultrapure blue thermally activated delayed fluorescence polymers. *Sci. China Chem.* 64: 547–551.

60 Xu, S., Duan, Y., and Liu, B. (2020). Precise molecular design for high-performance luminogens with aggregation-induced emission. *Adv. Mater.* 32: 1903530.

61 Xu, S., Liu, T., Mu, Y. et al. (2015). An organic molecule with asymmetric structure exhibiting aggregation-induced emission, Delayed fluorescence, and mechanoluminescence. *Angew. Chem. Int. Ed.* 54: 874–878.

62 Guo, J., Zhao, Z., and Tang, B.Z. (2018). Purely organic materials with aggregation-induced delayed fluorescence for efficient nondoped OLEDs. *Adv. Opt. Mater.* 6: 1800264.

63 Guo, J., Li, X., Nie, H. et al. (2017). Achieving high-performance nondoped OLEDs with extremely small efficiency roll-off by combining aggregation-induced emission and thermally activated delayed fluorescence. *Adv. Funct. Mater.* 27: 1606458.

64 Liu, H., Zeng, J., Guo, J. et al. (2018). High-performance non-doped OLEDs with nearly 100% exciton use and negligible efficiency roll-off. *Angew. Chem. Int. Ed.* 57: 9290–9294.

65 Tsujimoto, H., Ha, D., Markopoulos, G. et al. (2017). Thermally activated delayed fluorescence and aggregation induced emission with through-space charge transfer. *J. Am. Chem. Soc.* 139: 4894–4900.

66 Li, Q., Hu, J., Lv, J. et al. (2020). Through-space charge-transfer polynorbornenes with fixed and controllable spatial alignment of donor and acceptor for high-efficiency blue thermally activated delayed fluorescence. *Angew. Chem. Int. Ed.* 59: 20174–20182.

67 Peng, C., Yang, S., Li, H. et al. (2020). Highly efficient thermally activated delayed fluorescence via an unconjugated donor–acceptor system realizing EQE of over 30%. *Adv. Mater.* 32: 200388.

68 Wu, C., Liu, W., Li, K. et al. (2021). Face-to-face orientation of quasiplanar donor and acceptor enables highly efficient intramolecular exciplex fluorescence. *Angew. Chem. Int. Ed.* 60: 3994–3998.

69 Tang, X., Cui, L., Li, H. et al. (2020). Highly efficient luminescence from space-confined charge-transfer emitters. *Nat. Mater.* 19: 1332–1338.

70 Goushi, K., Yoshida, K., Sato, K., and Adachi, C. (2012). Organic light-emitting diodes employing efficient reverse intersystem crossing for triplet-to-singlet state conversion. *Nat. Photonics* 6: 253–258.

71 Zhang, T., Chu, B., Li, W. et al. (2014). Efficient triplet application in exciplex delayed-fluorescence OLEDs using a reverse intersystem crossing mechanism based on a ΔE_{S-T} of around zero. *ACS Appl. Mater. Interfaces* 6: 11907–11914.

72 Goushi, K. and Adachi, C. (2012). Efficient organic light-emitting diodes through up-conversion from triplet to singlet excited states of exciplexes. *Appl. Phys. Lett.* 101: 023306.

73 Sun, L., Hua, W., Liu, Y. et al. (2019). Thermally activated delayed fluorescence in an organic cocrystal: Narrowing the singlet–triplet energy gap via charge transfer. *Angew. Chem.* 131: 11433–11438.

74 Atsushi, K., Kengo, S., Toshitada, Y., and Seiji, T. (2010). Absolute measurements of photoluminescence quantum yields of 1-halonaphthalenes in 77 K rigid solution using an integrating sphere instrument. *Chem. Letter.* 39: 282–283.

75 Lower, S.K. and EI-Sayed, M.A. (1966). The triplet state and molecular electronic processes in organic molecules. *Chem. Rev.* 66: 199–241.

76 Yuan, W., Shen, X., Zhao, H. et al. (2010). Crystallization-induced phosphorescence of pure organic luminogens at room temperature. *J. Phys. Chem. C* 114: 6090–6099.

77 Cavallo, G., Metrangolo, P., Milani, R. et al. (2016). The halogen bond. *Chem. Rev.* 116: 2478–2601.

78 Bolton, O., Lee, K., Kim, H. et al. (2011). Activating efficient phosphorescence from purely organic materials by crystal design. *Nat. Chem.* 3: 205–210.

79 An, Z., Zheng, C., Tao, Y. et al. (2015). Stabilizing triplet excited states for ultralong organic phosphorescence. *Nat. Mater.* 14: 685–690.

80 Gong, Y., Zhao, L., Peng, Q. et al. (2015). Crystallization-induced dual emission from metal- and heavy atom-free aromatic acids and esters. *Chem. Sci.* 6: 4438–4444.

81 Chai, Z., Wang, C., Wang, J. et al. (2017). Abnormal room temperature phosphorescence of purely organic boron-containing compounds: the relationship between the emissive behavior and the molecular packing, and the potential related applications. *Chem. Sci.* 8: 8336–8344.

82 Yang, Z., Mao, Z., Zhang, X. et al. (2016). Intermolecular electronic coupling of organic units for efficient persistent room-temperature phosphorescence. *Angew. Chem. Int. Ed.* 55: 2181–2185.

83 Yang, J., Zhen, X., Wang, B. et al. (2018). The influence of the molecular packing on the room temperature phosphorescence of purely organic luminogens. *Nat. Commun.* 9: 840.

84 Yang, J., Gao, H., Wang, Y. et al. (2019). The odd-even effect of alkyl chain in organic room temperature phosphorescence luminogens and the corresponding in vivo imaging. *Mater. Chem. Front.* 3: 1391–1397.

85 Liu, Z., Tian, Y., Yang, J. et al. (2022). Direct demonstration of triplet excimer in purely organic room temperature phosphorescence through rational molecular design. *Light-Sci. Appl.* 11: 142.

86 Gu, L., Shi, H., Bian, L. et al. (2019). Colour-tunable ultra-long organic phosphorescence of a single-component molecular crystal. *Nat. Photonics* 13: 406–411.

87 Giese, A.C. and Leighton P.A. (1937). Phosphorescence of cells and cell products. *Science* 85: 428–429.

88 Zhou, Q., Cao, B., Zhu, C. et al. (2016). Clustering-triggered emission of nonconjugated polyacrylonitrile. *Small* 12: 6586–6592.

89 Fang, M., Yang, J., Xiang, X. et al. (2018). Unexpected room-temperature phosphorescence from a non-aromatic, low molecular weight, pure organic molecule through the intermolecular hydrogen bond. *Mater. Chem. Front.* 2: 2124–2129.

90 Wang, Y., Tang, S., Wen, Y. et al. (2020). Nonconventional luminophores with unprecedented efficiencies and color-tunable afterglows. *Mater Horiz.* 7: 2105–2012.

91 Ohkita, H., Sakai, W., Tsuchida, A., and Yamamoto, M. (1997). Charge recombination of electron-cation pairs formed in polymer solids at 20 K through two-photon ionization. *J. Phys. Chem. B* 101: 10241–10247.

92 Kabe, R. and Adachi, C. (2017). Organic long persistent luminescence. *Nature* 550: 384–387.

93 Alam, P., Leung, N.L.C., Liu, J. et al. (2020). Two are better than one: a design principle for ultralong persistent luminescence of pure organics. *Adv. Mater.* 32: 2001026.

94 Alam, P., Cheung, T.S., Leung, N.L.C. et al. (2022). Organic long-persistent luminescence from a single-component aggregate. *J. Am. Chem. Soc.* 144: 3050–3062.

95 Wang, Y., Gao, H., Yang, J. et al. (2021). High performance of simple organic phosphorescence host-guest materials and their application in time- resolved bioimaging. *Adv. Mater.* 33: 2007811.

96 Li, Y., Cai, Z., Liu, S. et al. (2020). Design of AIEgens for near-infrared IIb imaging through structural modulation at molecular and morphological levels. *Nat. Commun.* 11: 1255.

97 Li, C., Chen, G., Zhang, Y. et al. (2020). Advanced fluorescence imaging technology in the near-infrared-II window for biomedical applications. *J. Am. Chem. Soc.* 142: 14789.

5

Mechanoluminescence of Molecular Aggregates

Yujun Xie[1], Jinfeng Wang[1], and Zhen Li[1,2]**

[1] *Institute of Molecular Aggregation Science, Tianjin University, Tianjin, China*
[2] *Department of Chemistry, Wuhan University, Wuhan, China*

*Corresponding authors: xieyujun@tju.edu.cn; lizhen@whu.edu.cn

5.1 The Emission Mechanism of Mechanoluminescence

5.1.1 Brief Introduction to the History of Organic Mechanoluminescence

Mechanoluminescence (ML) is an ancient luminous phenomenon, referring to the process by which materials emit light in response to external mechanical stimuli.[1] Historically, ML was also termed triboluminescence or fractoluminescence; these designations reflect the specific mechanical conditions under which light is generated: "triboluminescence" implies light production via friction, while "fractoluminescence" indicates light emission upon material fracture (i.e. light is observable only when compounds are crushed by force). For instance, the term "triboluminescence" was coined by Wiedemann in 1895; its prefix "tribo-," meaning "to rub," is derived from the Greek word "tribein." Actually, ML can be triggered by a diverse range of mechanical forces, including grinding, crushing, deforming, cleaving, compressing, or treatment by a thermal/cold shock, crystallization, ultrasonic treatment, and infrared laser pulse.[2,3] The earliest recorded discovery of ML dates to 1605, when Francis Bacon documented the phenomenon in his work *Advancement of Learning*: he observed light emission from hard sugar when it was scraped.[1] Nowadays, ML remains easily observable in daily life, thanks to the widespread availability of sugar as a household item. A simple experiment can demonstrate this: fill a transparent plastic bottle to about a quarter of its volume with crystalline sugar, turn off the lights until the room is nearly dark, and then, shaking the bottle quickly, the glint of blue light will be visible. Notably, the faster the bottle is shaken, the brighter the emitted light.

ML, which seems to possess a "mystical power" to convert mechanical force into light, has long attracted sustained attention. In the nineteenth century, although numerous ML compounds were reported, systematic research on ML remained challenging. This difficulty stemmed from the fact that the discovery and observation of ML relied heavily on the researchers' keen observation in a dark room, an approach that lacked objectivity and reproducibility.[4] The predicament began to shift in the early twentieth century, as a series of studies uncovered correlations between the crystal structures and the ML properties. A pivotal milestone came in 1913, when Langevin proposed a foundational theory: he attributed ML origin to nitrogen discharge induced by the piezoelectric effect. In 1922, Longchambon further validated this theory via both experimental and theoretical approaches. He also became the first researcher to record ML spectra, achieved by exposing photographic film to continuous grinding of 5 kg of sugar for four hours. A breakthrough in ML research emerged in the 1950s with the invention of the photomultiplier tube (PMT) for spectroscopy characterization. This tool enabled convenient and detailed analysis of luminous properties: beyond the nitrogen discharge mechanism, ML was further categorized into three types: fluorescence, phosphorescence, and a combination of the two. During this period, a wide range of ML-active materials were identified, including organic/inorganic crystals, ZnS : Mn, alkali halides, rare-earth-doped ceramics, and metal-centered organic chelates. Concurrently, ML demonstrated immense application potential in fields such as pressure sensors, structure damage detection, displays, and bioimaging.[2,3,5] However, the development of organic ML lagged relatively behind. This slow progress was primarily attributed to three key limitations: low emission efficiency of organic emitters in the solid state, ambiguity surrounding their ML emission mechanisms, and insufficient exploration of practical applications.

Back in 2001, the Tang team introduced the concept of aggregation-induced emission (AIE).[6] Their findings illustrated that through deliberate molecular design, organic emitters can surmount the aggregation-caused quenching effect, thus achieving high luminous efficiency in the aggregated state. While the AIE concept initially focuses on organic emitters themselves, this research sparked renewed interest in the photophysical properties of aggregated systems. Since the ML of organic crystals falls under the category of aggregated-state luminescent phenomenon, the AIE concept also rekindled research interest in organic ML. A pivotal advance came in 2015: the Tang team discovered the first organometallic AIE chelate with ML properties,[7] and independently, the Chi group reported the first pure organic AIE emitter exhibiting ML properties.[8]

Concurrently, studies revealed that the triplet state plays a crucial role in regulating the luminous behavior. For instance, a narrow energy gap between the singlet excited state (S_1) and triplet excited state (T_1) facilitates thermally activated delayed fluorescence (TADF) by enhancing reverse intersystem crossing. In contrast, a wide energy gap between these two states tends to induce emissions associated with hybridized local and charge-transfer (HLCT) and triplet-triplet annihilation (TTA).[9–11] Meanwhile, the enhanced spin-orbit coupling (SOC) in certain organic crystals enables the stabilization of triplet excitons, which in turn gives rise to room-temperature phosphorescence (RTP).[12] Currently, a diverse range of organic compounds have been identified to exhibit ML properties,

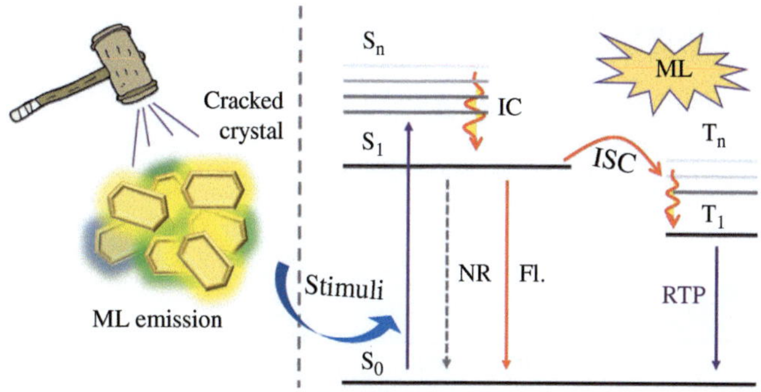

Figure 5.1 Schematic illustration of the production process of mechano-induced fluorescence (Fl) and room temperature phosphorescence (RTP), where the cracked crystal serves as the excitation source for organic compounds. IC: internal conversion, ISC: intersystem crossing. *Source*: Reproduced with permission from [14]/Royal Society of Chemistry.

accompanied by complex photophysical behavior. Notably, some organic ML systems have even been found to exhibit RTP characteristics with long emission lifetimes.[13] These discoveries not only expand the performance boundaries of organic ML materials but also open a new direction for exploring the intrinsic connections between ML and other aggregated-state luminescent phenomena.

The production of ML is remarkably simple: a mere mechanical impact is sufficient to trigger ML emission (Figure 5.1).[14] However, early research on ML phenomena encountered significant challenges. First, ML typically manifests as a transient flash, with an extremely short observation window that makes it difficult to capture the underlying process. Second, numerous interfering factors further complicate analysis. After extensive trial and error, researchers confirmed a key observation: ML in crystals can only be induced when the crystal is fractured by mechanical force. During the fracture process, the crystal first undergoes elastic and plastic deformation; once fracture occurs, a new interface is generated. At this fresh interface, an electric potential is formed due to the disruption of the original charge distribution. This potential is influenced by both charge dislocation within the crystal lattice and the contact potential of the grinding tool. Concurrently, the high pressure and temperature localized at the crack tip can induce phase transformations in the crystal's microregions. Additionally, impurities present in the crystal may alter its symmetry and charge distribution, further modulating ML behavior. Encouragingly, as research progresses, the intrinsic nature of ML is gradually being unraveled, with the development of ML mechanisms shedding light on the detailed process of ML generation.

5.1.2 ML Derives from the Gas Discharge of Organic Crystal Without Luminescent Center

The earliest reported ML compounds generally lack a luminescent center, such as sucrose and tartaric acid, and thus cannot emit fluorescence or phosphorescence under light source excitation. Nevertheless, these classic organic ML compounds

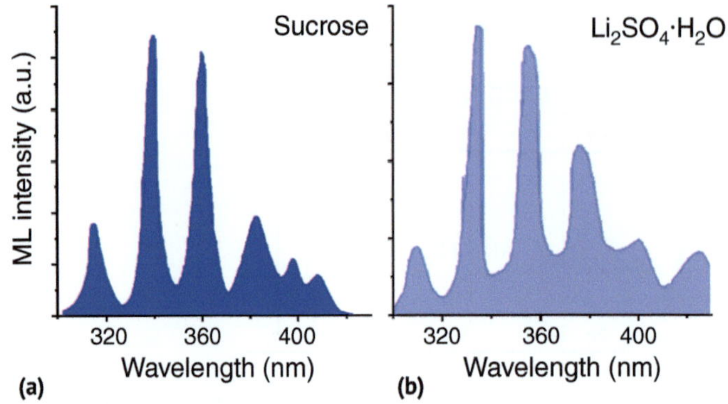

Figure 5.2 (a) The ML spectra of sucrose and (b) $Li_2SO_4{\cdot}H_2O$. *Source*: Reproduced with permission from [3]/Elsevier.

produce identical deep-blue ML (Figure 5.2a), whose emission is highly like a lightning flash. Notably, the ML of certain inorganic crystals (e.g. $Li_2SO_4{\cdot}H_2O$, Figure 5.2b) also exhibits this characteristic deep-blue emission, implying a shared ML mechanism across these compounds.[3] The first ML spectrum (recorded for sugar) was reported by Longchambon in 1922. This spectrum offers key evidence revealing that the ML emitted in air is indistinguishable from nitrogen's gas-discharge spectrum. It specifically corresponds to N_2' second and third positive groups, as well as the negative group of $N_2^+(^3\pi_u \rightarrow {}^3\pi_g)$. This hypothesis was further validated by modifying the surrounding gas and its pressure. In control experiments, Longchambon observed that reducing air pressure increased ML intensity, with the intensity peaking at a pressure range of 1–40 torr.[1] In contrast, when the same crystal was placed in a neon atmosphere, it emitted red ML instead of the typical deep-blue emission. These findings confirm beyond doubt that ML in compounds without a luminescent center originates from nitrogen (N_2) discharge.

A key question remains: how does this gas discharge occur, and what serves as its excitation source? In fact, as early as 1913, Langevin proposed that ML arises from the piezoelectric effect in noncentrosymmetric crystals. The asymmetric structure of these crystals enables the uneven arrangement of charges along the polarization direction. When pressure is applied, electrons migrate and accumulate at opposite ends of the crystal, creating a microscopic electric field and voltage drop inside the crystal. When the crystal fractures under mechanical force, new interfaces form, along with a strong electric field between the charged fracture surfaces. This electric field is sufficiently intense to trigger nitrogen discharge, ultimately producing ML.

To verify this theory, Zink et al. performed mathematical calculations to determine the electric field intensity at the cracked interface of $Li_2SO_4{\cdot}H_2O$ crystals. Three assumptions were put forward by them: (1) the piezoelectric constants of the crystal stay unchanged until the stress reaches the level that caused fracture; (2) prior to the excitation of ML, the stress in the vicinity of the crack tip does not undergo relaxation; (3) when it comes to charge development, the distribution along the polar

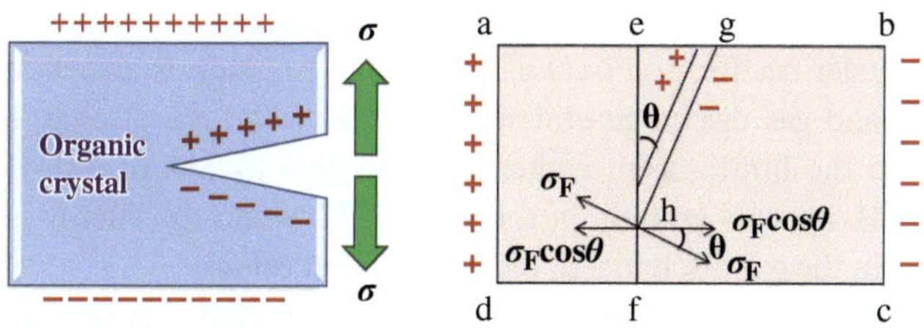

(a) **(b)**

Figure 5.3 (a) Under external stress, the charges of the piezoelectric crystal distribute along the polarization direction. (b) Piezoelectric field is generated at the cracked surface of the piezoelectric crystal. *Source*: Reproduced with permission from [3]/Elsevier.

direction exerts a more dominant influence than that along any other direction. As depicted by Figure 5.3, suppose the crack propagation orientation forms an angle θ with *ef* (the normal to the crystal's polar orientation *ab*). Under this condition, the charge density presented on the cracked surfaces is given by $d_{22}\sigma_f \cos\theta$, where d_{22} denotes the piezoelectric constant and σ_f refers to the stress necessary to pull apart the crystal surfaces. Additionally, the electric field existing between these cracked surfaces follows the expression $(1/\varepsilon_0)(d_{22}\sigma_f \cos\theta)$, in which ε_0 standing the permittivity of free space. For $Li_2SO_4{\cdot}H_2O$ crystals, the fracture stress σ (estimated from $^1/_{30}$th of the Young's modulus) equals 1.77×10^8 N·m^{-2}, while the value of d_{22} is 16.2×10^{-12} (C·N)$^{-1}$. The angle θ was determined to be $22 \pm 2°$ through an experiment: a $Li_2SO_4{\cdot}H_2O$ crystals was compressed between the jaws of a screw gauge, and this setup allowed for the determination of θ. When these parameters are substituted into the aforementioned electric field formula, the computed field strength is 1.26×10^8 V·m^{-1}. This value exceeds the 3×10^7 V·m^{-1} threshold required for gas discharge under atmospheric pressure. This approximate calculation thus verifies that gas discharge is capable of inducing ML. Specifically, the piezoelectric charging effect on the cracked crystal surfaces is sufficiently intense to initiate ML via nitrogen discharge.

5.1.3 ML from the Organic Compounds Contain Luminescent Center

Organic luminescent center is generally composed of aromatic chromophores with π-conjugated electronic structure. The ML emitted by organic emitters typically exhibits its nearly identical nature to their photoluminescence (PL). When mechanical stimuli serve as the excitation source, ML can manifest as fluorescence, phosphorescence, or a combination of the two. Inspired by the well-established mechanism of gas discharge-induced ML in piezoelectric crystals, Longchambon further extended piezoelectric theory to ML compounds containing luminescent centers.[1,15] He identified two components in the ML spectrum of uranyl nitrate hexahydrate ($UO_2(NO_3)_2{\cdot}6H_2O$): one matches the PL of uranyl, and the other, ~1% of the total intensity, stems from nitrogen discharge. He deduced that gas discharge is the primary ML event, whereas PL is a secondary process excited by the high energy of gas discharge. Four key conclusions follow: (1) no ML is observed if the electric field generated by crystal fracture is insufficient to trigger gas discharge; (2) in the absence of a luminescent center in the crystal, no energy transfer occurs, and only gas discharge emission is detected; (3) a combination of PL and gas discharge emission emerges when gas discharge transfers partial energy to the luminescent center, as illustrated by ML of uranyl nitrate hexahydrate; (4) ML appears entirely in the form of PL when the energy is completely transferred from the gas discharge to the luminescent center.

Yet the gas discharge mechanism fails to explain all ML phenomena. A case in point is the work by Hurt et al., who observed that europium chelates' ML is not initiated by gas discharge since their ML shows no reliance on either the type or pressure of the surrounding gas.[16] Beyond this, a notable contradiction emerges: many piezoelectric crystals do not exhibit ML properties, whereas a great number of nonpiezoelectric crystals do. This inconsistency is further reinforced by statistical data: among ML-active compounds, only 53% of sulfates have noncentrosymmetric structures. The corresponding proportions are 50% for nitrates and halides, 68% for organic aromatic emitters, and 73% for triclinic crystals.[17] Thus, the primary focus must be to clarify the process occurring during crystal fracture. During the fracture process, cracked surfaces emit fractoemission, including electron emission, positive ion emission, and neutral species, as observed by Dickinson et al. (Figure 5.4).[18] Drawing on this observation, Chandra et al. put forward a proposal: ML stems from electron bombardment produced during crystal fracture, and charge neutralization drives the recombination of electrons and ions on the fracture surface.[19] Of note, nitrogen

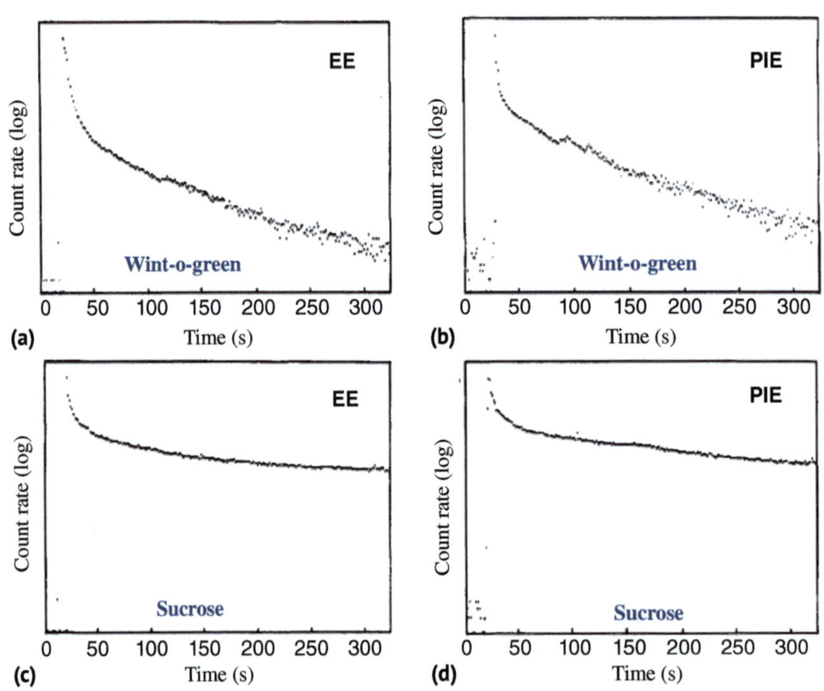

Figure 5.4 The electron emission (EE) and positive ion emission (PIE) of Wint-o-green (a and b) and sucrose (c and d). *Source:* Reproduced with permission from [3]/Elsevier.

discharge requires roughly 11 eV of excitation energy, whereas luminescent emitters need below 5 eV (corresponding to an excitation wavelength of about 250 nm). When a fracture occurs, electrons with energies exceeding 5 eV are detected; this finding confirms that electron bombardment provides sufficient energy to trigger ML.

Hence, the piezoelectric effect cannot be regarded as the key contributor to ML. For piezoelectric crystals, the higher charge density along the polarization direction leads to stronger electron bombardment when the crystal fractures, which in turn means a higher likelihood of ML production. While this theory can explain nearly all fracture-induced ML phenomena, it still fails to describe the light production process in detail. Excitons produced by mechanical stimuli exhibit features of both Frenkel and Wannier excitons; consequently, ML can manifest as fluorescence, phosphorescence, or even excimer emission. Moreover, ML spectra are highly dependent on the molecular conformation, stacking modes, and noncovalent interactions within the crystal. Exploring ML at the molecular level offers a novel perspective for unraveling its emission mechanism, a topic that warrants in-depth investigation. In the following section, we will classify organic ML into four categories: fluorescence-based, phosphorescence-based, host–guest doping systems, and special ML systems.

5.2 Organic Crystals with Fluorescent Mechanoluminescence

Fluorescence is the fundamental luminescence mode for organic compounds, arising from the radiative transition of singlet excitons. While mechanical and photo-stimuli differ in how they excite materials, the photophysical processes that drive light emission are nearly identical. For this reason, ML is generally classified as fluorescence. In 2015, Tang et al. identified the first AIE-active organic copper complex with ML properties.[7] Subsequently, a series of AIE-active and non-AIE organic ML compounds were reported in succession. In this section, various fluorescent organic ML compounds are categorized into three classes: triphenylamine (TPA) derivatives, tetraphenylethylene (TPE) derivatives, and nitrogen heterocyclic derivatives.

5.2.1 Triphenylamine Derivatives

Thanks to the strong conjugation, flexible conformation, and fine crystallinity, TPA is a prevalent aromatic unit in the design of opto-electronic functional materials as well as ML emitters. In 2017, by integrating an electron-withdrawing benzaldehyde into the TAP group, Li et al. developed the ML compound TAP-1BA. When subjected to grinding, this compound emits bright green ML with an emission peak at 500 nm (Figure 5.5a).[20] With well-developed crystallinity and high rigidity, TPA-1BA forms a noncentrosymmetric crystal structure that crystallizes in the $Pca2_1$ space group. At the molecular level, two types of molecular configurations, along with efficient C—H···O and C—H···π hydrogen bonds are observed in the crystal (Figure 5.5b). These structural elements work jointly to boost crystal rigidity, inhibit molecular

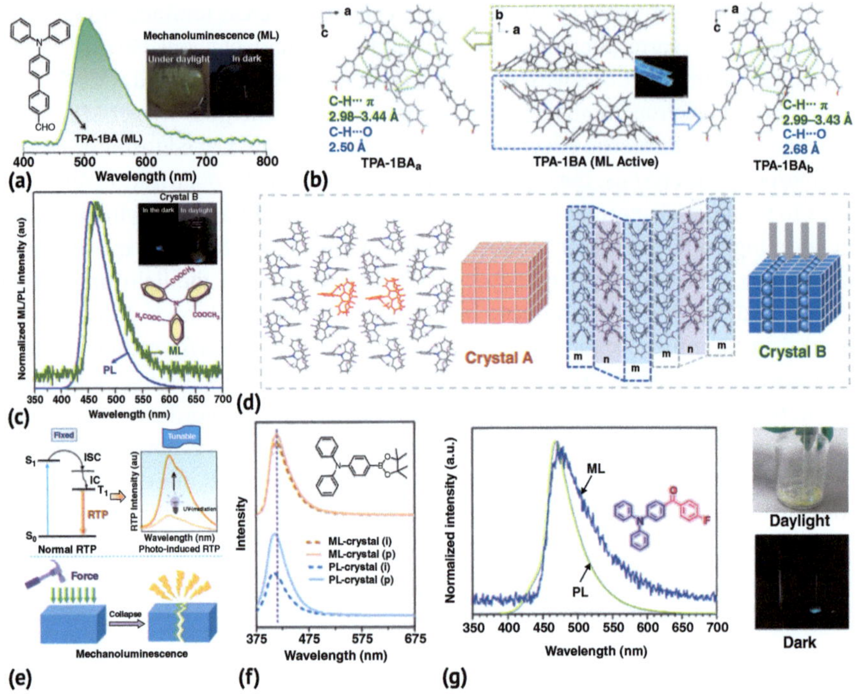

Figure 5.5 (a) Molecular structure of TPA-1BA and its ML spectrum, and photograph. (b) Molecular packing in the crystal of TPA-1BA. *Source*: Reproduced with permission from [20]/Royal Society of Chemistry. (c) PL and ML spectra, and a photograph of TPA-o-3COOMe in crystal B. (d) The polymorphism of TPA-o-3COOMe. *Source*: Reproduced with permission from [21]/John Wiley & Sons. (e) Schematic diagram of ML and photo-induced RTP of TPA-B. (f) ML and PL spectra of TPA-B in the initial state (crystal-i) and after UV irradiation (crystal-p). *Source*: Reproduced with permission from [22]/John Wiley & Sons. (g) ML spectrum and images of FCO-TPA. *Source*: Reproduced with permission from [23]/Royal Society of Chemistry.

slippage and collapse under external stress, and in turn contribute both to robust ML and to high PL efficiency.

Like the aldehyde group, the integration of an ester group into the TPA unit can also increase molecular polarity and strengthen the intermolecular interactions, specifically via hydrogen bonds formed by the oxygen atoms within the ester group. In 2019, Li et al. synthesized the compound TPA-o-3COOMe by introducing three ester groups at the ortho positions of the TPA unit.[21] This compound exhibits polymorphism, with its two polymorphs realizing RTP and ML, respectively. The two properties can be distinguished by crystal form: Crystal A crystallizes in the centrosymmetric space group P-1. Strong intermolecular interactions drive the formation of an ordered packing structure, which is more conducive to RTP generation; Crystal B is characterized by the noncentrosymmetric space group P_n (Figure 5.5c, d). It adopts an interlocked molecular packing mode that enhances the crystal's overall rigidity, thereby making it easier to trigger ML through the piezoelectric effect (Figure 5.5c, d). This result directly confirms that the luminescent properties of organic aggregates are highly correlated with molecular packing modes.

Reported in the same year, a TPA-based borate ester derivative (TPA-B) was found to simultaneously exhibit prominent ML and photoinduced RTP in the crystal state (Figure 5.5e).[22] Structural analysis reveals that TPA-B crystals belong to the noncentrosymmetric space group $Pna2_1$. Notably, no changes occur in either the crystal's space group or its packing mode when exposed to UV light; consequently, the ML properties are largely consistent before and after light treatment (Figure 5.5f). This observation further verifies that for both ML and photoinduced RTP, the molecular arrangement and noncovalent interactions of the aggregate serve as critical determinants of the luminescent response.

In 2020, by linking the p-fluorobenzoyl group to the TPA unit, FCO-TPA was designed with typical HLCT character (Figure 5.5g).[23] This compound can generate bright ML: on the one hand, the incorporation of F and O atoms can reduce nonradiation transition via enhancing intermolecular hydrogen bonds; on the other hand, its asymmetric structure is conducive to the formation of a noncentrosymmetric packing mode and generation of a piezoelectric effect, both of which play a role in promoting ML. Furthermore, the HLCT characteristics facilitate the efficient utilization of excited-state energy, which in turn aids in enhancing ML performance.

Halogen atoms, owing to their polar nature, can also enhance intermolecular interactions. In 2019, Li et al. carried out structural modification on TPA-CHO, which does not possess ML properties (Figure 5.6a).[24] By introducing halogen atoms (F, Cl, Br), three TPA derivatives (TPA-CHO-2F, TPA-CHO-2Cl, and TPA-CHO-2Br) exhibited altered electronic structure. This structural modification enabled tight molecular packing in the crystalline state, yielding prominent ML emission. Spectroscopic analysis revealed that the ML spectra of these three compounds were highly similar, with their emission peaks located at approximately 475 nm, 475 nm, and 473 nm, respectively.

By further inserting a phenyl group and decorating halogen atoms at the para- or meta-position of the phenyl group, Li et al. synthesized TPA-ph, TPA-oX, TPA-mX, and TPA-pX (X = F, Cl, Br) (Figure 5.6b), respectively.[25] However, the ortho-substituted derivatives did not show ML nature. TPA-ph and TPA-pF displayed weak ML and crystallized in the centrosymmetric space group of $P2_1/c1$. In contrast, the crystals of TPA-mX (X = F, Cl, Br) and TPA-pX (X = Cl, Br) exhibited prominent ML: their emission spectra showed distinct peaks at 431, 433, 435, 443, and 436 nm, respectively (Figure 5.6d). A key structural difference between the two groups emerged in their space groups; these strongly ML-active crystals all belonged to the noncentrosymmetric space group Clc1. From the PL and ML spectra, it's interesting to find that the ML spectra of these compounds resemble their PL spectra in the ground state (Figure 5.6c), implying their ML is generated from the cracked surface of the crystals. Based on these observations, for TPA derivatives, different halogen atoms induce the formation of diverse intermolecular interactions via variations in their electronegativity and atomic radius. These interactions

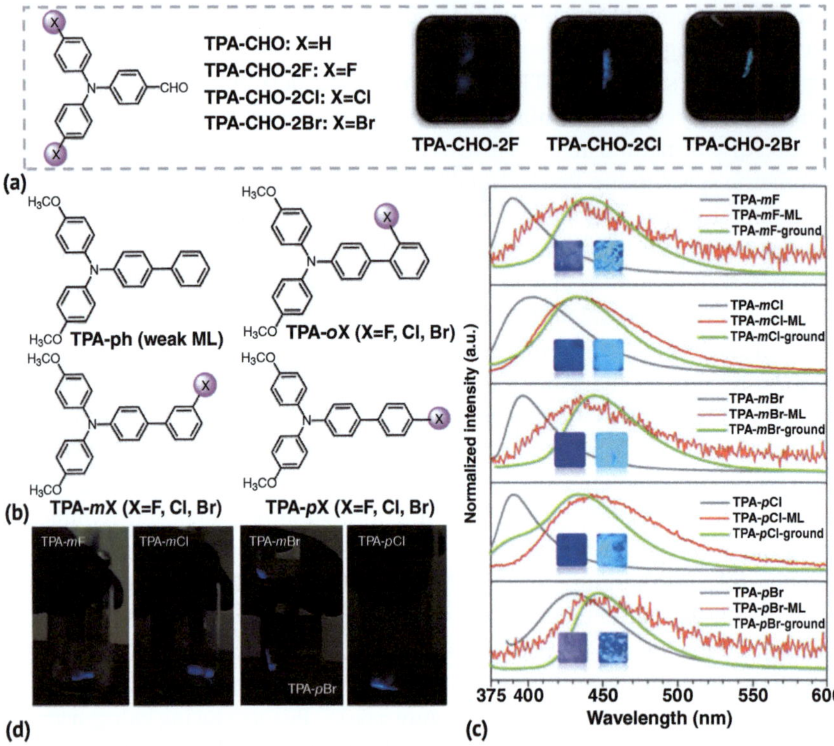

Figure 5.6 (a) Molecular structures and ML photographs of TPA-CHO derivatives. *Source*: Reproduced with permission from [24]/The Royal Society of Chemistry. (b) Molecular structures of TPA-ph derivatives. (c) PL and ML spectra of TPA-*m*X (X = F, Cl, Br) and TPA-*p*X (X = Cl, Br). (d) ML photographs of TPA-*m*X (X = F, Cl, Br) and TPA-*p*Cl. *Source*: Reproduced with permission from [25]/John Wiley & Sons.

effectively suppress the molecular slippage under mechanical stimulation, ultimately enhancing the generation of the ML.

5.2.2 Tetraphenylethylene Derivatives

TPE stands as one of the most representative AIE molecules and is widely employed as a building block in the field of organic luminescent materials. Owing to the flexible conformation and relatively loose packing within the crystalline structure, nearly all TPE derivatives exhibit mechanochromic behavior. The incorporation of polar groups (e.g., aldehyde, amino, ester, and methoxy) can effectively strengthen the intermolecular interactions and concurrently modulating the molecular crystallinity. Currently, a diverse range of TPE derivatives were reported to have ML activity. Leveraging their inherent AIE characteristics (high luminescent efficiency in the aggregated state), these materials typically display high ML brightness that can be clearly visualized even under daylight. This key attribute offers critical support for the practical applications of ML materials, such as visual stress sensing.

By incorporating a polar aldehyde group into TPE derivatives, Chi et al. constructed polymorphic systems to investigate the correlation between crystalline phases and ML behaviors. In 2015, they reported that the TPE derivative P_4TA forms two distinct polymorphs with contrasting ML behaviors: the block-shaped C_b crystal emits blue light (peak at 476 nm) and is ML inactive, while the rod-shaped C_g crystal emits green light (peak at 499 nm) and exhibits green ML emission (Figure 5.7a, b).[26] Crystallographic analysis revealed that both polymorphs belong to the noncentrosymmetric space group of P(2)1, while the C_g crystal possesses a stronger piezoelectric effect and a lower excitation energy, which collectively render it more prone to ML generation. Furthermore, a phase transition can occur between the two polymorphs: the C_b-phase crystal can be converted into the ML-active C_g phase via fumigation with dichloromethane. Afterward, they further directly appended an aldehyde group to the TPE and synthesized three ML materials (*p*-P_4A, *m*-P_4A, and *p*-P_4A_2), all of which crystallize in a noncentrosymmetric polar space group. This structural diversity was achieved by modulating both the numbers of aldehyde groups and their substitution position on the TPE scaffold (Figure 5.7c, d).[27]

In 2019, through the introduction of carboxyl or ester groups at different positions of the TPE unit, Li et al. prepared four TPE derivatives of *p*-COOH, *m*-COOH, *p*-ester, and *m*-ester (Figure 5.8a).[28] Among these derivatives, *p*-COOH

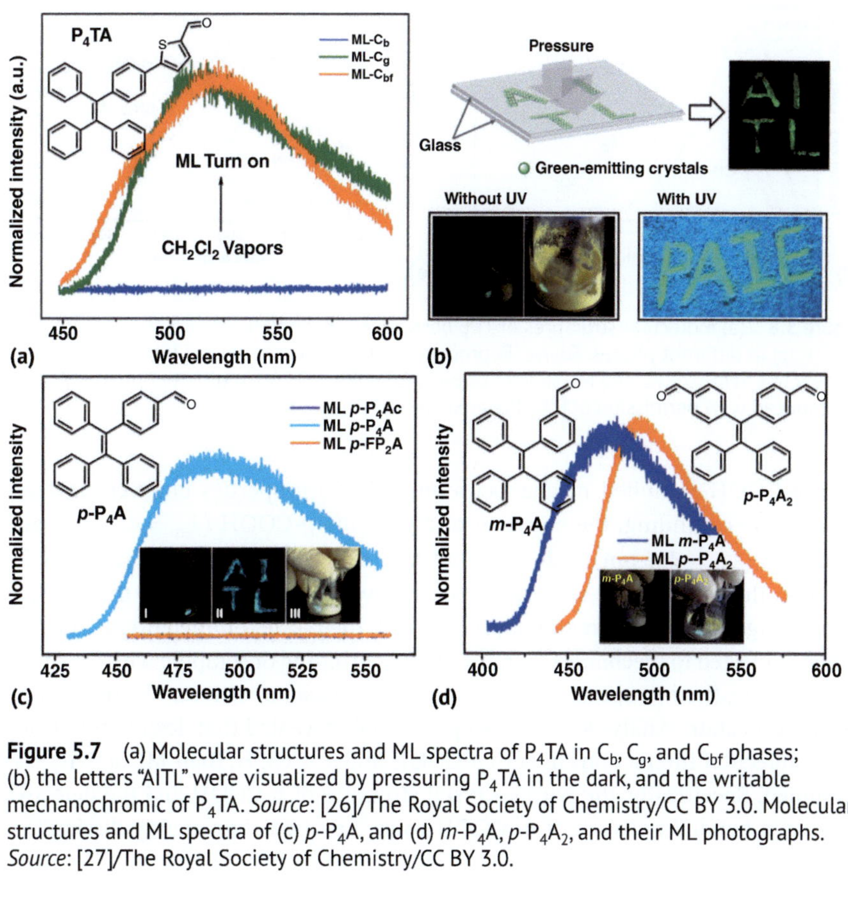

Figure 5.7 (a) Molecular structures and ML spectra of P_4TA in C_b, C_g, and C_{bf} phases; (b) the letters "AITL" were visualized by pressuring P_4TA in the dark, and the writable mechanochromic of P_4TA. *Source*: [26]/The Royal Society of Chemistry/CC BY 3.0. Molecular structures and ML spectra of (c) *p*-P_4A, and (d) *m*-P_4A, *p*-P_4A_2, and their ML photographs. *Source*: [27]/The Royal Society of Chemistry/CC BY 3.0.

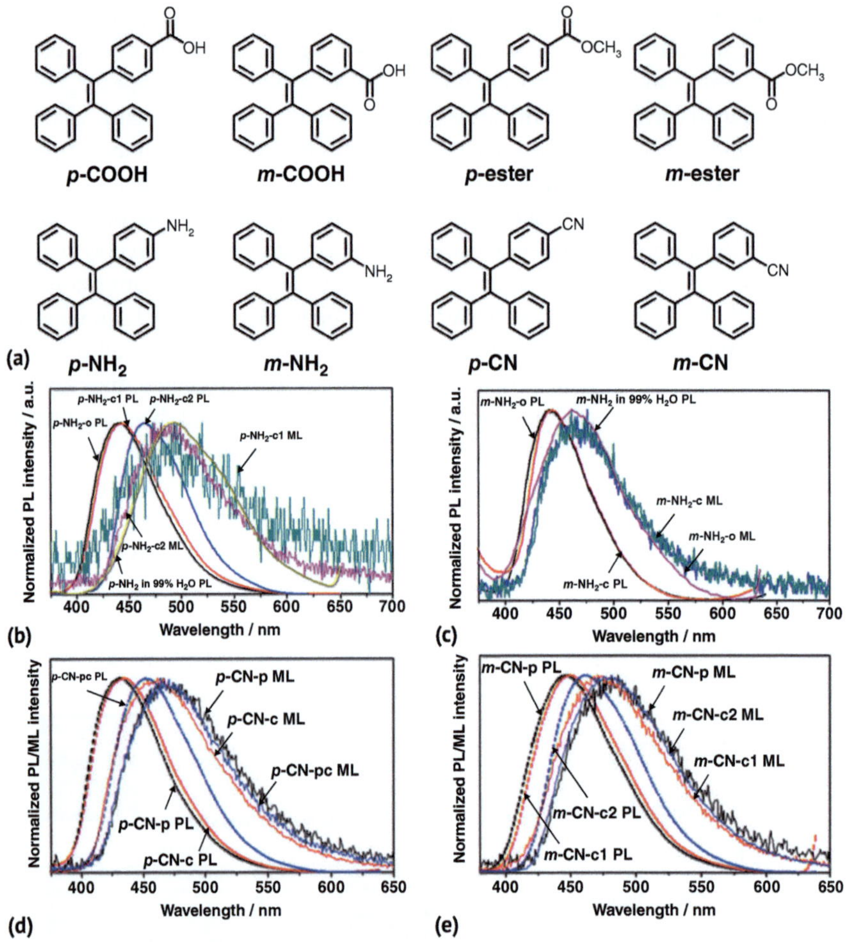

Figure 5.8 (a) Molecular structures of TPE derivatives. PL and ML spectra of *p*-NH_2 (b) and *m*-NH_2 (c) in different phases. *Source*: Reproduced with permission from [29]/The Royal Society of Chemistry. PL and ML spectra of *p*-CN (d) and *m*-CN (e) in different phases. *Source*: Reproduced with permission of [30]/Royal Society of Chemistry.

and *m*-COOH exhibited mechanochromism (MC) properties but were inactive in ML. After grinding, the emission of crystalline *p*-COOH (λ_{em} = 458 nm) and *m*-COOH (λ_{em} = 436 nm) underwent redshift to 482 nm and 450 nm, respectively. Notably, the ground state of these two derivatives could be recovered through solvent fumigation or heat treatment. In contrast, the *p*-ester emitted bright blue ML when subjected to mechanical stimuli, such as grinding or scraping with a spatula. Importantly, the ML spectrum of *p*-ester was consistent with its PL spectrum in the ground state. Analysis of their single crystals revealed that despite the strong intermolecular interactions in both *p*-COOH and *m*-COOH, their ML activity was inhibited. Specifically, the porous crystal structure of *p*-COOH and the defective hydrogen-bonded network of *m*-COOH caused excited-state energy dissipation, which accounted for their ML inactivity. In contrast, *p*-ester exhibited ML activity owing to its moderate packing density and intense hydrogen-bonded network.

Also in 2019, they substituted the ester group with an amino group, successfully synthesizing two TPE derivatives: *p*-NH_2 and *m*-NH_2.[29] Both compounds display ML activity (Figure 5.8b, c). Single-crystal structural analysis and complementary experimental data demonstrated that the ML properties of these derivatives are closely linked to molecular geometry and noncovalent interactions. Subsequently, when the amino group was replaced with a cyano group, the resulting derivatives, *p*-CN and *m*-CN, emitted bright blue ML under dim light conditions.[30] Notably, the ML activity of these compounds could be restored through a straightforward thermal treatment process (Figure 5.8d, e). Single-crystal analysis of *p*-CN and *m*-CN further revealed that the ML activity of these derivatives probably stems from the intact supramolecular interactions induced by the cyano groups.

Regarding molecular packing and ML properties, Li et al. conducted a systematic analysis using TPE derivatives as the research platform. In 2016, they introduced four methoxy groups into the TPE to obtain TMPE. This compound was found to exhibit two polymorphisms with distinct space groups: one a with $P2_1(c)$ space group (designated as the C_p form) and the other with a C2 space group (designated as the C_c form) (Figure 5.9a).[31] Experimental results demonstrated that only the Cp crystal exhibited bright deep-blue ML (with an emission peak at 460 nm, Figure 5.9b), whereas the Cc crystal showed no ML activity. From a macroscopic viewpoint, the piezoelectric effect of polar crystals drives charge separation along the polarization direction, leading to charge accumulation on the two sides of the crystal. This accumulation results in more intense electron bombardment when the crystal fractures, thereby facilitating the activation of mechanoluminescence. At the molecular packing level, the Cp crystal features stronger intermolecular interactions, including C—H···π and C—H···O hydrogen bonding, which enable denser molecular packing (Figure 5.9c). This specific crystal structure restricts intermolecular sliding and suppresses nonradiative transitions, ultimately promoting the occurrence of mechanoluminescence. In contrast, the ordered molecular arrangement in the Cc crystal is highly susceptible to sliding under external stress, precluding the formation of intense electron bombardment and thus the generation of ML.

Apart from introducing heteroatom-containing groups to enhance molecular polarity and noncovalent interactions, Li et al. also reported a pure hydrocarbon compound, TETPE. This compound crystallizes in the centrosymmetric space group C2 and exhibits notable ML activity (Figure 5.9d).[32] Theoretical calculations and single-crystal structure analyses reveal that the alkynyl moieties in TETPE molecules can induce the formation of hydrogen bonds and electrostatic interactions, leveraging the positronic character of terminal alkynyl hydrogen atoms. These interactions effectively facilitate the formation of excited states during crystal fragmentation, acting as the key driving force behind ML generation. In 2018, Li et al. constructed the *mm*-$TPE(PI)_2$ molecule by linking phenanthroimidazole, employed as the functional unit, to TPE at the meta-position (Figure 5.9a, e).[33] This molecule adopts a perpendicular conformation with

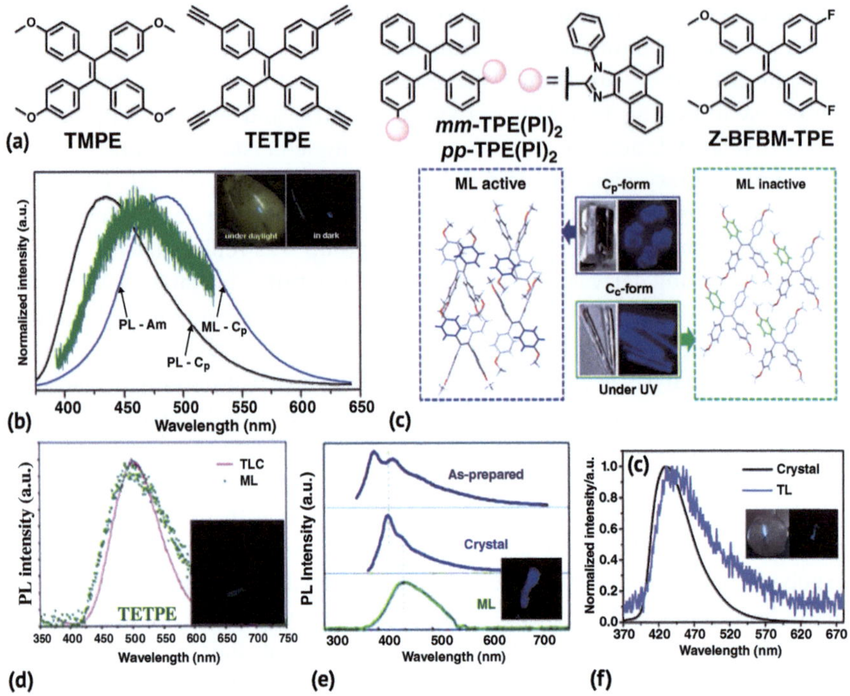

Figure 5.9 (a) Molecular structures of the TPE derivatives with ML properties. (b) PL and ML spectra of TMPE. (c) The molecular packing of TMPE. *Source*: Reproduced with permission from [31]/The Royal Society of Chemistry; (d–f) PL and ML spectra of TETPE, *mm*-TPE(PI)$_2$, and Z-BFBM-TPE; the insets were the corresponding ML photographs. *Source*: Reproduced with permission from [32]/The Royal Society of Chemistry. Reproduced with permission from [33]/The Royal Society of Chemistry. Reproduced with permission from [34]/John Wiley & Sons.

weak intramolecular conjugation, and adjacent molecules stack in a perpendicular orientation. Notably, this interlocking molecular arrangement gives rise to strong noncovalent interactions; these interactions, in turn, facilitate the formation of a dense packing structure, ultimately enabling the molecule to exhibit blue ML. Li et al. further investigated the influence of *cis-trans* isomerism on ML performance. Specifically, when two methoxy and fluorine substituents were introduced at the *cis*-positions of TPE, the dipole moment and polarity of the molecule were significantly enhanced (Figure 5.9a, f).[34] The resulting compound, Z-BFBM-TPE, thus exhibited ML activity. In contrast, the corresponding *trans*-substituted analog displayed no ML properties, attributed to its smaller dipole moment. This finding confirms that the polarity differences induced by *cis-trans* isomerism in molecules exert a significant impact on their packing modes and macroscopic photophysical properties.

Triphenylethylene is a structural variant of TPE. It retains AIE properties and has an asymmetric structure, which facilitates the formation of tight molecular packing and enables the generation of ML. In 2018, Li et al. incorporated a thiophene unit into the triphenylethylene scaffold, synthesizing the compound tPE-5-Meth.[35]

This compound exists in two distinct crystal forms (Crystal-P1 and Crystal-P2) with markedly divergent ML behaviors: the ML-active Crystal-P1 exhibits an emission peak at 440 nm, whereas the non-ML-active Crystal-P2 displays its emission peak at 472 nm (Figure 5.10a, d). Notably, leveraging the supramolecular self-assembly properties of the compound, reversible recovery ML performance can be achieved via heat treatment, an advancement that opens a novel pathway for the cyclic application of ML materials. From the perspective of molecular packing, molecules in Crystal-P1 adopt a twisted conformation, and strong intermolecular interactions drive the formation of an ordered and highly compact packing structure. This ordered packing architecture serves as the core rationale behind its efficient ML performance. In contrast, within the centrosymmetric Crystal-P2, molecules arrange in a pairwise cross configuration, leading to an antiparallel packing pattern in the crystal. This packing mode fails to induce the charge separation essential for ML, ultimately resulting in the absence of ML activity. The discrepancy between the two crystal forms further validates the critical regulatory role of molecular packing modes in aggregates with respect to ML properties.

Subsequent studies have uncovered an additional key attribute of the thiophene units: beyond their self-assembly functionality, they possess a unique "structural

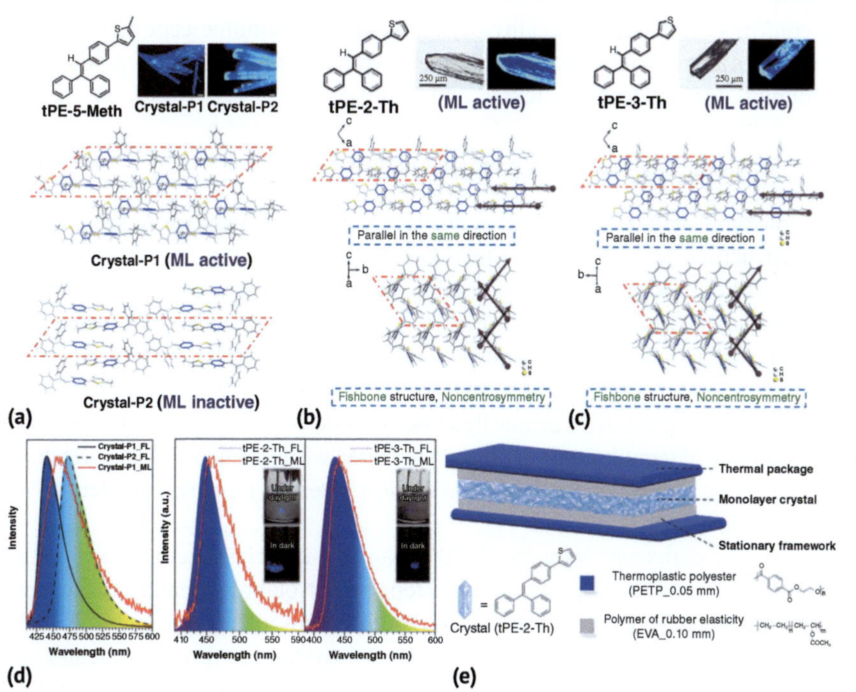

Figure 5.10 (a) Molecular structures of tPE-5-Meth and its molecular packing of polymorphism (b, c) Molecular structures of tPE-2-Th and tPE-3-Th, and their corresponding molecular packing. (d) PL and ML spectra of crystal P1 of tPE-5-Meth, tPE-2-Th, and tPE-3-Th. (e) Schematic structure of ML devices based on tPE-2-Th. *Source*: Reproduced with permission from [35]/The Royal Society of Chemistry. Reproduced with permission from [36]/with permission of Elsevier.

repair" capability. When external mechanical stimuli cause localized damage to the crystal structure, the self-assembly properties of the thiophene units enable the reorientation of molecules and the reconstruction of ordered packing. This repair mechanism provides a vital safeguard for the sustained generation of ML and the long-term stability of its performance.

When introducing thiophene unit at distinct positions, the resultant compounds, tPE-2-Th and tPE-3-Th, are found to crystallize in noncentrosymmetric lattices.[36] Their molecular packing adopts a layered herringbone pattern, and both compounds exhibit substantial dipole moments, measuring 9.49 D and 9.29 D, respectively (Figure 5.10b–d). This unique structural profile offers two key advantages: first, the herringbone arrangement strengthens intermolecular interactions, thereby minimizing energy loss through nonradiative transitions when the material is subjected to mechanical stress; second, the combination of a noncentrosymmetric structure and large dipole moments promotes charge separation and accumulation via the piezoelectric effect, laying a favorable groundwork for ML generation. Employing tPE-2-Th's outstanding ML performance, characterized by high brightness and exceptional sensitivity, Li et al. developed a flexible ML device featuring a sandwich architecture (Figure 5.10e). Through comprehensive testing, they further established a quantitative correlation between applied pressure and luminescence intensity. This work not only confirms the practical viability of organic mechanoluminescent materials but also provides a crucial technical basis for their seamless application in areas including communications, information storage, and healthcare.

5.2.3 N-containing Heterocyclic Derivatives

Carbazole represents a classic class of nitrogen-containing heterocyclic compounds for constructing opto-electronic functional materials, owing to its high triplet energy and luminescent efficiency. Notably, carbazole derivatives also hold significant potential in the field of ML. In 1981, 9-ethyl-3-vinylcarbazole (ECz) was found to exhibit ML activity in its crystalline state.[37,38] Subsequently, in 1983, Samoc et al. reported the ML properties of n-isopropylcarbazole (Figure 5.11a).[38] Their study proposed that ML generation primarily stems from the formation of cracked surfaces upon mechanical stimulation, which are associated with the material's polarization. Together, these two cases constitute the earliest research on carbazole derivatives with ML properties.

In 2015, the carbazole derivative PCDOS, an asymmetric-structured compound integrating AIE and TADF properties, was reported.[8] Notably, when the as-prepared sample was scratched, intense green ML could be observed even under daylight (Figure 5.11b). This remarkable ML behavior is attributed to three key factors: the molecule's asymmetric structural features, its crystallization in the noncentrosymmetric space group Pna21, and its intricate "bagua"-like molecular packing (a traditional Chinese geometric pattern). In this packing arrangement, the planar carbazole moiety of one molecule interlocks with the nonplanar phenothiazine moiety of an adjacent molecule, forming a tightly packed, highly ordered structure, with no π–π stacking interactions present.

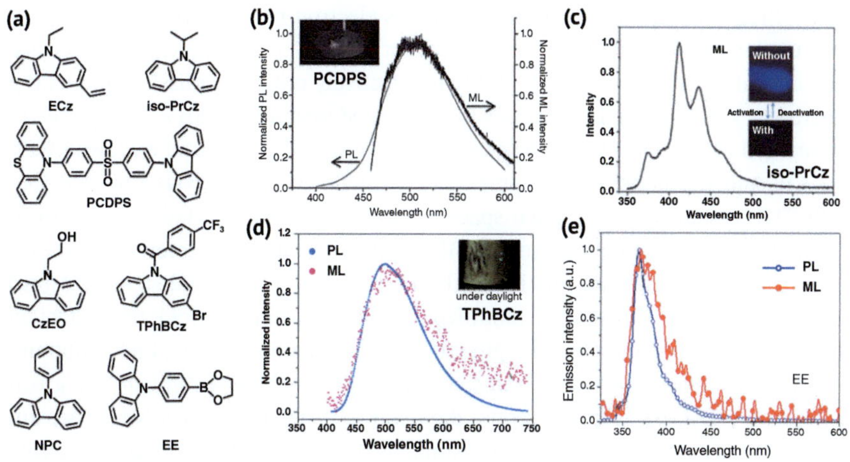

Figure 5.11 (a) Molecular structures of carbazole derivatives. (b) PL and ML spectra of PCDPS. *Source*: Reproduced with permission from [8]/John Wiley & Sons. (c) ML spectrum for the photo-activated CZEO. Inset: ML photograph of CZEO with and without activation. *Source*: Reproduced with permission from [39]/The Royal Society of Chemistry. (d) PL and ML spectra of TPhBCz. Inset: ML photograph. *Source*: Reproduced with permission from [40]/Elsevier. (e) PL and ML spectra of EE. Inset: ML photograph. *Source*: Reproduced with permission from [42]/Royal Society of Chemistry.

Subsequently, in 2019, Lin et al. discovered that 2-(9H-carbazol-9-yl)ethanol (CZEO) exhibits both ML activity and photoinduced RTP.[39] The switching between ML and photoinduced RTP can be achieved via grinding or UV activation (Figure 5.11c). Specifically, after UV irradiation for 2 minutes, CZEO's ML is deactivated while its RTP is activated; this switching is primarily driven by UV-induced changes in molecular packing. Photoactivation elongates the C—H⋯O hydrogen bond length and shortens the C—H⋯π distance within molecular dimers, rendering the dimers in the crystal more compact. This enhanced dimer compactness restricts molecular motion, thereby activating RTP and suppressing ML. Conversely, grinding reverses this state: the dimers become loosely packed, and increased dimer motion dissipates triplet-state energy, switching off RTP and reactivating ML.

In 2020, Xu et al. reported compound TPhBCz, which combines three functional properties: AIE, TADF, and ML (Figure 5.11d).[40] Detailed crystal structure analysis revealed that the crystal's compact molecular packing and strong noncovalent interactions are responsible for its efficient ML. That same year, Yang et al. also reported a series of carbazole derivatives with ML properties.[41,43] Notably, even though compound NPC is a nonpolar molecule, it crystallizes in a piezoelectric space group and forms a hydrogen-bonding network, which is a key attribute that contributes to its ML emission. Building on this work, Yang et al. further developed a series of borated derivatives based on NPC. Among these, the organic crystal EE stands out: it adopts a piezoelectric space group and features strong intermolecular interactions, enabling bright ML emission (Figure 5.11e).

Phenothiazine, a moiety distinguished by its electron-rich nature and nonplanar conformation, is an excellent building block for designing opto-electronic materials. A key advantage of this unit is its ability to effectively mitigate emission quenching triggered by π–π stacking interactions in aggregated systems. In 2017, Thilagar et al. documented ML in DMPBPTZ, a phenothiazine-derived compound endowed with both AIE and mechanochromic properties.[43] Single-crystal analysis revealed that DMPBPTZ crystallizes in the polar space group R3c; when subjected to mechanical stimuli, it emits bright greenish-yellow ML. While ML emission (centered at 530 nm) undergoes a slight red shift relative to the PL spectrum of its solid state (513 nm) (Figure 5.12a). Later, Ghosh et al. modified the phenothiazine scaffold: they attached a diarylboryl group to the phenyl segment of phenothiazine, yielding the compound BMBPTZ. This derivative was found to exhibit both mechanochromism and green ML (Figure 5.12a, b).[44] Spectroscopic data showed that BMBPTZ's ML spectrum (λ_{em} = 520 nm) is red-shifted by approximately 30 nm, and this shift aligns closely with the PL spectrum of the compound in its ground state. Notably, BMBPTZ's single crystal adopts an asymmetric space group of *P*31, which is believed to be the core factor enabling its ML property.

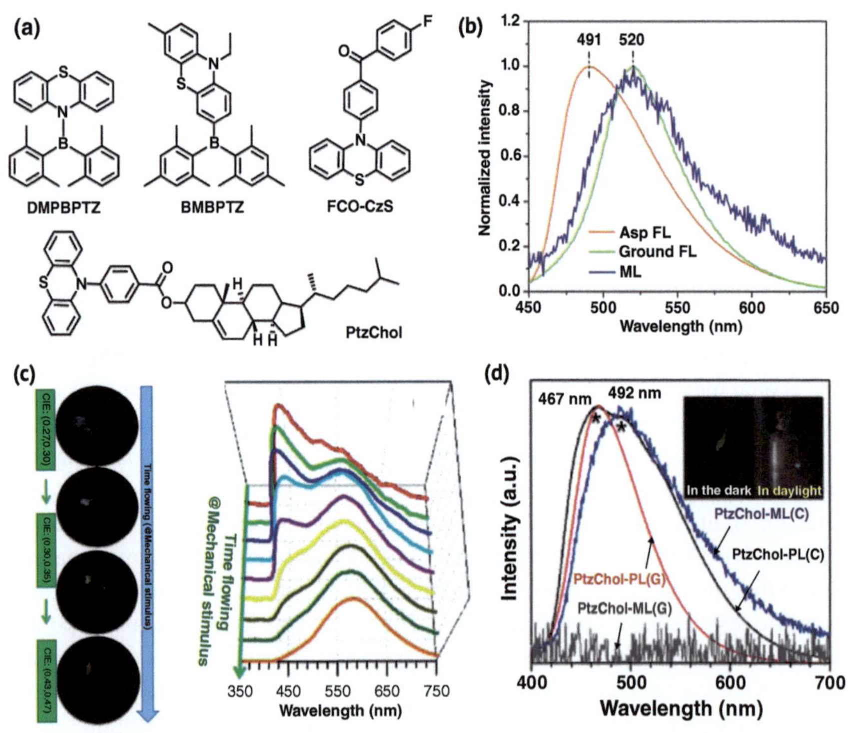

Figure 5.12 (a) Molecular structures of phenothiazine derivatives. (b) PL and ML spectra of BPBPTZ. *Source*: Reproduced with permission from [44]/John Wiley & Sons. (c) The ML pictures and normalized ML spectra of FCO-CzS with different grinding times. *Source*: Reproduced with permission from [45]/John Wiley & Sons. (d) PL and ML spectra of PtzChol. *Source*: Reproduced with permission from [46]/Springer Nature.

In 2018, Li et al. synthesized a phenothiazine derivative by the coupling of phenothiazine with 4-fluorophenyl benzophenone. This compound exhibits an exceptionally unique ML property: upon exposure to external mechanical force, its ML color undergoes a gradual transition from blue to white and eventually shifts to yellow after prolonged grinding (Figure 5.12a, c, d).[45] Crystal structure analysis reveals that phenothiazine adopts two distinct conformations, quasi-axial and quasi-equatorial. The quasi-axial conformation features a higher molecular energy, while the quasi-equatorial counterpart possesses superior conjugation characteristics. During the application of external mechanical stimulation, the molecular conformation of FCO-CzS gradually transforms from the quasi-axial to the quasi-equatorial form. This conformational transition directly induces changes in the molecular electronic structure, which in turn triggers the shift in ML color, thereby giving rise to distinctive dynamic electronic features and luminescent behavior. Additionally, the fluorine atoms within the molecule can strengthen intermolecular interactions, effectively suppressing nonradiative decay processes: a role that contributes positively to enhancing both PL intensity and ML intensity.

In the effort to expand phenothiazine derivatives, Li et al. further reported in 2021 the incorporation of cholesterol, a nonconjugated chiral group, into the phenothiazine molecular scaffold, enabling the design and synthesis of the compound PtzChol (Figure 5.12a, d).[46] This compound not only exhibits similar ML and RTP properties analogous to those of FCO-CzS but also acquires additional chiral characteristics due to the introduction of cholesterol, marking one of the earlier investigations into chiral organic room-temperature phosphorescence. On one hand, the cholesterol group constructs a rigid crystalline environment; on the other hand, it induces the formation of a chiral molecular packing structure. Results from crystal structure characterization and circular dichroism (CD) measurements demonstrate that PtzChol crystals generate detectable CD signals. However, following severe grinding, the ordered chiral packing structure of the crystals is disrupted, and the CD signals vanish almost entirely. This phenomenon provides direct evidence for the correlation between the ordering of crystal packing and chiral signals.

Given that ML properties are closely associated with the piezoelectric effect, and materials exhibiting piezoelectric properties typically possess a dipolar structure and a noncentrosymmetric molecular arrangement. Imide derivatives represent a class of acceptors inherently bearing a large dipole moment. In 2012, Yamashita et al. designed several N-phenylimide derivatives via introducing the polar groups of CF_3, SF_5, and p-$CF_3C_6H_4$. The resulting derivatives, ImCF_3, ImSF_5, and Im-*p*-$CF_3C_6H_4$, exhibiting remarkable ML properties (Figure 5.13a).[47] The three compounds show an obvious blue ML in the dark (taking ImCF_3 as an example, as shown in Figure 5.13b), which remarkably increases at low temperature due to the suppressed radiationless deactivation process. Furthermore, they increased the conjugation of the molecule and obtained imide derivatives NImCF_3 and NIm-*p*-$CF_3C_6H_4$ with blue ML emission. These two compounds have larger dipole moments and twisted conformations. The experimental findings enable the deduction of the ML mechanism: electron-withdrawing groups (CF_3 and SF_5) augment the molecular dipole moment

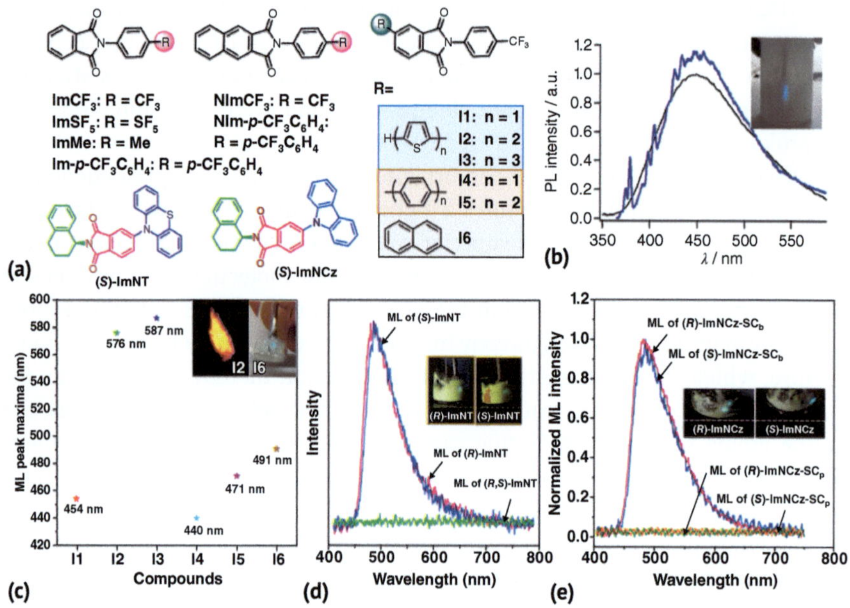

Figure 5.13 (a) Molecular structures of the imide derivatives. (b) PL and ML spectra of compound $ImCF_3$. *Source*: Reproduced with permission from [47]/American Chemical Society. (c) The ML peak maxima of compounds I1–I6. *Source*: Reproduced with permission from [48]/American Chemical Society. (d) ML spectra of enantiomers and their racemate of ImNT. *Source*: Reproduced with permission from [49]/The Royal Society of Chemistry. (e) ML spectra of polymorphic compound ImNCz; inset: ML images of compounds. *Source*: Reproduced with permission from [50]/Elsevier.

and promote the formation of piezoelectric crystals. Mechanical scraping induces fracture of these piezoelectric crystals, generating electrified surfaces at the fracture sites. This, in turn, triggers electronic discharge at the fractured interfaces, which excites both the compounds and ambient N_2 gas to their excited states, thereby producing luminescence.

In 2016, the research team synthesized a library of imide derivatives (labeled I1–I6, Figure 5.13a)[48] by functionalizing a phthalimide core, itself modified with a trifluoromethylphenyl group, with various π-extended substituents, including thienyl, phenyl, and naphthyl groups. A key structural feature shared by all these derivatives is their noncentrosymmetric molecular packing, and this structural trait is accompanied by a striking ability to emit mechanoluminescence (ML) across a range of colors. Detailed ML characterization revealed distinct emission profiles for each compound: I1 emits intense greenish-blue ML, with its emission maximum at 454 nm; I2 shows strong orange ML, peaking at 576 nm; I3 exhibits weak reddish-orange ML, with a peak centered at 587 nm; I4 produces bright blue ML, whose maximum lies at 440 nm; I5 emits intense greenish-blue ML, with an emission peak at 471 nm; I6 displays exceptionally strong light-blue ML, peaking at 491 nm (Figure 5.13c). Collectively, these findings confirm that the ML color of phthalimide-derived materials can be precisely modulated by selecting and incorporating appropriate π-extended groups onto the phthalimide core.

In 2019, Xu et al. conducted a study on chiral AIE enantiomers and their racemic mixture.[49] The molecular design centered on a phthalimide core (serving as the electron acceptor) and phenothiazine (acting as the electron donor); crucially, chirality was imparted to the compounds via the introduction of a phthalimide-functionalized tetrahydronaphthalene segment. Notably, the individual enantiomers, (*R*)-ImNT and (*S*)-ImNT, displayed clear ML activity, emitting vivid bluish-green light (Figure 5.13a, d). By sharp contrast, the formation of the racemate (*R*,*S*)-ImNT led to the total loss of ML activity. This trend mirrors the compounds' circularly polarized luminescence (CPL) behavior, highlighting a critical insight: chirality not only triggers CPL but also serves as an activator for ML in the AIEgens (*R*)-ImNT and (*S*)-ImNT. Crystal structure analysis shed light on the structural origins of this performance difference. For the enantiomers, a combination of extensive intermolecular interactions, including C—H···O, C—H···S, and C—H···π contacts, and a large dipole moment acts in synergy, which is responsible for their bright ML emission. In the case of the racemate (*R*,*S*)-ImNT, however, ML is fully quenched. This is attributed to two key factors: weaker intermolecular interactions and a centrosymmetric crystal arrangement, even though the racemate still maintains its AIE property.

In 2021, Xu and colleagues replaced the phenothiazine unit with carbazole, focusing their research on ImNCz, a polymorphic compound that exhibits both afterglow and ML properties.[50] They found that the enantiomer of ImNCz forms two distinct polymorphs, emitting intense blue and bluish-green fluorescence, respectively. A key observation was that the bluish-green-emitting crystals (SCb form) displayed striking ML activity, with emission peaks centered at 482 nm (Figure 5.13a, e). The blue-emitting crystals (SCp form) showed no ML activity, yet they produced photo-induced afterglow, consisting of TADF and ultralong phosphorescence. Molecular packing arrangements uncovered notable differences in intermolecular interactions: the SCp crystals featured a rich noncovalent interaction, including C—H···O and C—H···π; the SCb crystals, by contrast, had a more limited hydrogen bond interaction. Critical structural distinctions emerged, however: the SCb crystals adopted the chiral, polar space group $P2_1$, which correlates closely with the helical arrangement of chiral molecules that enhances structural dissymmetry. Moreover, the SCb crystals exhibited a significantly larger net dipole moment compared to their SCp counterparts. These findings imply that the strong noncovalent interactions and AIE properties play crucial roles in realizing ML performance.

Phenanthroimidazole is also a fused-ring compound with a large dipole moment. In 2019, Li et al. focused on designing and investigating ML compounds using phenanthroimidazole as the core building unit. As illustrated in Figure 5.14a, b, the boronate ester-substituted derivative tPTI-Bpin exists in two crystalline polymorphs that exhibit opposing ML behaviors.[51] Crystal structure analyses revealed that the boronate ester groups enhance intermolecular interactions through hydrogen bonding and coordination, while the incorporation of tert-butyl groups improves crystallinity and optimizes molecular

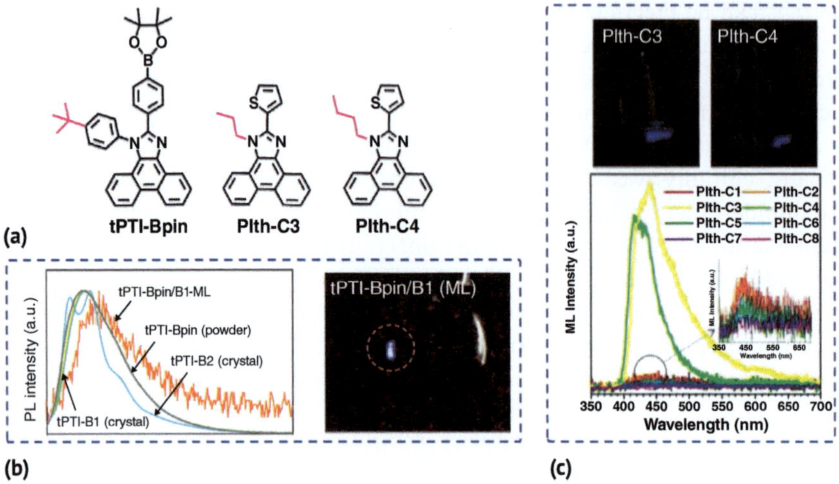

Figure 5.14 (a) Molecular structures of tPTI-Bpin, PIth-C3, and PIth-C4. (b) ML and PL properties of two crystal forms (tPTI-B1 and tPTI-B2) of tPTI-Bpin. *Source*: Reproduced with permission from [51]/The Royal Society of Chemistry. (c) ML spectra of PIth-C3 and PIth-C4, with their ML photographs. *Source*: Reproduced with permission from [52]/Royal Society of Chemistry.

packing, collectively providing a structural foundation for ML generation. In the tPTI-B1 polymorph, molecules adopt a highly twisted conformation and form a unique triangular-like packing arrangement. This configuration strengthens intermolecular forces and creates a favorable structural environment for charge separation under mechanical stimulation, thereby facilitating ML emission. In contrast, molecules in the tPTI-B2 polymorph arrange in parallel. This packing lacks the constraint of strong intermolecular interactions, leading to facile intermolecular slippage under external force. Such slippage causes energy dissipation via nonradiative transitions within the crystal, which fails to satisfy the structural requirements for ML and results in the absence of mechanoluminescent activity.

To expand this class of ML materials, Li et al. introduced thiophene units into the phenanthroimidazole scaffold, constructing a novel aromatic core structure. Based on this design, they synthesized a series of derivatives with alkyl chain lengths ranging from methyl to octyl (C1–C8) (Figure 5.14a, c).[52] The strategic introduction of thiophene aimed to leverage its electronic properties and spatial configuration (electron-rich, planar structure), inducing diverse molecular conformations while regulating crystal packing modes. Among these derivatives, only PIth-C3 (propyl-substituted) and PIth-C4 (butyl-substituted) exhibited distinct ML activity; no ML phenomenon was observed in the others. Single-crystal X-ray diffraction analyses of PIth-C3 and PIth-C4 revealed that their molecular dimers adopt a highly twisted packing mode. This nonparallel arrangement reduces nonradiative transition losses under mechanical stimulation. Additionally, strong C—H⋯S interactions within the dimers enhance crystal structural rigidity, creating favorable conditions for charge separation upon crystal fragmentation. These

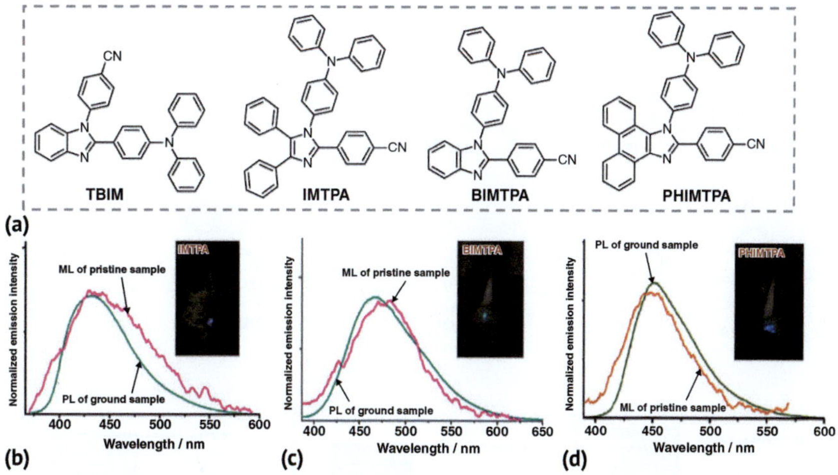

Figure 5.15 (a) Molecular structures of imide derivatives with ML properties. (b–d) ML spectra of IMTPA, BIMTPA, and PHIMTPA, and corresponding ML images. *Source*: Reproduced with permission from [54]/Royal Society of Chemistry/CC BY-NC 3.0.

specific packing modes and intermolecular forces effectively stabilize the crystal structure, ensuring PIth-C3 and PIth-C4 possess both ML activity and efficient luminescence.

Xia et al. designed a series of ML compounds by integrating imidazole derivatives with TPA units (Figure 5.15a). In 2019, they reported two polymorphs of TBIM: $TBIM_G$ and $TBIM_B$, which exhibit deep-blue fluorescence (at 435 nm) and green fluorescence (at 505 nm), respectively.[53] Upon mechanical stimulation, $TBIM_B$ emitted blue ML at 432 nm, and $TBIM_G$ produced green ML at 500 nm. In 2020, they further developed three imidazole-based ML molecules, IMTPA, BIMTPA, and PHIMTPA, by strategically introducing 4-cyanophenyl and TPA groups into the imidazole, respectively.[54] Despite these compounds crystallizing in centrosymmetric space groups, they retained ML activity (Figure 5.15a–d). Structural analyses revealed that molecules in the crystals assemble into molecular pairs, and these pairs adopt a nonantiparallel arrangement. This specific configuration gives rise to substantial dipole moments and facilitates intermolecular charge transfer. These observations collectively indicate that the presence of molecular pairs with sufficiently large dipole moments is the critical factor underlying the ML activity of these centrosymmetric crystals.

Subsequently, they further introduced the imidazole units to the TPA, synthesizing compounds IMTPA-1, IMTPA-2, and IMTPA-3 (Figure 5.16a).[55] For the IMTPA-1 crystal, a faint emission was barely observable even under heavy grinding. In contrast, intense ML was observed when gently grinding IMTPA-2 and IMTPA-3 crystals (Figure 5.16b–d). The enhanced force responsiveness of IMTPA-2 and IMTPA-3 stems from two key structural features: the "engine-shaped" geometry of the TPA core and the blade-like aryl groups on the imidazole units. Together, these characteristics endow them with substantial dipole moments that drive the formation of highly polar molecular pairs, ultimately

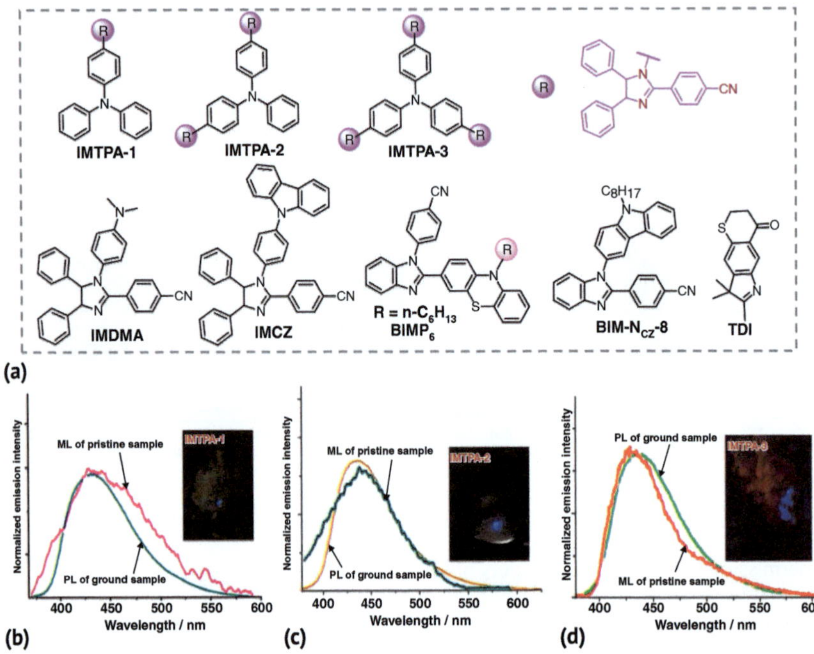

Figure 5.16 (a) Molecular structures of N-containing heterocyclic compounds with ML properties. (b–d) PL and ML spectra of IMTPA-1, IMTPA-2, and IMTPA-3, and corresponding ML images. *Source*: Reproduced with permission from [55]/Elsevier.

leading to the remarkable ML performance. Later, they introduced dimethylaminophenyl and N-carbazolyphenyl groups onto the imidazole core, respectively, yielding two additional ML-active materials (IMDMA, IMCZ, Figure 5.16a).[56] In the crystals of these compounds, the significantly higher dipole moments of the dimers contribute substantially to the polarizability of the corresponding crystals, which is regarded as the intrinsic origin of their ML activity. Separately, Xia et al. incorporated phenothiazine into the imidazole as an electron donor.[57] By adjusting the alkyl chain length on the phenothiazine moiety, they synthesized a series of imidazole derivatives. Among these, the BIMP6 (Figure 5.16a) crystal exhibited ML activity: when lightly pressed, it emitted a bright greenish-yellow ML that was clearly visible to the naked eye.

5.3 Organic Crystals with Phosphorescent Mechanoluminescence

Pure organic compounds typically exhibit a small SOC coefficient, which forbids the transition between their singlet and triplet states. Meanwhile, triplet excitons in organic compounds show remarkably poor stability: they tend to dissipate via nonradiative transitions and are readily quenched by atmospheric oxygen. Consequently, phosphorescence from organic compounds is barely detectable under room-temperature conditions. However, in the aggregated state, nonradiative molecular motion is suppressed and oxygen is isolated. Further introduction of groups with lone-pair electrons or heavy atoms can significantly promote the generation of triplet excitons, thus enabling the emergence of RTP. From the perspective of its generation mechanism, ML also belongs to the category of aggregated-state luminescence. External mechanical stimulation induces electron bombardment at the fractured interface of the crystal, generating excitons at the organic luminescent centers, ultimately yielding mechanophosphorescence.

Actually, in 1974, Zink et al. first reported organic phosphorescent ML compounds phthalic anhydride and acenaphthene,[58] whose ML spectra consisted of both fluorescence and phosphorescence. Recently, lots of organic phosphorescent ML compounds were reported with novel emission properties. Li et al. found an organic compound of DPP-BO in 2017, exhibiting dual ML bands consisting of fluorescence (around 350 nm) and phosphorescence (around 450 nm) (Figure 5.17a).[59] DPP-BO crystal is in the noncentrosymmetric space group of *Cmc*21, and abundant intermolecular hydrogen bond interaction was found in the crystal, which contributes to enhance the ISC effect. However, the phosphorescence cannot be observed by photo irradiation at room temperature. Phenothiazine derivatives of CzS-CH_3 and CzS-C_2H_5 (Figure 5.17b) exhibited more remarkable phosphorescence properties, and their RTP can be detected by UV light irradiation.[60] Their ML were also composed of fluorescence and phosphorescence, but the phosphorescence cannot be observed after ceasing mechanical stimuli because of the short phosphorescence lifetime.

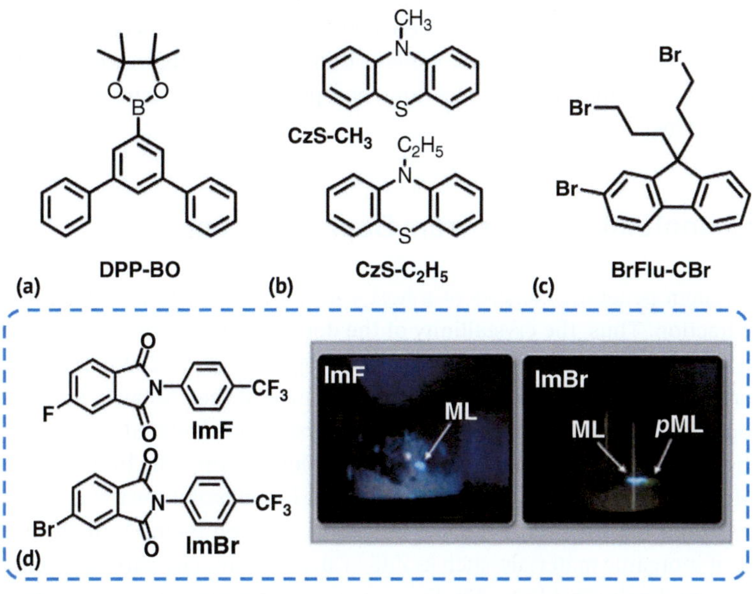

Figure 5.17 (a) The Molecular structures of DPP-BO; (b) CzS-CH_3 and CzS-C_2H_5; (c) BrFlu-CBr; (d) ImF and ImBr as well as corresponding ML photographs. *Source*: Reproduced with permission from [13]/John Wiley & Sons.

Organic compounds with relatively long RTP lifetimes would produce persistent ML at room temperature. In 2018, Li et al. introduced the Br atom to BrFlu-CBr to enhance the SOC effect.[61] As a result, intense RTP located at 480 nm with an efficiency of 4.56% was found for the BrFlu-CBr crystal. Upon mechanical grinding, cyan ML, composed of both fluorescent and phosphorescent components, was initially observed; this ML color subsequently shifted to green-white, implying the presence of phosphorescent ML. After continuous grinding of BrFlu-CBr powder, only dim blue ML was observed (Figure 5.17c); this further indicates that the molecular packing within the crystal plays a crucial role in regulating the phosphorescence intensity. Trifluoromethylphenyl-substituted phthalimide derivatives were reported to have bright ML, due to the dipolar geometry and the asymmetric crystal structure. In 2018, Chi et al. found simultaneously RTP and ML properties in two phthalimide derivatives, namely ImF and ImBr. ImF exhibited blue fluorescence (peaks at 455 nm) and RTP (peaks at 555 nm) with a lifetime of 35.3 ms.[13] For ImBr, dual PL bands were also observed: one corresponds to blue fluorescence (centered at 457 nm, with a lifetime of 2.6 ns), and the other to intense RTP (with an ultralong lifetime of 102.1 ms). More interestingly, under UV light irradiation, the RTP intensity of ImBr gradually increased to a maximum within approximately 0.8 s, accompanied by a color shift from blue to white (Figure 5.17d). After removing the UV light, yellow RTP emission was detected, confirming the occurrence of tricolor emission. Under mechanical grinding, ImF exhibited blue mechanoluminescence (ML), and its ML spectrum was analogous to its PL spectrum. For ImBr, blue ML was observed at the force-applied site; notably, a yellow afterglow ribbon persisting for 1 s was detected at the postgrinding position, indicating the presence of persistent ML (p-ML). The dynamic ML spectra were in good agreement with these phenomena: the ML spectrum at the force-applied site was like that of ImF, while a transition from fluorescence-only emission to dual emission (fluorescence and phosphorescence) occurred within approximately 60 ms.

5.4 Mechanoluminescence from Doping System

The doping system generally consists of a host with a major fraction and a guest with a trace fraction. Thus, the crystallinity of the doping system is like that of the host, which determines the basic photophysical properties of ML. Whereas the guest plays a crucial role in the local microstructure phase of the crystal and photophysical properties of PL and ML. When the doping system is composed of a nonluminescent host and a luminescent guest, the guest would become the luminescent center for ML. As for a doping system composed of a luminescent host and guest, ML property would rely on energy transfer efficiency between them. Actually, the ML property of inorganic materials, such as ZnS : Cu, ZnS : Mn, rare earth element-doped ceramic materials ($Sr_3Al_2O_6$: Eu, Dy), is mainly determined by the dopants. The role of guest on organic ML was also investigated in the form of impurities.

Wint-O-Green Lifesavers, which consist of sucrose and the additive ethyl salicylate, can produce brighter blue ML rather than the deep-blue ML of pure sucrose.[1] Harvey et al. further investigated the ML properties of sugars with different flavors. Specifically, wintergreen-flavored sugar emitted bright greenish ML; in contrast, no ML was detected for chocolate-, licorice-, and cinnamon-flavored sugars, while only faint ML was observed for lemon-, lime-, and clove-flavored sugars.

Owing to the lack of systematic design strategies, organic emitters exhibiting bright ML remain relatively scarce. This scarcity not only results in limited ML color diversity but also restricts their practical applications. Host-guest doping offers a feasible solution to this dilemma, as it circumvents the need for complex molecular design and synthesis. Specifically, the host molecules with deep-blue ML can act as an excitation source, while guest molecules with distinct emission colors enable the generation of ML with varied hues. In 2018, Yang et al. observed intense blue ML from crystals of N-phenylcarbazole (NPC) (Figure 5.18a) under mechanical stimuli.[41,62] Bright ML was also detectable from bulk crystals prepared via melting and sublimation. The ML property of NPC is attributed to two key factors: its piezoelectric space group Fdd2 and intense noncovalent interactions within the crystal lattice. Given NPC's high thermal stability, high crystallinity, and moderate melting point (91 °C), they further explored its application as an ML host. In its molten liquid state, NPC can act as a nonpolar organic solvent, capable of dissolving or dispersing organic fluorescent dyes. Four dyes with green, yellow, orange, and red PL were doped into NPC at 110 °C at a concentration of 4–5 wt%; subsequent cooling yielded small nanocrystals. The absorption spectra of the four dyes overlapped well with the ML spectrum of NPC, indicating that ML-induced energy transfer between NPC (host) and the dyes (guests) is feasible. Notably, the host-guest blends exhibited bright green, orange, and red ML, with emission profiles analogous to their respective PL spectra.

In 2018, Chi et al. also realized host-guest ML with a similar approach. They reported two alkyl-substituted carbazole derivatives (Cz-alkyl-6 and Cz-alkyl-7, Figure 5.18a): although the Cz-alkyl-6 crystal has a noncentrosymmetric structure and the Cz-alkyl-7 crystal adopts a centrosymmetric structure, both can emit bright deep-blue ML.[63] The long alkyl chains in the crystals enhance the material's sensitivity to pressure, effectively lowering the pressure threshold required for ML generation. Meanwhile, their low melting points (68.8 °C and 35.9 °C, respectively) endow them with excellent crystallinity, allowing their ML performance to be easily recovered via heating-cooling cycles. Notably, Cz-alkyl-6 even exhibits intense ML during the crystallization process; this phenomenon is known as crystalloluminescence. In further studies, various organic dyes with different luminescent colors were doped into Cz-alkyl-6. The resulting composite systems showed ML covering the entire visible light range from blue to red. Moreover, the ML spectra of these composites exhibited dual emission peaks, consisting of the blue light from the host Cz-alkyl-6 and the fluorescence from the guest dye molecules.

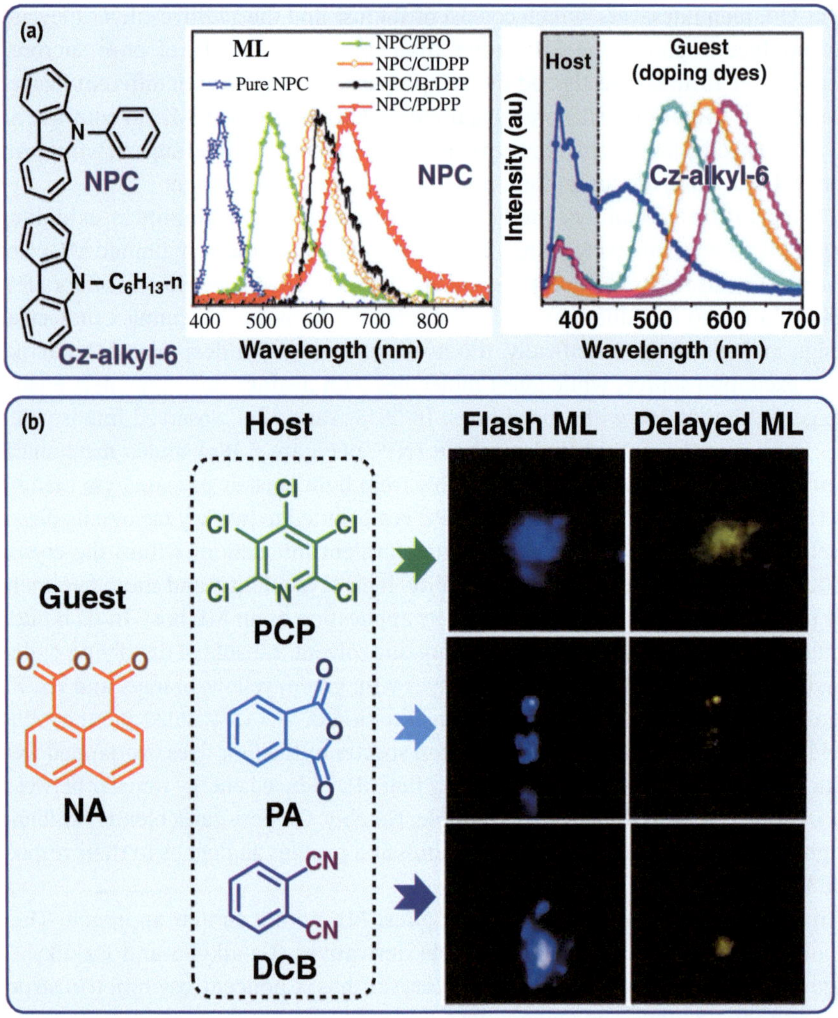

Figure 5.18 (a) The molecular structures of NPC and Cz-alkyl-6/7 and corresponding ML and doped ML spectra with different dyes. *Source*: Reproduced with permission from [62]/The Royal Society of Chemistry. *Source*: Reproduced with permission from [63]/John Wiley & Sons. (b) Molecular structures of the guest of NA and the host of PCP, PA, and DCB, as well as photographs of flash ML and delayed ML. *Source*: [64]/Springer Nature/CC BY 4.0.

In 2019, Tang et al. reported a novel and interesting doping system with ML property. They found bright green, deep-blue, and blue ML from host compounds of PCP, PA, and DCB, as well as RTP of PCP and DCB around 515 and 550 nm (Figure 5.18b).[64] When the guest molecule of NA was doped into PCP, PA, and DCB with the ratio of 1 wt% via melt-casting, grinding, or evaporation, similar yellow RTP around 540 nm was observed from the three blend systems with lifetimes of 363, 492, and 603 ms, respectively. Under the crushing process of solid NA/PCP, NA/PA, and NA/DCB obtained by melt-casting, a prompt bright ML with delayed yellow ML that can last seconds was found. Thus, the RTP of three blends was excited by ML, demonstrating long lifetime phosphorescent ML.

5.5 Special Mechanoluminescence Phenomena

The excitation source for ML is mechanical stimulus, but the stimulus is not limited to grinding, bending, cracking, extrusion, hot/cold shock, crystallization, etc. The noncontact excitation for ML can also produce ML. For example, ML caused by rapid crystallization mentioned above is generated by noncontact excitation, but the high requirements on crystallization speed limit the number of such compounds. Whereas, it has been found that ML can be excited by multiple nonmechanical stimuli, such as ultrasound and laser. Suslick et al. found that when liquids are irradiated by high-intensity ultrasonics, the bubbles generated by acoustic cavitation are formed in the liquids, followed by growing and collapsing sharply, creating extreme conditions of thousands of degrees Celsius and hundreds of atmospheres in the bubbles.[65,66] A strong shock wave from the collapsed bubble spreads outward at speeds faster than the speed of sound. Hence, micrometer-scale solid particles suspended in the liquid are accelerated to very high speeds by the shock wave, even reaching the order of 100 m/s. Compared to manual grinding, which can only produce a speed of tens of meters per second, the high speed generated by ultrasonics can cause the strong collision of solid particles. The ML intensity produced by ultrasonic treatment of sucrose powder in dodecane solution was an order of magnitude higher than ML produced by grinding sucrose in dodecane solution, and their ML spectra were the same as that produced by grinding sucrose in air. When the ML compound of resorcinol was crushed in air, the ML spectrum consisted of nitrogen discharge and PL of resorcinol. In contrast, in a hexadecane solution saturated with a helium and argon atmosphere, the suspension of resorcinol produced strong helium and argon discharge spectral lines and weak crystal luminescence under ultrasonic action. Ultrasonic is undoubtedly a stronger excitation source for ML than mechanical grinding. Accordingly, Xu et al. have verified that the ultrasonic wave induced ML (USML) for nondestructive and noninvasive mechanical stimulation in the body, which can act as an excitation source for TiO_2 in photodynamic therapy, exhibiting the wide application prospect of USML.[67]

Another effective noncontact excitation for ML is a laser. Under the irradiation of a high-power laser, the absorption of the laser energy will lead to an increase in the temperature of the materials. The rapid variation in temperature would promote the formation of thermal stress; the faster the temperature increase, the greater the stress. In 1979, Hardy et al. successfully reported ML from coumarin, sodium chloride, and potassium chloride by excitation of a laser pulse of Nd^{3+} : YAG laser (1060 nm at a time of 20 ns, with the pulse energy of 0.5~4 J cm^{-2}).[68] In 2003, Kitamura et al. further developed the laser-driven shock wave (LISW) method, which uses an Nd^{3+}:YAG laser (1064 nm, 10 ns) to irradiate N-isopropyl carbazole (NIPCz) crystal to generate shock waves.[69] ML was generated when the crystal was broken, with the ML spectrum identical to the PL spectrum of the NIPCz crystal. The LISW method was proved to effectively improve the ML production efficiency. Compared to manually grinding glucose crystals, the ML intensity produced by the LISW method increased 10^5 times (Figure 5.19a).[70] In addition, due to the short laser pulse time, ML can be utilized for analysis of the dynamic characteristics of

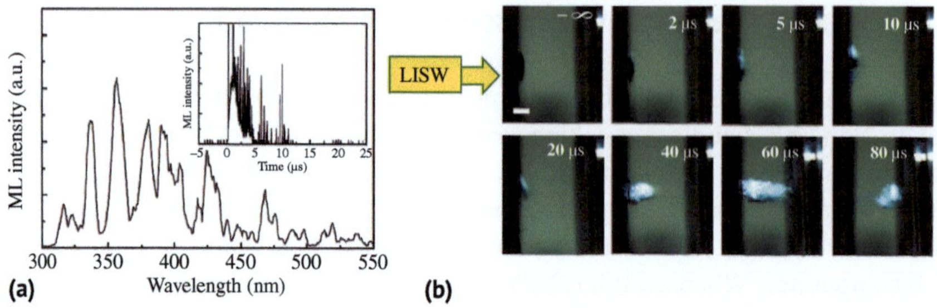

Figure 5.19 (a) ML spectrum of amorphous glucose excited by a laser pulse of Nd^{3+}:YAG laser (1064 nm, 10 ns); the inset shows the temporal profile ML spectrum. (b) The images of fracture dynamics caused by laser-driven shock wave (LISW), the $-\infty$ means before application of the shock wave. *Source*: Reproduced with permission from [70]/with permission of American Chemical Society.

crystal fracture, which was found closely related to the transient ML intensity. It is found that ML is generated after 200 ns laser irradiation (Figure 5.19b), and the ML intensity can be quantitatively controlled by the laser power, which provides an effective method for noncontact semiquantitative study of the ML phenomenon.

There is also a very fancy way to produce ML, that is, the glow can be observed by peeling tape. The daily use tape is attached to the surface of the polyethylene film by a viscous acrylic adhesive. Peeling off the pressure-sensitive tape requires the overcoming of the Van der Waals interaction between the interfaces. From the perspective of bond energy, the Van der Waals force is the weakest among chemical bonds, which is nearly one hundredth of the energy required to generate visible light. However, Camara et al. reported that peeling tape can produce light that is visible to the naked eye.[71] It was found that the transient emission of visible light and X-ray photons were generated during the process of peeling tape. The blue glow was generated by the scintillator after absorbing high-energy X-rays, while the red glow at the tape stripping point was the enhanced ML under a neon atmosphere. When the pressure of the device is reduced to 10^{-3} Torr, the enhanced high-energy rays can generate a glow that even illuminates the entire scintillator. Based on the piezoelectric theory, Camara et al. proposed a reliable mechanism: during the peeling process, the viscous acrylic adhesive was positively charged, while the polyethylene tape was negatively charged. Hence, the electric field was generated by peeling tape. When the accumulated positive charges collide with the negative charges at low pressure, bremsstrahlung X-rays are produced.

5.6 Summary

In summary, organic ML has undergone rapid development in recent years, with the continuous discovery of organic ML compounds featuring diverse photophysical properties. The generation of ML strongly depends on molecular packing modes and intermolecular interactions—such as hydrogen bonding, π–π stacking, and electrostatic interactions—while the ML properties of these compounds bear significant similarity to the PL of their crystalline states. Based on luminescence characteristics, ML can be categorized into fluorescence, phosphorescence, and long-lived phosphorescence. In this chapter, we have briefly reviewed the mechanisms of ML, organic ML compounds with fluorescent and phosphorescent properties, ML derived from doped systems, and special ML phenomena. In-depth understanding of the fundamental mechanisms underlying organic ML not only advances basic luminescence theory but also enhances our comprehension of the photophysical properties of organic compounds in the aggregated state. As more ML compounds with distinct photophysical behaviors are discovered, mechanical stimuli are expected to demonstrate prospects comparable to those of light, electrical, and bioenergy excitation sources.

References

1 Walton, A.J. (1977). Triboluminescence. *Adv. Phys.* 26: 887–948.

2 Chen, B., Zhang, X., and Wang, F. (2021). Expanding the toolbox of inorganic mechanoluminescence materials. *Acc. Mater. Res.* 2: 364–373.

3 Xie, Y. and Li, Z. (2018). Triboluminescence: Recalling interest and new aspects. *Chem* 4: 943–971.

4 Jha, P. and Chandra, B P. (2014). Survey of the literature on mechanoluminescence from 1605 to 2013. *Luminescence* 29: 977–993.

5 Peng D., Chen B., and Wang, F. (2015). Recent advances in doped mechanoluminescent phosphors. *ChemPlusChem* 80: 1209–1215.

6 Luo, J., Xie, Z., Lam, J.W.Y. et al. (2001). Aggregation-induced emission of 1-methyl-1,2,3,4,5-pentaphenylsilole. *Chem. Commun.* 1740–1741.

7 Mei, J., Leung, N.L., Kwok, R.T. et al. (2015). Aggregation-induced emission: Together we shine, United we soar! *Chem. Rev.* 115: 11718–11940.

8 Xu, S., Liu, T., Mu, Y. et al. (2015). An organic molecule with asymmetric structure exhibiting aggregation-induced emission, delayed fluorescence, and mechanoluminescence. *Angew. Chem. Int. Ed.* 54: 874–878.

9 Yang, J., Fang, M., and Li, Z. (2021). Stimulus-responsive room temperature phosphorescence materials: Internal mechanism, design strategy, and potential application. *Acc. Mater. Res.* 2: 644–654.

10 Wang, C., Li, X.-L., Gao, Y. et al. (2017). Efficient near-infrared (NIR) organic light-emitting diodes based on donor-acceptor architecture: An improved emissive state from mixing to hybridization. *Adv. Opt. Mater.* 5: 1700441.

11 Wong, M.Y. and Zysman-Colman, E. (2017). Purely organic thermally activated delayed fluorescence materials for organic light-emitting diodes. *Adv. Mater.* 29: 1605444.

12 An, Z., Zheng, C., Tao, Y. et al. (2015). Stabilizing triplet excited states for ultralong organic phosphorescence. *Nat. Mater.* 14: 685–690.

13 Li, J.A., Zhou, J., Mao, Z. et al. (2018). Transient and persistent room-temperature mechanoluminescence from a white-light-emitting AIEgen with tricolor emission switching triggered by light. *Angew. Chem. Int. Ed.* 57: 6449–6453.

14 Xie, Y. and Li, Z. (2020). The development of mechanoluminescence from organic compounds: breakthrough and deep insight. *Mater. Chem. Front.* 4: 317–331.

15 Zink, J.I. (1975). Triboluminescence of uranyl nitrate. *Inorg. Chem.* 14: 555–558.

16 Hurt, C.R., McAvoy, N., Bjorklund, S., and Filipescu, N. (1966). High intensity triboluminescence in europium tetrakis (dibenzoylmethide)-triethylammonium. *Nature* 212: 179–180.

17 Chandra, B.P., Chandra, V.K., and Jha, P. (2013). Models for intrinsic and extrinsic fracto-mechanoluminescence of solids. *J. Lumin.* 135: 139–153.

18 Dickinson, J.T., Brix, L.B., and Jensen, L.C. (1984). Electron and positive ion emission accompanying fracture of Wint-o-green lifesavers and single-crystal sucrose. *J. Phys. Chem.* 88: 1698–1701.

19 Chandra, B.P., Chandra, V.K., Jha, P. et al. (2012). Fracto-mechanoluminescence and mechanics of fracture of solids. *J. Lumin.* 132: 2012–2022.

20 Fang, M., Yang, J., Liao, Q. et al. (2017). Triphenylamine derivatives: Different molecular packing and the corresponding mechanoluminescent or mechanochromism property. *J. Mater. Chem. C* 5: 9879–9885.

21 Wang, J., Chai, Z., Wang, J. et al. (2019). Mechanoluminescence or room-temperature phosphorescence: Molecular packing-dependent emission response. *Angew. Chem. Int. Ed.* 58: 17297–17302.

22 Dang, Q., Hu, L., Wang, J. et al. (2019). Multiple luminescence responses towards mechanical stimulus and photo-induction: The key role of the stuck packing mode and tunable intermolecular interactions. *Chem. Eur. J.* 25: 7031–7037.

23 Li, D., Yang, J., Wang, Y. et al. (2020). Bright mechanoluminescent luminogens even in daylight through close intermolecular interaction with the characteristic of hybridized local and charge transfer (HLCT). *J. Mater. Chem. C* 8: 10852–10858.

24 Tu, J., Fan, Y., Wang, J. et al. (2019). Halogen-substituted triphenylamine derivatives with intense mechanoluminescence properties. *J. Mater. Chem. C* 7: 12256–12262.

25 Yu, Y., Wang, C., Wei, Y. et al. (2019). Halogen-containing TPA-based luminogens: Different molecular packing and different mechanoluminescence. *Adv. Optical Mater.* 18: 1900505.

26 Xu, B., He, J., Mu, Y. et al. (2015). Very bright mechanoluminescence and remarkable mechanochromism using a tetraphenylethene derivative with aggregation-induced emission. *Chem. Sci.* 6: 3236–3241.

27 Xu, B., Li, W., He, J. et al. (2016). Achieving very bright mechanoluminescence from purely organic luminophores with aggregation-induced emission by crystal design. *Chem. Sci.* 7: 5307–5312.

28 Huang, G., Jiang, Y., Wang, J. et al. (2019). The influence of intermolecular interactions and molecular packings on mechanochromism and mechanoluminescence – a tetraphenylethylene derivative case. *J. Mater. Chem. C* 7: 12709–12716.

29 Jiang, Y., Wang, J., Huang, G. et al. (2019). Insight from the old: mechanochromism and mechanoluminescence of two amine-containing tetraphenylethylene isomers. *J. Mater. Chem. C* 7: 11790–11796.

30 Jiang, Y., Chang, X., Xie, W. et al. (2021). Cyano-containing tetraphenylethene isomers: similar bright mechanoluminescence, but diverse recoverable processes. *Mater. Chem. Front.* 5: 885–892.

31 Wang, C., Xu, B., Li, M. et al. (2016). A stable tetraphenylethene derivative: aggregation-induced emission, different crystalline polymorphs, and totally different mechanoluminescence properties. *Mater. Horiz.* 3: 220–225.

32 Xie, Y., Tu, J., Zhang, T. et al. (2017). Mechanoluminescence from pure hydrocarbon AIEgen. *Chem. Commun.* 53: 11330–11333.

33 Liu, F., Tu, J., Wang, X. et al. (2018). Opposite mechanoluminescence behavior of two isomers with different linkage positions. *Chem. Commun.* 54: 5598–5601.

34 Tu, J., Liu, F., Wang, J. et al. (2019). Fluorine-substituted tetraphenylethene isomers with different triboluminescence properties. *ChemPhotoChem* 3: 133–137.

35 Wang, C., Yu, Y., Chai, Z. et al. (2019). Recyclable mechanoluminescent luminogen: Different polymorphs, different self-assembly effects of the thiophene moiety and recovered molecular packing via simple thermal-treatment. *Mater. Chem. Front.* 3: 32–38.

36 Wang, C., Yu, Y., Yuan, Y. et al. (2020). Heartbeat-sensing mechanoluminescent device based on a quantitative relationship between pressure and emissive intensity. *Matter* 2: 181–193.

37 Inoue, T. and Tazuke, S. (1981). Triboluminescence and tribopolymerization of 9-ethyl-3-vinylcarbazole. *Chem. Lett.* 10: 589–592.

38 Nowak, R., Krajewska, A., and Samoc, M. (1982). Efficient triboluminescence in N-isopropylcarbazole. *Chem. Phys. Lett.* 94: 270–271.

39 Huang, Q., Mei, X., Xie, Z. et al. (2019). Photo-induced phosphorescence and mechanoluminescence switching in a simple purely organic molecule. *J. Mater. Chem. C* 7: 2530–2534.

40 Liu, C., Chen, J., Xu, C. et al. (2020). AIEgens with bright mechanoluminescence and thermally activated delayed fluorescence derived from (9H-carbazol-9-yl)(phenyl)methanone. *Dyes Pigm.* 174: 108093.

41 Zhang, K., Sun, Q., Zhang, Z. et al. (2018). Touch-sensitive mechanoluminescence crystals comprising a simple purely organic molecule emit bright blue fluorescence regardless of crystallization methods. *Chem. Commun.* 54: 5225–5228.

42 Zhang, K., Sun, Q., Tang, L. et al. (2018). Cyclic boron esterification: Screening organic room temperature phosphorescent and mechanoluminescent materials. *J. Mater. Chem. C* 6: 8733–8737.

43 Neena, K.K., Sudhakar, P., Dipak, K., and Thilagar, P. (2017). Diarylboryl-phenothiazine based multifunctional molecular siblings. *Chem. Commun.* 53: 3641–3644.

44 Arivazhagan, C., Maity, A., Bakthavachalam, K. et al. (2017). Phenothiazinyl boranes: A new class of AIE luminogens with mega stokes shift, Mechanochromism, and mechanoluminescence. *Chem. Eur. J.* 23: 7046–7051.

45 Yang, J., Qin, J., Geng, P. et al. (2018). Molecular conformation-dependent mechanoluminescence: Same mechanical stimulus but different emissive color over time. *Angew. Chem. Int. Ed.* 57: 14174–14178.

46 Tian, Y., Yang, X., Gong, Y. et al. (2021). The initial attempt to reveal the emission processes of both mechanoluminescence and room temperature phosphorescence with the aid of circular dichroism in solid state. *Sci. China Chem.* 64: 445–451.

47 Nakayama, H., Nishida, J.-i., Takada, N. et al. (2012). Crystal structures and triboluminescence based on trifluoromethyl and pentafluorosulfanyl substituted asymmetric N-phenyl imide compounds. *Chem. Mater.* 24: 671–676.

48 Nishida, J., Ohura, H., Kita, Y. et al. (2016). Phthalimide compounds containing a trifluoromethylphenyl group and electron-donating aryl groups: Color-tuning and enhancement of triboluminescence. *J. Org. Chem.* 81: 433–441.

49 Chen, Y., Xu, C., Xu, B. et al. (2019). Chirality-activated mechanoluminescence from aggregation-induced emission enantiomers with high contrast mechanochromism and force-induced delayed fluorescence. *Mater. Chem. Front.* 3: 1800–1806.

50 Li, J.-A., Song, Z., Chen, Y. et al. (2021). Colour-tunable dual-mode afterglows and helical-array-induced mechanoluminescence from AIE enantiomers: Effects of molecular arrangement on formation and decay of excited states. *Chem. Eng. J.* 418: 129167.

51 Yu, Y., Fan, Y., Wang, C. et al. (2019). Phenanthroimidazole derivatives with minor structural differences: crystalline polymorphisms, different molecular packing, and totally different mechanoluminescence. *J. Mater. Chem. C* 7: 13759–13763.

52 Yu, Y., Fan, Y., Wang, C. et al. (2021). Achieving enhanced ML or RTP performance: alkyl substituent effect on the fine-tuning of molecular packing. *Mater. Chem. Front.* 5: 817–824.

53 Liu, X., Jia, Y., Jiang, H. et al. (2019). Two polymorphs of triphenylamine-substituted Benzo[d]imidazole: Mechanoluminescence with different colors and mechanofluorochromism with emission shifts in opposite direction. *Acta Chim. Sinica* 77: 1194–1202.

54 Liu, X.J., Gao, G.L., Jiang, H. et al. (2020). Deep insight into reasons for the mechanoluminescence phenomenon of triphenylamine-substituted imidazoles and their distinct mechano-fluorochromic behaviours. *RSC Adv.* 10: 23187–23195.

55 Jia, Y.-R., Jiang, H., Gao, G.-L. et al. (2021). Derivation-density-engineered molecular packing, emission property, mechanofluorochromism and mechanoluminescence of triarylimidazole-substitued triphenylamines. *Dyes Pigm.* 194: 109541.

56 Gao, G.-L., Jia, Y.-R., Jiang, H., and Xia, M. (2021). Bright mechanoluminescence and dramatically different mechanofluorochromic behaviors of tetraarylimidazoles by slight modulation of substituent structure. *Dyes Pigm.* 186: 109030.

57 Jiang, H., Liu, X.J., Jia, R.R. et al. (2019). Side chain effects on the solid-state emission behaviours and mechano-fluorochromic activities of 10H-phenothiazinylbenzo[d]imidazoles. *RSC Adv.* 9: 30381–30388.

58 Zink, J.I. (1974). Tribophosphorescence from nonphotophosphorescent crystals. *J. Am. Chem. Soc.* 96: 6775–6777.

59 Yang, J., Ren, Z., Xie, Z. et al. (2017). AIEgen with fluorescence-phosphorescence dual mechanoluminescence at room temperature. *Angew. Chem. Int. Ed.* 56: 880–884.

60 Yang, J., Gao, X., Xie, Z. et al. (2017). Elucidating the excited state of mechanoluminescence in organic luminogens with room-temperature phosphorescence. *Angew. Chem. Int. Ed.* 56: 15299–15303.

61 Wang, J., Wang, C., Gong, Y. et al. (2018). Bromine-substituted fluorene: Molecular structure, Br-Br interactions, Room-temperature phosphorescence, and tricolor triboluminescence. *Angew. Chem. Int. Ed.* 57: 16821–16826.

62 Sun, Q., Zhang, K., Zhang, Z. et al. (2018). A simple and versatile strategy for realizing bright multicolor mechanoluminescence. *Chem. Commun.* 54: 8206–8209.

63 Li, W., Huang, Q., Mao, Z. et al. (2018). Alkyl chain introduction: In situ solar-renewable colorful organic mechanoluminescence materials. *Angew. Chem. Int. Ed.* 57: 12727–12732.

64 Zhang, X., Du, L., Zhao, W. et al. (2019). Ultralong UV/mechano-excited room temperature phosphorescence from purely organic cluster excitons. *Nat. Commun.* 10: 5161.

65 Eddingsaas, N.C. and Suslick, K.S. (2006). Mechanoluminescence: Light from sonication of crystal slurries. *Nature* 444: 163.

66 Nathan C. and Eddingsaas, K.S.S. (2007). Intense mechanoluminescence and gas phase reactions from the sonication of an organic slurry. *J. Am. Chem. Soc.* 129: 6718–6719.

67 Terasaki, N., Yamada, H., and Xu, C.-N. (2013). Ultrasonic wave induced mechanoluminescence and its application for photocatalysis as ubiquitous light source. *Cataly. Today* 201: 203–208.

68 Hardy, G.E., Chandra, B.P., and Zink, J.I. (1979). Pulsed infrared laser induced visible luminescence. *J. Am. Chem. Soc.* 101: 2787–2788.

69 Tsuboi, Y., Seto, T., and Kitamura, N. (2003). Laser-driven shock wave-induced triboluminescence of an organic crystal: Toward a semiquantitative study. *J. Phys. Chem. B* 107: 7547–7550.

70 Tsuboi, Y., Seto, T., and Kitamura, N. (2008). Laser-induced shock wave can spark triboluminescence of amorphous sugars. *J. Phys. Chem. A* 112: 6517–6521.

71 Camara, C.G., Escobar, J.V., Hird, J.R., and Putterman, S.J. (2008). Correlation between nanosecond X-ray flashes and stick–slip friction in peeling tape. *Nature* 455: 1089–1092.

6

Molecular Aggregates as Charge Transport Layers in Perovskite Solar Cells

Shuyan Shao[1] and Zhen Li[1,2]*

[1] *Institute of Molecular Aggregation Science, Tianjin University, Tianjin, China*
[2] *Department of Chemistry, Wuhan University, Wuhan, China*

*Corresponding author: shuyan_shao@tju.edu.cn

6.1 Introduction

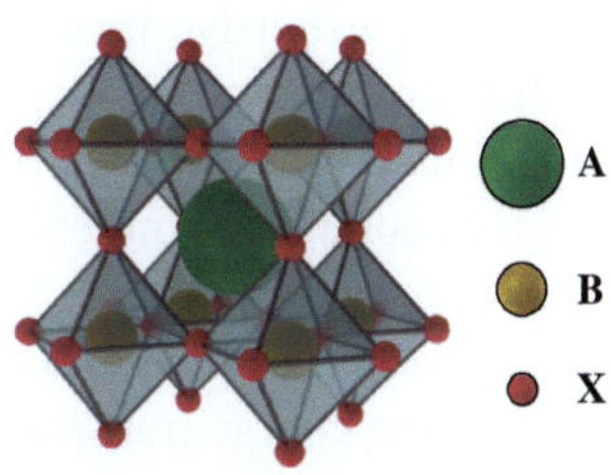

Figure 6.1 The crystal structure of the 3D metal halide perovskite.

Metal halide perovskites (HP) form a corner-shared ABX_3 (3D) crystal structure as shown in Figure 6.1, when the size of the A, B, X ions collectively meets the requirements defined by the empirical tolerance factor and the octahedra factor. To date, very limited elements with appropriate size have been found to fit the A, B, X sites of the 3D metal halide perovskites. The monovalent organic or inorganic cation such as Cs^-, $CH_3NH_3^+$, or $CH(NH_2)^+$ sits at the cavity (A site); the divalent metal cation such as Pb^{2+}, Sn^{2+}, or Ge^{2+} occupies the body center (B site) of the octahedra; and halide anions (Cl^-, Br^-, I^-) sit at the corner (X site) of the octahedra.[1–6] These perovskite materials have excellent photophysical and electrical properties, such as high absorption coefficient, tunable optical bandgaps, small exciton binding energy, long charge carrier diffusion length, and balanced charge carrier transport. Moreover, these materials are compatible with the low cost and massive production owing to their good solution processability. These characteristics of metal halide perovskites trigger intensive research interest to explore their potential as light absorbing layers for high-efficiency and low-cost photovoltaic technology.

In 2009, Miyasaka and coworkers carried out the pioneering research in using the perovskite materials ($CH_3NH_3PbBr_3$ and $CH_3NH_3PbI_3$) as the sensitizer in liquid dye-sensitized solar cells (DSSCs).[2] However, the devices were unstable and lasted for just a few minutes owing to the dissolution of the perovskites in the liquid electrolytes.[2] In 2011, Park et al. further improved the PCE of the liquid-based DSSCs to 6.5% by modifying the surface of TiO_2 and the deposition method for $MAPbI_3$.[7] Their results further pointed to the common instability issue of the metal halide-based DSSCs, which persists as long as the polar liquid electrolytes were used. However, the same group achieved a milestone result one year later by replacing the liquid electrolyte with a solid-state 2,2′,7,7′-tetrakis(N,N-di-p-methoxyphenylamine)-9,9′-spirobifluorene (Spiro-OMeTAD) hole-transporting layer (HTL), which significantly improved not only the efficiency, but also the stability of the perovskite solar cells (HPSCs).[8] These first solid-state mesoscopic heterojunction HPSCs using sub-micrometer thick mesoporous TiO_2 film (0.6–1 μm) as electron transport layer (ETL), $MAPbI_3$ as light harvesting layer, and Spiro-OMeTAD as HTL delivered a PCE of 9.7% and survived 500 h in air. In a similar period, Snaith and coworkers independently reported a PCE of over 10% using Al_2O_3 as the scaffold. These achievements have triggered great passion of the researchers to advance the HPSC technology from 2012 onward.[9] To date, lead-based single junction HPSCs have achieved a high efficiency of 27.02% by extensive investigation in the new device architectures, deposition techniques, design and optimization of the perovskite materials, passivation techniques, hole and electron transport materials.[10]

An important lesson learned from the history of HPSCs is that the solid-state hole and electron transport layers are indispensable for almost all the high-performance HPSCs. Fundamentally, the chemical structures and the aggregated state of the charge transport materials determine the intra- and inter-molecular electrical couplings and, therefore, the optical and electrical properties. Therefore, these interfacial materials and the corresponding interfaces play a very important part in the operation of the HPSCs, determining their efficiency and lifetime. The absence of the ETL or HTL on the other side quite often causes low efficiency and fast degradation of the HPSCs due to the poor charge extraction and facile oxygen/water invasion.

In the past several years, great efforts have been made to explore both organic and inorganic charge transport layers in HPSCs. To date, only very limited options have been found for the inorganic hole (NiO, CuI, CuSCN) and electron (TiO_2, ZnO, SnO_2) charge transport materials.[11–16] Compared to the inorganic materials, organic materials have abundant structural diversity tailored by the chemical synthetic tools, thus fine-tuning of the optical, electrical and chemical properties of the charge transport materials. Organic molecules possess many intermolecular interactions in the solid state such as van der Waals force, hydrogen bonding, dipole–dipole interactions, or π–π interactions. Their molecular conformation and stacking behaviors in thin films thus directly determine the intra- and intermolecular charge transfer. Planar conjugated systems can enhance intermolecular electronic coupling through effective p-orbital overlapping, thereby facilitating charge transfer between the molecules. To meet the need for cost-effective solution-processing techniques, researchers have been keen to develop organic semiconductors with careful molecular engineering on conjugate backbone, side chains, or linkage to modulate the intermolecular interactions and thus their molecular orientation and aggregation behaviors in solid states. Importantly, the introduction of heteroatoms into conjugated systems has proved effective in inducing noncovalent interactions for π–π stacking in thin films, which is beneficial for the charge transport in opto-electronic devices. The molecular orientation, crystallinity, and solubility of organic semiconductors should be carefully controlled since they are closely related to the hole mobility and film-forming ability.

Organic transport materials, including small molecules and polymers, have also been developed. Spiro-OMeTAD and PEDOT: PSS are the first and still most frequently used solid-state organic HTLs in conventional and inverted structures, respectively. Stimulated by the success of the spiro-OMeTAD, a lot of organic materials including small molecule and polymers have been developed. Since the p-type dopants are often hydrophobic and thus harmful for the lifetime of the HPSCs, intensive research attention has shifted to improve the hole mobility/conductivity of the organic molecules by tailoring their structures to avoid the use of p-dopants. Quite some dopant-free organic hole transport materials have been used to fabricate the inverted HPSCs, reaching a highest PCE over 24%.[17] Unlike the inorganic ETLs, organic ETLs are mostly used in inverted structures, in which fullerene C_{60} and phenyl-C_{61}-butyric acid methyl ester (PCBM) are the most popular organic ETLs. Since C_{60} is insoluble in most of the organic solvents, some fullerene-based derivatives with different functional side chains such as $PC_{70}BM$, 1′,1″,4′,4″-tetrahydro-di[1,4]methanonaphthaleno[5,6]fullerene-C60 (ICBA), a fulleropyrrolidine with an electron-donating triethylene glycol monoethyl ether side chain (PTEG-1) have also been incorporated into the HPSCs.[18–20] However, the hydrophilic PCBM cannot fully inhibit the invasion of water into the perovskite film and therefore produces unstable HPSCs. To overcome the disadvantages of the fullerene-based ETLs, non-fullerene derivatives ETLs with higher water resistance have also been developed for improving the stability of the solar cells in a humid environment.

This chapter provides an overview of the recent advances in the chemical engineering of the small molecule-based HTLs and ETLs (Figure 6.2), and at the same time summarizes the fundamental understanding of how the structures and aggregated states of the organic hole/electron transport materials influence the opto-electronic properties as well as the performance of the HPSCs. It begins with an introduction of the device structures and the working principle of the HPSCs, and then reviews the previous studies on the organic HTL and ETLs. Last, it highlights the key challenges for designing good-performing charge transport layers toward highly efficient and stable HPSCs, and provides an outlook for future directions.

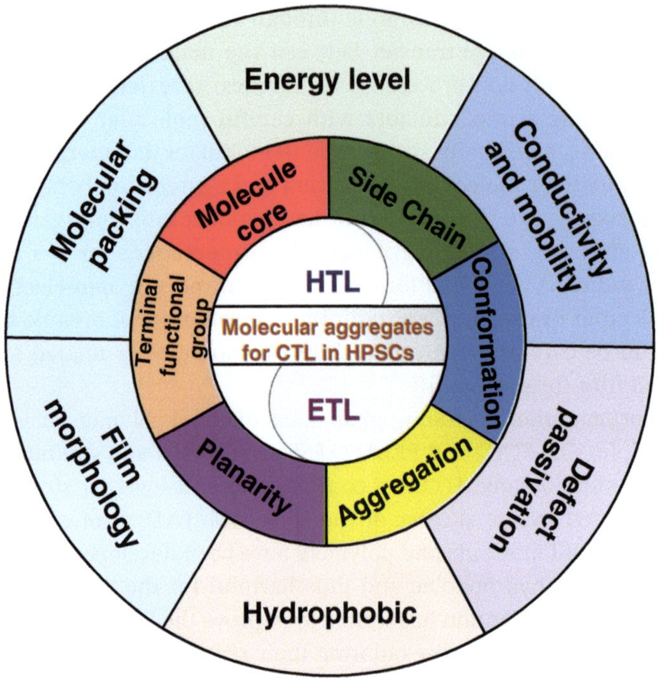

Figure 6.2 An illustrated summary of this chapter: molecular aggregates for charge transport layer (CTL: hole transport layer HTL and electron transport layer ETL) and the required opto-electronic and film properties.

6.2 Construction and Work Principle

6.2.1 Device Structure

HPSCs normally have a sandwich architecture with perovskite materials sitting between the hole and electron transport materials (Figure 6.3). In general, the architectures of the HPSCs are categorized into mesoporous and planar ones depending on the nanostructure of the electron and hole transport materials. Each of the two architectures can be further categorized into p-i-n and n-i-p structures, depending on the stacking sequence of the three layers relative to the transparent conducting electrode.

In a mesoscopic nanostructure (Figure 6.3a, b), a mesoporous metal oxide scaffold with the thickness of several hundred of nanometers, which is normally put on top of a compact semiconducting metal oxide layer (NiO or TiO_2) covered transparent conducting electrode, is used as the hole/electron transport layer or just template materials depending on the materials used. P-type NiO, n-type TiO_2, and an insulating dielectric oxide Al_2O_3 have often been used as the scaffold. Since the scaffold often requires a high-temperature sintering around 500 °C, fluorine-doped tin oxide (FTO) is more appropriate as the transparent electrode due to its robustness at high temperatures. Perovskite materials are infiltrated into a mesoporous metal oxide scaffold with an additional perovskite layer (capping layer) covering the top of the

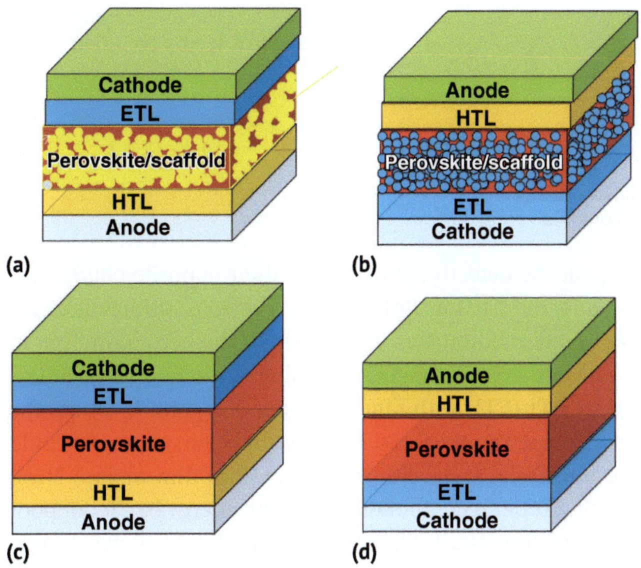

Figure 6.3 (a) mesoporous p-i-n, (b) mesoporous n-i-p, (c) planar p-i-n, and (d) planar n-i-p device structures.

scaffold completely, which is critical for preventing the leakage current in the solar cells. Finally, organic or inorganic HTL/ETL is deposited on top of the perovskite capping layer and then followed by a metal electrode. The scaffold aids the formation of a homogeneous film and acts as a hole or electron transporter if the scaffold is semiconducting.

The ambipolar nature and high charge carrier mobility, as well as the long charge carrier lifetime of the perovskite materials, enable them to function as efficient hole/electron transporters without the aid of a mesoporous scaffold layer at all. By avoiding the high-temperature sintered scaffold, all the layers in a planar heterojunction configuration (Figure 6.3c, d) are compact and flat films, which are compatible with the low-cost and massive production techniques. When it comes to the p-i-n and n-i-p configurations, the stacking sequence of the layers including the HTL, perovskite film, and ETL relative to the transparent conducting electrode in the planar architecture, is the same as that of the mesoporous ones.

6.2.2 Work Principle

Figure 6.4 depicts the photophysical processes that determine the conversion process of the light to electricity including the charge generation, transport, collection, and recombination in HPSCs.[21] First, electrons in the conduction band (CB) and holes in the valence band (VB) of the perovskite are generated upon light excitation. Then, perovskite materials conduct these photo-generated holes/electrons to the perovskite/HTL or perovskite/ETL interfaces under a built-in electric field. Before reaching the interfaces, charge carriers could recombine following first order (k_R) or/and second order (k_2) fashion in the bulk of the perovskite film,

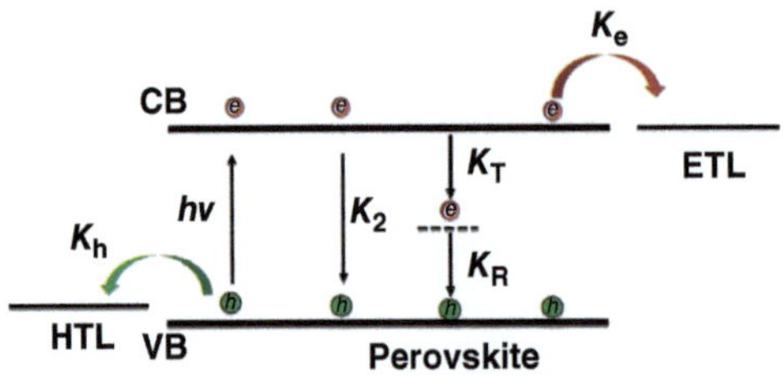

Figure 6.4 Schematic diagram of photophysical events occurring in the HPSC.

when they are tapped (k_T) at the defective sites or find their opposite charge carriers with a distance around the exciton Bohr radius. Previous studies indicate that the first-order recombination dominates in the bulk of the perovskite film in HPSCs operated under one sun condition. After crossing the active layer, holes/electrons are transferred to HTL/ETL with rate constants of k_h and k_e across the perovskite/HTL and perovskite/ETL interfaces. Here, partial charge carriers could also recombine, because interfacial traps either in perovskite film or interfacial materials could cause the first order recombination, and the energy barrier or poor charge transport capability of the ETL/HTL causes accumulation of the charge carriers at the interfaces, which then undergo the subsequent second-order charge recombination. Finally, those surviving charge carriers are extracted into external circuits across the HTL/anode and ETL/cathode interfaces.

6.3 Small Molecules as Hole Transport Layers in HPSCs

As compared to the inorganic counterparts, organic small molecules attracted a surge of interest due to the structural versatility by tailoring the building blocks and peripheral groups, which leads to fine-tuned electrical and optical properties, cost-effective synthesis as well as high batch-to-batch material reproducibility. Therefore, organic small molecules become the promising candidates for scalable manufacturing of the HPSCs. In the past several years, researchers have devoted themselves to manipulating the opto-electronic properties and film forming properties of the HTL by optimizing the molecular structures and conformation/orientation. The common strategies, including designing different cores, bridging groups/terminating groups, as well as the substitutional groups have been developed. As the most popular HTL used in HPSCs, spiro-OMeTAD (Figure 6.6) features an orthogonal molecular conformation with two rigid molecular halves linked by the central carbon atom. Such a twisted structure confers a high glass-transition temperature $\left(T_g\right)$, suitable highest occupied molecular orbital (HOMO) energy levels (5.2 eV), matching well with most perovskites, good solubility in commonly used organic solvents as well as excellent capability to form transparent films. These properties are ideal for fabricating the devices via solution-processing techniques. However, the non-planar structure induces very weak intermolecular electronic coupling and leads to relatively low intrinsic hole mobility (ca. 10^{-5} cm^2 V^{-1} s^{-1}). When used as HTL in HPSCs, spiro-OMeTAD is always doped with p-type dopants to achieve high device efficiencies. The commonly-used p-type dopants, including tris(2-(1H-pyrazol-1-yl)-4-tert-butylpyridine)cobalt(III)tri(bis(trifluoromethane)sulfonimide) (FK209), lithium bis(trifluoromethanesulfonyl)imide (LiTFSI), and 4-tert-butylpyridine (t-BP) are very hydrophobic, leading to unstable HPSCs in the ambient environment.[17, 22–24] Moreover, high-purity spiro-OMeTAD is too costly ($\approx$ 500 €g^{-1})for massive production due to its tedious synthesis, costly sublimation, as well as toxic and corrosive reagents. To achieve cost effective, highly efficient, and stable HPSCs in the absence of the hydrophobic dopants, researchers devoted themselves to developing the highly conductive small-molecule-based HTLs by carefully designing their molecular structures. In next sessions, we will summarize the structural design and properties of organic small molecules HTLs either with or without p-type dopants, and the corresponding performance achieved in HPSCs.

6.3.1 Chemically Doped Small Molecule HTLs

In most cases of the n-i-p structures, perovskite layers deposited on top of the inorganic metal oxide electron transport layer often have a very rough and defective surface, which requires thick HTLs to cover them completely. In this regard, HTLs should have sufficiently high hole mobility or conductivity to guarantee efficient charge transport. Strong intermolecular π–π stacking of the planar core structure is a significant parameter for achieving high hole mobilities. However, the planar molecules are normally prone to aggregate and crystallize, forming bad film morphology and harming the device performance. To get good-quality films, reducing the planarity and rigidity of the molecules or incorporating alkyl chains are commonly used, sacrificing the charge transport capability of the HTLs. Therefore, the planarity and solubility of the molecules should be considered in the molecule design. In addition, the electron withdrawing units or electron rich units are often used to tune the HOMO level and mobility of the organic molecules. However, most small molecules developed at the early stage are not conductive enough, which causes large series resistance and inhibits the charge collection by the electrodes. In this case, the devices suffered from severely low short-circuit current density (J_{SC}) and fill factor (FF). Chemical (molecular) doping represents a powerful strategy for tuning the electronic properties of organic semiconductors.[25] It often involves a charge transfer process with a strong oxidizing agent (p-dopant) or a reducing agent (n-dopant) (as schematically shown in Figure 6.5a). The doped HTLs with improved conductivities are able to eliminate piling up of charge carriers at the perovskite/HTL interface (Figure 6.5b), which suppresses the charge recombination and increases the charge collection efficiency. Thus far, p-type dopants including 2,3,5,6-Tetrafluoro-7,7,8,8-tetracyanoquinodimethane (F_4TCNQ), FK209, LiTFSI combined with 4-tert-butylpyridine (tBP) have been developed (or used) to dope HTLs. In the meantime, researchers have developed many other small molecules by modifying the cores, the terminating groups, as well as the side chains. In the following, we will discuss the small HTLs developed in the past several years, which are used along with the p-dopants to achieve highly efficient HPSCs.

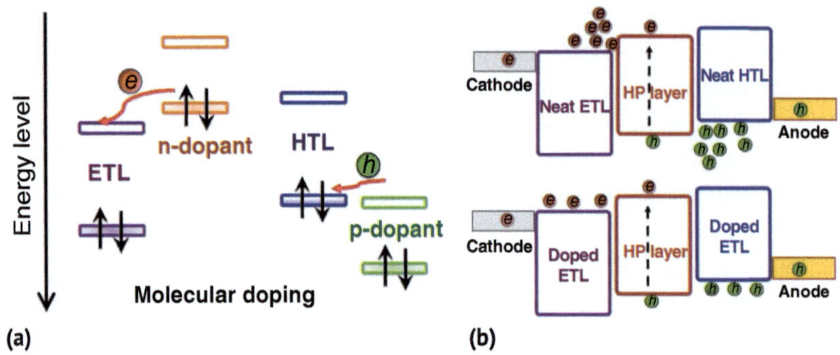

Figure 6.5 (a) The schematic of chemical (molecular) doping of organic semiconductors (HTL and ETL); (b) photo-generated charge carrier accumulation at perovskite/HTL (or ETL) interfaces in the case of neat charge transport and doped charge transport layer.

Currently, the twisted spirobifluorene-based bulky molecule spiro-OMeTAD and its derivatives (Figure 6.6) have been proven to be the most promising HTLs for HPSCs devices due to the merits of high thermal stability, good solubility in common organic solvents, and non-planar 3D architecture. Modifying the spirobifluorene core has been one of the most used strategies to design alternative HTLs. Since the distribution and density of the electron clouds around the core of the HTLs are critical in determining their electronic band structure and properties, researchers have devoted themselves to modifying the substitutional electron-donating and electron-withdrawing groups of the spirobifluorene-based derivatives in the past several years. From a structural point of view, the methoxy groups (–OCH_3) could be either electron withdrawing (meta) or electron donating (para) depending on their substitution position in the aromatic ring of the HTL structure. In addition to such electronic effects, substitution at the ortho position influences the oxidation potential by steric effects as well. It has been demonstrated by previous works that varying the substitution position of the -OMe group in arylamine-based HTMs is a simple and effective strategy to finely tune the electronic properties of the HTMs. In 2014, Seok et al. investigated the effects of replacing p-OMe groups in spiro-OMeTAD with o- and m-OMe groups. The resulting compounds were labeled as pm-, pp-, and po-spiroOMeTAD, respectively.[26] All three compounds have suitable optical properties, good solubility, and wide bandgap. The ortho-methoxy substituted derivative shows a high LUMO energy level. This could play a well role in blocking the electrons, which would lead to higher FF and PCE. Their electrochemical properties including the HOMO and LUMO energy levels can be tuned via changing substitution site of the methoxy groups. As a result, PCEs of 13.9%, 14.9%, and 16.7% have been achieved in HPSCs based on pm-, pp-, and po-spiro-OMeTAD as the HTL, respectively.

Fluorination of the conjugated materials is an effective strategy to realize multiple benefits to the opto-electronic properties of the materials for solar cell applications (e.g., lowering energy levels, enhancing intermolecular interaction/molecular packing, and hydrophobicity through the induced dipole along the C—F bond). Saliba et al. found that the fluorinated dithiophene core enabled the 2′,7′-bis(bis(4-methoxyphenyl) amino)spiro[cyclopenta[2,1-b:3,4-b′] dithiophene-4,9′-fluorene] (FDT) to stack in a slipped fashion through CH/π hydrogen bonds and π–π interactions.[27] Interestingly,

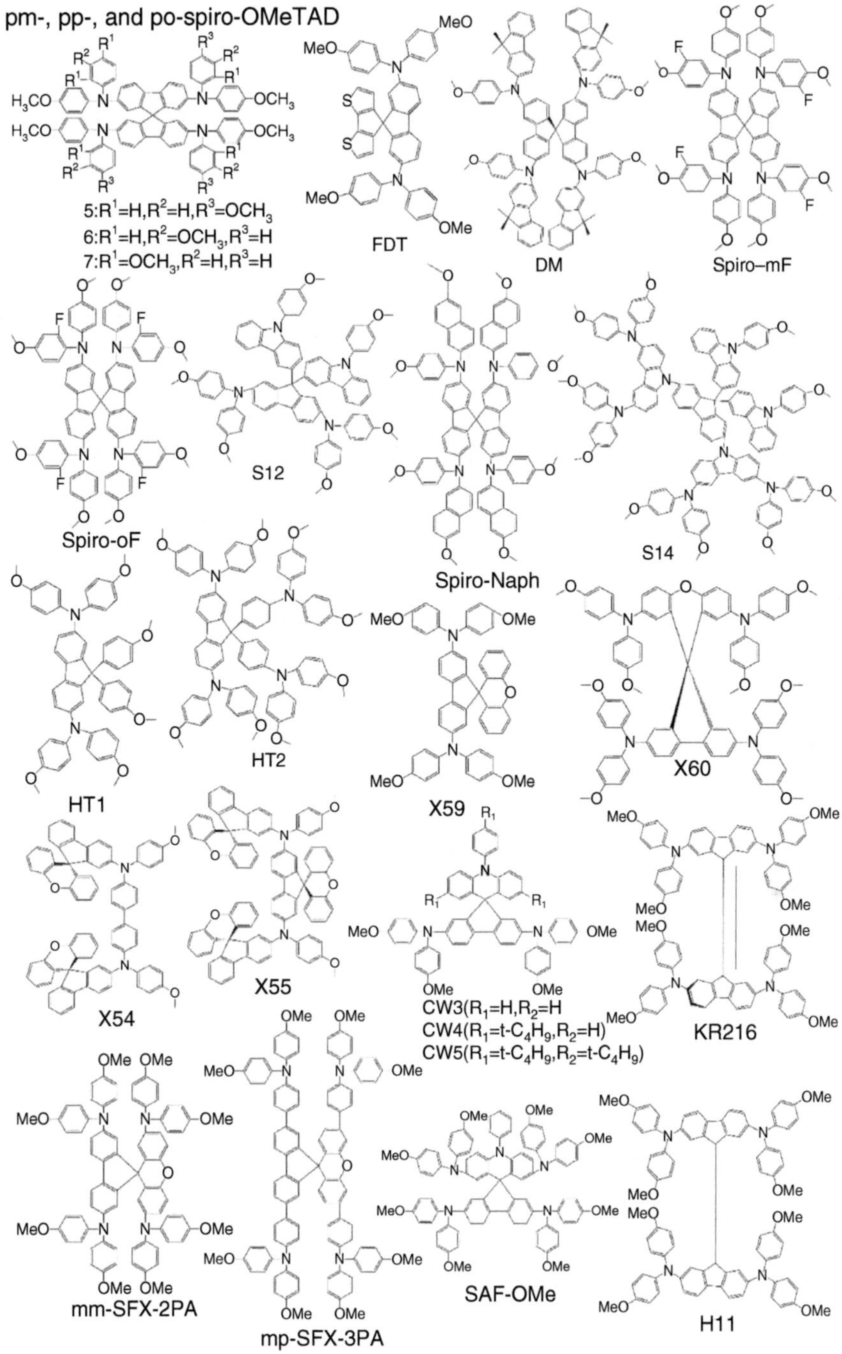

Figure 6.6 Chemical structures of the hole transporting materials based on the spiro or modified spiro cores.

the production cost of the FDT is only one-fifth of spiro-OMeTAD. A stronger adhesion between FDT and perovskite enhanced interfacial coupling through thiophene-iodine interaction, improving hole transfer at the FDT/perovskite interfaces. The HPSC based on FDT HTL showed an impressive PCE of 20.3% with a high V_{OC} value (1.15 V), which is superior to that of spiro-OMeTAD-based counterpart under the same conditions. Seok and coworkers reported a fluorene-terminated spiro-OMeTAD (N^2,$N^{2\prime}$,N^7,$N^{7\prime}$-tetrakis(9,9-dimethyl-9H-fluoren-2-yl)-N^2,$N^{2\prime}$,N^7,$N^{7\prime}$- tetrakis(4-methoxyphenyl)-9,9′-spirobi[fluorene]-2,2′,7,7′-tetraamine) (DM) with an extraordinary efficiency of 23.2% with J_{SC} = 24.91 mA cm^{-2}, V_{OC} = 1.144 V, and FF = 0.81.[28] When exposed to the thermal stress at 60 °C for more than 500 h, the device using DM as HTL exhibited much better stability than the reference device using spiro-OMeTAD as HTL, retaining 95% of its initial performance.

Jeong et al. revealed the important impact of the conformation and stacking mode of the fluorinated isomeric analogs (spiro-mF and spiro-oF) of spiro-OMeTAD on the performance of the HPSCs.[29] The inductive electron-withdrawing effect of the fluorine atoms on the aromatic rings of the conjugated backbone down-shifts the HOMO level to −5.19, −5.06 eV for spiro-mF, and spiro-oF relative to −4.97 eV of spiro-OMeTAD, which improves the charge transfer at the HTL/perovskite interfaces. In-depth atomistic simulations of the DFT and molecular dynamics (MD) revealed that the conformation and the packing mode of the HTLs determine their hole transport capability. Spiro-mF and spiro-oF, which adsorbs closer to the perovskite surface, enables faster hole transport than spiro-OMeTAD. Spiro-mF, which exhibits a more unfolded structure regarding the angle between the centers of spirobifluorene and two fluorinated methoxyphenyl groups, adsorbs with fluorene and phenyl groups together onto the perovskite surface and exhibits higher hole mobility than the spiro-oF due to denser solid-state packing through noncovalent inter- and intra-interactions. In contrast, the fluorine atoms at the ortho-sites of the latter cause steric hindrance regarding the free movement of the methoxyphenyl groups, hindering the intermolecular interaction and the charge hopping. HPSCs fabricated with fluorinated HTLs in the conventional n-i-p configuration delivered a significantly improved PCE as compared to those using spiro-OMeTAD HTL. Especially, spiro-mF accelerated the hole transport and the hole transfer rate at the interfaces, achieving a certified PCE of 24.64% in a small area device and a PCE of 22.31% in a large area (1 cm^2) device. The small loss (0.3 V) in V_{OC} is close to the theoretical minimum value based on the radiative limit defined by the Shockley–Queisser theory. Moreover, the fluorine atoms create barriers that slow down the intrusion of O_2 and H_2O and therefore enhance the ambient stability of the HPSCs.

Introducing asymmetric subunits into small organic semiconductors is also one of the effective strategies to suppress the intermolecular interactions and adjust their energy levels. Very recently, Jeong et al. demonstrated that replacing the symmetric diphenylamines with asymmetric phenylnaphthylamine within the spiro-OMeTAD would reinforce its π-electron delocalization, effective conjugation length, and structural robustness.[17] The symmetric spiro-OMeTAD molecules exhibited a near-isotropic geometry, while the asymmetric spiro-Naph molecules showed strong dependence on their orientations. Empirical van der Waals correction-assisted calculations revealed that an optimal spiro-Naph configuration with the phenylnaphthylamine edges aligned in parallel increases the intermolecular van der Waals interactions owing to the higher directionality and larger area of the π orbitals of the phenylnaphthylamine. The doped spiro-Naph exhibited better conductivity and reduced series resistance, more efficient charge extraction and transport occurring at the perovskite–HTL interfaces. Moreover, spiro-Naph has significantly higher thermal stability, and the fused-ring naphthyl moieties help to prevent moisture. The superior electrical properties of the spiro-Naph enhanced the PCE of the resultant HPSC, which not only achieved a high PCE (24.43%) but also improved device stability at elevated temperatures and humid environments (>78% PCE retention at 60 °C without encapsulation after 400 h; >89.4% PCE retention in air at ~25% relative humidity after 2000 h). These results highlighted the key role of the tighter molecular packing on the excellent performance of the spiro-Naph-based HPSCs.

The carbazole moiety has a significant impact on geometrical alignment and hydrophobicity, which determines the stability as well as efficiency of the HPSCs. In another work, Sheibani et al. developed a new class of nonfused 3D asymmetric structures based on carbazole-coded S14 and S12.[30] It was found that the π-conjugation effective length in nonfused 3D structures is of significance for the hole mobility. S14 forms a compact amorphous film with long-term morphology stability due to suppressed crystallization and high-glass-transition temperature ($T_g = 165\,°C$). Carbazole moiety extended the π-conjugated system, and, thus, enhanced intermolecular interaction on the nonfused structure of the S14, which consequently improved PCE of the HPSCs. The device using S14 as HTL was stable for 1000 h at atmospheric condition without encapsulation, while those using S12 and spiro-OMeTAD decayed to 75% and 69% under the same conditions.

Sun et al. improved the hole mobility by an order of magnitude when they broke one of the carbon bonds of fluorenes in spiro-OMeTAD to synthesize HT1 and HT2.[31] The facile two-step reaction from cheap starting materials with a total yield higher than 90% enabled HT2 to be an ideal candidate for the large-scale industrial production. The twisted 3D structure enabled HT2 to form good film and yield efficient HPSCs. The more extended 3D structure of the HT2 benefited the hole extraction and led to a HPSC with an impressive PCE of 18.04%, superior to HT1-based cells. In a similar period, the researchers from the same group synthesized another two low cost spiro[fluorine 9,9′-xanthene] (SFX)-based HTMs, X59 and X60, via a two-step procedure, which offered remarkably high PCEs of 19.8% in HPSCs.[32,33] Based on these works, the authors made further modifications to SFX-based 3D oligomers to synthesize X54 and X55, by using a one-pot Buchwald–Hartwig cross-coupling reaction with very cheap starting materials.[34] X55 stood out as HTL in terms of much deeper HOMO level, higher hole mobility, and thermal stability as well as film-forming property, than that of spiro-OMeTAD. These features enabled the HPSC devices based on X55 HTL to deliver an impressively high PCE of 20.8%, and excellent stability, which retained 93% of the initial efficiency after 6 months aging in a controlled atmosphere, outperforming the spiro-OMeTAD based HPSCs. Clearly, the facile one-pot synthesis of HTLs holds great potential for replacing the commonly used spiro-OMeTAD in HPSCs. By modifying the positions of bis(4-methoxy phenyl)aniline or bis(4-methoxyphenyl) amine groups in SFX core, Zhan et al. synthesized three small molecule compounds including mp-SFX-3PA, mmSFX-2PA, and mp-SFX-2PA. The mp-SFX-2PA possesses higher hydrophobicity than that of spiro-OMeTAD, protecting the perovskite more effectively.[35]

Guo et al. revealed the important role of the steric interactions of the three small molecule HTLs based on spiro-acridine-fluorene (SAF) core named as CW3, CW4, and CW5, on the film forming properties.[36] With proper alkyl chains, CW4 formed a compact layer on top of the perovskite layer, leading to a PCE of 16.56% in conventional mesoporous structure.

Bifluorene (BF) core, which is formed by insertion of carbon–carbon bonds or other functional groups in the spiro, allows a gradual change in the rigidity of the 3D molecular arrangement. By simply modifying the structural linkage between two fluorene units, diverse molecules could be obtained. The rotation or rigidity of the linkage between the fluorene planes determines their torsion angle and molecular properties. Nazeeruddin et al. obtained 4,4′-dimethoxydiphenylamine-substituted 9,9′-bifluorenylidene (KR216) from a straightforward two-step synthetic procedure by replacing the sp3 carbon atom in the spiro-OMeTAD with a C=C double bond.[37] The competition between the planarization and repulsive steric hindrance led to diversified torsion angles among the four bulky propeller-shaped diphenyl amine units, and a smaller dihedral angle between the two double-bond connected fluorenes than that of spiro-OMeTAD. The four independent molecules in one unit cell are stacked in a highly slipped fashion through $\pi \cdots \pi$, C ··· C, and C ··· O short contacts as well as CH/π (CH/O) hydrogen bonds, forming abundant intermolecular interactions. However, eight fluorenes have three different types of orientation, and the absence of the face-to-face overlap due to the highly sterically hindered geometry led to low conductivity of the HTL. Hagfeldt et al. revealed the effect of the 3D molecular structure on the opto-electronic properties of the HTLs and the performance of the solar cells.[38] They designed H11 by changing a "single atom" in spiro-OMeTAD to a "C—C" configuration. Density functional calculations (DFT) using the Gaussian 09 program package at the B3LYP/6-31G(d)* level of theory indicated the dihedral angles between the two fluorene planes in the order of spiro-OMeTAD (90°) > H11 (81°) > H12 (52°). The central linkage suppressed the delocalization of the electron density distribution (EDD) of the HOMO of H11 and spiro-OMeTAD to the other fluorene plane and isolated them electronically, whereas the EDD of the HOMO and LUMO of the H12 was delocalized throughout the core to the different biphenyl groups and the central bifluorene planes, respectively. Therefore, the charge transport capability exhibited an opposite trend to that of the torsion angle. Benefitted from faster charge extraction at the HTL/perovskite interface, the device using H11 as HTL outperformed that using H12 as HTL.

In addition to the spiro core, alternative core units such as triphenylamine (TPA), carbazole, phenoxazine, acene derivatives, azomethine, triazatruxene, etc., have also been developed to construct small molecule HTLs. The TPA unit represents one of the most used building blocks (either as terminal or core units) for the small molecule HTLs (Figure 6.7) due to its capability to transport positive charges. The nitrogen atoms of the TPA units interact with the aromatic rings by attracting electrons from the latter (inductive effect) and simultaneously

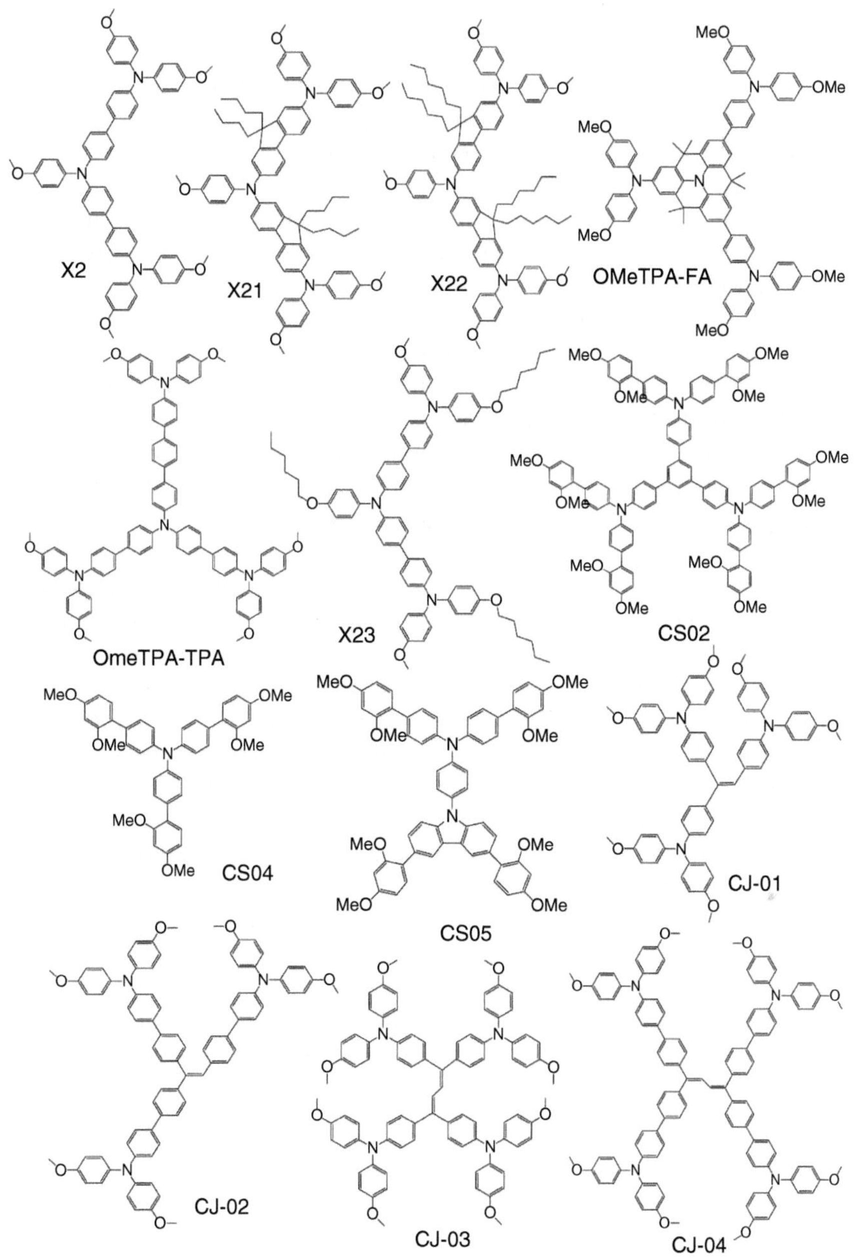

Figure 6.7 Chemical structures of the hole transporting materials based on the TPA and multiphenylethylene cores.

supplying their unshared electrons to the latter (π–π conjugated effect). As the conjugated effect outweighs the induction effect, the electron cloud is widely distributed along the electron rich triphenylamine moiety. Therefore, it is prone to lose electrons to form a positively charged vacancy, facilitating the holes transmission. The propeller-like structure of the TPA core gives rise to the non-planar geometry that regulates the disordered packing of the molecules in the solid state and their opto-electronic properties. The TPA-based small HTLs possess low ionization potentials, good film-forming property, and light stability, which are desirable properties for good performing HTLs. However, the non-planar structures of the TPA-based compounds result in weaker intra- and intermolecular interactions, and, thus, low hole mobility (10^{-6}–10^{-5} $cm^2 V^{-1} s^{-1}$). Thus, p-type dopants are required to enhance hole mobility and regulate their energy levels. A few novel-structure cores have been reported to construct TPA-based small molecule HTLs.

The alkyl chains have constructive effects on the molecular structures and opto-electronic properties, which are critical for the hole transfer kinetics and hole transport abilities in HPSCs. In addition, the fluorene unit on the biphenyl group significantly affects the structural planarity, charge delocalization property, and the hole transport properties of the triphenylamine-based small molecules. Hagfeldt et al. designed four simple triphenylamine-based HTLs, namely X2, X21, X22, and X23, by grafting and tuning the substituted position and length of the alkyl chains on the skeleton of triphenylamine core.[39] They revealed that the length of the alkyl chains influences the solubility of the HTL in organic solvent, and thereby the molecular ordering during the spin coating process. More importantly, the alkyl chain affected the geometrical configuration and electronic distribution, which control the structural planarity, redox potentials, and electronic conductivity. The alkyl chains on the triphenylamine organic molecules played crucial roles in the charge carrier mobility of molecules and solid film self-assembly. The absence of alkyl chain led to severely aggregated X2 film with pin holes, leading to poor device performance. The two alkyl chains on the fluorene of X21 facilitated not only homogenous and complete surface coverage, but also faster hole transfer at the interfaces and hole transport, leading to the best device performance.

Ko et al. synthesized a series of star-shaped hole-transporting materials with a fused triphenylamine or triphenylamine core and diphenylethenyl side arms.[40] All the materials have high transparency in the region of the solar spectrum where perovskite layers absorb, and suitable HOMO levels for extracting holes from perovskite. For example, the star-shaped triphenylamine core leads to a non-planar structure of the tris{N,N-bis(4- methoxyphenyl)-N-phenyl}amine quinolizino acridine (OMeTPA-FA), yielding low hole mobility (1.08×10^{-4}). In the absence of the p-type dopants, the less conductive OMeTPA-FA harms the J_{SC} and FF of the devices using OMeTPA-FA as HTL, leading to a low efficiency of 11%. When tBP and LiTFSI dopants are used to dope the OMeTPA-FA, the J_{SC} and FF of the devices are significantly improved due to the improved hole extraction. OMeTPA-TPA has more planar configuration with a smaller torsion angle of the diphenyl units, exhibiting three times faster hole transport capability (3.67×10^{-4} $cm^2 v^{-1} s^{-1}$). It is worth mentioning that the devices become more efficient when OMeTPA-FA and OMeTPA-TPA are doped, meaning that the hole transport properties of the pristine materials are still not optimal.

Methoxy-substituted triphenylamine (TPA) moiety is a well-known unit, where methoxy groups not only act as Lewis bases to passivate the defect sites at the interface between HTL and the perovskite layer, but also augment the molecular solubility for superior film quality. The Lewis base character is associated with the nitrogen atom in TPA, which acts as the doping site and is stabilized through conjugation with neighboring phenyl rings. The non-planar configuration of TPA induces amorphous nature and good film forming property of the HTL layer. Three TPA-derivatives comprising CS02, CS04, and CS05 enabled triple-cation mixed halide perovskite-based HPSCs to achieve PCEs of 15.40%, 18.10%, and 19.4%, respectively.[41] CS05 outstood in the whole series of the HTLs due to the combined advantages of a more rigid molecular structure, better π–π stacking, and a deeper HOMO level.

Chen et al. reported a family of asymmetric propeller-shaped HTLs with a highly twisted conformation, which contain a multiphenylethylene core and three diphenylamine-/TPA-pending groups.[42] CJ-01 with a diphenylamine moiety produced HPSCs with a similar PCE of about 18% to that of the spiro-OMeTAD, whereas CJ-02 with TPA moieties produced inferior efficiency (15.5%). This type of structure possesses decent hole transport properties and film forming properties. The authors also reported HTLs based on dimethoxydiphenylamine- or TPA-substituted tetraphenylbutadiene fragments.[43] In the absence of the long alkyl chains, CJ-03 with the four-blade screw-like structure formed loose but ordered molecular stacking with homogenous thin film. Moreover, CJ-03 exhibited a deeper HOMO level and high hole extraction efficiency. Aided by the faster charge transport and hole extraction, CJ-03 contributed to a higher J_{SC} and FF than the counterpart CJ-04 in the HPSCs.

Figure 6.8 shows the carbazole-based HTL materials. Zhang et al. studied the impact of linking topology on the optical and electronic properties of the carbazole-based HTLs.[44,45] The 2,7-substituted carbazole-based HTL-X18 and X50 display higher hole mobility and conductivity than that of the 3,6-substituted carbazole-based HTL-X19 and X51, producing better HPSCs. Rakstys et al. systematically investigated a family of isomeric structural HTLs, where two diphenylamine-substituted carbazole units were linked by a nonconjugated benzyl spacer.[46] The larger π-conjugated system through additional carbazole unit amended the charge transport in the film and facilitated the hole extraction. The devices using V1209, V1221, V1225, and V1207 as HTLs achieved average PCEs of 17 and 15%, respectively.

The rigidity of the HTLs significantly influences the hole transport capability and HOMO level. Compared to SAF-5, SCZF-5 possesses a more rigid backbone and faster hole transport capability due to the combined merits of the carbazole unit and spiro-core, obtaining promising HPSCs with a PCE of 20.1%.[47] Ahmad and coworkers developed two HTLs by linking the pyridine core and methoxy diphenylamine-substituted carbazole arm, which allows the continuous π-conjugation by tuning the dihedral angles between core and arm.[48] The authors found that the substitutional position of the arm group on the pyridine core strongly impacts the charge transfer and uniform film-forming ability. 3,5PyDANCBZ formed a very rough and incompact film with pinholes and voids. In contrast, 2,6PyDANCBZ formed smooth film and led to more intimate contact between the perovskite and HTL layer, suppressing the leakage current in the devices. The hole transport properties of the methoxydiphenylamine-substituted carbazole (MODPACz) have heavy dependence on the linkage between

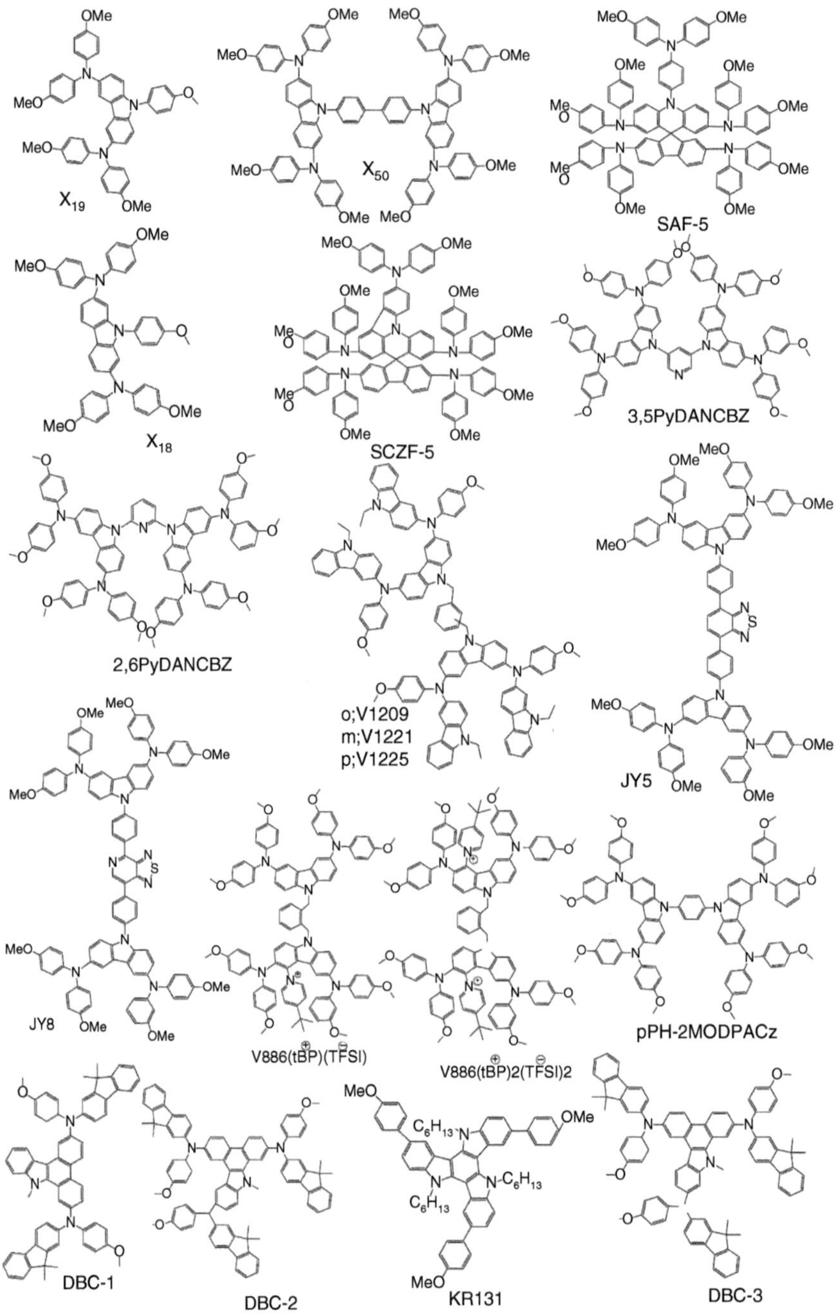

Figure 6.8 Chemical structures of the hole transporting materials based on the carbazole cores.

the MODPACz units.[49] Using phenyl as the bridge between the MODPACz units, the molecules exhibited reasonable hole mobility and conductivity, and good film-forming capability, leading to a PCE over 20% in planar HPSCs. In contrast, MODPACz with the nonconjugated benzyl linkage produced a less efficient solar cell. V886 containing the pyridine in the backbone enabled to form V886(tBPþ)($TFSI^-$) and V886($tBPþ)_2(TFSI^-)_2$ material when it reacts with tert-butylpyridine and lithium bis(trifluoromethane)sulfonamide reagents.[50] These radical-cation species of HTL solution allow the removal of the additives. Moreover, pyridine unit passivated the traps and the charge recombination at the HTL/perovskite interfaces. Unfortunately, this strategy was not effective in achieving high-performance HPSCs. The carbazole-based materials, JY8 and JY5, adopting the same N^3,N^3,N^6,N^6-tetrakis(4-methoxyphenyl)-9-phenyl-9H-carbazole-3, 6-diamine peripheral groups, but different cores, led to distinct photovoltaic performance with PCEs of 19.14 and 17.13%, respectively.[51] Both molecules have the same N^3,N^3,N^6,N^6-tetrakis(4-methoxyphenyl)-9-phenyl-9H-carbazole-3,6-diamine peripheral groups, but they have the thiadiazolopyridine (PT) and benzothiadiazole as the core structures, respectively. Compared to JY5 with benzothiadiazole core, JY8 has not only higher intermolecular dipole–dipole interaction, but also more planar configuration in the central segment due to the robust electron-withdrawing ability of thiadiazolopyridine (PT) core, which is beneficial for charge-transporting capacity and hole extraction.

Recently, Liu et al. developed three small molecules by decorating a new dibenzo[a,c]-carbazole (DBC) core with different peripheral donor groups.[52] Among the Y-shaped structures, DBC-2 contains π-extended phenanthrene group with twisted C12 linkage peripheral positions. The devices based on DBC-2 showed the highest conversion efficiency of 20.02%, performing better than that of prototype spiro-OMeTAD (18.18%) under the same circumstances. One of the main reasons for getting high efficiency for these carbazole derivatives is the tighter molecular packing in the crystal structure, resulting in high hole mobility and hole extraction.

Four center symmetric star-shaped HTLs comprising planar triazatruxene core and electron-rich methoxy-engineered side arms have been synthesized and successfully employed in $(FAPbI_3)_{0.85}(MAPbBr_3)_{0.15}$ perovskite solar cells.[53] These HTLs were obtained from relatively cheap starting materials by adopting a facile preparation procedure, without using expensive and complicated purification techniques. These compounds have suitable HOMO with respect to the valence band level of the perovskite, and time-resolved photoluminescence (TRPL) spectra indicated that hole injection from the valence band of perovskite into the HOMO of triazatruxene-based HTLs is relatively more efficient as compared to that of well-studied spiro-OMeTAD. A remarkable PCE over 18% was achieved in HPSCs using 5,10,15-trihexyl-3,8,13-tris(4-methoxyphenyl)-10,15-dihydro-5H-diindolo[3,2-a:3′,2′-c]carbazole (KR131) as HTL. This result proved triazatruxene-based compounds as a new class of HTL for the fabrication of highly efficient HPSCs.

Electron-rich thiophene-containing segments have been widely incorporated into the skeletons of HTLs to modulate the energy level and carrier mobility (Figure 6.9). Previous studies revealed that the sublinear small molecules based on dithieno[3,2-b:2′,3′-d]thiophene (DTT) core outperform those based on either fluoroene such as N^2,N^2,N^7,N^7-tetrakis (4-methoxyphenyl)-9,9-dimethyl-9H-fluo- rene-2,7-diamine (Z1) and 4,4′-(9,9-dimethyl-9H-fluorene-2,7-diyl)bis(N,N-bis(4-methoxyphenyl)aniline) (Z2) or bifluorene (spiro-OMeTAD).[54] Even though the amorphous Z2 and spiro-OMeTAD formed smooth film, the highly disordered molecular stacking hindered the charge intermolecular hopping,

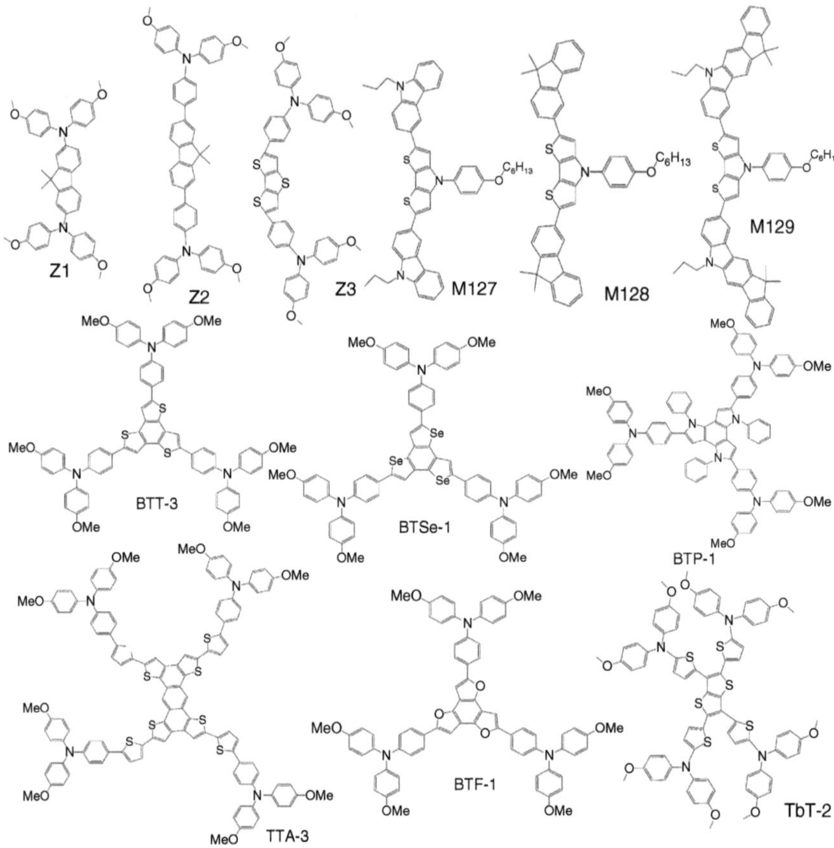

Figure 6.9 Chemical structures of the hole transporting materials based on the thiophene-containing segments.

resulting in lower charge mobility than the semicrystalline Z1 and Z3. HPSCs using Z3 as HTL achieved a PCE of 20.84%.

Although the methoxy group lowers the oxidation potential and favors the hole transport, it causes inherent problems such as a cathodic shift of the oxidation energy level and lower V_{OC} as well as hydrophilicity of HTL layer. Therefore, researchers decorated the HTLs with alternative donor groups. Methoxy-free M129 molecules integrated with indeno[1,2-b]carbazole building block exhibited a promising PCE over 20.35 and 17.50%, respectively, in the presence and absence of the p-dopants in a device configuration of ITO/SnO_2/$Cs_{0.05}(FA_{0.83}MA_{0.17})_{0.95}Pb(Br_{0.17}I_{0.83})_3$/HTLs/Au[55], while HTLs with only carbazole or fluorene donor groups such as M127 and M128 produced inferior devices.

The molecular structure of HTLs influences the π-orbital overlap and face-to-face π-stacking, determining the intermolecular charge hopping rate. The heteroatom-containing central scaffold has significant impact on the electrochemical and photophysical properties, as well as the photovoltaic performance. Previous work studied the intermolecular charge hopping in a series of star-shaped fused HTLs including BTP-1, BTF-1, BTSe-1, and BTT-1, which contain benzotripyrrole, benzotrifuran, benzotriselenophene, and benzotrithiophene central cores linked with three-armed TPA moieties, respectively.[56] BTP-1, having a crowded core structure, suffered from severe steric hindrance and poor intermolecular interactions, exhibiting a low intermolecular hopping rate. The poor charge transport capability inhibited charge extraction and therefore caused obvious hysteresis in the JV curves. In contrast, BTF-1, which has stronger molecular π–π interactions than those in BTT-3 and BTSe-1, produced a PCE of 18%.

Thieno[3,2-b]thiophene central scaffold decorated with conjugated arylamine arms produces a series of TbT molecules with favorable properties, which were employed as HTLs in mixed $(FAPbI_3)_{0.85}(MAPbBr_3)_{0.15}$-based HPSCs. In particular, TbT-3 has good film-forming property, decent hole mobility, and a low-lying HOMO level due to its good solubility and the more extended π-conjugation, yielding superior HPSCs to the spiroOMeTAD based counterparts under the same circumstances.[57] Nazeeruddin et al. used a saddle-like, π-conjugated skeleton composed of four fused thiophene rings to synthesize a cyclooctatetrathiophene (CoTh) with four triphenylamines (CoTh-TTPA).[58] The new HTL showed a bright red color stemming from a direct conjugation between the TPA groups and the central CoTh scaffold. This leads to a charge transfer band, due to the combination of the weak acceptor moiety, the CoTh unit, and the electron-donating *p*-methoxytriphenylamine groups. CoTh-TTPA exhibited a suitable HOMO level matching well with the valence band edge of the perovskite, leading to efficient hole extraction at the perovskite/HTL interface. It has been applied as the HTL in combination with a mixed perovskite $([FAPbI_3]_{0.85}[MAPbBr_3]_{0.15})$ and a state-of-the-art triple cation perovskite $([(FAPbI_3)_{0.87}(MAPbBr_3)_{0.13}]_{0.92}[CsPbI_3]_{0.08})$ reaching decent PCEs of 16.3 and 15.9%, respectively. These values are slightly lower than those measured for the benchmark spiro-OMeTAD HTL. The authors further studied the face-to-face π-interaction and intermolecular π-conjugation effect on the charge hopping in the HTLs incorporating sulfur-rich nonfused tetrathienylbenzene (TTB) and more π-extended sulfur-rich core tetrathienoanthracene (TTA) as novel scaffolds.[59] The triple-cation mixed perovskite $(Cs_{0.1}FA_{0.74}MA_{0.13}PbI_{2.48}Br_{0.39})$-based HPSCs benefitted from the homogenous TTA-3 films, delivering an impressive PCE of 18.76%. In contrast, HPSCs suffered from the poor surface coverage of the TTA-1 and TTA-2 films, exhibiting PCEs of 16.54 and 14.09%, respectively.

Two novel pyrene core structures, TEAP and Py-C, were used to construct the HTLs.[60] Replacing one p-methoxyphenyl unit of the amine substituent with the 9,9′-dimethylfluorene group improved the solubility and the film quality of TFAP. More importantly, the fluorene unit not only extended the conjugation length but also possessed a deeper HOMO level, facilitating the charge transfer and extraction. The conventional HPSC with $ITO/SnO_2/MAPbI_3$/HTM/Au architecture using TEAP HTL delivered a PCE of 12.04% as compared to the device using Py-C HTL. The nonfused tetraphenylethylene (TPE) limited the PCE of the HPSC to 15.4% due to its poor morphology. In contrast, π-extended fused tetraphenylethylene (F-TPE) produced a more homogenous surface, high hole mobility, and deeper oxidation potential, thus leading to a PCE of 18.30%.[61]

A simple one-pot imine approach was used to synthesize V950 with enamine structure by the condensation of 3-amino-9-ethylcarbazole and 2,2-bis(4-methoxyphenyl) acetaldehyde. This simple synthetic route shrunk the price of the final product by one order of magnitude, in comparison with the commercial spiro-OMeTAD.[62] Simple chemical reactions were also carried out to investigate the impact of the structures on thermal and electrochemical behaviors of the spiro[fluorene-9,9′-xanthene] (SFX)-based HTMs bearing TPA groups.[63] Linking TPA groups to the conjugated fluorene moiety led to a significant change in hole extraction and electrochemical redox properties by more than 300 meV, while TPA pending groups on the xanthene units increased not only the hole mobility but also the Tg values by more than 30 °C.[64] The molecular structures of the HTM-FX and HTM-FX0 determined the stability of the films as well as the photophysical properties, yielding PCEs of 19.5 and 20.8%, respectively. Wang et al. judiciously designed a new 3D-twisted structure, spirobisindane core-coded Spiro-I with lower symmetry HTLs, which produced more stable HPSCs with a PCE around 18.57% due to the absence of the phase transition in the range from 25 to 320 °C for the spiro-I.[64]

The traps located at the surface of the perovskite film are among the critical factors reducing the performance of the HPSCs. HTLs play a key role in determining the quality of the perovskite films. Mei et al. reported an efficient bifunctional and cost effective HTL by performing a dopant exchange treatment between the electrochemically deposited polyaniline (PAN), which was doped by dodecyl benzene sulfonic acid (DBSA), and the HCl.[65] The chlorine-doped PAN HTL increased the hole conductivity by a factor of 4 as compared to DBSA-doped PAN layer, facilitating the hole collection. More importantly, the hydrophilic chlorine terminal group not only lowered the surface energy of the PAN but also coordinated with the Pb by filling the iodide vacancy of the perovskite surface. The Cl-PAN substrate led to substantially larger grain sizes (~ 540 nm) and higher crystallinity of the $Cs_{0\cdot05}FA_{0\cdot9}MA_{0\cdot05}PbI_{2\cdot55}Br_{0.45}$-based perovskite films than that on DBS-PAN (~ 360 nm) substrate due to the increased hydrophilicity and improved wettability of the surface. The multiple functionalities such as nucleation sites and defect passivator of the chlorine-based PAN HTL suppressed the trap-assisted nonradiative recombination, improved the charge transfer at the perovskite/cl-PAN interface, and enabled faster hole transport. As the consequence, the device using cl-PAN HTL delivered efficiencies of 20.7 and 18.4% for small-area and large-area (1 cm^2) inverted devices, respectively. Besides, the Cl-PAN HPSCs showed enhanced stability as compared to the untreated PAN based PSCs with 89% initial efficiency retained after stored for 30 days in an inert N_2 atmosphere, thus demonstrating the promising application of this low-cost HCl-doped PAN HTL in HPSCs.

6.3.2 Dopant-free Small Molecule HTLs

The hydrophilic p-type dopants enable the water molecules to invade facilely into the perovskite layer and accelerate the degradation of the perovskite film as well as the device performance. Aiming at enhancing the stability of the HPSCs by avoiding the hydrophilic p-dopants, many dopant-free small-molecular HTLs have been reported based on energetic considerations for charge transfer and criteria for charge transport. In this section, we review previous studies working on synthesis of the dopant-free HTLs and present the advancement in the performance of the HPSCs.

6.3.2.1 Undoped HTLs in n-i-p Structure

Finely tuning the planarity of the HTLs is a critical strategy to realize balanced charge transport and film forming properties of the HTLs. Figure 6.10 shows the dopant-free HTLs used in n-i-p structure of HPSCs. TTE is a good candidate for the core of the HTL to achieve tunable planarity of HTLs.[66] It forms an orthogonal conformation in the aggregated state, in which the two thiophene rings are almost perpendicular to the plane formed by the two trans thiophene rings and the ethene double bond. Interestingly, the cis thiophene rings can be semi- or fully locked by a feasible synthetic route, leading to a gradual increase in molecular planarity. Zhu and coworkers reported the first hybrid orthogonal and planar conformation by adopting a conformation tunable tetrathienylethene (TTE) as the core for HTLs. TTE-1, with unlocked TTE as the core and end-capped with four triphenylamine units, possessed high-lying HOMO level (−5.01 eV) and poor hole transport capability, limited the charge collection and the PCE (around 13%) of the HPSCs. Though the fully locked TTE has high planarity favorable for charge transport, it is unsuitable for solar cell applications due to the poor solubility and solution processability, and thus the incompact film. By using the semi-locked TTE as the core, TTE-2 takes the advantages of both unlocked and fully locked conformations, that is, a hybrid conformation of the orthogonal and planar ones. Such a unique semi-locked TTE core with controllable planarity is expected to synergistically enhance the energetic, electronic, and morphological properties of TTE-2 as the hole-transporting layer in HPSCs. The weak electronic conjugation between the semi-locked TTE core and TPA arms leads to the lower lying HOMO (−5.3 eV) level, promoting the hole transfer at the HTL/perovskite interface. Moreover, such a hybrid conformation enables this molecule to have high hole mobility and form a compact and smooth film. The HPSC with a conventional planar structure based on the dopant-free TTE-2 HTL showed a very promising PCE of 20.04%, with a V_{oc} of 1.11 V, a J_{sc} of 23.26 mA cm^{-2}, and an FF of 0.77. As a consequence of the superior electronic and processing properties of the TTE-2, this device outperformed the one using dopant-free TTE-1 particularly in terms of the significantly improved *J*sc and FF. An overall improvement of 50% in the PCE of the device was obtained by using the dopant-free TTE-2 HTL due to the improved hole mobility and faster hole transfer rate between TTE-2 and the perovskite film. Importantly, TTE-2 has a T_g value of 158.8 °C, much higher than that of spiro-OMeTAD (126.8 °C). The relatively high T_g value favors the formation of a uniform thin film and morphology stability during the operation of the solar cell at elevated temperatures.

Xu et al. reported carbazole based molecules of 9-(2-ethylhexyl)-N^3,N^3,N^6,N^6-tetrakis(4-(methylthio)phenyl)-9H-carbazole-3,6-diamine (CMT), that is, (2-ethylhexyl)-9H-carbazole equipped with N,N-di-p-methylthiophenylamine as terminal groups.[57] In the absence of the p-dopants, CMT produced HPSCs

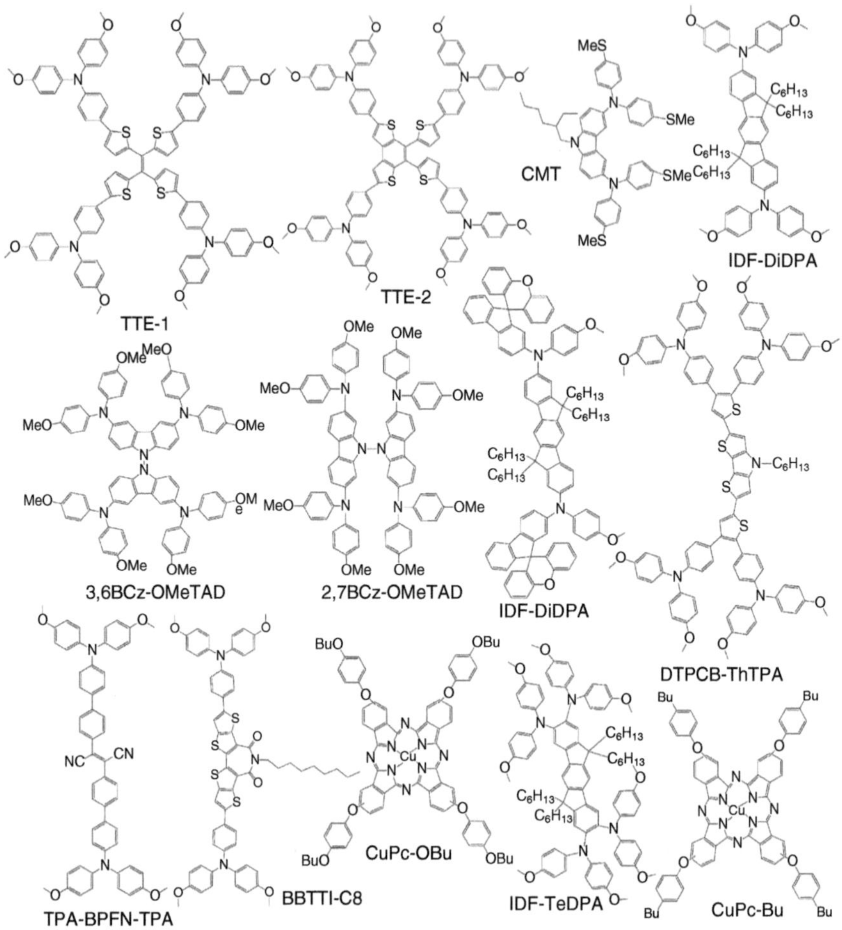

Figure 6.10 Chemical structures of the dopant-free hole transporting materials used in n-i-p segments.

with an optimal PCE of 13.05%. Simple and cheap carbazole-based molecules, 3,6 BCz-OMeTAD and 2,7 BCz-OMeTAD, were used as dopant-free HTLs to fabricate large-area HPSCs.[67] They have the same donor groups but with different substitutions of carbazole location. The HPSCs based on the dopant-free 3,6 BCz-OMeTAD and 2,7 BCz-OMeTAD demonstrated PCEs of 17.0 and 17.6%, respectively, outperforming the corresponding undoped spiro-based device. In the absence of the hydrophilic dopants, these hydrophobic HTLs produced good device durability, maintaining 90% of the initial efficiency after 2000 h without encapsulation. Similarly, dopant-free polycyclic heteroaromatic hydrocarbon PAC-1 showed encouraging efficiencies of 15.59 and 18.17% in the n-i-p planar and mesoscopic device architectures, respectively.[68] Star-shaped molecule with thiophene bridge DTPC$_8$-ThDTPA produced a device with a PCE of 19.42%. Donor–acceptor–donor (D–A–D) molecules with symmetric geometry, such as 3- bis(4′-(bis(4-methoxyphenyl) amino)-[1,1′-biphenyl]-4-yl)-fumaronitrile

(TPA–BPFN–TPA) containing fumaronitrile electron-deficient moiety and TPA electron-donating moiety, was employed as HTL in a mesoporous $MAPbI_3$ architecture without the assistance of any p-dopants. The corresponding HPSCs showed the highest efficiency of 18.4% among dopant-free D–A–D molecules.[69] Unlike D–π–D, A–D–A, or D–A–π types of molecular structures, D–A–D-based molecules tend to lower the bandgap and the HOMO level, leading to higher V_{OC} and PCE of HPSCs. 6,12-Dihydroindeno[1,2-b]fluorine (IDF) core equipped with three different bulky periphery groups, that is, IDF-DiDPA, IDF-SFXPh, and IDFTeDPA, were used as HTLs in planar n-i-p device configuration with a structure of $FTO/C_{60}/CH_3NH_3PbI_{3x}Cl_x)/HTM/MoO_3/Ag$. IDF-SFXPh containing twisted SFX structures possessed higher hole mobility benefiting from good intermolecular π–π stacking of the pentacyclic plane structure center. The HPSCs using dopant-free IDF-SFXPh HTL showed a PCE of 17.6% as well as excellent device stability.

Among the strategies to design dopant-free HTLs, highly dipolar electron donor (D)–acceptor (A) scaffolds impart the intramolecular charge transfer, enforce self-doping, and reinforce the potential for charge separation.[70] The side chains of the molecules significantly impact the molecular packing, film morphology, and consequently device performance. Three dipolar small molecules 2,8-bis(4-(bis(4-methoxyphenyl) amino)phenyl)-5-alkyl-4H-thieno[2′,3′:4,5]thieno[3,2-c]thieno[2′,3′:4,5]thieno[2, 3-e]azepine-4,6(5H)-dione (BTTI) with different alkyl chain lengths were applied in the planar device with the structure of ITO/SnO_2/CsFAMA/HTLs/Au. BTTI-C_6 with the shortest alkyl chain exhibits better hole mobility and film forming ability, leading to a superior PCE of 19.69% for those of the BTTI-C_8 and BTTI-C_{12} based devices. Decorating the peripheral position of the copper (II) phthalocyanines (CuPc) with alkyl chain could increase the solubility of the materials. More importantly, manipulating the peripheral substituents and the metal center could tune the electronic properties of phthalocyanine. The structural changes from butyl to butoxy group in the substituents of the Pc ring strongly impacted the molecular ordering and hole mobility of the two new CuPc derivatives, namely CuPc-Bu and CuPc-OBu. CuPc-OBu with butoxy groups exhibited ordered molecular stacking, leading to high hole mobility.[71] In addition, CuPc-OBu-based HPSC proved to be more stable when stored in ambient conditions without encapsulation.

6.3.2.2 Undoped HTMs in p-i-n Structure

In addition to the normal device structure, hydrophobic organic small molecules or 2D materials have also been used in the inverted structure as bottom HTLs, whose surface energy is critical for the quality of the perovskite film.[72] Figure 6.11 summarizes the representative dopant-free small molecule HTLs employed in inverted HPSCs.

Meng et al. obtained Spiro-E and Spiro-S via a 67 three-step synthetic routine by replacing para-methoxy substituent in spiro-OMeTAD with ethyl and methylsulfanyl groups.[73] The heteroatom substitution in spiro-bifluorene lowers the HOMO energy levels of the Spiro-E and Spiro-S, minimizing the energy loss at the HTL/perovskite interface and extracting the holes more efficiently. Once being employed in the inverted HPSCs, Spiro-E and Spiro-S layers yielded high quality perovskite films due

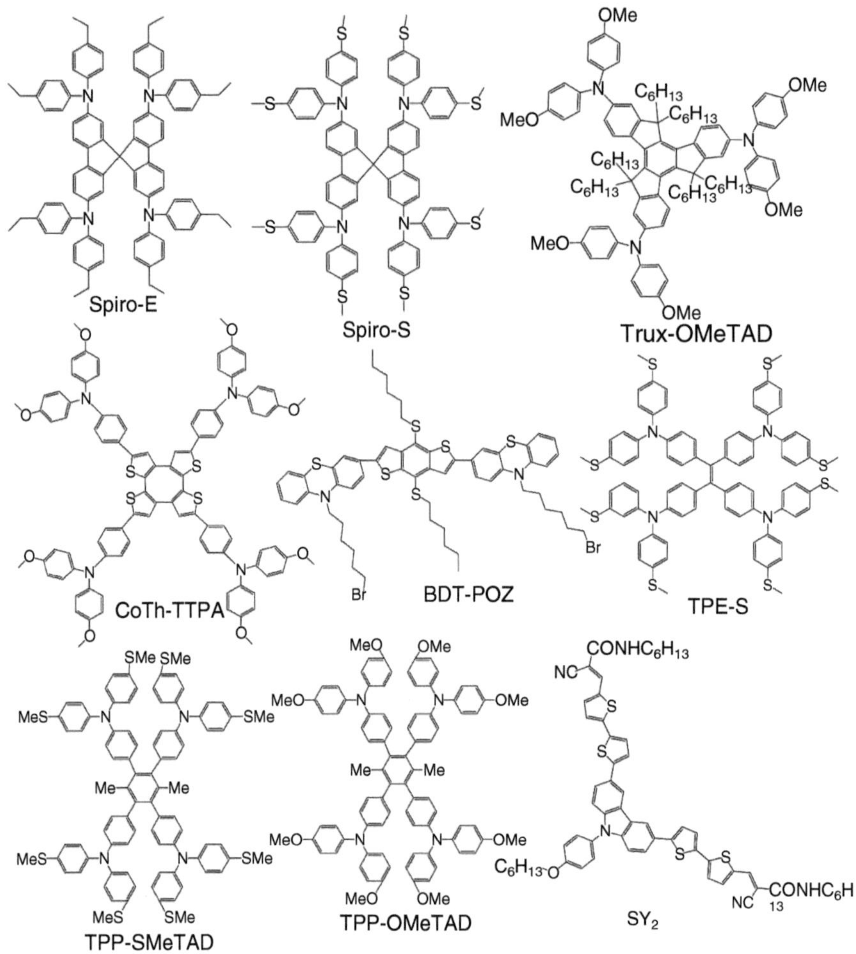

Figure 6.11 Chemical structures of the dopant-free hole transporting materials used in p-i-n devices.

to the benefits of high hydrophobicity. Though those heteroatom-substituted HTLs have comparable hole mobilities to that of spiro-OMeTAD, their thin layers of them without any dopants significantly improved the fill factor and J_{SC} of the devices.

Trux-OMeTAD, consisting of a C3h Truxene-core with arylamine terminals and hexyl side-chains, adopts a planar, rigid, and fully conjugated molecular geometry.[74] These structural features lead to excellent optical and electrical properties of the material including high hole mobility, hydrophobicity, transparency, and matched frontier energy level with perovskite. Consequently, the transfer dynamics of the holes is fast at the Trux-OMeTAD/perovskite interface. Moreover, the hydrophobic TruxOMeTAD surface facilitates the nucleation and growth of larger grains as compared to the hydrophilic p-doped spiro-OMeTAD and PEDOT:PSS, reducing the defect density and the charge recombination centers in perovskite films. The resulting solar cells showed superior PCE (18%) with considerably reduced hysteresis compared to reference devices.

Dopant-free $N^{2\prime}$, $N^{2\prime}$, $N^{7\prime}$, $N^{7\prime}$-tetrakis(4-methoxyphenyl) spiro[[1,3]dioxolane-2,9′-fluorene]-2′,7′-diamine (DFH) as HTL offered a superior PCE of 20.06% in an inverted HPSC with very simple device configuration.[75] Even though DFH lacks of large π-stacking structures or D–A moieties, it transforms from amorphous to paracrystalline dimers when heated at temperatures > 120 °C with the assistance of the 1,3-dioxolane polar group, leading to high intrinsic hole mobility (1×10^{-3} cm^{-2} V^{-1} s^{-1}).[76] D–π–D-type conjugated molecules, featuring with high hole mobility, appropriate energy level, and lower optical absorption in the visible region, have also been used to design dopant-free HTLs. Two *D*-π–D type HTLs (BDTPSZ and BDT-POZ), using 4,8-di(hexylthio)-benzo[1,2-b:4,5-b′]dithiophene (BDT) as the π-conjugated linker and N-(6-bromohexyl) phenothiazine (PTZ)/N-(6-bromohexyl) phenoxazine (POZ) as donor blocks produced p-i-n HPSCs with PCEs of 18.26 and 19.16%, respectively.[77] The short N-N distance of 3.850 Å between the closest POZ units facilitated the hole hopping between adjacent molecules, accounting for the exceptional high FF of 81.7% of the device based on BDT-POZ HTL. Jiang et al. reported a novel twisted 3D dopant-free HTL, which was composed of the tetraphenylethene and bis(4-(methylthio)phenyl)amine groups (TPE-S).[78] When TPE-S was used as HTL in conventional HPSCs, a remarkable efficiency of 21.0% was obtained. The TPE-S based devices showed much better stability than the device using PEDOT:PSS HTL. Beside the superior mobility, the stronger coordination bonding of the S atoms of TPE-S with Pb 2þ ions of the perovskites passivated the surface defects more effectively. Recently, Chen et al. used TPP-OMeTAD and TPP-SMeTAD containing a tetraphenyl core and aryamine pending groups as HTLs to fabricate the p–i–n HPSCs in the absence of the dopants.[79] The variation from the oxygen (TPP-OMeTAD) to the sulfur atom (TPP-SMeTAD) in the HTL, lowers effectively the HOMO of the latter and passivated more defects by the stronger Pb–S interaction with perovskites, promoting the hole collection. The deeper HOMO level and the defect passivation capability of the TPP-OMeTAD resulted in a better PCE of 16.6% with negligible hysteresis.

A–D–A conjugated molecules composed of the carbazole building block and ester/amide electron acceptor unit were used as dopant-free HTLs in inverted HPSCs.[80] When Y2 was used as the bottom HTL, the inverted device with the structure of ITO/NiO_X/interfacial layer (Y2)/perovskite/PCBM/BCP/Ag exhibited a decent PCE of 18.96% (J_{SC} = 21.76 mA cm^2, V_{OC} = 1.102 V, and FF = 0.79), and maintained 87% of the original efficiency after 1000 h in Ar atmosphere. In contrast, amide-based SY2 as the dopant-free HTL produced an inferior HPSC with a PCE of 13.59%. More attention needs to be paid to optimize the energy level and the interaction at the perovskite/HTL interface in order to minimize the interfacial charge recombination.

6.4 Small Molecules as ETLs in HPSCs

6.4.1 Fullerene and Derivatives

Fullerenes and their derivatives are the organic ETLs most widely employed in inverted HPSCs (Figure 6.12).[81] C_{60} and its derivatives, such as PCBM, $IC_{60}BA$ and $PC_{71}BM$, are popular ETLs due to their low temperature fabrication, suitable energy level alignment, and decent electron mobility.[18,82] Despite the high mobility of C_{60}, it requires thermal

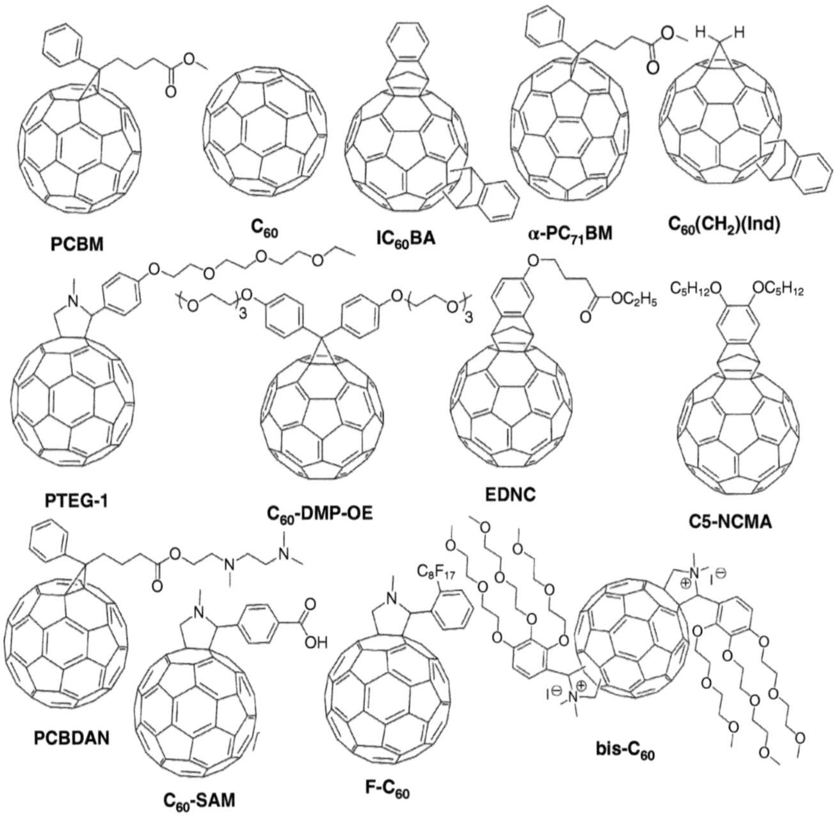

Figure 6.12 Fullerene-based ETLs employed in HPSCs.

evaporation in high vacuum due to negligible solubility in the common organic solvents. To achieve solution processing, the soluble PCBM with alkyl chain is quite often used as ETL in HPSCs. The first p-i-n HPSC using PCBM as ETL delivered a PCE of 3.9%, which is mainly limited by the deficient optimization in device structure, interfacial materials and perovskite and PCBM films. When the device is fully optimized by tuning composition, deposition methods, interfacial materials as well as the device structures, the HPSCs using PCBM ETL achieved a PCE higher than 20%.

The traps at the surface or in the bulk of the perovskite layer has caused the notorious hysteresis and light soaking phenomena in the performance of the HPSCs, leading to unreliable power output. In addition to transport electrons to the cathode, PCBM played an important role in eliminating the hysteresis in the *J–V* curves of the HPSCs by suppressing the ion migration and passivating the defects in the perovskite layers through halide-π noncovalent interaction.[83] Thermal annealing enabled PCBM to diffuse into the bulk of the perovskite layer and passivates the bulk defects, improving the J_{SC} and FF of the p–i–n HPSC.

However, PCBM as ETL failed to eliminate the light soaking effect of the HPSCs when the perovskite film contained plenty of defects. Since the pristine PCBM film exhibits a typically low conductivity of 1.6×10^{-9} S cm^{-1} at room temperature, thick films of PCBM caused big series resistance and inhibited charge collection at the cathode. The conductivity of PCBM decreased monotonously with the descending temperature, creating higher series resistance and significant bimolecular recombination in the HPSCs. Shao et al. found that the PCE of the HPSCs using PCBM as ETL dropped sharply from 4.2% at 298 K to around 1.4% at 140 K.[84] A fulleropyrrolidine with an electron donating triethylene glycol monoethyl ether side chain PTEG-1 as ETL could passivate the traps more effectively by coordinating with the halide vacancies and therefore successfully eliminated the light soaking effect in the HPSCs. Devices based on PTEG-1 exhibited negligible changes in PCE before and after light soaking (15.2% before light soaking vs 15.7% after light soaking).[16] The electron-rich ethylene glycol side chain enabled PTEG-1 to possess higher conductivity due to the self-doping effect, yielding HPSCs that maintains high PCE in a broad low temperature window due to the fast charge extraction.

The processing techniques impact the crystallinity and stacking order as well as the conductivity of PCBM, and therefore the charge collection in HPSCs. Solvent annealing of PCBM by dichlorobenzene (DCB) eliminated the energy disorder in the bulk of the PCBM film, minimized the charge recombination, and improved the V_{OC} of the HPSCs.[85] Another important way to improve the conductivity of PCBM is to introduce the n-type dopants into PCBM.[84] Shao et al. improved the room temperature conductivity of PCBM by more than four orders of magnitude by using LiF and PFN as n-type dopants due to the reduced activation energy for the charge transport, which is 0.5 eV smaller than the pristine PCBM. Consequently, these doped PCBM ETLs keep high conductivity and fast charge transport when the temperature declined from room temperature to about 140 K. Such efficient charge transport enabled the HPSCs to work efficiently from room temperature to about 140 K.

Since the low-lying LUMO level of C_{60} and PCBM results in deficient capability in blocking holes, fullerene derivatives with smaller electron affinity have been used as ETLs to improve the V_{OC} of the p–i–n PSCs.[86,87] $IC_{60}BA$ with smaller electron affinity matches better with the conduction band of $MAPbBr_3$ perovskite and blocks backflowing of the holes more efficiently, enhancing the V_{OC} of the HPSCs.[88] The attaching units decrease the electron mobility of the fullerene derivatives and the electron mobility follows the order of C_{60} > PCBM > $IC_{60}BA$, yielding HPSCs with PCE of 8.06%, 13.37%, and 15.44%, respectively. Five fullerene derivatives including two indene fullerene derivatives (IPB or IPH) and three methanofullerene derivatives (PCBM, PCBH, or $PC_{61}BB$) were compared in p–i–n HPSCs.[89] Indene fullerene derivatives enabled the HPSCs to deliver higher V_{OC} and PCE than PCBM based counterparts, as a result of their higher LUMO energy levels. Similarly, the V_{OC} was improved from 1.05 to 1.13 V by using another indene fullerene bisadduct $C_{60}(CH_2)$(Ind), which possesses a higher LUMO level (3.66 eV) than PCBM.[90]

Cao and coworkers synthesized two groups of fullerene derivatives by attaching oligomer ether (OE) onto phenyl group of the diphenylmethanofullerene (DPM) moiety, where C_{60} and C_{70} were used as the core, respectively.[91] The authors assumed that the electron-rich OE groups not only passivate the surface traps of perovskite layer, but also lower the work function of the Ag electrode, facilitating the charge extraction. C_{60}-DPM-OE, C_{60}-$(DPM\text{-}OE)_2$, C_{70}-DPM-OE, and C_{70}-$(DPM\text{-}OE)_2$

enhanced the photovoltaic behavior of the p-i-n devices due to more efficient charge extraction than the PCBM and $PC_{70}BM$ based counterparts.

Echegoyen and coworkers investigated the impact of the terminal alky groups of the fullerene derivatives on the performance of the inverted HPSCs. Replacing the methyl group with 2-ethylhethyl group increased the solubility of the PCBEH and resulted in smoother film with full coverage. As such, the improvement in the morphology of the PCBEH eliminated the leakage current in the inverted HPSCs, which outperformed the PCBM based counterpart in terms of the J_{SC}, FF, and PCE.[92] Therefore, the conductivity or mobility is an important parameter needing to be considered for rational designing good ETLs. The charge transport and film-forming properties of fullerene-based ETLs also played important role in the photovoltaic performance.[93] Benefitted from good film forming properties, C_{60} EDNC as ETL yielded superior device performance (12.64%) than that of BDNC-based device.

The hydrophobicity of the fullerene derivatives plays a critical role in the ambient stability of HPSCs. One of the major concerns that limit the practical applications of HPSCs lies in their short lifetime in the ambient environment due to the fast degradation of the perovskite layer as well as the perovskite/HTL or perovskite/ETL interfaces upon water ingression. In addition to the charge transport properties, the interfacial layers physically serve as the shell to protect the perovskite layer. To inhibit the ingress of the water molecules, the ETL ideally should possess high hydrophobicity. Previous studies demonstrated that hydrophobic fullerene derivatives as ETLs prevented efficiently the penetration of the water molecules into perovskite layer and enhanced significantly the stability of the inverted HPSCs.[94,95] A new fullerene derivative C5-NCMA as ETL yielded a more stable and efficient HPSC than PCBM due to its merits of higher hydrophobicity, higher LUMO energy level and stronger ability of self-assembly.[95] Devices using C5-NCMA ETL showed a PCE of 17.6% with negligible hysteresis, much higher than that of PCBM (16.1%). After aging for 30 days under a controlled humidity of 30% in ambient at room temperature without encapsulation, the HPSCs with C5-NCMA ETL retained 70% of the initial PCE, which is superior to that using PCBM ETL.

Crosslinking the ETLs with hydrophobic groups, which improves significantly not only the morphological stability, but also the ambient stability, is an effective strategy to improve the stability of the HPSCs. A C_{60}-substituted benzoic acid self-assembled monolayer containing cross-linkable and hydrophobic silane side chains (C_{60}-SAM) was employed as ETL to improve the water resistance.[96] To prevent water ingression, a relatively thick and compact C_{60}-SAM layer was needed to cover the perovskite layer, but it caused big series resistance, inhibits the charge transport due to low conductivity. Methylammonium iodide (MAI) was used to dope the crosslinked C_{60}-SAM ETL to improve its electron conductivity. Consequently, HPSCs exhibited a PCE of 19.5%, and retained nearly 90% of the original high efficiency after exposing to an ambient environment for 30 days.

Jen and coworkers reported a self-doped fullerene derivative, fulleropyrrolidinium iodide (FPI). The self n-doping process, which is driven by the electron transfer between iodide and the fullerene core, endowed the FPI with a high conductivity of 2.0 S m^{-1}.[97] Li and coworkers developed a series of fullerene derivatives (FPPI,

Bis-FPPI, Bis-FIMG, and Bis-FITG) using a quaternary ammonium halide moiety as the common backbone, driving electron transfer from halides to fullerene core and resulting in anion-induced n-doping of the fullerenes derivatives. These derivatives, as the cathodic interfacial layer, yielded HPSCs with a PCE of 19%.[98] Yu and coworkers replaced PCBM with ultrathin fulleropyrrolidinium iodide (bis-C_{60}) in an inverted HPSC, extracting the electrons efficiently and inhibiting the backflow of the holes at the perovskite/bis-C_{60} interface. Consequently, the *J*sc of the HPSC was improved from 17.91 to 20.71 mA cm^{-2}.[99] Dumb-belled PCBM (D-PCBM), consisting of two fullerene balls, helps to improve the quality of the ETL when blended with PCBM as ETLs.[100] PCBM:D-PCBM blends suppressed the charge recombination at the perovskite/ETL interface and enhanced the exciton dissociation.

Previous studies demonstrated that alternative composite ETLs formed by fullerene derivatives and C_{60} reduce the energy barrier at the ETL/cathode interface and enhance the extraction and transportation of the electrons. DMAPA-C_{60} dipolar interfacial layer significantly improved the V_{OC} and FF of the HPSCs. PCBDAN serving as a hydrophobic interlayer between PCBM and Ag electrode enhanced the stability of the HPSCs by providing higher water resistance.[101] The composite ETL composted of two fullerene derivatives, N-methyl-2-(2-perflurooctylphenyl)-3,4-fullero-pyrrolidine (F-C_{60}) and bis-C_{60} surfactant (N-methyl2-(2,3,4-tris(2-(2-methoxyethoxy)ethoxy) phenyl)-3,4-fulleropyrrolidium iodide bisadduct), combined the advantage of the high hydrophobicity of F-C_{60} and the tunable work function of bis-C_{60}, improving the PCE and stability of the HPSCs effectively.[102]

6.4.2 Non-fullerene Small Molecules

6.4.2.1 Perylene Diimides (PDIs) as ETLs in HPSCs

In the past several years, PDIs and their derivatives (Figures 6.13 and 6.14) have attracted a lot of attention as the non-fullerene ETLs due to their advantages of easy synthesis, good thermal and chemical stability as well as their suitable LUMO and HOMO levels.[103–105] However, the PDI-based devices still lag behind the PCBM based counterparts in terms of the PCE. The poor device performance is mainly associated with the strong aggregation behavior and the bad film morphology caused by the strong intermolecular π–π interactions between the PDI molecules. In order to solve this issue, a common strategy adopted by the research community is to disrupt the planarity of PDIs by introducing functional groups.

Halogen substitution significantly influences the intermolecular force, determining the arrangement and solubility of the molecules. Consequently, PDI films with different halogen substituents show different morphologies and electron mobilities. Chen et al. studied the effect of the halogen atoms on the opto-electronic properties of the EH-PDI and the performance of the HPSCs. Br-PDI has weaker intermolecular interactions, as the larger atomic radius of the Br atom suppresses the aggregation of the molecules compared to the EH-PDI and F-PDI. Thus, Br-PDI has the highest solubility (>10^{-1} mol L^{-1}) in chloroform and forms the smoothest film among the three PDIs. In contrast, the intermolecular interactions of the F-PDI molecules are much stronger due to the π–π, F–π, and F–F interactions, yielding

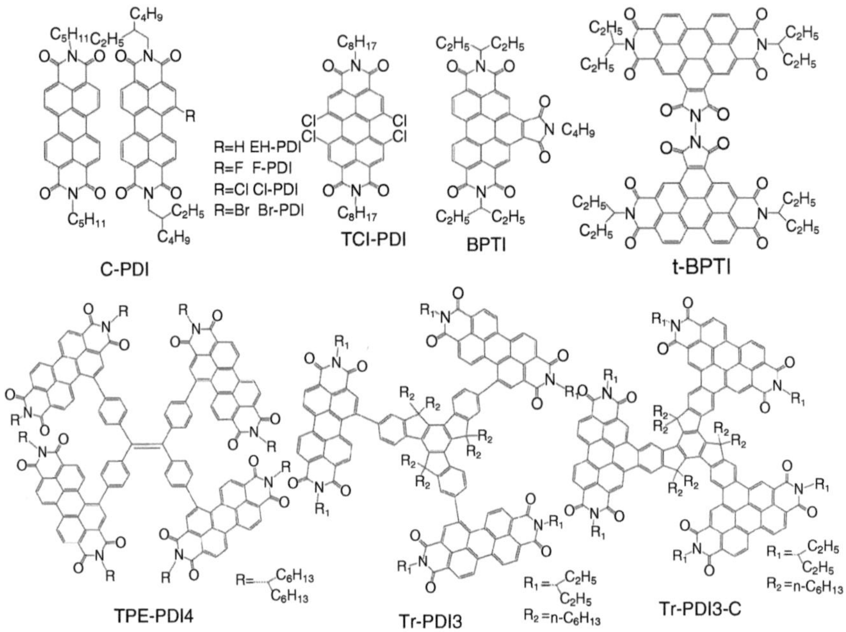

Figure 6.13 Chemical structures of 3D structure PDI derivatives.

the worst film morphology. Besides the morphology, the intermolecular interactions also impact the electron mobility, which increases with the size of the halogen atoms. Br-PDI exhibited an electron mobility (1.08×10^{-3} $cm^2\ V^{-1}\ s^{-1}$) at least one order of magnitude higher than the EH-PDI (1.12×10^{-4} $cm^2\ V^{-1}\ s^{-1}$) and F-PDI (8.31×10^{-6} $cm^2\ V^{-1}\ s^{-1}$). Br-PDI as HTL produced the best device performance due to faster charge transport and lower leakage current.[106] Tetrachloro-substituted perylene diimide (TCl-PDI) as an ETL yielded a highly crystalline perovskite film and forms a barrier-free ETL/perovskite interface. These desirable properties led to superior conventional planar HPSCs compared to those based on TiO_2 ETL.[107]

The strong intermolecular π–π interactions of the PDI molecules are one of the main factors for the poor solubility and film morphology. A common strategy to avoid aggregation is to lower the planarity of the PDI molecules by increasing the steric hindrance. Some twisted dimers or 3D PDIs formed denser and more uniform ETL, reducing the direct contact between the perovskite and the cathode and promoting the charge transfer.[108] The dimer of Benzo[ghi] perylenetriimide (BPTI), t-BPTI as novel ETL overcomes the self-aggregation of PDI molecules and forms smooth and continuous films on the top of the perovskite layer, yielding better HPSCs.[108]

PDI containing a 3D core is an alternative type of stereo molecule, which inhibits its aggregation more effectively and has isotropic charge transport. Jiang et al. synthesized a 3D molecule by grafting the PDI units onto a tetraphenyl ethylene core (TPE). The resulting $TPE\text{-}PDI_4$ achieved good electron mobility, excellent solution processability, favorable film morphology, and suitable energy levels.[109] The inverted HPSCs using TPE-PDI4 as ETL achieved a PCE of 16.29%, which was

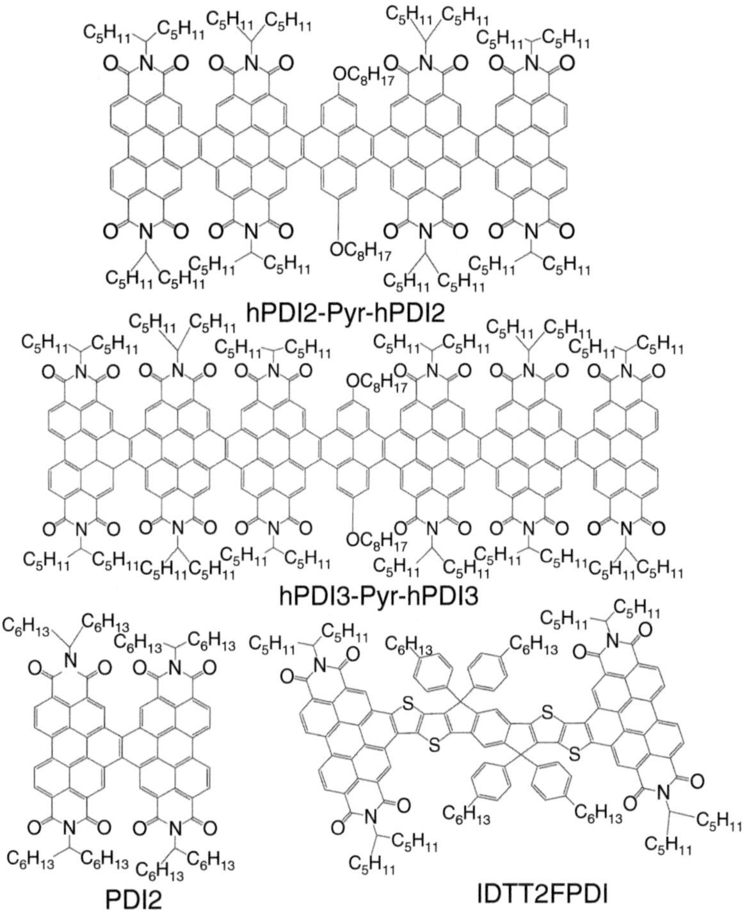

Figure 6.14 The chemical structures of PDIs with a conjugate plane.

slightly higher than that of the PCBM-based devices. A 3D molecule SFX-PDI4 based on the Spiro[fluorene-9,9′-xanthene] (SFX) core exhibited low-lying HOMO and LUMO levels, high electron mobility, and good film-formation properties.[109] Based on such an ETL, a high PCE of 15.3% without obvious hysteresis was achieved in the inverted HPSCs. Wang et al. found that the connection fashion of the two PDI trimers with the truxene core influences the geometry of Tr-PDI$_3$ and Tr-PDI$_3$-C. The TrPDI$_3$ molecule with the three PDI units connected to the truxene core by C—C bond has a twisted structure. In contrast, Tr-PDI$_3$-C, which was connected by a fused ring, has a planar structure. Tr-PDI$_3$ as the ETL produced a superior device than the device based on the Tr-PDI$_3$-C ETL.[110]

It is an effective method to improve the electron transport or modify the energy level by extending the π-conjugation length of the ETL materials. The extended conjugate plane of the PDIs leads to helical chemical structures. A balance needs to be achieved between the aggregation and electron mobility to fabricate high performance HPSCs. Nuckolls et al. used two graphene nanoribbons, hPDI2-Pyr-hPDI2 and hPDI3-Pyr-hPDI3, as ETLs in inverted HPSCs.[111] HPDI3-Pyr-hPDI3 has a higher mobility than hPDI2-Pyr-hPDI2 due to its larger conjugate plane. The more conductive hPDI3-Pyr-hPDI3 ETL produced a better HPSC with a PCE of up to 16.5%, which was higher than that of hPDI2-Pyr-hPDI2. The small molecule PDI_2 with extended π-conjugation length accelerated the electron transport and suppressed the carrier recombination in HPSCs. The efficient charge transport capability of PDI_2 ETL alleviated not only hysteresis, but also light soaking effect of the HPSCs, when it was used as the interfacial modifier between TiO_2 and the perovskite layer.[112]

A ladder-type conjugated PDI dimer containing an IDT unit (IDTT2FPDI) benefits from the multiple fused thiophene ring structure and exhibits high electron mobility and suitable energy levels. The hydrophobic IDTT2FPDI ETL not only facilitated efficient electron transport, but also protected the perovskite layer from water strike.[113] A double layered ETL formed by IDTT2FPDI and PCBM yielded a device with a PCE of 19.4%.

Amino and sulfur modified PDIs serve as good interfacial layers in HPSCs, as they reduce the work function of metal electrodes and optimize energy level alignment. Moreover, the lone electron pairs of the nitrogen and sulfur atoms enhance the electron transport of the ETL and passivate the electron traps of the perovskite film. Zhang et al. made the first attempt to apply PDIN as an ETL in inverted HPSCs and obtained a PCE of 17.66%.[114] Miao et al. first extended the application of PDIN as ETL in inverted HPSCs, achieving a PCE of 15.28%.[115] The oxidized forms of PDIN and PDINO exhibited higher conductivity and lower HOMO level. PDINO as a buffer layer between PCBM and Ag helps to extract the electrons more efficiently than the PCBM only device. More importantly, the uniform and smooth PDINO film improved the stability of the device, as it effectively blocked the diffusion of I or Ag between the perovskite and cathode.[116] A discotic coronene diimide derivative (CDIN), which has an expanded p-conjugated plane and amino terminal substituents at the imide-positions, could increase the charge mobility of ETLs and affect the grain size of the perovskite. Consequently, CDIN as ETL improved the PCE of HPSCs from 12.0% of the TiO_2 based device to 17.1% with smaller hysteresis.[117] Cheng et al. synthesized two PDI dimers (FA^{2+}-PDI2 and (FA^{2+}-PDI2)Br_2) by adding the amino or ammonium groups at bay positions. The charged ammonium moiety endowed (FA^{2+}-PDI2)Br_2 a higher electronic conductivity and better hydrophobicity than FA^{2+}-PDI2, leading to a better HPSC.[118]

Modification of the PDIs with sulfur atom also serves as an effective strategy to tune the energy levels, improve the conductivity of PDIs, and passivate the perovskite layer. Li et al. obtained a series of ETLs including 2PDI-0S, 2PDI-2S, 2PDI-3S, and 2PDI-4S by decorating the PDI dimers with annulated sulfur atoms.[119] As the number of sulfur atoms increases, the LUMO of the PDIs downshifts from 3.7, 3.9 to 4.0 eV for 2PDI-2S, 2PDI-3S, and 2PDI-4S, respectively. The chemical interaction between sulfur atoms of 2PDI-4S and the perovskite layer led to a more stable device than PCBM based counterpart, which showed negligible attenuation at the maximum power for 400 hours.[120]

6.4.2.2 NDIs as ETLs in HPSCs

NDIs (Figure 6.15) that replace the perylene core of PDIs by the naphthalene core have also received increasing attention as ETLs in HPSCs. NDIs exhibit a larger optical bandgap and are optically more transparent than PDIs, minimizing the loss of the incident light in a conventional device configuration. The facile accessibility of NDIs significantly reduces the production cost. However, NDIs suffer from the strong intermolecular π–π interactions and severe aggregation, leading to poor solubility and film quality. Several strategies such as N- or side-substitution on mono-NDIs and NDI dimers have been developed to control the intermolecular interactions. Introduction of the heteroatoms (e.g., O, S, N atoms, etc.) onto the skeleton of the NDIs helps to passivate interfacial defects.

The substituents attached to the nitrogen atom of NDIs play a critical role in the opto-electronic and processing properties of the materials. Decoration of the NDI derivatives with the flexible N-substituted alkyl chains leads to high conformational freedom, which is harmful to the stability of HPSCs.[121] On the other hand, the

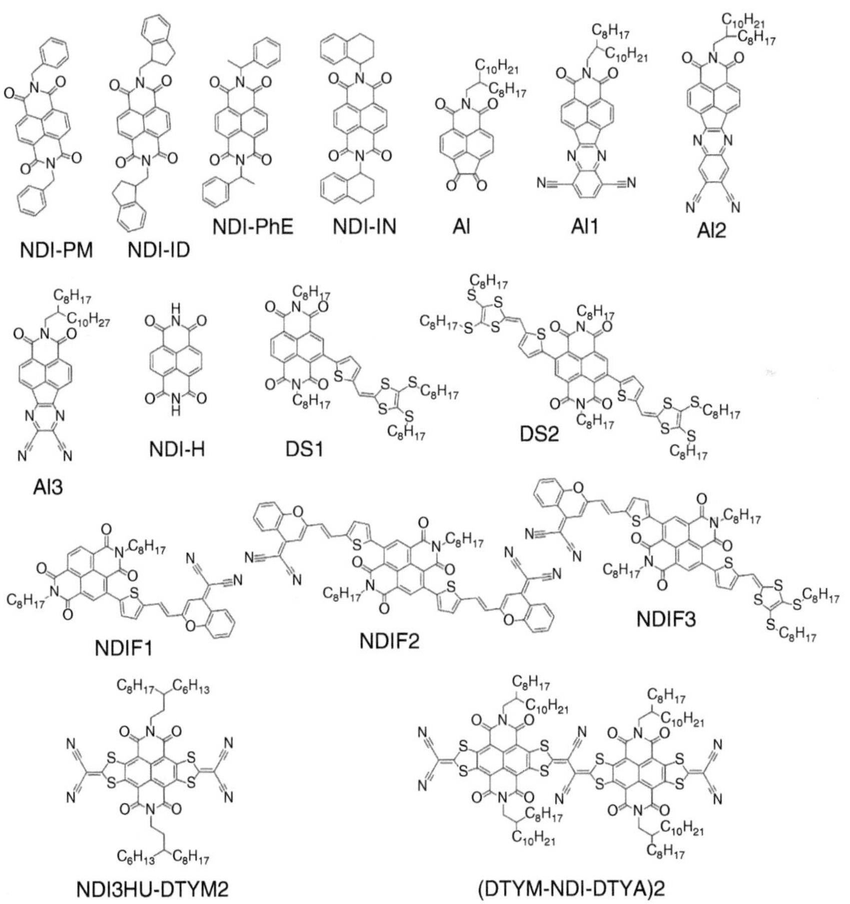

Figure 6.15 The chemical structures of NDIs.

N-substituted aromatic groups enhanced the long-term temporal stability, but sacrificed the solubility of NDIs in organic solvents.[122] Im et al. introduced N-substituted phenylmethyl into the NDI core (NDI-PM) and reported the first inverted HPSC using NDI-PM as ETL.[123] Even though NDI-PM ETL produced HPSCs with PCE comparable to the PCBM based counterparts, it enhanced the thermal stability of HPSCs due to the strong hydrogen bonding between NDI-PM molecules. Since NDI-PM exhibited electron mobility and solubility comparable to PCBM, it yielded HPSCs with PCEs similar to the PCBM based counterparts. Kwon et al. studied the photovoltaic behavior of the HPSCs using a series of N-substituted PDIs, including 1-indanyl (NDI ID(RR)), (R)-1-phenylethyl (NDI-PhE), 1,2,3,4-tetrahydro-1-naphthyl (NDI-TN(RR)), and their corresponding heterochiral counterparts (NDI-ID(RS) and NDI-TN(RS). NDI-ID has excellent low temperature processability and high stability due to the combined alicyclic and aromatic characteristics of the indane moiety. Consequently, NDI-ID-based devices delivered a higher PCE (up to 20.2%) and stability than that of the PCBM-based ones.[124] Since the two asymmetric-shaped chiral (R)-1-phenylethyl (PhE) groups reduced the molecular symmetry, adding a methyl group to the structure of the NDI-PM yielded NDI-PhE with higher solubility and better film-forming ability. As a consequence, NDI-PhE as ETL yielded inverted HPSCs with a higher PCE than PCBM.[125] The same group systematically investigated deeper into how the substituents with chiral stereoisomerism influence the physical and electronic properties of NDIs.[126] The non-centrosymmetric geometry of the enantiomeric NDIs with a homochiral pendant group endowed them with higher solubility and thermal stability than the homochiral NDIs. The highly uniform NDI-ID (RR) and NDI-TN(RR) ETLs yielded HPSCs with PCEs of 20.4 and 17.9%, respectively, outperforming those of the NDI-ID (RS) and NDI-TN(RS) based devices (20 and 16.5%). The more stable NDI-ID (RR) slowed down the degradation rate (less than 10% at 100 °C in air for 1000 h) of the performance of the HPSCs. Control of the structural symmetry of the NDIs is a typical approach to realize high solution processability and significant temporal stability of these materials. The J_{SC} and FF of HPSCs were significantly improved by inserting a series of asymmetrical small molecules based on acenaphthylene-imide (AI) including AI1, AI2, and AI3 as the buffer layer between TiO_2 and the perovskite film due to their high electron mobility ($>10^4$ cm^{-2} V^{-1} $s^{-1)}$ and solution processability.[127]

Naphthalene-1,8,4,5-tetracarboxylic diimide (NDI-H), which adopts a specific face-to-face orientation relative to the substrate, was used as ETLs in HPSCs.[128] Introducing side substituents onto NDI core could also suppress the aggregation of NDIs, resulting in a smooth and uniform film. Among the side chains, those containing thiophene groups not only passivated the uncoordinated Pb traps of the perovskite film but also impacted the stacking of NDIs.[129] The dithiafulvenyl NDIs with two thiophene groups (DS2) could passivate more efficiently the uncoordinated Pb traps at the surface of the perovskite film than DS1. DS2 as ETL yields efficient HPSCs with a PCE (11.4%) as comparable to that of PCBM-based devices (13.5%).[130] The dithiafulvenyl and 2-(2-methyl-4H-chromen-4-ylidene)malononitrile (CMCN) groups have big influence on the aggregation behavior of the NDIs. Modifying the position and the number of the side chains on the skeleton of NDIs could finely tune the extent of the molecular interaction and aggregation. NDIF1 formed a smooth film and a low series resistance, yielding a HPSC with a PCE of 15.6%. However, NDIF2 and NDIF3 with either too strong or too weak intermolecular interaction produced very bad morphology due to strong aggregation in the former and pin holes in the latter, reducing the device performance.[131] It is necessary to thoroughly investigate the relationship between the side chains and the molecular interactions to design good ETLs.

The highly planar aromatic rings of NDIs caused very strong aggregation of the molecules, producing poor solution processability and film quality.[128] The twisted NDI dimers inhibited the aggregation of the NDI molecules, improving the film forming property and inducing more isotropic charge transport.[132] The twisted NDI dimer $(DTYM\text{-}NDI\text{-}DTYA)_2$ formed good film morphology and higher electron mobility due to weaker aggregation of the molecules and the isotropic charge transport. Moreover, the LUMO level of $(DTYM\text{-}NDI\text{-}DTYA)_2$ aligned well with the conduction band of the perovskite, promoting the electron transfer efficiency. $(DTYMNDI\text{-}DTYA)_2$ yielded a HPSC with a superior PCE of 12.9% than the HPSC based on NDI3HU-DTYM2.[133] The piperazine-1,4-diyl bridged NDI dimers exhibited stronger defect passivation capability with the increased nitrogen atoms. As compared to PDPT, PMDPT containing a larger number of nitrogen atoms enabled the inverted HPSCs to deliver higher V_{OC} by passivating more defects at the surface of the perovskite film.[132]

6.4.2.3 Azaacene-based Small Molecules as ETLs in HPSCs

Azaacene derivatives (Figure 6.16) have also been used as ETLs in HPSCs due to their excellent charge transport properties.[134] The electron-deficient pyrazine unit on azaacene provided good n-type electron transport. However, azaacenes have been limited by their relatively high LUMO and poor solution processability. Recent research tried to tackle these problems by introducing different substituents onto the azaacenes.

Manipulating the alkylsulfanyl chains length or the sulfur oxidation states of azaacene-based ETLs (sulfide-$HATNAS_3C_4$, $HATNAS_3C_7$, $HATNAS_3C_7\text{-}C_{3h}$, $HATNAS_3C_7$-Cs; sulfoxide-$HATNASOC_7$-Cs; and sulfone-$HATNASO_2C_7$-Cs) could improve the solution processability as well as defect passivation capability. However, the electron-donating alkylsulfanyl chains have downshifted the LUMO level significantly and created an energy barrier at the ETL/perovskite interface. In contrast, the LUMO level could be elevated by replacing the sulfide (S) with sulfoxide (SO) or sulfone (SO_2). Such modifications in the structure of the ETL improved the PCE from 4.69% of the inverted device using HATNA-F6 as ETL to 17.62% of the device using $HATNASOC_7$-Cs as ETL. The fluorine group increased the hydrophobicity of the $HATNASOC_7$-Cs ETL and enhanced the stability of the HPSCs, which maintained 80% of the initial PCE in air for 30 days.[135] Though a novel azaacene-based QCAPZ has an electron mobility of 4.7×10^4 cm^2 V^{-1} s^{-1} as comparable to that of PCBM, it produced an inferior inverted HPSC with a PCE of only 10.26% due to the mismatched energy levels with the perovskite layer.[136] The sulfur atoms of the azaacene ETLs have a noticeable impact on the performance of the HPSCs. Gu et al. found that the interaction between the sulfur atom and the perovskite surface improved

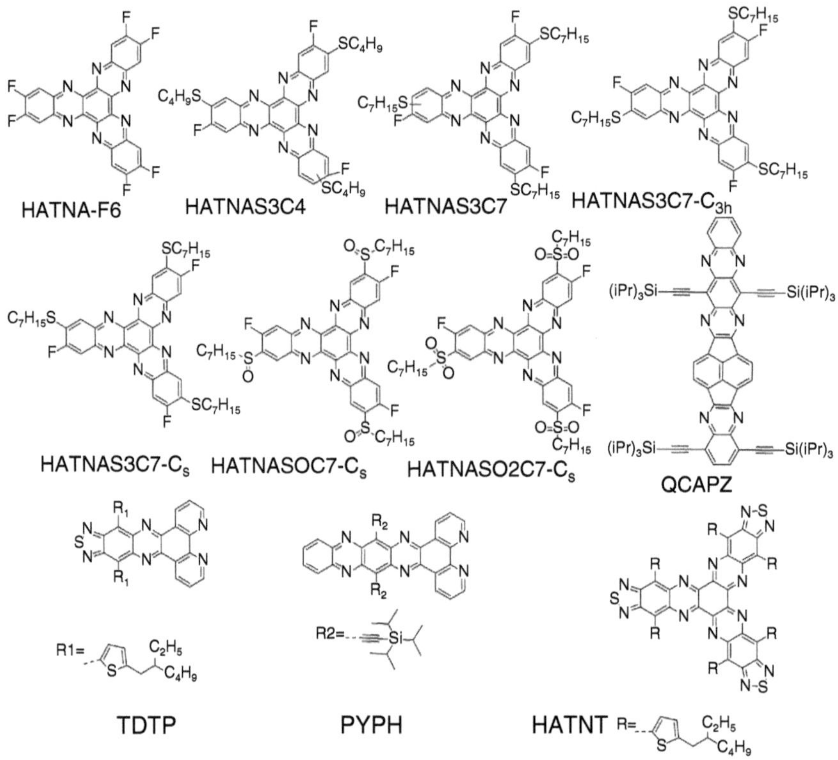

Figure 6.16 The chemical structures of the azaacene-based small molecules.

the performance of HPSCs. The device using TDTP as ETL showed a PCE of 18.2%, outperforming that based on PYPH (9.5%) due to the stronger interaction between TDTP and the perovskite surface.[137] A new sulfur-containing n-type azaacene-based small molecule, hexaazatrinaphtho[2,3-c][1,2,5]thiadiazole (HATNT), was demonstrated to suppress the photogenerated charge recombination at the perovskite/ETL interface due to its suitable energy levels, high electron mobility, as well as good defect passivation property.[138] HATNT as an ETL produced better HPSCs with a PCE of 18.1% as compared to the PCBM-based counterpart (16.2%).

6.4.2.4 IDT-based Small Molecules as ETLs in HPSCs

Recently, IDTs (Figure 6.17) have been used as ETLs in HPSCs due to their good charge transport properties and excellent film filming properties. Manipulating the electron-donating fused ring core and the side chains could modify the structures and the corresponding opto-electronic properties of the PDIs.[139]

Functionalization of the terminal groups of IDTs with halogen atoms or Me groups is an effective strategy to manipulate the LUMO level, electron transport, and the hydrophobicity of the materials.[140,141] Zhan group employed a series of the fluorinated IDTs containing different numbers of fluorine atoms (INIC, INIC-1F, and INIC2F) as ETLs in the inverted HPSCs. The authors found that increasing

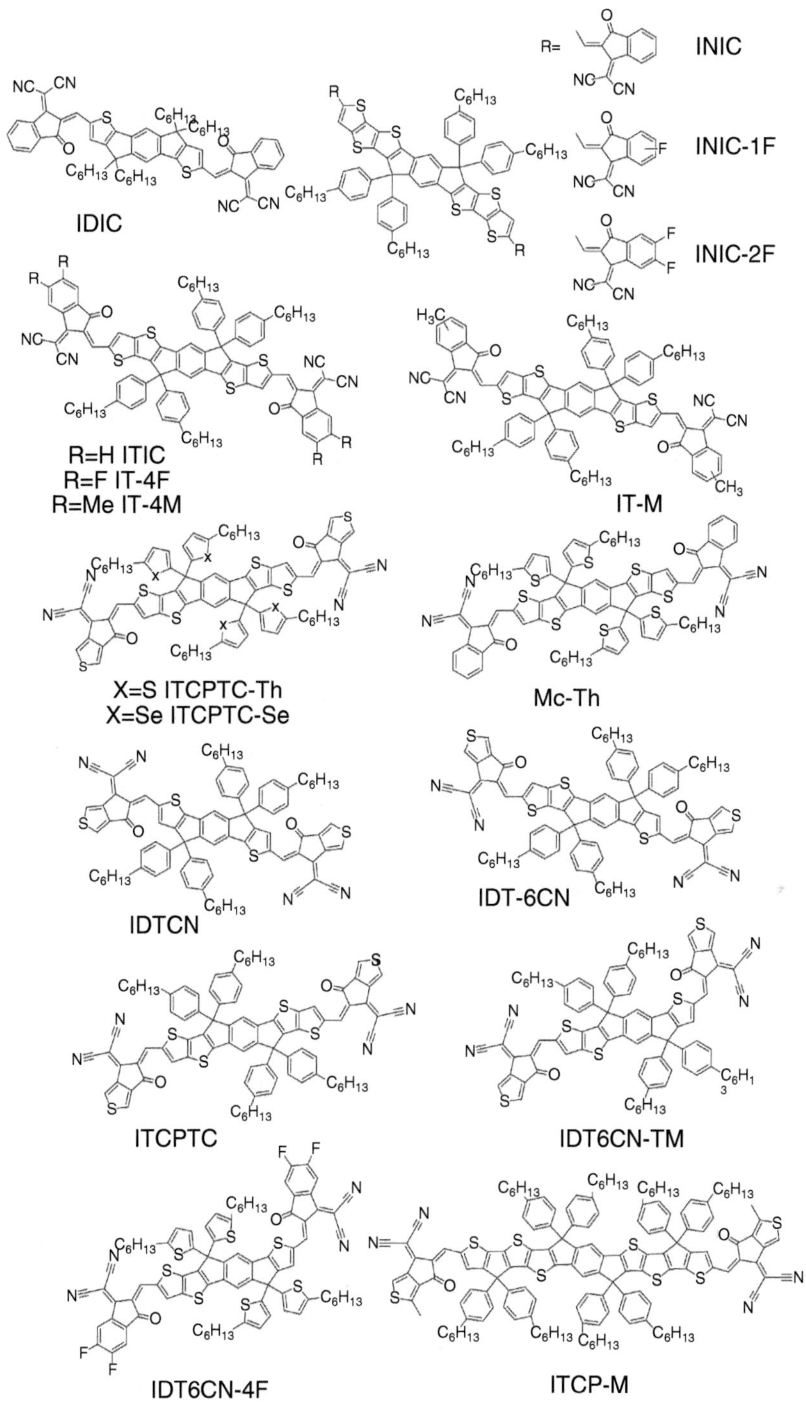

Figure 6.17 The chemical structures of IDTs with different terminal groups.

the number of fluorine atoms enhanced not only the electron mobility but also the hydrophobicity of the fluorinated IDTs. Consequently, INIC-2F as ETL improved the charge extraction and ambient stability of HSCs.[140] The electronegativity of the terminal groups has a vital impact on the energy levels of IDTs. Yang et al. obtained IT4F and IT-4M by modifying the terminal groups of the ITIC with fluorine atoms and methyl groups.[141] The LUMO levels of these materials drop from −3.96, −4.04 to −4.19 eV following the electron-withdrawing strength of the terminal groups of the IT-4M, ITIC, and IT-4F, respectively. IT-4M as ETL produced the best HPSCs with a PCE of 17.62% due to the well-aligned energy levels at the interfaces.

Introducing the chalcogen atoms, such as S, N, and O, into the terminal groups of the IDTs passivated the defects of the perovskite layer.[142,143] IDIC containing 1,1-dicyanomethylene-3- indanone terminal groups effectively passivated the Lewis acid traps of $MAPbI_3$. Moreover, this material has a high mobility up to 1.1×10^{-3} cm^2 V^{-1} s^{-1} and a suitable energy level, producing a HPSC with a PCE up to 19.5%.[142] A gradient IDIC intermediate layer formed by the antisolvent method passivated the defects at the grain boundaries and surface of the perovskite layer, promoting the charge extraction and leading to a PCE of 19.86%.[143] Interestingly, IDIC could also function well as ETL in inverted HPSCs, as its hydrophobic surface improved the crystallinity of the perovskite. This is further evidenced by the fact that IDIC as ETL yielded a HPSC with a PCE of 19.1% superior to TiO_2-based devices (17.4%).[144]

The side chains of the IDTs also play an important role in the charge transport and the film forming properties of the materials. Wu et al. employed two conjugated small molecules (ITCPTC-Th and ITCPTC-Se) composed of the fused rigid ladder type backbone as the model system and investigated the effect of the side chains on the electron transport properties and the performance of the inverted HPSCs.[145] Modification of the side chains of IDTs with alkyl chains and thiophenes improved the charge transport and morphology of IDTs. ITCPTC-Th with n-hexylthiophene side chain formed a smoother film, and transported the electrons faster than ITCPTC-Se with selenophene. A PCE of 17.11% was achieved in ITCPTC-Th-based device, which is higher than that of the ITCPTC-Se based device (16.12%). Changing the benzene rings of the side chains of the ITIC to the thiophenes of the ITIC-Th significantly enhanced the electron mobility of the latter owing to the intermolecular sulfur–sulfur interaction.[146]

Manipulating the symmetry and the conjugation length of IDT core could improve the film morphology and the electron mobility and extend the photo-response to the NIR absorption.[147] In 2018, Yang et al. synthesized three IDTs, IDTCN, IDT_6CN, and ITCPTC with five, six, and seven fused rings as the core. The larger conjugated core enhanced the electron transport capability. Moreover, ITCPTC adopted dominant "face-on" orientation in the films, facilitating the charge transport in the solar cells. Therefore, ITCPTC as ETL produced HPSCs with a highest PCE of 17.42% due to the enhanced charge collection. ITCPTC also functioned well as the intermediate layer to reduce the energy barrier, improving the PCE (19.51%) further.[148] ITCP-M with larger fused core extended the light absorption of the inverted HPSCs to the NIR region. Moreover, ITCP-M has suitable energy levels, smooth film morphology and decent electron mobility, producing inverted HPSCs with an excellent PCE of 19.15%.[147]

Manipulating the symmetry of the molecules could adjust the dipole–dipole and π–π interactions of IDT-based molecules. Yang et al. employed two asymmetric molecules, the methyl-substituted IDT6CN-TM and the fluorinated IDT6CN-4F, as ETLs in inverted HPSCs.[149] Overall, these two materials exhibit high electron mobility, efficient charge collection, and good surface coverage on the perovskite film, leading to inverted HPSCs with PCEs of 18.36 and 17.75%, respectively. The superior device performance of the IDT6CN-TM based device has been associated with the higher electron mobility and stronger π–π interaction of the material.

6.4.2.5 Other Small Molecules as ETLs

Some other small molecules (Figure 6.18) have also been employed as ETLs in HPSCs beyond the materials discussed in the previous sections. Both the PCE and the stability of the inverted HPSCs were improved by using hexadecafluorophthalocyaninatocopper ($F_{16}CuPc$) based ETL as compared to the fullerene-based HPSCs.[150] A small molecule B2F based on electron-accepting benzobis(thiadiazole) unit has a LUMO of −4.02 eV matching well with the conduction band of the perovskite layer. Furthermore, it could passivate the defects of the perovskite film due to the interaction between the sulfur and the uncoordinated lead. The inverted HPSCs using B2F as the ETL showed a PCE of 12.35%.[151] The locations of the sulfur atoms significantly influenced the energy level, the mobility, and the passivation capability of the two small molecules based on pyrene and thieno[3,4-b]thiophene(TT).[152] The inverted HPSC employing PTTI-1 as ETL showed a higher PCE of 15.37% than

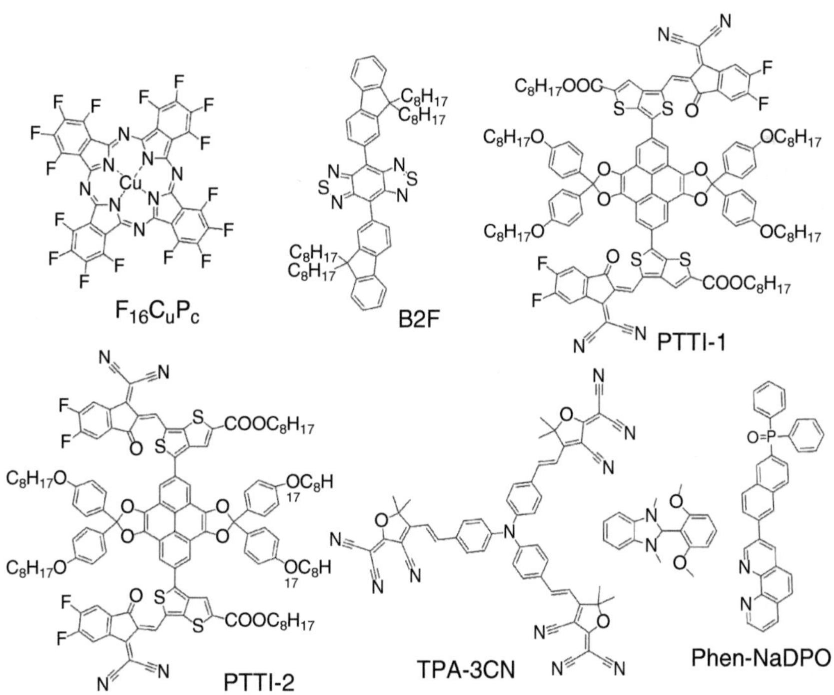

Figure 6.18 The chemical structures of other small molecules.

that of PTTI-2 (11.07%). TPA-3CN, containing a TPA core and 3CN capping groups, aligned well with the energy level of the perovskite layer, as the strong electron-withdrawing 3CN group down shifts the HOMO and LUMO levels of the material. The HPSCs employing the doped TPA-3CN as ETL exhibited a PCE of 19.2% comparable to that of PCBM based counterpart (19.1%).[153]

6.5 Conclusion

As the most promising next generation photovoltaic technology, HPSCs have been drawing increasing attention from academy and industry. Although this type of solar cells has witnessed sky-rocketing increase in the PCE, they are still far from the commercial criteria due to the poor stability. HTL and ETL are two critical components that determine the PCE and lifetime of the HPSCs. Organic HTL and ETL have great potential to produce low cost and high performance HPSCs due to the structural diversity and their low-cost production. Therefore, it is paramount to develop excellent HTLs and ETLs to improve the performance of the HPSCs.

This chapter summarizes in detail the advancement in the organic hole and electron transport materials for HPSCs. In the meantime, it discusses how the structural parameters of these materials (i.e., the backbone, side chain as well as the terminal and functional groups), and the aggregation behavior (i.e., conformation, crystallinity as well as orientation) influence their physical, chemical, and mechanical properties as well as the performance of the HPSCs. Here, it reveals a clear fact that the performance of HPSCs increases with the improved chemical structures of the charge transport materials. More importantly, it provides a deep understanding of the relationship between the structures of the charge transport materials and the device performance to guide the research community for rational design of the better charge transport materials for HPSCs.

Despite the progress discussed in the previous sections, the state-of-the-art organic HTL and ETL materials are still suffering from low charge mobility and resistance to the strikes from water, heat, as well as light. In order to realize the practical applications, future research needs to address the poor stability issue and the large-scale processability of the HPSCs. Besides developing robust perovskite materials, designing organic HTL and ETLs will persist as one of important research directions to improve the lifetime of the HPSCs. More specifically, future attention should be paid to the following aspects when it comes to the design of the charge transport materials: (1) developing simple synthetic routes of organic hole/electron transport materials in order to reduce the production cost; (2) manipulating the energy levels of the organic charge transport materials to minimize V_{OC} loss; (3) improving the thermal stability, hydrophobicity as well as the light stability of the HTL and ETL; (4) improving the charge mobility of the HTL and ETL. From the structural point of view, there is an intrinsic trade-off between some of the physical and electrical properties of the organic charge transport materials. It is very challenging to design ideal HTLs that possess simultaneously all the positive photophysical and electrical properties.

A more delicate control in the chemical structures of the charge transport layers is the key to achieve good performing HPSCs.

References

1 Green, M.A., Ho-Baillie, A., and Snaith, H.J. (2014). *Nature Photonics* 8: 506.
2 Kojima, A., Teshima, K., Shirai, Y., and Miyasaka, T. (2009). *Journal of the American Chemical Society* 131: 6050.
3 Pitaro, M., Tekelenburg, E.K., Shao, S., and Loi, M.A. (2022). *Advanced Materials* 34: 2105844.
4 Noel, N.K., Stranks, S.D., Abate, A. et al. (2014). *Energy & Environmental Science* 7: 3061.
5 Hao, F., Stoumpos, C.C., Cao, D.H. et al. (2014). *Nature Photonics* 8: 489.
6 Sheibani, E., Yang, L., and Zhang, J. (2020). *Solar RRL* 4: 2000461.
7 Im, J.H., Lee, C.R., Lee, J.W. et al. (2011). *Nanoscale* 3: 4088.
8 Kim, H.S., Lee, C.R., Im, J.H. et al. (2012). *Scientific Reports* 2: 591.
9 Lee, M.M., Teuscher, J., Miyasaka, T. et al. (2012). *Science* 338: 643.
10 Guixiang, L., Zuhong Z., Benjamin A-T. et al. (2025). *Nature Photonics*. https://doi.org/10.1038/s41566-025-01791-1
11 Bai, Y., Chen, H., Xiao, S. et al. (2016). *Advanced Functional Materials* 26: 2950.
12 Liu, D., Yang, J., and Kelly, T.L. (2014). *Journal of the American Chemical Society* 136: 17116.
13 Ke, W., Fang, G., Liu, Q. et al. (2015). *Journal of the American Chemical Society* 137: 6730.
14 Burschka, J., Pellet, N., Moon, S.J. et al. (2013). *Nature* 499: 316.
15 Takahashi, K. and Suzuki, Y. (2017). *Japanese Journal of Applied Physics* 56: 08MC04.
16 Shao, S., Abdu-Aguye, M., Qiu, L. et al. (2016). *Energy & Environmental Science* 9: 2444.
17 Jeong, M., Choi, I.W., Yim, K. et al. (2022). *Nature Photonics* 16: 119.
18 Jeng, J.Y., Chiang, Y.F., Lee, M.H. et al. (2013). *Advanced Materials* 25: 3727.
19 Sun, S., Salim, T., Mathews, N. et al. (2014). *Energy & Environmental Science* 7: 399.
20 Lin, Y., Chen, B., Zhao, F. et al. (2017). *Advanced Materials* 29: 1700607.
21 Shao, S. and Loi, M.A. (2020). *Advanced Materials Interfaces* 7: 1901469.
22 Cheng, T., Goddard, W.A., An, Q. et al. (2017). *Physical Chemistry Chemical Physics* 19: 2666.
23 Bach, U., Lupo, D., Comte, P. et al.(1998). *Nature* 395: 583.
24 Wang, S., Huang, Z., Wang, X. et al. (2018). *Journal of the American Chemical Society* 140: 16720.
25 Méndez, H., Heimel, G., Opitz, A. et al. (2013). *Angewandte Chemie International Edition* 52: 7751.
26 Jeon, N.J., Lee, H.G., Kim, Y.C. et al. (2014). *Journal of the American Chemical Society* 136: 7837.

27 Saliba, M., Orlandi, S., Matsui, T. et al. (2016). *Nature Energy* 1: 15017.
28 Jeon, N.J., Na, H., Jung, E.H. et al. (2018). *Nature Energy* 3: 682.
29 Jeong, M., Choi, I.W., Go, E.M. et al. (2020). *Science* 369: 1615.
30 Sheibani, E., Heydari, M., Ahangar, H. et al. (2019). *Solar Energy* 189: 404.
31 Hua, Y., Zhang, J., Xu, B. et al. (2016). *Nano Energy* 26: 108.
32 Xu, B., Bi, D., Hua, Y. et al. (2016). *Energy & Environmental Science* 9: 873.
33 Bi, D., Xu, B., Gao, P. et al. (2016). *Nano Energy* 23: 138.
34 Xu, B., Zhang, J., Hua, Y. et al. (2017). *Chem* 2: 676.
35 Liu, K., Yao, Y., Wang, J. et al. (2017). *Materials Chemistry Frontiers* 1: 100.
36 Li, M.H., Hsu, C.W., Shen, P.S. et al. (2015). *Chemical Communications* 51: 15518.
37 Rakstys, K., Saliba, M., Gao, P. et al. (2016). *Angewandte Chemie International Edition* 55: 7464.
38 Zhang, J., Hua, Y., Xu, B. et al. (2016). *Advanced Energy Materials* 6: 1601062.
39 Zhang, J., Xu, B., Johansson, M.B. et al. (2016). *Advanced Energy Materials* 6: 1502536.
40 Choi, H., Paek, S., Lim, N. et al. (2014). *Chemistry – A European Journal* 20: 10894.
41 Rodríguez-Seco, C., Méndez, M., Roldán-Carmona, C. et al. (2020). *Angewandte Chemie International Edition* 59: 5303.
42 Chen, J., Xia, J., Yu, H.J. et al. (2019). *Chemistry of Materials* 31: 5431.
43 Chen, J., Xia, J., Gao, W.J. et al. (2020). *ACS Applied Materials & Interfaces* 12: 21088.
44 Wang, L., Sheibani, E., Guo, Y. et al. (2019). *Solar RRL* 3: 1900196.
45 Qiu, J., Liu, H., Li, X. et al. (2019). *Solar RRL* 3: 1900202.
46 Rakstys, K., Paek, S., Drevilkauskaite, A. et al. (2020). *ACS Applied Materials & Interfaces* 12: 19710.
47 Zhu, X.D., Ma, X.J., Wang, Y.K. et al. (2019). *Advanced Functional Materials* 29: 1807094.
48 Huang, P., Manju, S.K., Sivakumar, G. et al. (2020). *ACS Applied Materials & Interfaces* 12: 22881.
49 Yu, W., Yang, Q., Zhang, J. et al. (2019). *ACS Applied Materials & Interfaces* 11: 30065.
50 Magomedov, A., Kasparavičius, E., Rakstys, K. et al. (2018). *Journal of Materials Chemistry C* 6: 8874.
51 Ji, Y., He, B., Lu, H. et al. (2019). *ChemSusChem* 12: 1374
52 Liu, F., Wu, F., Ling, W. et al. (2019). *ACS Energy Letters* 4: 2514.
53 Rakstys, K., Abate, A., Dar, M.I. et al. (2015). *Journal of the American Chemical Society* 137: 16172.
54 Fang, L., Zheng, A., Ren, M. et al. (2019). *ACS Applied Materials & Interfaces* 11: 39001.
55 Wang, J., Zhang, H., Wu, B. et al. (2019). *Angewandte Chemie International Edition* 58: 15721.
56 García-Benito, I., Zimmermann, I., Urieta-Mora, J. et al. (2018). *Advanced Functional Materials* 28: 1801734.
57 Urieta-Mora, J., García-Benito, I., Zimmermann, I. et al. (2020). *The Journal of Organic Chemistry* 85: 224.

58 Urieta-Mora, J., García-Benito, I., Zimmermann, I. et al. (2019). *Journal of Materials Chemistry C* 7: 6656.

59 Rojas, D.E.M., Cho, K.T., Zhang, Y. et al. (2018). *Advanced Energy Materials* 8: 1800681.

60 Shao, J.Y., Yang, N., Guo, W. et al. (2019). *Chemical Communications* 55: 13406.

61 Zhang, X., Zhou, Z. a., Ma, S. et al. (2020). *Chemical Communications* 56: 3159.

62 Daskeviciute-Geguziene, S., Zhang, Y., Rakstys, K. et al. (2022). *Angewandte Chemie International Edition* 61: e202113207.

63 Chiykowski, V.A., Cao, Y., Tan, H. et al. (2018). *Angewandte Chemie International Edition* 57: 15529.

64 Wang, X., Zhang, J., Yu, S. et al. (2018). *Angewandte Chemie International Edition* 57: 12529.

65 Mei, Y., Shen, Z., Kundu, S. et al. (2021). *Chemistry of Materials* 33: 4679.

66 Shen, C., Wu, Y., Zhang, H. et al. (2019). *Angewandte Chemie International Edition* 58: 3784.

67 Yin, C., Lu, J., Xu, Y. et al. (2018). *Advanced Energy Materials* 8: 1800538.

68 Li, Y., Scheel, K.R., Clevenger, R.G. et al. (2018). *Advanced Energy Materials* 8: 1801248.

69 Pham, H.D., Jain, S.M., Li, M. et al. (2020). *Advanced Electronic Materials* 6: 1900884.

70 Pham, H.D., Hayasake, K., Kim, J. et al. (2018). *Journal of Materials Chemistry C* 6: 3699.

71 Wu, S., Chen, C., Liu, Q. et al. (2018). *ACS Applied Energy Materials* 1: 5539.

72 Chen, S. and Shi, G. (2017). *Advanced Materials* 29: 1605448.

73 Hu, Z., Fu, W., Yan, L. et al. (2016). *Chemical Science* 7: 5007.

74 Huang, C., Fu, W., Li, C.Z. et al. (2016). *Journal of the American Chemical Society* 138: 2528.

75 Cao, Y., Li, Y., Morrissey, T. et al. (2019). *Energy & Environmental Science* 12: 3502.

76 Zhang, J., Sun, Q., Chen, Q. et al. (2019). *Advanced Functional Materials* 29: 1900484.

77 Chen, Y., Xu, X., Cai, N. et al. (2019). *Advanced Energy Materials* 9: 1901268.

78 Jiang, K., Wang, J., Wu, F. et al. (2020). *Advanced Materials* 32: 1908011.

79 Chen, H., Fu, W., Huang, C. et al. (2017). *Advanced Energy Materials* 7: 1700012.

80 Chiu, Y.L., Li, C.W., Kang, Y.H. et al. (2022). *ACS Applied Materials & Interfaces* 14: 26135.

81 Liu, J., Qiu, L., and Shao, S. (2021). *Journal of Materials Chemistry C* 9: 16143.

82 Docampo, P., Ball, J.M., Darwich, M. et al. (2013). *Nature Communications* 4: 2761.

83 Shao, Y., Xiao, Z., Bi, C. et al. (2014). *Nature Communications* 5: 5784.

84 Shao, S., Liu, J., Fang, H.H. et al. (2017). *Advanced Energy Materials* 7: 1701305.

85 Shao, Y., Yuan, Y., and Huang, J. (2016). *Nature Energy* 1: 15001.

86 Yan, W., Li, Y., Sun, W. et al. (2014). *RSC Advances* 4: 33039.

87 Shao, S., Liu, J., Portale, G. et al. (2018). *Advanced Energy Materials* 8: 1702019.

88 Wu, C.G., Chiang, C.H., and Chang, S.H. (2016). *Nanoscale* 8: 4077.

89 Momblona, C., Gil-Escrig, L., Bandiello, E. et al. (2016). *Energy & Environmental Science* 9: 3456.

90 Xue, Q., Bai, Y., Liu, M. et al. (2017). *Advanced Energy Materials* 7: 1602333.
91 Xing, Y., Sun, C., Yip, H.L. et al. (2016). *Nano Energy* 26: 7.
92 Tian, C., Betancourt-Solis, G., Nan, Z. et al. (2021). *Science Bulletin* 66: 339.
93 Dai, S.M., Deng, L.L., Zhang, M.L. et al. (2017). *Inorganica Chimica Acta* 468: 146.
94 Tian, C., Kochiss, K., Castro, E. et al. (2017). *Journal of Materials Chemistry A* 5: 7326.
95 Meng, X., Bai, Y., Xiao, S. et al. (2016). *Nano Energy* 30: 341.
96 Bai, Y., Dong, Q., Shao, Y. et al. (2016). *Nature Communications* 7: 12806.
97 Li, C.Z., Chueh, C.C., Ding, F. et al. (2013). *Advanced Materials* 25: 4425.
98 Yan, K., Liu, Z.X., Li, X. et al. (2018). *Organic Chemistry Frontiers* 5: 2845.
99 Huang, J., Yu, X., Xie, J. et al. (2016). *ACS Applied Materials & Interfaces* 8: 34612.
100 Liu, J., Guo, X., Qin, Y. et al. (2012). *Journal of Materials Chemistry* 22: 1758.
101 Xie, J., Yu, X., Sun, X. et al. (2016). *Nano Energy* 28: 330.
102 Zhu, Z., Chueh, C.C., Lin, F., and Jen, A.K.Y. (2016). *Advanced Science* 3: 1600027.
103 Nielsen, C.B., Holliday, S., Chen, H.Y. et al. (2015). *Accounts of Chemical Research* 48: 2803.
104 Liu, S.Y., Wu, C.H., Li, C.Z. et al. (2015). *Advanced Science* 2: 1500014.
105 Liu, Y., Mu, C., Jiang, K. et al. (2015). *Advanced Materials* 27: 1015.
106 Wu, J.l., Huang, W.K., Chang, Y.C. et al. (2017). *Journal of Materials Chemistry A* 5: 12811.
107 Zou, D., Yang, F., Zhuang, Q. et al. (2019). *ChemSusChem* 12: 1155.
108 Karuppuswamy, P., Chen, H.C., Wang, P.C. et al. (2018). *ChemSusChem* 11: 415.
109 Jiang, K., Wu, F., Yu, H. et al. (2018). *Journal of Materials Chemistry A* 6: 16868.
110 Wang, R., Jiang, K., Yu, H. et al. (2019). *Materials Chemistry Frontiers* 3: 2137.
111 Castro, E., Sisto, T.J., Romero, E.L. et al. (2017). *Angewandte Chemie International Edition* 56: 14648.
112 Yang, L., Wu, M., Cai, F. et al. (2018). *Journal of Materials Chemistry A* 6: 10379.
113 Wang, H., Yang, F., Xiang, Y. et al. (2019). *Journal of Materials Chemistry A* 7: 24191.
114 Zhang, H., Xue, L., Han, J. et al. (2016). *Journal of Materials Chemistry A* 4: 8724.
115 Akbulatov, A.F., Frolova, L.A., Griffin, M.P. et al. (2017). *Advanced Energy Materials* 7: 1700476.
116 Hou, Y., Chen, W., Baran, D. et al. (2016). *Advanced Materials* 28: 5112.
117 Zhu, Z., Xu, J.Q., Chueh, C.C. et al. (2016). *Advanced Materials* 28: 10786.
118 Chen, C., Cheng, M., Li, H. et al. (2018). *Materials Today Energy* 9: 264.
119 Wang, K., Xia, P., Wang, K. et al. (2020). *ACS Applied Materials & Interfaces* 12: 9528.
120 Nakayama, H., Zheng, Y., Schneider, J.A. et al. (2019). *Journal of Materials Chemistry C* 7: 11126.
121 Jung, B.J., Lee, K., Sun, J. et al. (2010). *Advanced Functional Materials* 20: 2930.
122 Guha, S., Goodson, F.S., Corson, L.J., and Saha, S. (2012). *Journal of the American Chemical Society* 134: 13679.
123 Heo, J.H., Lee, S.C., Jung, S.K. et al. (2017). *Journal of Materials Chemistry A* 5: 20615.

124 Jung, S.K., Heo, J.H., Lee, D.W. et al. (2018). *Advanced Functional Materials* 28: 1800346.

125 Jung, S.K., Heo, J.H., Lee, D.W. et al. (2019). *ChemSusChem* 12: 224.

126 Jung, S.K., Heo, J.H., Oh, B.M. et al. (2020). *Advanced Functional Materials* 30: 1905951.

127 Shao, J., Guo, X., Shi, N. et al. (2019). *Science China Materials* 62: 497.

128 Nakamura, T., Shioya, N., Shimoaka, T. et al. (2019). *Chemistry of Materials* 31: 1729

129 Shaikh, D.B., Ali Said, A., Wang, Z. et al. (2019). *ACS Applied Materials & Interfaces* 11: 44487.

130 Shaikh, D.B., Said, A.A., Bhosale, R.S. et al. (2018). *Asian Journal of Organic Chemistry* 7: 2294.

131 Liu, W., Shaikh, D.B., Rao, P.S. et al. (2020). *Chemistry – An Asian Journal* 15: 112.

132 Said, A.A., Wagalgave, S.M., Xie, J. et al. (2019). *Journal of Solid State Chemistry* 270: 51.

133 Ge, C., Wu, W., Hu, L. et al. (2018). *Organic Electronics* 61: 113.

134 Li, J. and Zhang, Q. (2015). *ACS Applied Materials & Interfaces* 7: 28049.

135 Zhao, D., Zhu, Z., Kuo, M.Y. et al. (2016). *Angewandte Chemie International Edition* 55: 8999.

136 Gu, P.Y., Wang, N., Wu, A. et al. (2016). *Chemistry – An Asian Journal* 11: 2135.

137 Gu, P.Y., Wang, N., Wang, C. et al. (2017). *Journal of Materials Chemistry A* 5: 7339.

138 Wang, N., Zhao, K., Ding, T. et al. (2017). *Advanced Energy Materials* 7: 1700522.

139 Yan, C., Barlow, S., Wang, Z. et al. (2018). *Nature Reviews Materials* 3: 18003.

140 Liu, K., Dai, S., Meng, F. et al. (2017). *Journal of Materials Chemistry A* 5: 21414.

141 Liu, X., Li, X., Zou, Y. et al. (2019). *Journal of Materials Chemistry A* 7: 3336.

142 Lin, Y., Shen, L., Dai, J. et al. (2017). *Advanced Materials* 29: 1604545.

143 Ma, J., Zheng, M., Chen, C. et al. (2018). *Advanced Functional Materials* 28: 1804128.

144 Zhang, M., Zhu, J., Liu, K. et al. (2017). *Journal of Materials Chemistry A* 5: 24820.

145 Wu, F., Gao, W., Yu, H. et al. (2018). *Journal of Materials Chemistry A* 6: 4443.

146 Lin, Y., Zhao, F., He, Q. et al. (2016). *Journal of the American Chemical Society* 138: 4955.

147 Wu, F., Gao, W., Zhu, L. et al. (2020). *Solar RRL* 4: 1900565.

148 Zhu, L., Gao, W., Wu, F. et al. (2018). *Journal of Materials Chemistry A* 6: 18044.

149 Gao, W., Wu, F., Liu, T. et al. (2019). *Journal of Materials Chemistry A* 7: 2412.

150 Jin, F., Liu, C., Hou, F. et al. (2016). *Solar Energy Materials and Solar Cells* 157: 510.

151 Zhu, L., Song, C., Li, X. et al. (2017). *Journal of Materials Chemistry C* 5: 10777.

152 Chen, W., Said, A.A., Wang, Z. et al. (2019). *ACS Applied Energy Materials* 2: 5716.

153 Chen, C., Li, H., Ding, X. et al. (2018). *ACS Applied Materials & Interfaces* 10: 38970.

7

Molecular Aggregates as Active Layers in Organic Solar Cells

Jin He[1], *Yibin Li*[1], *Zhong'an Li*[1*], *and Zhen Li*[2]

[1] *School of Chemistry and Chemical Engineering, Huazhong University of Science and Technology, Wuhan, China*
[2] *Department of Chemistry, Wuhan University, Wuhan, China*

*Corresponding author. lizha@hust.edu.cn

7.1 Introduction

To date, it is quite urgent to develop renewable energy to reduce the proportion of fossil fuels due to the increasingly serious climate change and environmental pollution. Undoubtedly, the solar energy has been considered as the most available renewable energy sources in both academic and industrial fields because of its abundant and inexhaustible characteristics. And great efforts have been continuously devoted on developing new photovoltaic (PV) technologies that can directly convert the sunlight into electricity, which have experienced several generations of development since 1950s. At present, the first and second generation of PV technique, i.e., crystalline silicon and inorganic thin-film solar cells still dominate the commercial market, however, their sophisticated, high-cost manufacture is unfavorable from both the economic and environmental perspective. As the new generation of PV technologies (third generation), organic solar cells (OSCs), perovskite solar cells (PVSCs), quantum dot solar cells (QDSCs) and dye-sensitized solar cells (DSSCs) have attracted more and more attentions due to the unique advantages of low cost, solution processing ability, light weight, potential of making flexible and/or semitransparent devices, etc., and impressive advance on the power conversion efficiency (PCE) has been made in the past several years.[1]

The photoactive layer is the key component of any type of solar cells, which convert the absorbed light into the charge carriers. For OSCs, the photoactive layer is composed of organic polymeric materials, so distinct device functions such as colorful, flexible, stretchable, semi-transparent, and even suitable for indoor light can be realized their diversity of molecular functionalities, easily tunable structure and opto-electronic properties.[2] Indeed, the PV effect of organic materials has been observed as early as the 1900s,[3] and the first OSCs was demonstrated by Pope et al. in 1959 with a single-layer device structure, in which anthracene crystals were used as the photoactive layer sandwiched by two metal electrodes.[4] However, the main driving force for dissociating photogenerated excitons into free charge carriers in single-layer OSCs is the built-in electric field generated by the work function difference between two metal electrodes, so it is quite difficult to fully dissociate the photogenerated excitons, making the PV performance very poor (PCE of ~0.1%). Until 1986, a monumental breakthrough was achieved by Deng, who developed a bilayer-type OSC by stacking p-type and n-type layers via vacuum deposition method. Such device structure engineering enables a striking (PCE of 0.95% at that time, thanks to the effectively improved charge separation at the interface.[5] Since then, OSCs have gradually attracted more interests, but not too much progress has been realized due to poor charge transport in bilayer OSCs.

In 1995, Heeger and Friend *et. al.* individually achieved another breakthrough in OSCs by inventing the bulk heterojunction (BHJ, Figure 7.1a) device structure,[6,7] in which a p-type donor is mixing with a n-type acceptor through simple solution

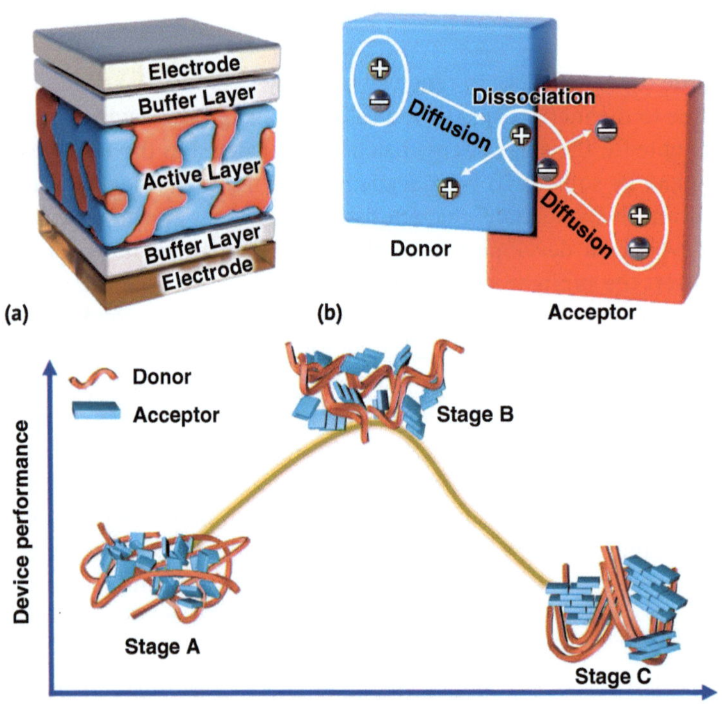

Figure 7.1 Illustration of (a) device structure and (b) operation mechanism of BHJ-OSCs; (c) schematic diagram of three stage evolution of BHJ film morphology. Stage A: Highly amorphous film morphology; Stage B: Crystalline film morphology with suitable nanoscale phase separation; Stage C: Highly crystalline film morphology with large phase separation. *Source*: Zhong'an Li.

processing method to constitute as a binary photoactive layer. Such BHJ structure can simultaneously improve the charge separation and charge transport by forming bicontinuous and interpenetrated nanoscale network, then can not only enable bulk p–n junction interfaces but also achieve ultrafast photoinduced charge transfer ($t < 100$ *fs*) (Figure 7.1b).[8] Therefore, in the following two decades, OSCs have achieved significant PCE improvement, which is mainly due to the innovative design of various p-type donor materials.

Before 2015, fullerene derivatives have dominated n-type acceptor materials in OSCs, and the highest PCE of fullerene-based binary OSCs has approached ~12%.[9] However, fullerene derivatives have some intrinsic shortcomings such as weak light absorption in the visible and NIR region, large recombination loss, and self-aggregation behaviors, badly limiting the further improvement of device performance. Since 2015, non-fullerene acceptors (NFAs) have attracted more and more interest due to the emergence of fused ring electron acceptors (FREAs),[10,11] which have significantly elevated the performance of OSCs, with PCEs exceeding 19% achieved recently.[12–14] In comparison with the fullerene acceptors, NFAs in particular, FREAs can be designed with strong and intense near-infrared (NIR) absorption, which significantly enhances the light-harvesting ability of OSCs to increase the photocurrent. What's more, the driving force of exciton dissociation required for NFA-based BHJ blend is much smaller than that for fullerene-based counterpart, which thus can effectively boost the device efficiency by decreasing the non-radiative recombination losses.[15]

In addition to the innovative development of new photoactive materials, the optimization of BHJ film morphology is another critical driving force of the impressive performance improvement of OSCs.[8,16] However, the formation of a BHJ film with multiple components indeed is a very complicated process, which can be affected by a variety of factors such as the chemical structures of photoactive components, the blend compositions, the deposition method, the processing solvents and/or additives, post-treatments, etc.[17,18] Nonetheless, it is widely recognized that the BHJ film morphology is highly related with the aggregation behaviors of photoactive components during BHJ formation (Figure 7.1c). In Stage A, amorphous film morphology can be generated when donor and acceptor material have a great compatibility and both of them have a low crystallinity. This morphology will increase p-n junction interface, however also negatively affect the charge transport channel due to the lack of bicontinuous interpenetrating network. Stage B describes a crystalline film morphology with nanoscale phase separation (10~20 nm), if donor and acceptor materials exhibit good compatibility and suitable molecular crystallinity with favorable face-to-face packing. This is the optimal BHJ morphology due to simultaneous improvement of p-n junction interface and charge transport. Stage C shows a highly crystalline film morphology with large phase separation, caused by low compatibility between donor and acceptor materials and/or their self-aggregations. This will not only induce significantly decreased p-n junction interface, but also a rough surface that is harmful to the interface contact. As the result, the molecular aggregates in BHJ film play a key role in determining the device performance of OSCs, however the information on the relationship of molecular structure-molecular aggregates-device performance has not been well understood. In this chapter, we thus have comprehensively summarized the effects of molecular aggregates of both donors and acceptors in the device performance.

7.1.1 Effects of Donor Aggregates

For donor materials in OSCs, small molecule donors have attracted particular research interests due to their definite chemical structures and molecular weights, no batch-to-batch difference and good reproducibility in device performance.[19] Besides, a striking property of small molecule donors is their ability to form aggregates in appropriate conditions due to strong intermolecular interactions.[20] The aggregates demonstrate distinctly different optical and electrical properties compared to single molecules, making critical effects on the photophysical and electronic properties as well as the film morphology of photoactive films.[21] As a result, significant efforts have been devoted to studying molecular aggregates forming in BHJ films and their effects on the photovoltaic performance,[22] and many types of small molecule donors have been investigated including squaraines (SQs),[23] boron dipyrromethenes (BODIPYs),[24] porphyrin derivatives,[25] and so on. In general, the strategies for tuning the aggregation behaviors of molecules could be divided into two categories: (i) modification of molecular structures; (ii) control of processing methods/conditions.

The first study, which presented the effect of J-aggregates on the OSC performance, was reported by Würthner and co-workers in 2009.[26] As shown in Figure 7.2, a squaraine dye, namely SQ1, was found to form a densely slip-stacked packing arrangement with J-type coupling in the crystal lattice with enhanced π–π contacting, which is thus beneficial for efficient exciton and charge-carrier transport. Results from organic field-effect transistors (OFETs) showed that SQ1 thin films exhibited a hole-carrier mobility (μ_h) of 1.5×10^{-5} cm^2 V^{-1}s^{-1}, which can be further increased to 1.3×10^{-3} cm^2 V^{-1}s^{-1} after annealing due to the brickwork arrangement in films. Owing to the well-packed J-type domains in the SQ1:PC$_{61}$BM blend film after annealing at 110 °C for 15 min, the solution-processed BHJ OSCs achieved extended light-harvesting ability, balanced free charge-carrier transport, and high photoresponse in the NIR region, thereby giving rise to a PCE of 1.79% and a high short-circuit current (J_{SC}) value of 12.6 mA cm^{-2}. However, the improved J_{SC} was

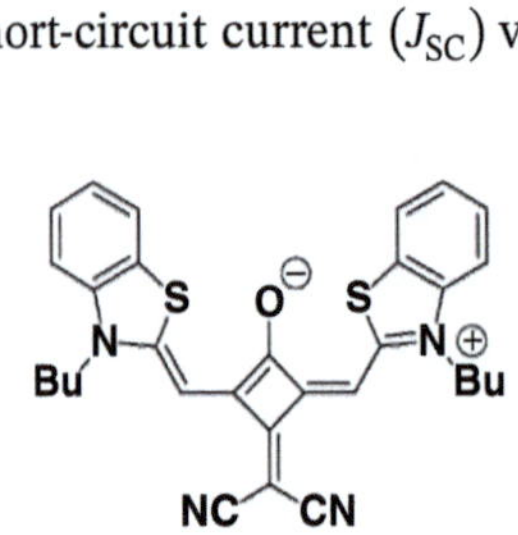
(a) SQ1

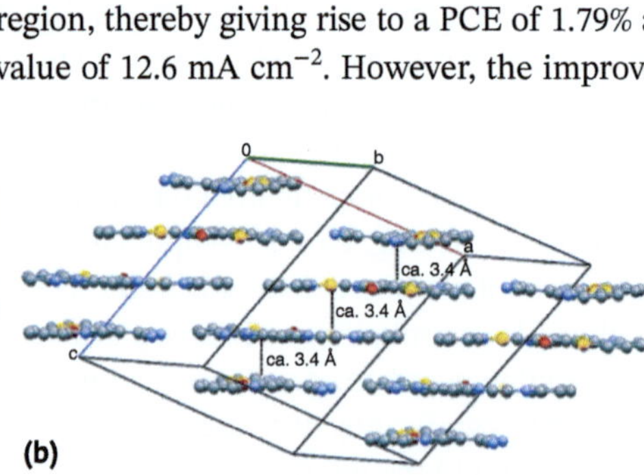

(b)

Figure 7.2 (a) Molecular structure of SQ1. (b) Spatial slip-stacked packing arrangement of SQ1 in the crystal lattice. *Source*: Reproduced with permission from [26]/John Wiley & Sons.

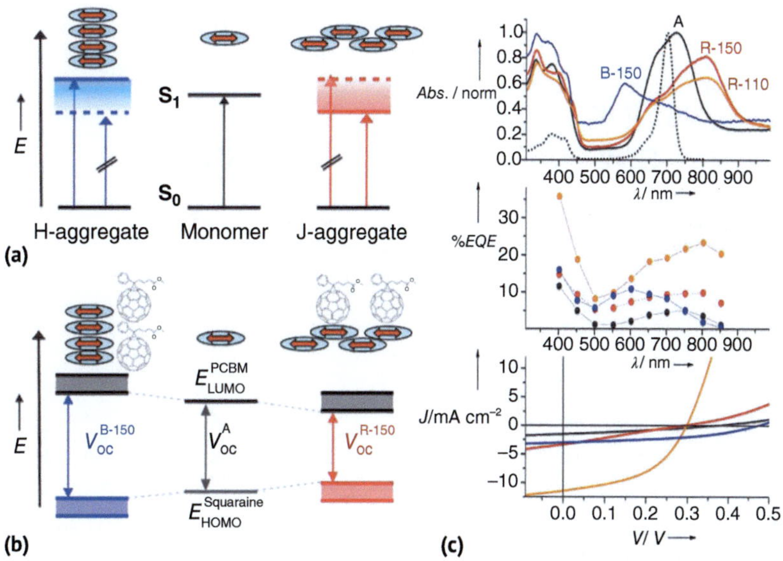

Figure 7.3 (a) Schematic representation of the ground and excited state in a monomer (middle, black) and H-aggregates (left, blue) and J-aggregates (right, red). (b) Qualitative schematic representation of the HOMO level of the SQ donor and LUMO level of the $PC_{61}BM$ acceptor in amorphous (middle, black), H-aggregate (left, blue), and J-aggregate (right, red) of SQ1:$PC_{61}BM$ blends. (c) Absorption spectra, EQE spectra, and J–V characteristic of the SQ1:$PC_{61}BM$ BHJ OSCs. *Source*: [27]/Reproduced with permission of Royal Society of Chemistry.

achieved upon annealing at 110 °C for 15 min, which conversely slightly reduced the open-circuit voltage (V_{OC}). Therefore, further investigations were demanded to improve the electronic properties of molecular aggregates toward the simultaneous achievement of high V_{OC} and high J_{SC}.

Soon after, Meerholz and Würthner et al. further carefully investigated the effects of molecular aggregation type of SQ1 on device performance.[27] They found that SQ1 molecules were easy to form aggregates due to π–π stacking (Figure 7.3a), either H-aggregates or J-aggregates in the BHJ layer when blended with $PC_{61}BM$ acceptor, which can be controlled by thermal annealing (TA) conditions (Figure 7.3b). As shown in Figure 7.3c, different aggregation states in the SQ1:$PC_{61}BM$ blend films could be obtained depending on the annealing temperature. In the absorption spectra, the band A represented the amorphous state from an as-cast film, and the band R and B represented red-shifted (J-aggregate) and blue-shifted (H-aggregate) domains obtained after annealing, respectively. The H-aggregate showed an increase of excitonic bands, thus resulting in a blue-shift in the absorption maximum and an increase in the V_{OC}. On the contrary, the J-aggregate exhibited a red-shift of the absorption band that resulted in significant improvement in the external quantum efficiency (EQE) and J_{SC} but a decrease in the V_{OC}.

Apart from TA, modification of molecular structures has been recognized as the most effective and common method to tune the aggregation behaviors of photoactive materials in solid films. As shown in Figure 7.4a, Kido et al.

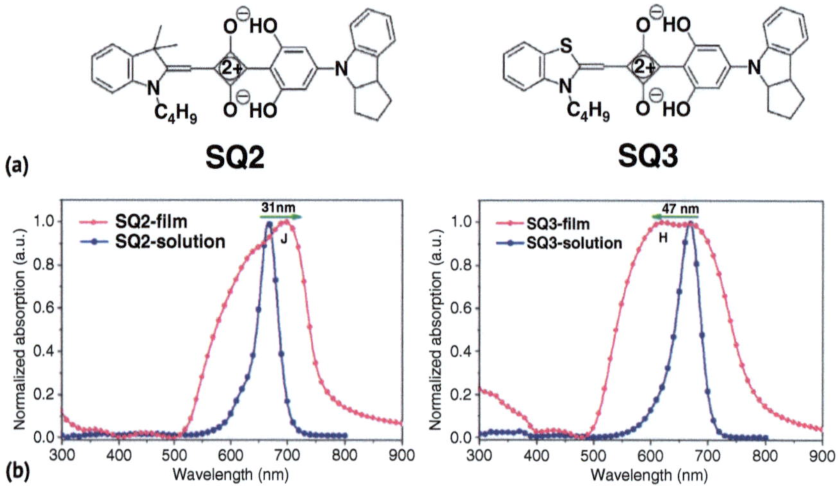

Figure 7.4 (a) Molecular structures of SQ2 and SQ3. (b) Absorption spectra of SQ2 and SQ3 in solution and thin films. *Source*: Reproduced with permission from [28]/Elsevier.

synthesized two unsymmetrical SQs with different electron-rich subunits (SQ2 and SQ3).[28] The two branched methyl groups on the indolenine group of SQ2 can hinder the π-π stacking, resulting in J-aggregates in thin film with a red shift of 31 nm (Figure 7.4b, left). On the contrary, the benzothiazole segment of SQ3 showed a more planar structure, enabling H-aggregates in thin film with a blue shift of 47 nm (Figure 7.4b, right). As a result, the J-aggregates of SQ2 led to a significant increase in the J_{SC}. In addition, although the μ_h of the SQ2 neat film was lower than that of the SQ3, the μ_h of the SQ2:$PC_{71}BM$ blend film (2.03×10^{-5} cm^2 V^{-1} s^{-1}) was nearly 4 times higher than that of the SQ3:$PC_{71}BM$ (5.10×10^{-6} cm^2 V^{-1} s^{-1}), which was ascribed to the J-aggregates of SQ2 maintaining a sufficiently larger interface with the acceptor $PC_{71}BM$. Benefiting from the simultaneously enhanced V_{OC}, J_{SC}, and fill factor (*FF*), the BHJ-OSCs based on SQ2:$PC_{71}BM$ exhibited a much higher PCE of 4.27% than that of the SQ3:$PC_{71}BM$ (2.78%).

In 2014, Loo et al. also investigated the structure-aggregation-property relationships of a series of isoindigo-based derivatives (Figure 7.5a) and found that the chemical structure modification at the different positions of the core with electron-rich or electron-poor substituents played important roles in the molecular organization and final device performance of OSCs.[29] Absorption results shown in Figure 7.5a indicated that 6,6′-derivatives generally formed J-aggregates and well-organized lamellar structures in the solid state, particularly when the substituent was electron-rich benzothiophene (BT). This is because the BT unit could couple strongly to the electron-poor isoindigo core on neighboring compounds, resulting in head-to-tail type stacking. On the contrary, a weaker charge transfer band was observed in the 5,5′-substituted derivatives, accompanied by the formation of H-aggregates in the solid state. Among these molecules, the device with the incorporation of 6,6′-BT as the electron donor showed the highest PCE of 0.47% compared to others (0.01%), because of the strongest and broadest light absorption and

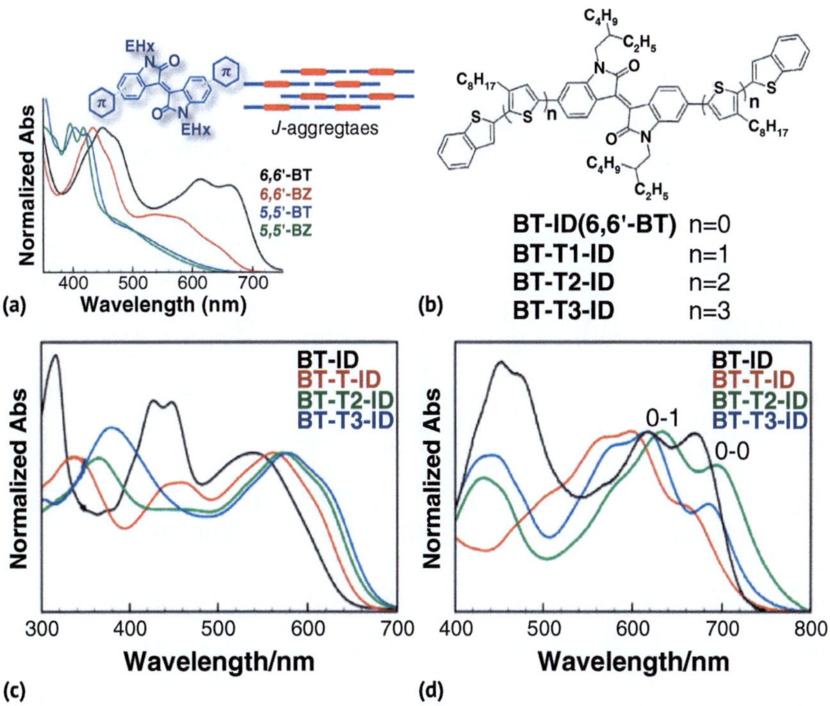

Figure 7.5 (a) Molecular structures and absorption spectra of 6,6′- and 5,5′-substituted isoindigo-based compounds. *Source*: [29]/American Chemical Society. (b) Molecular structures of 6,6′-BT derivatives. Absorption spectra of the isoindigo-based derivatives (c) in solution and (d) in annealed thin films. *Source*: [30]/American Chemical Society.

well-organized film morphology derived from J-aggregation. As an extension of that work, the same group further reported the second-generation isoindigo-based molecular semiconductors based on 6,6′-BT (Figure 7.5b), in which the conjugation length was extended through the incorporation of octyl-thiophene moieties at the 6,6′-position. Although the resulting isoindigo derivatives showed a red-shift in solution absorption with increasing number of octyl-thiophene units due to the increase in the conjugation length (Figure 7.5c), the measured thin film spectra suggested that the extended isoindigo derivatives formed H-aggregates in the solid state (Figure 7.5d). In particular, BT-T2-ID with the most crystalline and ordered state showed different molecular packing motif from other dyes, achieving the best PCE up to 3.4% when blended with $PC_{61}BM$.[30]

In 2017, Kido et al. also found that the processing solvents can control the aggregation behavior of squaraines without modifying the molecular structure.[31] It has been well known that a J-aggregation state is kinetically meta-stable, while a H-aggregation state is thermodynamically stable.[32] In this work, they demonstrated low-boiling solvent with fast evaporation process allows molecules to realize a kinetically stable state, i.e., J-aggregation, whereas high-boiling point solvent leads to H-aggregation (Figure 7.6a). There are three solvents such as chloroform (CF), chlorobenzene (CB), and o-dichlorobenzenes (*o*-DCB) used, whose boiling point follows a trend of

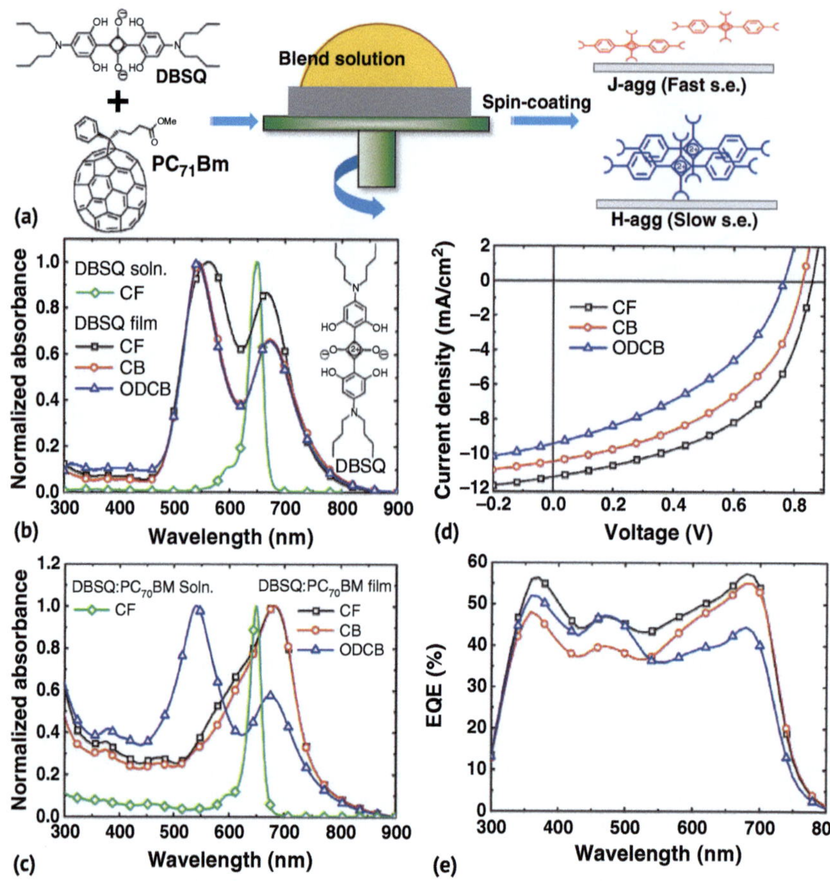

Figure 7.6 (a) Schematic representations of the structural characteristics and aggregation behavior of DBSQ in the DBSQ: $PC_{71}BM$ blend films spin-coated from various solvents (fast s.e.: fast solvent evaporation; slow s.e.: slow solvent evaporation). UV–vis absorption spectra of (b) DBSQ and (c) DBSQ: $PC_{71}BM$ blend in CF solution and as thin solid-state films cast from CF, CB, and *o*-DCB, respectively. (d) *J–V* characteristics under AM 1.5G solar spectrum at 100 mW cm^{-2} illumination; (e) EQE spectra measured at the AM 1.5G condition of DBSQ: $PC_{71}BM$ BHJ cells processed from various solvents. *Source*: Reproduced with permission from [31]/Elsevier.

CF < CB < *o*-DCB. Although both H- and J-aggregates of DBSQ molecules formed in the neat DBSQ films coated from all three processing solvents (Figure 7.6b), the aggregation behaviors of DBSQ molecules in the DBSQ: $PC_{71}BM$ blend films were found to be significantly different in comparison with those in neat films. When DBSQ: $PC_{71}BM$ blend films were deposited by CF or CB with fast solvent evaporation rates, there was only J-aggregates presented, while H-aggregates were more pronounced by using high-boiling-point *o*-DCB as the processing solvent (Figure 7.6c). Due to the stronger absorption in the NIR range and a deeper HOMO level from J-aggregates, the device performance of DBSQ:$PC_{71}BM$-based BHJ devices processed by CF and CB can be effectively improved (Figure 7.6d, e), showing much enhanced J_{SC} and V_{OC} values, and thus resulting in higher PCEs of 5.02% and 4.03%, respectively, compared to that derived from the H-aggregate active layer (2.87%).

Solvent vapor annealing (SVA) is also a common method to tune the molecular aggregation behaviors in blend films, which allows solvent vapor to penetrate into the film and downshifts the glass transition temperature, softening the blend films and endowing the molecules with mobility to crystallize. Related research by Peng and co-workers on a porphyrin derivative electron donor of ZnP2-DPP (Figure 7.7a) showed that the homogeneous morphology with unapparent nano-scale phase separation in the as-cast ZnP2-DPP: $PC_{61}BM$ blend film (Figure 7.7b) can cause severe bimolecular recombination, leading to an extremely low J_{SC} of 6.88 mA cm^{-2} and FF of 31.02%, as shown in Figure 7.7c.[33] Interestingly, after being treated with CS_2 SVA, an optimized blend film with a fiber-like structure can be achieved as shown in Figure 7.7d due to the formation of J-aggregates, which effectively improved the hole transport of BHJ films. Consequently, a significantly enhanced PCE of 9.47% was achieved, attributed to the simultaneous improvement of J_{SC} (21.40 mA cm^{-2}) and FF (69.14%). This work further indicates that the formation of J-aggregates could be an effective approach for tuning the BHJ film morphology toward high performance OSCs.

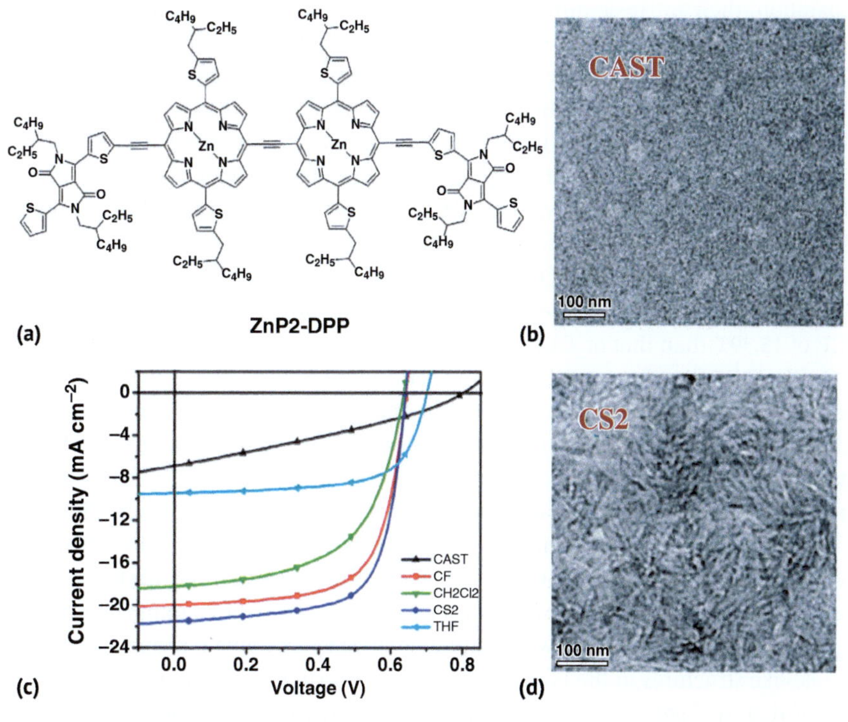

Figure 7.7 (a) Molecular structure of ZnP2-DPP. (b) TEM images of the as-cast ZnP2-DPP: $PC_{61}BM$ blend film. (c) The J–V curves of the ZnP2-DPP: $PC_{61}BM$ based solar cells under different solvent vapor annealing processing conditions. (d) TEM images of the CS_2 vapor treated ZnP2-DPP: $PC_{61}BM$ blend films. *Source*: Reproduced with permission [33]/The Royal Society of Chemistry.

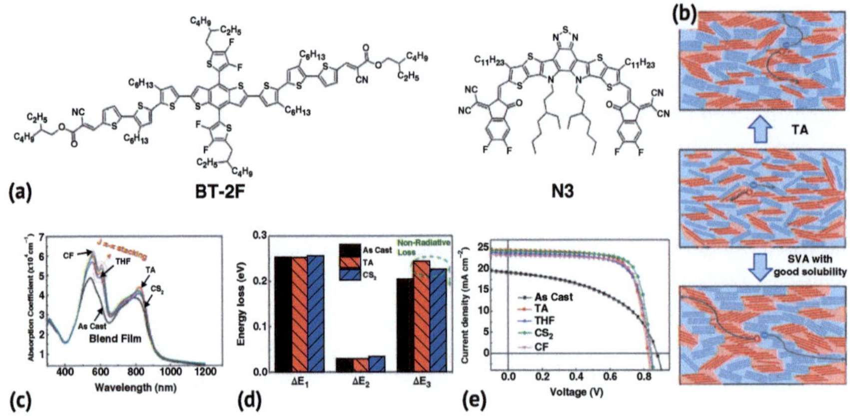

Figure 7.8 (a) Molecular structures of BT-2F and N3. (b) Schematic diagram of morphology evolution. (c) Absorption spectra of BT-2F:N3 films with and without annealing treatments. (d) Summary of energy loss for radiative and non-radiative recombination. (e) *J*–*V* curves of the devices with and without annealing treatments. *Source*: Reproduced with permission from [34]/John Wiley & Sons.

Although both TA and SVA can tune the molecular crystallization and phase separation, their intrinsic differences have been rarely mentioned until the investigation from Ge et al.[34] They systematically investigated the influences of TA and SVA on BT-2F:N3 active layer (Figure 7.8a, b) through a careful characterization of molecular stacking and photovoltaic performance. As shown in Figure 7.8c, although both of TA and SVA treatments can enhance the J-aggregates of BT-2F in blend films, CS_2 SVA treatment particularly induced a stronger J-aggregation and better photoresponse compared to the TA treatment. In addition, CS_2 SVA treatment showed donor-selective dissolution and little effect on N3, and thus more decreased non-radiative recombination (Figure 7.8d) has been achieved to increase V_{OC} and *FF* simultaneously (Figure 7.8e). Finally, the CS_2 treated devices delivered a higher PCE of 15.39% than that of TA treatment (14.66%).

On the other hand, a recent work further indicated that the J-aggregation might also be a promising method for reducing the energy loss of OSCs. As shown in Figure 7.9a, Leo et al. reported two BODIPY molecules, BDP-H and BDP-J, and applied them in both planar heterojunction (PHJ) and BHJ OSCs.[35] For BDP-H, H-aggregation is the preferred molecular aggregation type in the neat film (Figure 7.9b, left). However, when two H atoms on the peripheral phenyl rings of BDP-H were substituted by F atoms, the molecular packing of BDT-J tended to form J-aggregation, due to the modified electronic static potential by the strong electronegativity of F atoms (Figure 7.9b, right). In addition, BDP-J showed a more homogeneous domain shape and clearer needle-like structures in the layers on top of C_{60}, due to the higher crystallinity than that of BDP-H (Figure 7.9c). In PHJ devices, the device based on J-aggregated BDP-J showed a significantly higher electroluminescent EQE (EQE_{EL}), as shown in Figure 7.9d, thus possessed a lower non-radiative voltage loss of 0.35 V than that of 0.49 V in the H-aggregated BDP-H device. Although the aggregation was greatly disturbed in the BHJ blends, the BHJ device of BDP-J still showed higher EQE_{EL} than BDP-H, mainly because of a pronounced radiative and a depressed non-radiative decay rate.

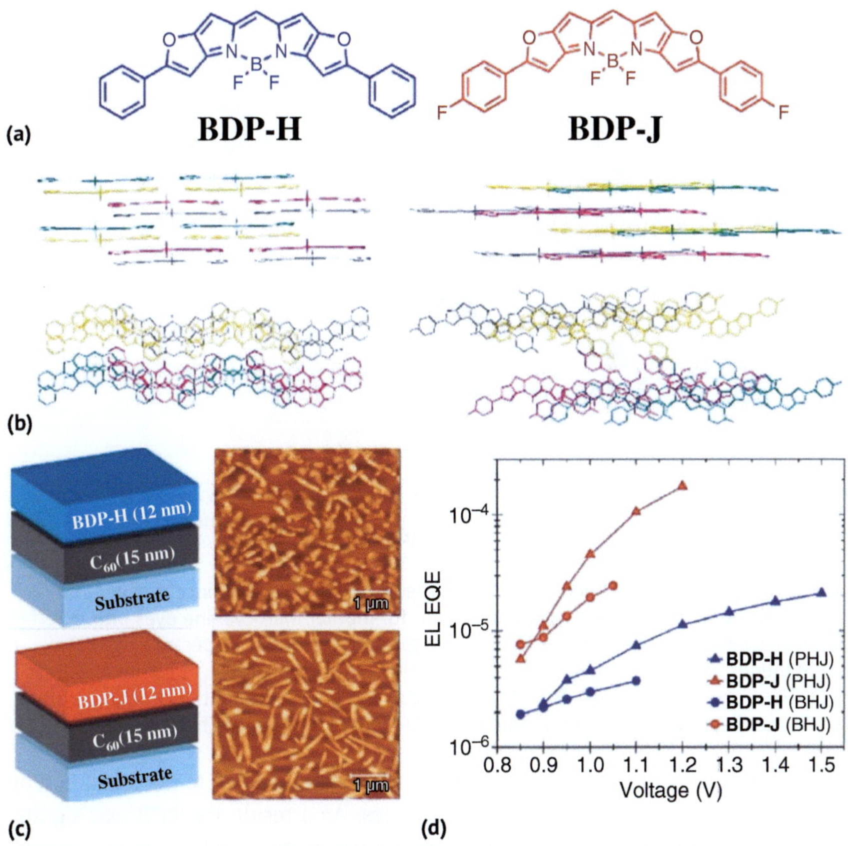

Figure 7.9 (a) Molecular structures and (b) packing diagrams of BDP-H and BDP-J in single crystal. (c) The schematic illustrations and AFM images of neat BDP-H (top) and BDP-J (bottom) layers on top of C_{60}. (d) Electroluminescent EQE of both PHJ and BHJ devices at different applied voltage. *Source*: Reproduced from [35]/with permission of American Chemical Society.

A quite general observation has been realized that J-aggregates can increase both J_{SC} and PCE, while the positive effects of H-aggregates on photovoltaic performance have been less investigated. In 2013, Meerholz and Würthner et al. synthesized a bis(merocyanine) dye (Figure 7.10a), and found that a nearly twofold increase in the J_{SC} was stemmed from the H-aggregated dye in the photoactive layer.[36] In a nonpolar environment, the dipolar dyes formed H-aggregation via intramolecular folding (Figure 7.10b), and exhibited a blue-shifted absorption (Figure 7.10c). Furthermore, incorporation of such H-aggregate domains into the dye:$PC_{61}BM$ BHJ solar cells induced a substantial increase of the EQE between 400 and 500 nm (Figure 7.10d) for improving the percolation pathways for excitons and charge carriers, finally resulting in a significant increase of the J_{SC} from 2.26 to 3.92 mA cm^{-2}. In 2016, Han et al. also indicated that a fine adjustment of the ratio of H-/J-aggregation state in blend films would be crucial to further improve the device performance.[37] In the devices, the J_{SC} can be increased from 8.29 to 15.0 mA cm^{-2}, accompanied by the FF increased from 40% to 58%, when the H/J ratio in the p-DTS($FBTTh_2$)$_2$: $PC_{71}BM$ blend thin film

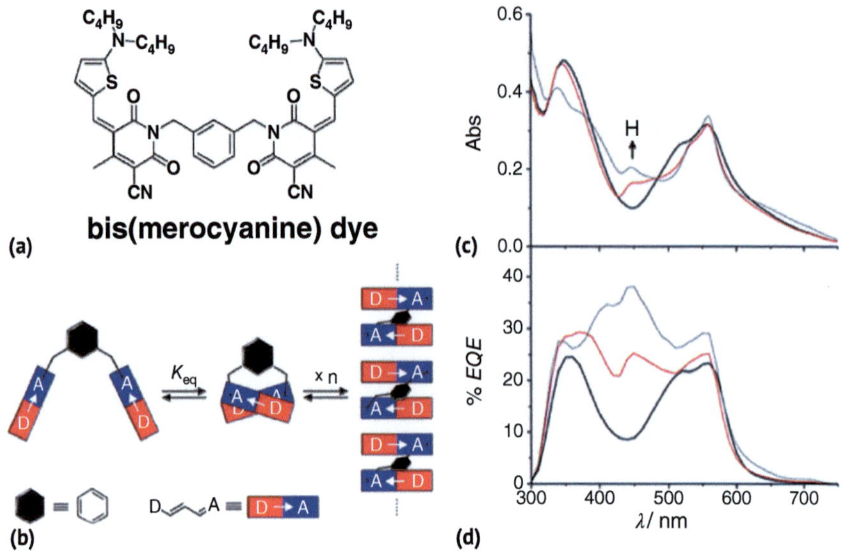

Figure 7.10 (a) Chemical structure of the bis(merocyanine) dye. (b) Concept of a hierarchical supramolecular polymerization path for dipolar merocyanine dyes into cofacially stacked H-aggregates. (c) Thin film absorption and (d) EQE spectra of dye: $PC_{61}BM$ blends on the casting solvent mixture $CHCl_3$ (black line), $CHCl_3$–toluene (red line), and $CHCl_3$–o-xylene (grey line). The arrow indicates the emergence of the H-aggregate band. *Source*: Reproduced with permission from [36]/Royal Society of Chemistry.

increased from 0.71 to 1.47. Such enhancements were attributed to the formation of better BHJ morphology induced by H-aggregates. As a result, the PCE was significantly boosted from 2.64% to 6.63%, corresponding to an amazing 150% enhancement.

The aggregation behaviors of conjugated polymers are quite different from small molecules, which usually easily form π–π stacking or side chain layer stacking through interchain interactions.[38] Short-range aggregates or long-range crystallites can be generated, depending on the structure of the polymer, processing conditions, and the strength of interchain interaction.[39] The highly π electron delocalization along the polymer backbone would give rise to interchain π–π stacking to afford charge-transporting channels, whose direction is largely related to the stacking orientation.[40] For OSCs with a vertical device structure, the face-on orientation is highly desired, because it can form straightforward charge-transporting channels between the electrodes.[41] Currently, utilizing the aggregation behaviors of polymer donors with strong temperature-dependent aggregation (TDA) property has been proven as an effective approach to control BHJ film morphology for high performance OSCs.[42]

In 2012, Hou et al. reported a conjugated polymer named PBDTBDD (Figure 7.11a) with a strong TDA effect in solution state.[43] As shown in Figure 7.11b, PBDTBDD had strong J-aggregation tendency in *o*-DCB solution under 30 °C, and such aggregation could be eliminated by further heating up, which is a reversible thermochromic process. During the spin-coating process of device fabrication, the aggregation size in the solution was very tiny and stable under ambient temperature, so small fibril domains formed in the PBDRBDD:$PC_{61}BM$ blend film, which permeated evenly.

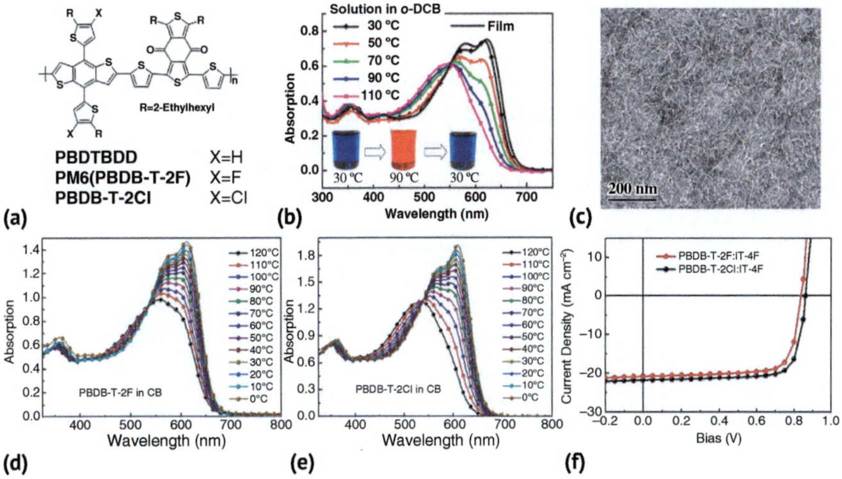

Figure 7.11 (a) Chemical structure of PBDTBDD, PM6, and PBDB-T-2Cl. (b) UV–vis spectra of PBDTBDD as a film and in *o*-DCB solution under different temperatures (inset: the color of the solution under different temperatures. *Source*: Reproduced from [43]/with permission of American Chemical Society. (c) TEM images of PM6:PC71BM blend films with 1% DIO. Source: Reproduced with permission from [44]/John Wiley & Sons. Temperature-dependent UV–vis absorption spectra of the solutions of (d) PM6 and (e) PBDB-T-2Cl in CB as the temperature decreased from 120 to 0 °C. (f) *J*–*V* curves of the devices based on PM6:IT-4F and PBDBT-2Cl:IT-4F. *Source*: Reproduced from [45]/with permission of John Wiley & Sons.

Therefore, the OSCs based on PBDRBDD:$PC_{61}BM$ BHJ exhibited a remarkable PCE of 6.67% and a high *FF* of 72.27%. Soon after, Hou et al. introduced the fluorine atom into BDD unit, and the resulting new polymer donor, namely PM6 (Figure 7.11a), exhibited a deeper HOMO level of −5.45 eV and strong crystallinity with a dominant face-on packing in neat film.[44] However, the π–π stacking of PM6 in the blend films was disrupted by the $PC_{71}BM$ aggregation. After adding 1,8-diiodooctane (DIO) as the additive, the lamellar diffraction and π–π stacking diffraction of PM6 can be significantly increased, and in particular, a fibrillar network with a small length scale was obtained with 1% DIO added (Figure 7.11c), which could provide better charge transport and reduced recombination. As a result, a high PCE of 9.2% with a high V_{OC} of 0.98 V was achieved in the PM6-based inverted OSCs. In 2018, the same group further replaced the critical fluorine with chlorine (Figure 7.11a). PBDB-T-2Cl showed similar strong aggregation effects in CB solution with those observed from PM6 (Figure 7.11d, e), and also formed nanoscale fibrillar aggregations in the BHJ film when blended with an NFA (IT-4F). Finally, the PBDB-T-2Cl:IT-4F based devices delivered an obviously enhanced PCE, up to 14.4% (Figure 7.11f).[45]

In 2017, Yan et al. reported a family of conjugated polymers with strong TDA property, and found that the second-position branched alkyl chains, as well as the fluorination on the polymer backbone, are two key structural features that can enable the strong TDA property.[46] Taking the PffBT4T-2OD polymer as an example, the steric hindrance effect of the second-position branched alkyl chain resulted in a highly twisted conformation between the two thiophene units at high temperatures, producing sufficient steric hindrance for polymer disaggregation. Meanwhile, the fluorination on the backbones of conjugated polymers could enhance the intermolecular interaction and π–π stacking between polymer chains by increasing the coplanarity of the polymer backbones, affording the force to overcome the steric hindrance caused by the second branched alkyl chains and thus resulting in strong aggregation in solution at room temperature.

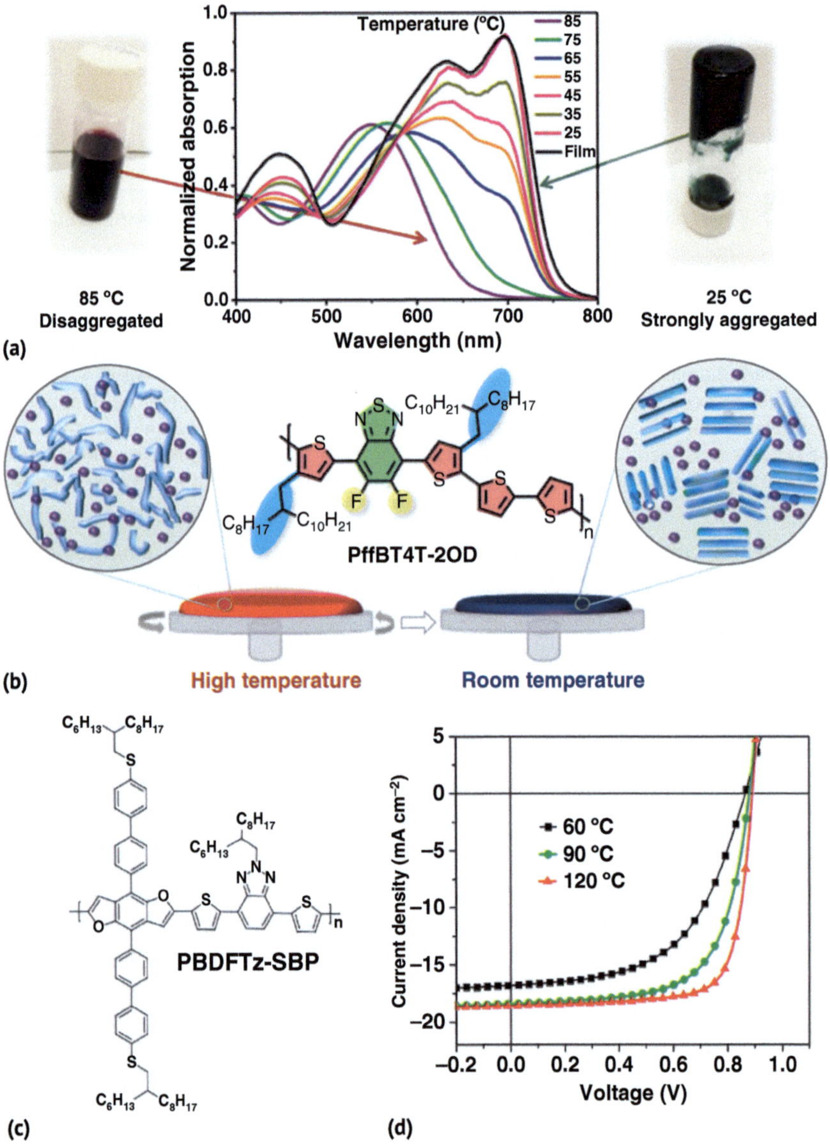

Figure 7.12 (a) Temperature-dependent aggregation property of PffBT4T-2OD. (b) Schematic illustration of the spin-coating and morphology forming process. *Source*: Reproduced with permission from [46]/American Chemical Society. (c) The chemical structure of the polymers PBDFTz-SBP. (d) The *J*–*V* curves PBDFTz-SBP:ITIC at different temperatures. *Source*: Reproduced from [48]/with permission of John Wiley & Sons.

As shown in Figure 7.12a, the absorption profile of the solution at room temperature was nearly identical to that observed in film state, confirming the presence of highly ordered aggregates in low temperature solution. The unique aggregation properties offered various processing windows to control and optimize the blend morphology (Figure 7.12b), such as temperature, spin-rates, concentration, etc. The polymer chains could disaggregate into random polymer chains like "spaghetti," when the active-layer solution was preheated to high temperature. Along with the solution temperature gradually decreasing to room temperature during the spin coating process, these polymer chains had enough time to aggregate again and form the right-sized crystalline domains. As a result, these polymers based BHJ OSCs achieved high PCEs over 10%. Therefore, this hot processing technique, coupled with the unique property of TDA polymers, has enabled precise control over the blend morphology for high-efficiency devices.[47]

In 2019, Yang et al. further applied this aggregation-breaking strategy to inhibit the trend of self-aggregation of the benzo[1,2-b:4,5-b'] difuran (BDF) based polymer PBDFtz-SBP (Figure 7.12c) with TDA property, and obtained encouraging photovoltaic performance for BDF-based polymers.[48] When the solution was heated to 120 °C before spin-coating, PBDFTz-SBP:ITIC blends formed a bicontinuous interpenetrating network in the active layer, and resulted in a favorable charge extraction as well as an enhanced PCE of 12.42% with a high *FF* of 75.19%, which was much higher than that of the devices (7.95%) prepared at a low temperature of blend solution, as shown in Figure 7.12d.

In addition, polymer aggregation also plays an important role in the device performance of all-polymer solar cells (all-PSCs). In 2020, Kim et al. synthesized three polymer donors with the same polymer backbone but different side chains, i.e., ethylhexyl (EH), ethylhexylthio (SEH), and tripropylsilyl (Si) side chains, and named them as P-EH, P-SHE, and P—Si (Figure 7.13a), respectively.[49] Among three polymers, P—Si showed the strongest aggregation behaviors in solution at the processing temperature, owing to the proper steric hindrance from the three-way branched trialkyl silyl side chains. However, in all-PSCs, the P—Si polymer donor with the strongest aggregation suffered from excessive phase separation in the blend films (Figure 7.13b), resulting in poor charge generation and collection, and consequently the lowest PCE of 3.14%. In sharp contrast, when blended with the fashion small molecule Y6 acceptor, the resulting P—Si/Y6 BHJ achieved a high PCE of 12.02%, attributed to the well-mixed BHJ morphology with suitable nanoscale phase separation (Figure 7.13b).

7.1.1.1 Effects of Acceptor Aggregates

As introduced previously, before 2015, the choice of electron acceptors focused on the fullerene derivatives like PCBM and ICMA (Figure 7.14a).[50] Most fullerene acceptors exhibit their high electronic mobility and can form dense and isotropic packing due to their spherical shape, leading to good charge separation and charge transport properties.[9,51] However, the best PCE of fullerene-based binary OSCs is limited to ~12%, due to several inherent shortcomings of fullerene acceptors, such as weak absorption, undesired energy levels, and large recombination loss.[52,53] The emergence of NFAs has greatly promoted the development of OSCs, as their absorption spectra and

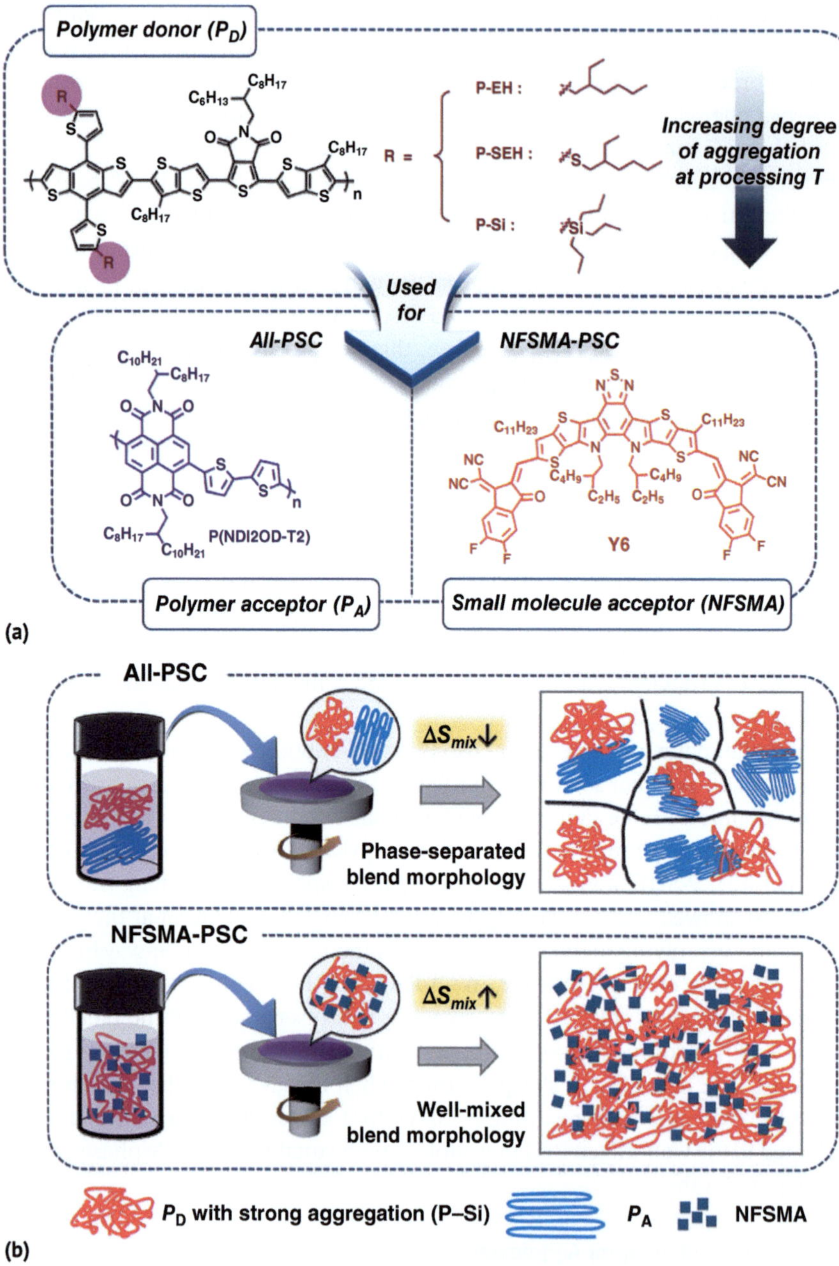

Figure 7.13 (a) Chemical structures of polymer donors with different side chains (P-EH, P-SEH, and P–Si) and representative acceptors (P(NDI2OD-T2) and Y6) used for a model study. (b) Illustration of different effects of polymer aggregation on blend morphology. *Source*: Reproduced from [49]/with permission of American Chemical Society.

energy levels can be easily tailored to boost the PCE of OSCs significantly.[54,47] The best NFA-based OSCs have achieved PCEs exceeding 19%.[12,55] There is no doubt that NFA-based OSCs represent the future direction of organic photovoltaics.

The early small molecule NFAs studied in OSCs are mainly rylene diimides such as perylene diimides (PDIs), which exhibit advantages of remarkable electron mobility and affinity, high absorption coefficients, and outstanding oxidative/thermal stability (Figure 7.14b).[56,57] However, these NFAs often show too strong aggregation tendency to achieve desired domain sizes for efficient exciton dissociation, even though several strategies have been adopted to decrease the aggregation extent via chemical structure modification (Figure 7.14c).[58,59] So far, the PCE of PDI-based OSCs has just surpassed 11%,[60] which is still lower than that of fullerene-based OSCs. Therefore, it is important to control the aggregation behaviors of NFAs for achieving a high device performance.

Since 2015, the emergence of a new generation of FREAs has greatly improved device efficiency.[8] FREAs typically adopt an acceptor–donor–acceptor (A–D–A) structure, wherein an electron-donating aromatic core is end-capped by two strong electron-accepting end groups via ethylene bridge. Such innovative structure design can enable a highly coplanar structure, leading to strong push–pull electronic effect to induce a strong intramolecular charge transfer (ICT) process for opto-electronic properties. For example, ITIC (Figure 7.14d), the first milestone of A-D-A FREAs, exhibits a strong absorption peak at 664 nm in dichloromethane solution. Relative to its solution, the absorption peak of the ITIC film exhibits an obvious redshift of 38 nm, suggesting that there is J-aggregation occurring in the film state (Figure 7.14e).[10] Molecular dynamics simulations revealed

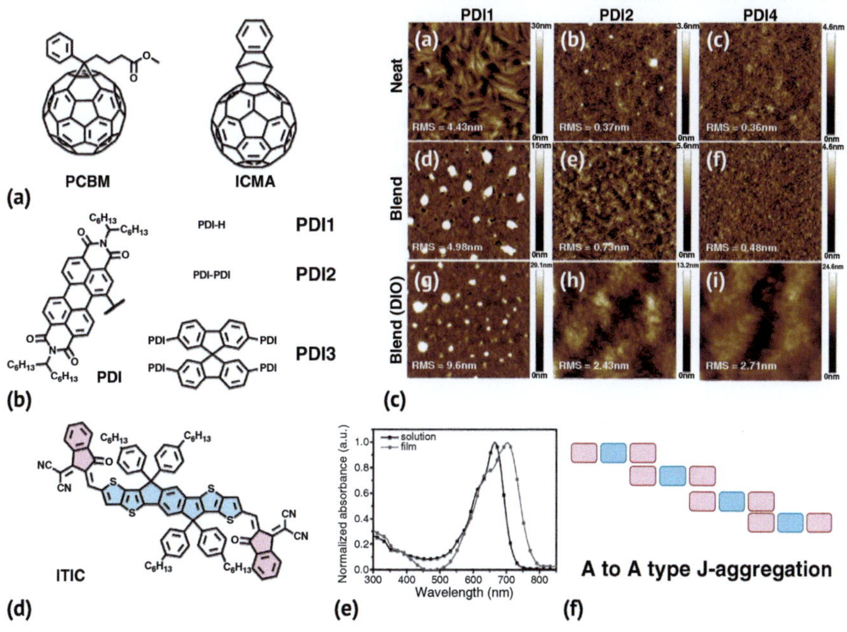

Figure 7.14 (a) Molecular structures of PCBM and ICMA. (b) Chemical structures of PDI series. (c) AFM height images of the PDI series pure and blend films with PTB7-Th without DIO and with DIO. *Source*: Adapted from [59]. (d) Molecular structure of A–D–A type ITIC. (e) UV–vis absorption spectra of ITIC in dichloromethane solution and in thin film. *Source*: Reproduced from [10]/with permission of John Wiley & Sons. (f) Schematic presentation of A to A stacking induced J-aggregation.

that ITIC can form a two-dimensional molecular packing by forming A-to-A type J-aggregation for efficient charge transport (Figure 7.14f).[61] With a broad absorption and a good BHJ morphology of the active layers, PTB7-TH:ITIC based OSCs showed a higher PCEs of 6.80% than that (6.05%) of PCBM-based control OSCs.[10] Later on, more and more studies demonstrated that the opto-electronic properties and aggregation behaviors of A–D–A FREAs are easily tunable by optimizing the chemical structures as well as the processing technologies, making great progress achieved in a short time.[11,60,62] It is worth noting that the aggregation control of polymer acceptors based on FREAs has achieved much less progress than that of small molecule FREAs so far, which is highly dependent on the structural units and molecular weights.[63] Thus here we will focus on discussing the representative progress of the aggregation control of small molecule FREAs.

The planarity of the core units of small molecule FREAs is directly correlated with the aggregation behaviors of the A–D–A NFAs. A planar core unit can achieve good crystallinity, dense π–π packing, and favorable face-on orientation, leading to higher charge mobility. Currently, a common strategy of the aromatic core unit engineering of FREAs is to increase the number of fused rings. As shown in Figure 7.15a, the central benzene ring can be extended to naphthalene[64] and further to naphthodithiophene,[65] affording IDIC1 to IHIC1/IHIC2 to IOIC2. It was shown that IOIC2 with a more extended core exhibited stronger lamellar and π–π packing in films than IHIC2 (Figure 7.16a). The improved π–π stacking in IOIC2 benefited the charge transport to enable a much higher electron mobility than that of IHIC2. As a result, significantly enhanced PCE can be achieved for the resulting IOIC2-based OSCs, which can be further improved when adding DIO as the addi-

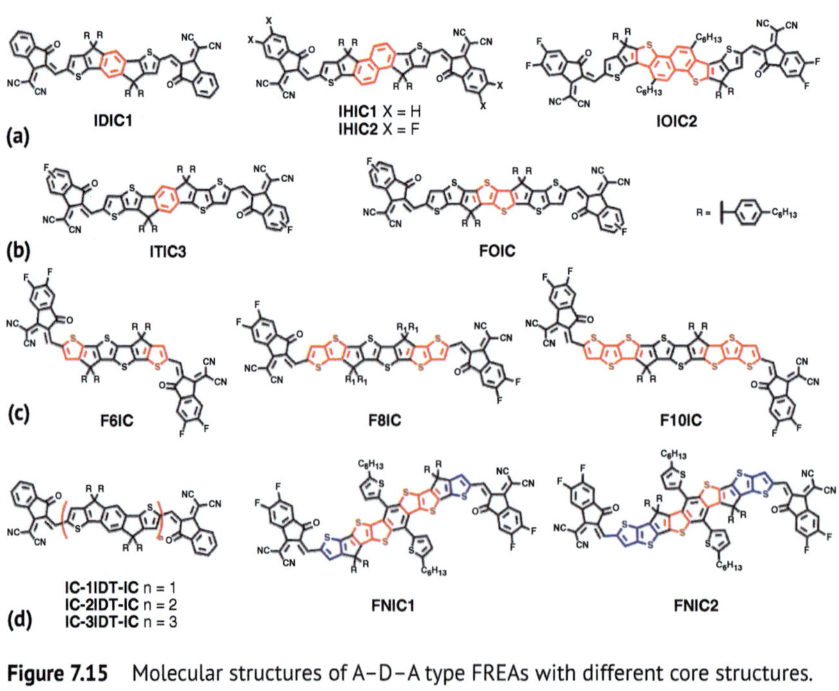

Figure 7.15 Molecular structures of A–D–A type FREAs with different core structures.

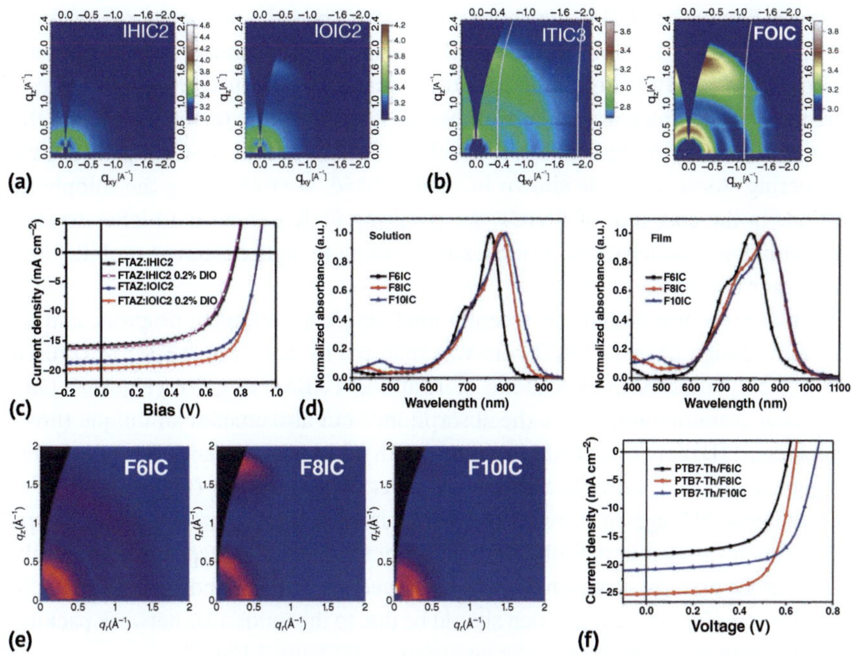

Figure 7.16 (a) 2D GIWAXS patterns of the IHIC2 and IOIC2 films. *Source*: Reproduced with permission from [65]/John Wiley & Sons. (b) 2D GIWAXS patterns of the ITIC3 and FOIC films. *Source*: Reproduced with permission from [66]/John Wiley & Sons. (c) *J*–*V* curves of the devices based on FTAZ:IHIC2 and FTAZ:IOIC2. *Source*: Reproduced with permission from [65]/John Wiley & Sons. (d) The UV–vis absorption spectra of the FnIC series in dichloromethane solution and in film. (e) 2D GIWAXS patterns of the FnIC series film. (f) *J*–*V* curves of the devices based on PTB7-Th:FnIC series. *Source*: Reproduced with permission from [67]/John Wiley & Sons.

tive (Figure 7.16c). Replacing the benzene unit with a TT unit can draw a similar conclusion. Figure 7.15b shows the molecular structure of resulting FREAs (ITIC3 and FOIC), and their 2D GIWAXS patterns (Figure 7.16b) clearly demonstrated that the scattering peaks of pure FOIC and blend films are much stronger and sharper than those from ITIC3 films, due to the higher crystallinity and more ordered face-on orientation of the former.[66]

Furthermore, FREAs of F6IC, F8IC, and F10IC (Figure 7.15c) were synthesized by systematically extending the core groups from six to eight, and then to ten fused rings, which showed gradually red-shifted absorption spectra with the absorption maxima changing from 714 to 798 nm in solution (Figure 7.16d).[67] However, F8IC film exhibited a similar absorption profile as that of F10IC film, indicating the former had a much stronger J-aggregation. Besides, F8IC exhibited a more obvious face-on orientation, dense π–π packing, leading to a more desired phase separation in BHJ films (Figure 7.16e). Thus, as presented in Figure 7.16f, the PTB7-Th:F8IC based device achieved the best J_{SC} (25.12 mA cm^{-2}) and the highest PCE (10.9%).

On the other hand, repeating the core unit is another design strategy to extend the backbone length, based on which, IC-nIDT-IC were successfully designed as shown in Figure 7.15d. Among them, IC-1IDT-IC had the strongest π–π packing and the highest crystallinity because of the most planar backbone,[68] suggesting the importance of carefully designing the core units for handling aggregate behaviors. Moreover, besides increasing the number of fused rings in the core units of the NFAs, isomerizing the fused rings is another effective design strategy for core unit engineering. As an example shown in Figure 7.15d, after changing the thiophene positions in the core unit of FNIC1, the resulting FNIC2 showed a higher crystallinity with dense lamellar and π–π packing, leading to higher electron mobility and thus higher PCEs.[69]

The core structure design has recently made more encouraging progress, and an innovative design of using DA'D core was reported by Zou et al.[70] The DA'D fused core not only endows the backbone of resulting FREAs, e. g. Y6 (Figure 7.17a), with better planarity to improve the absorption,[71] but also enables an unique three-dimensional (3D) network packing structure with multi-packing modes between cores (H-aggregation) and between terminal A units (J-aggregation) (Figure 7.17b).[72,73] The synergetic H/J aggregations can make the intermolecular interactions of Y-type FREAs much stronger in the solid state than from those of ITIC-type acceptors, making the device performance significantly improved (PCE>15%) and insensitive to the film thickness (Figure 7.17c), which should be due to the unique 3D network packing that can facilitate the charge transfer and transport (Figure 7.17d).[70]

Homogeneous elements in heterocycles have also been found to play an important role in regulating the aggregation behaviors. For instance, compared to sulfur atom, selenium atom is more polarizable and has a larger atomic radius and d-orbitals.[74] Through simply substituting the thiophene ring in Y6 by selenophene ring, the resulting Y6-Se and Y6-2Se (Figure 7.17a) can exhibit a much denser π–π packing than Y6 and further form more suitable nanoscale phase separation in BHJ films (Figure 7.17e), due to the induced strong intermolecular Se-Se non-covalent interactions.[75,76] The optimized molecular packing and BHJ morphology produced a more efficient exciton dissociation, better charge transportation, and lower charge recombination compared to Y6 acceptor.

Constructing non-fused NFAs have also attracted much attention due to the relatively simple synthesis and low material cost.[77] One of typical examples is PTIC acceptor (Figure 7.18a) reported by Li et al., which showed rotatable conformation in the solution state.[78] However, such rotation can be restrained into a planar and stackable conformation in the condensed solid state (Figure 7.18b), thereby reinforcing the intramolecular interactions among the core unit to enable dense molecular packing. Although PTIC and its fused counterpart IT-4F processed similar bandgap and energy levels as shown in Figure 7.18c, PTIC had a much better photostability than IT-4F (Figure 7.18d). The ^{1}H NMR spectra shown in Figure 7.18e demonstrated that two new proton peaks (denoted as H_a' and H_a'') of IT-4F can rise up subsequently, accompanied by the decrease of origin vinyl protons (H_a) under light illumination, revealing that the vinyl groups of IT-4F suffered photoisomerization and photooxidation. In sharp contrast, such undesired effects were absent for PTIC, suggesting that the steric hindrance between thiophene and A units due to

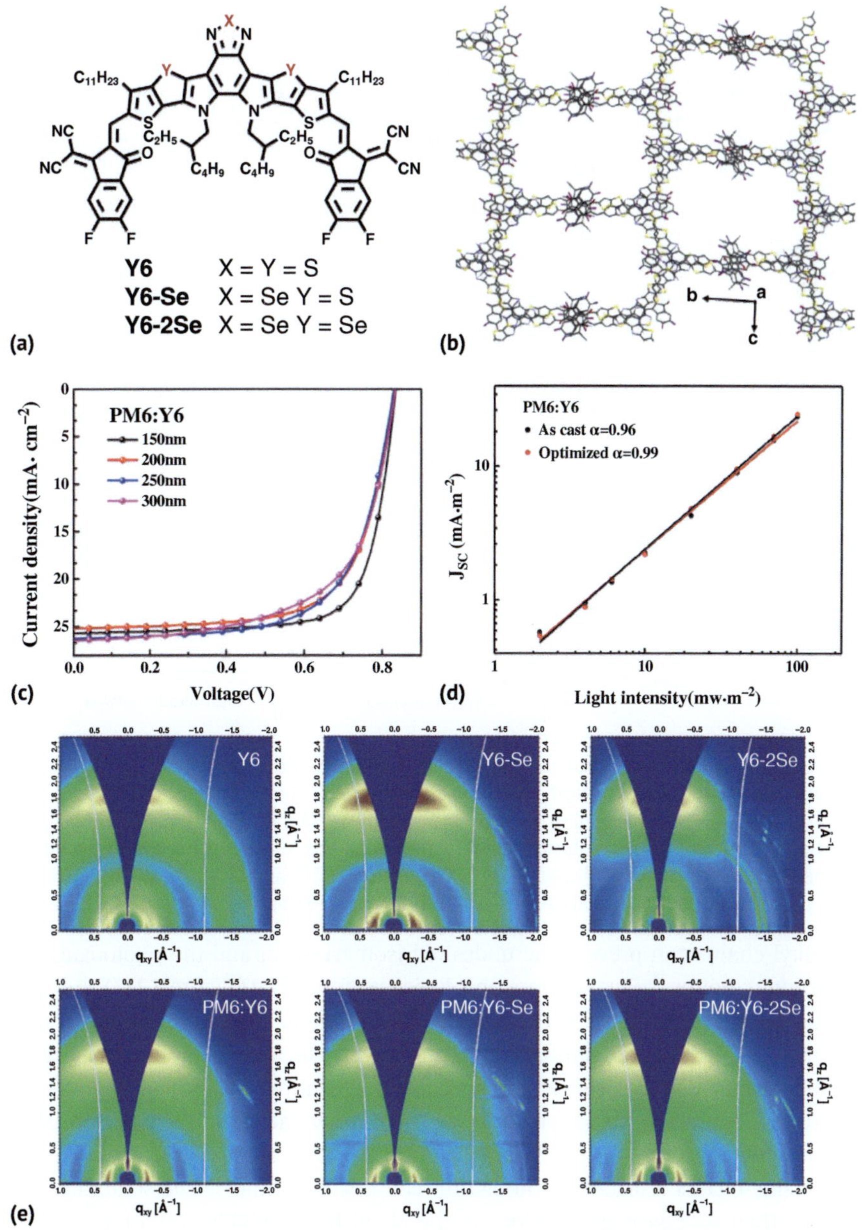

Figure 7.17 (a) Molecular structures of Y6 series FREAs. (b) Molecular packing sketch map of Y6 according to single crystal data. *Source*: Reproduced from [72]/with permission of American Chemical Society. (c) *J–V* curves of the devices based on PM6:Y6. (d) Photocurrent versus light intensity of the as-cast and optimized devices based on PM6:Y6. *Source*: Reproduced with permission from [70]/Elsevier. (e) 2D GIWAXS patterns of the Y6, Y6-Se, and Y6-2Se pure films and blend films with PM6. *Source*: Reproduced with permission from [75]/The Royal Society of Chemistry.

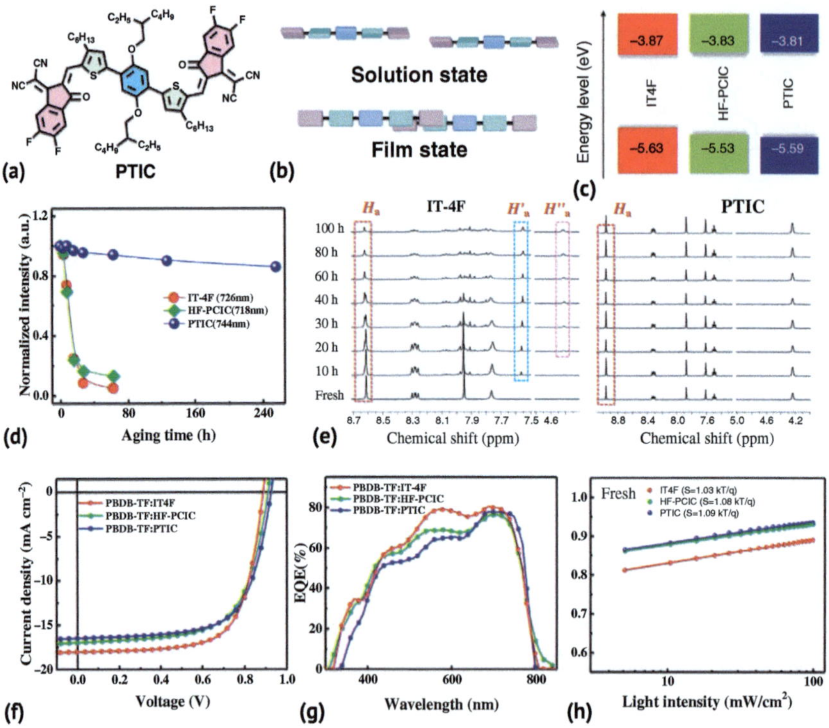

Figure 7.18 (a) Molecular structure of PTIC. (b) The schematic presentation of PTIC conformation rearrangement from solution to solid state. (c) Energy levels for IT4F and PTIC. (d) The change of remaining absorption intensity at 64 h illumination. (e) ^{1}H NMR spectra of IT-4F and PTIC in solutions under continuous one-sun-equivalent illumination. (f) The *J*–*V* curves and (g) EQE spectra of OSCs based on PTIC and IT-4F. (h) V_{OC} versus light intensity of fresh devices based on PTIC or IT-4F. *Source*: [78]/Springer Nature / CC BY 4.0.

the alkyl chains can prevent the undesired isomerization and the photooxidation of terminal groups.[79] Furthermore, the device performance (Figure 7.18f) and EQE (Figure 7.18g) of PTIC-based OSCs were comparable to those of IT-4F-based one. Nonetheless, it is worth noting that the relatively disordered molecular packing of non-fused NFAs also caused enhanced reorganization energy and poor charge transport (Figure 7.18h), thereby limiting the improvement of device performance to some degree.[80]

The side-chains attached to the D unit can flexibly enhance the solubility, and prevent their excessive aggregations in the blend films, which would mitigate the undesired charge recombination.[81,82] Even with the same backbone, the film morphology of NFAs with different side-chains could be quite different. Besides, the trade-off between solubility and conductivity also requires careful consideration when designing side-chains. For instance, IDTBR acceptors with n-octyl (O-IDTBR) and 2-ethylhexyl (EH-IDTBR) shared a similar absorption property in solution (Figure 7.19a, b), whereas O-IDTBR exhibited stronger J-aggregation with more red-shifted absorption relative to EH-IDTBR, which can be explained by enhanced intermolecular interactions (Figure 7.19c) induced by the linear alkyl chain.[83] As

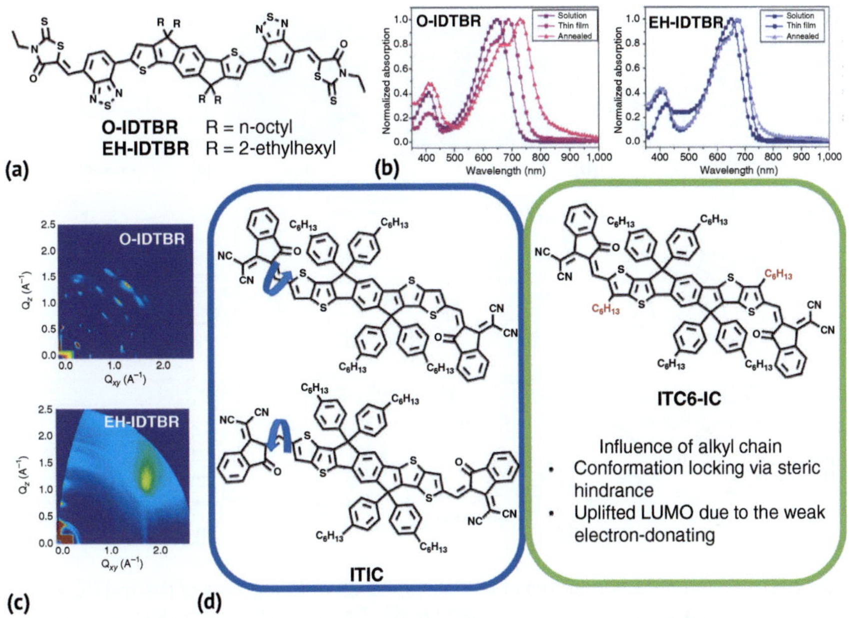

Figure 7.19 (a) Molecular structures of O-IDTBR and EH-IDTBR. (b) UV–vis–NIR absorption spectra of EH-IDTBR and O-IDTBR in solution, thin films with or without TA. (c) 2D GIWAXS patterns of the O-IDTBR and EH-IDTBR films. *Source*: [83]/Springer Nature / CC BY 4.0. (d) The schematic presentation of ITC6-IC.

a result, a more suitable phase separation in the P3HT:O-IDTBR blend film was achieved for efficient charge dissociation and transport, leading to an increased J_{SC} from 12.1 mA cm^{-2} to 13.9 mA cm^{-2}.

The side chains attached to the terminal β-position of core units are so-called terminal side chains, which can influence the aggregation behaviors of resulting acceptors significantly. Tang et al. introduced terminal hexyl chains into the fused IT core of ITIC to successfully obtain ITC6-IC (Figure 7.19d),[84] which led to an increased steric hindrance, and consequently a planar structure and better solubility.[84] Furthermore, the LUMO level of ITC6-IC was slightly uplifted (0.09 eV) due to the weak electron-donating ability of alkyl chains. Thus, PBDB-T: ITC6-IC-based OSCs achieved a higher PCE with a higher V_{OC} than that of ITIC. Similar influences of terminal side chains were also observed on the A-DA'D-A series.[85–87]

The site modification of alkyl side-chains is another way to tune the side-chain effects. As an isomeric counterpart of ITIC with para-alkyl-phenyl substitution, m-ITIC with meta-alkyl-phenyl substitution (Figure 7.20a) showed a larger crystalline coherence and more predominant face-on orientation (Figure 7.20b).[88] As the result, an improved electron mobility was obtained for m-ITIC, resulting in a higher PCE than that of ITIC (Figure 7.20c). What's more, m-ITIC-based OSCs showed less thickness-dependent photovoltaic performance than ITIC-based OSCs in the active-layer thickness range of 80–360 nm, which will be beneficial for large area device fabrication. Besides, another popular FREA named as

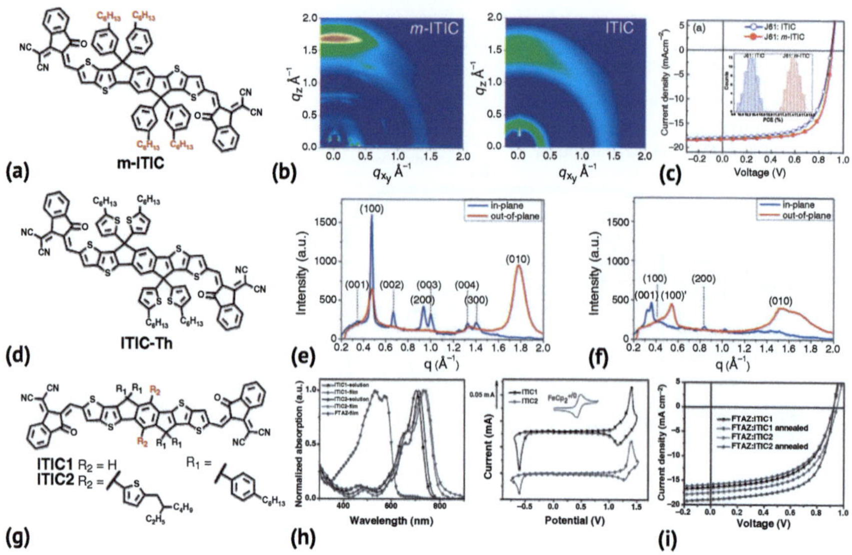

Figure 7.20 (a) Chemical structures of m-ITIC. (b) 2D GIWAXS patterns of the m-ITIC and ITIC films. (c) *J–V* curves of the devices based on J61:ITIC or m-ITIC. *Source*: Reproduced with permission from [88]/American Chemical Society. (d) Chemical structures of ITIC-Th. The intensity profiles of 2D GIWAXS patterns of the ITIC-Th (e) and ITIC (f) films in the in-plane and out-of-plane directions. *Source*: Reproduced with permission from [89]/John Wiley & Sons. (g) Chemical structures of ITIC1 and ITIC2. (h) UV–vis absorption spectra and cyclic voltammograms of ITIC1 and ITIC2. (i) *J–V* curves of the devices based on FTAZ:ITIC1 or ITIC2. *Source*: Reproduced with permission from [91]/John Wiley & Sons.

ITIC-Th was designed by replacing the phenyl side-chains in ITIC with the thiophene derivative (Figure 7.20d), which possessed much higher crystallinity and denser π–π packing (Figure 7.20e, f).[89,90] Furthermore, as shown in Figure 7.20g, the conjugated side chains can be introduced at the center of conjugated core to construct 2D-conjugated structures, and the resulting ITIC2 exhibited stronger molecular interactions in the aggregate state, leading to redshifted absorption and deeper LUMO level (Figure 7.20h).[91] Besides, TA treatment can further improve the electron mobility of FTAZ:ITIC2 blend films, thereby enhancing the device J_{SC} and *FF* (Figure 7.20i).

Terminal units generally contain electrophilic atoms, such as oxygen and nitrogen, which can facilitate intramolecular charge transport to tune the energy levels. To date, most of FREAs were constructed by employing 2-(3-oxo-2,3-dihydroinden-1-ylidene)malononitrile (IC) and its derivatives as the terminal groups, due to the strong electron-withdrawing ability and high flatness.[54] A small difference in terminals might result in a large change in aggregation behaviors. For instance, IT-M and IT-DM were synthesized by introducing one or two methyl groups to modify the IC units (Figure 7.21a), and the results showed that a much denser π–π packing was achieved in the IT-M films (Figure 7.21b), leading to higher electron mobility.[92] Moreover, space units are sometimes also inserted between core and terminal units to achieve strong and tight π–π packing.[93]

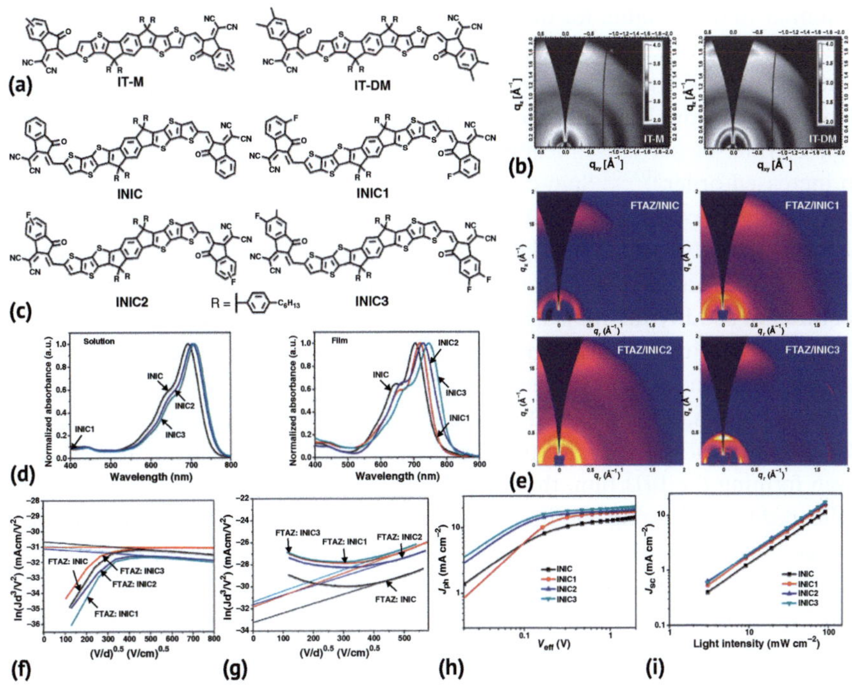

Figure 7.21 (a) Molecular structures of IT-M and IT-DM. (b) 2D GIWAXS patterns of the IT-M and IT-DM films. *Source*: Reproduced with permission from [92]/John Wiley & Sons. (c) Molecular structures of INIC series. (d) UV–vis absorption spectra of the INIC series. (e) 2D GIWAXS patterns of the INIC series blend films with FTAZ. The *J*-*V* characteristics in the dark for hole-only (f) and electron-only (g) devices based on FTAZ: INIC series. J_{ph} versus V_{eff} characteristics (h) and J_{SC} versus light intensity (i) of the optimized devices based on FTAZ:INIC series. *Source*: Reproduced from [96]/with permission of American Chemical Society.

In addition, many other electron-withdrawing groups have been recently employed as the terminal units of FREAs,[94] demonstrating that subtle modification of the terminal units can effectively modulate the energy level alignment and the BHJ phase separation for charge separation.[95] Among various structural modification strategies, halogenation is a feasible way to modify the terminal units to lower the energy levels, while the induced strong nonclassical H bonding also affects the aggregation behaviors of resulting FREAs.[96] Among a series of fluorinated FREAs (Figure 7.21c), INIC3 with two fluorine atoms attached at different positions of terminal units showed the most redshifted absorption, suggesting the formation of strongest J-aggregation as shown in Figure 7.21d. Moreover, INIC3 further processed the strongest face-on orientation and co-crystallite with the donor polymer FTAZ (Figure 7.21e), resulting in enhanced electron mobility and thus more balanced hole/electron transport (Figure 7.21f, g). Thus, the FTAZ:INIC3-based OSCs achieved the highest PCE among these fluorinated FREAs, due to more efficient charge dissociation and collection (Figure 7.21h) and negligible bimolecular charge recombination (Figure 7.21i).

Similar to those found for the donors, processing techniques also have significant impacts on the aggregation behaviors of acceptors. Processing solvents,[97,98] temperatures,[99,100] as well as solvent additives[101] can all affect the aggregation dynamics and the film morphology during processing, while post-treatments, e.g., thermal or solvent annealing, can reorganize the phase domains, leading to increased order.[102,103] Spin-coating and blade-coating are two most common fabrication process (Figure 7.22a). During the spin-coating process, the films can form rapidly due to the high spinning speed of the substrate with a strong centrifugal effect. However, in the case of blade coating, a much longer time is required to solidify the films as well as promote the aggregates for active layers. Thus, spin-coating and blade-coating techniques always produce films with different nanoscale morphological features.[104]

In some cases, the films of FREAs were found to exhibit enhanced crystallinity when the film formation time is prolonged, like FNIC3 as shown in Figure 7.22b, c.[105] Even forming the BHJ film, the crystallinity (Figure 7.22d) of FNIC3 can still be enhanced by prolonging the film formation time, which was found to improve the charge transport. Unfortunately, excessive aggregation was induced to reduce the interface area and thus to increase non-radiative energy loss, leading to lower J_{SC} and V_{OC}. Similarly, BTP-BO-4Cl (Figure 7.20e) also showed significantly distinct aggregation behaviors when changing the processing solvents or times.[98] As shown

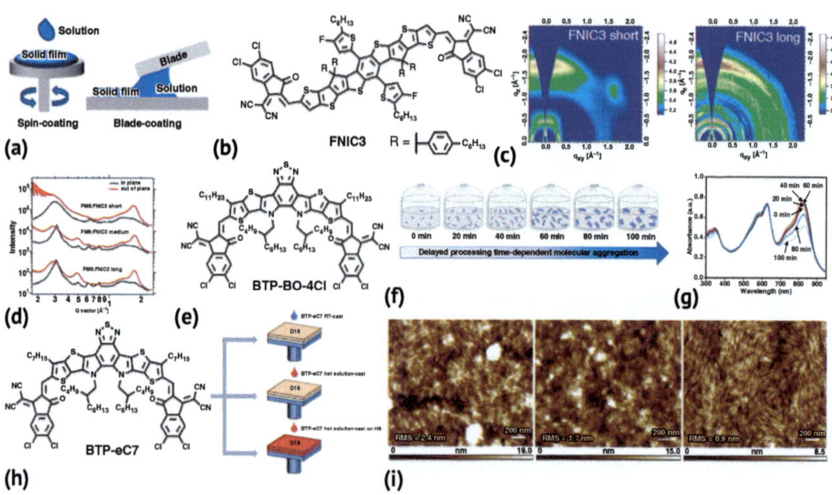

Figure 7.22 (a) The schematic presentation of spin-coating and blade-coating. *Source*: [104]/John Wiley & Sons/CC BY 4.0. (b) Molecular structures of FNIC3. (c) 2D GIWAXS patterns of the FNIC3 films with different film formation times. (d) 2D GIWAXS profiles of PM6:FNIC3 blended films with short, medium, and long film formation time. *Source*: [105]/John Wiley & Sons/CC BY 4.0. (e) Molecular structures of BTP-BO-4Cl. (f) The schematic presentation of BTP-BO-4Cl with different delay times. (g) UV–vis absorption spectra of BTP-BO-4Cl films with different delay times. *Source*: Reproduced with permission from [98]/Royal Society of Chemistry. (h) Molecular structures of BTP-eC7 and the schematic presentation of BTP-eC7 from RT cast, hot-solution cast, and hot-solution cast on the hot substrate. (i) AFM images of D18/BTP-eC7 sequential-cast films by RT cast, hot-solution cast, and hot-solution cast on the hot substrate. *Source*: [106]/John Wiley & Sons / CC BY 4.0.

in Figure 7.22f, the pre-aggregation in the solution was enhanced when prolonging the delay time, resulting in an increased absorption of around 840 nm (Figure 7.22g). Finally, the devices achieved a significantly improved PCE of 17.33%, attributed to the optimized morphology with efficient charge separation and transport.

Temperature also plays an important role in the aggregation control during film processing process. For instance, heating-induced aggregation behaviors of BTP-eC7 during sequential casting on D18 fibrillar layers were demonstrated in Figure 7.22h.[106] Compared to the films coated at room temperature, BTP-eC7 films coated from hot solution exhibited significantly reduced aggregation tendency, which can be further eliminated by using the hot substrate. Such processing temperature control led to an appropriate aggregation of BTP-eC7 and nanoscale phase separation in the photoactive layer (Figure 7.22i), both of which are beneficial to the exciton

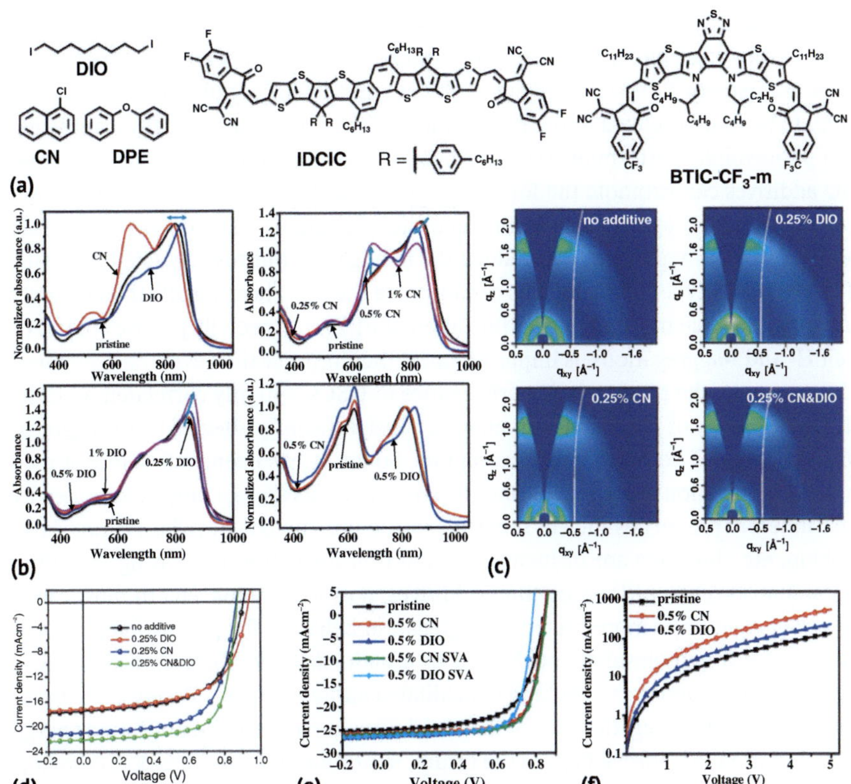

Figure 7.23 (a) Molecular structures of DIO, CN, IDCIC, and BTIC-CF_3-m. (b) UV–vis absorption spectra of PBDBTF: BTIC-CF_3-*m* films with or without different additives. *Source*: Reproduced with permission from [108]/Royal Society of Chemistry. (c) 2D GIWAXS patterns of the IDCIC films with or without different additives. (d) *J*–*V* curves of FTAZ:IDCIC-based OSCs without or with an additive. *Source*: Reproduced with permission from [107]/John Wiley & Sons. (e) *J*–*V* curves of devices based on PBDB-TF:BTIC-CF_3-m. (f) Dark *J*–*V* curves of electron-only devices based on PBDB-TF:BTIC-CF_3-m. *Source*: Reproduced with permission from [108]/Royal Society of Chemistry.

dissociation and charge transport. Thus, the resulting OSCs achieved a high PCE of 15.9%, nearly twice that of OSCs from room-temperature sequential casting.

Among various pre-treatment methods, adding solvent additives has been widely proven as a common but efficient method to modify the resulting film morphology. As shown in Figure 7.23a, DIO, 1-chloronaphthalene (CN), and diphenyl ether (DPE) are the most frequently used additives.[101] For instance, using DIO as the additive can effectively promote the aggregate of IDCIC molecules (Figure 7.23a) and further enhance the domain size and domain purity in the FTAZ:IDCIC blend film (Figure 7.23c).[107] As a result, an enhanced V_{OC} was achieved for OSCs using DIO as the additive, leading to a high PCE of 9.76% (Figure 7.23d). In contrast, the addition of CN as the additive was found to inhibit the aggregation of IDCIC, and further facilitate the acceptor molecules to diffuse into the FTAZ donor matrix, resulting in reduced domain size and purity. Thus, the FTAZ: IDCIC-based OSC with 0.25% CN&DIO treatment achieved the best PCE of 13.58%, due to the formation of desired BHJ film morphology.

Solvent additive can also regulate the aggregation behaviors of A-DA'D-A series NFAs. As an example, a trifluoromethyl-substituted Y6 acceptor was reported by He et al., namely BTIC-CF_3-*m* (Figure 7.23a), and its aggregation behaviors were found to be highly related to the processing additives, resulting in significantly distinct photovoltaic performance (Figure 7.23e).[108] As shown in Figure 7.23b, CN and DIO additives can promote the formation of H- and J-aggregates, respectively, and they found that H-aggregates can broaden the bandgap of BTIC-CF_3-*m* to increase the V_{OC}, while J-aggregates can reduce the bandgap to improve J_{SC} (Figure 7.23f). Moreover, both types of aggregates can improve the *FF*. These results thus strongly suggest that subtle morphology optimization of FREAs, especially precise control of their H- and J-aggregations, is important to further improve the device performance.

In summary, the photovoltaic performance of OSCs is highly correlated with the BHJ morphology of photoactive layer, which is significantly affected by the aggregation behaviors of both donor and acceptors. So far, it has been well demonstrated that the aggregation behaviors of photovoltaic materials are not only determined by their inherent properties such as molecular structure, molecular polarity, molecular packing, etc., but also are influenced by the film formation processing including processing solvents, processing time, additives, post-treatments, etc. In most cases, J-type aggregation is more preferred aggregation behavior for photovoltaic materials, because it can simultaneously lead to a bathochromic absorption to achieve higher J_{SC} and a dense π–π packing to facilitate charge transport. Besides, an appropriate control of aggregation size in the blend films can reduce the undesired charge recombination to improve V_{OC}. In the future, to fulfill the requirements of commercialization of OSCs, more efforts have to be devoted to developing effective strategies to precisely control the molecular aggregates in large areas and flexible BHJ films, which are processed via advanced processing techniques such as blade coating and roll-to-roll printing. To achieve this target, new molecular designs of photoactive materials as well as new treatment methods are highly demanded.

References

1 Green, M.A., Dunlop, E.D., Hohl-Ebinger, J. et al. (2022). Solar cell efficiency tables (Version 60). *Prog. Photovolt.: Res. Appl.* 30 (7): 687–701.

2 Yeh, N. and Yeh, P. (2013). Organic solar cells: Their developments and potentials. *Renew. Sust. Energ. Rev.* 21: 421–431.

3 Spanggaard, H. and Krebs, F.C. (2004). A brief history of the development of organic and polymeric photovoltaics. *Sol. Energy Mater. Sol. Cells* 83 (2–3): 125–146.

4 Kallmann, H. and Pope, M. (1959). Photovoltaic effect in organic crystals. *J. Chem. Phys.* 30 (2): 585–586.

5 Tang, C.W. (1986). Two-layer organic photovoltaic cell. *Appl. Phys. Lett.* 48 (2): 183–185.

6 Yu, G., Gao, J., Hummelen, J.C. et al. (1995). Polymer photovoltaic cells: Enhanced efficiencies via a network of internal donor-acceptor heterojunctions. *Science* 270 (5243): 1789–1791.

7 Halls, J.J.M., Walsh, C.A., Greenham, N.C. et al. (1995). Efficient photodiodes from interpenetrating polymer networks. *Nature* 376 (6540): 498–500.

8 Huang, Y., Kramer, E.J., Heeger, A.J. et al. (2014). Bulk heterojunction solar cells: morphology and performance relationships. *Chem. Rev.* 114 (14): 7006–7043.

9 Zhao, J., Li, Y., Yang, G. et al. (2016). Efficient organic solar cells processed from hydrocarbon solvents. *Nat. Energy* 1 (2): 15027.

10 Lin, Y., Wang, J., Zhang, Z.G. et al. (2015). An electron acceptor challenging fullerenes for efficient polymer solar cells. *Adv. Mater.* 27 (7): 1170–1174.

11 Wang, J. and Zhan, X. (2021). Fused-ring electron acceptors for photovoltaics and beyond. *Acc. Chem. Res.* 54 (1): 132–143.

12 Cui, Y., Xu, Y., Yao, H. et al. (2021). Single-junction organic photovoltaic cell with 19% efficiency. *Adv. Mater.* 33 (41): e2102420.

13 Zhu, L., Zhang, M., Xu, J. et al. (2022). Single-junction organic solar cells with over 19% efficiency enabled by a refined double-fibril network morphology. *Nat. Mater.* 21 (6): 656–663.

14 Gao, W., Qi, F., Peng, Z. et al. (2022). Achieving 19% power conversion efficiency in planar-mixed heterojunction organic solar cells using a pseudosymmetric electron acceptor. *Adv. Mater.* 34 (32): e2202089.

15 Qian, D., Zheng, Z., Yao, H. et al. (2018). Design rules for minimizing voltage losses in high-efficiency organic solar cells. *Nat. Mater.* 17 (8): 703–709.

16 Roncali, J. (2009). Molecular bulk heterojunctions: An emerging approach to organic solar cells. *Acc. Chem. Res.* 42 (11): 1719–1730.

17 Cui, C. and Li, Y. (2021). Morphology optimization of photoactive layers in organic solar cells. *Aggregate* 2 (2): e31.

18 Yoon, S., Shin, E.-Y., Cho, N.-K. et al. (2021). Progress in morphology control from fullerene to nonfullerene acceptors for scalable high-performance organic photovoltaics. *J. Mater. Chem. A* 9 (44): 24729–24758.

19 Kan, B., Kan, Y., Zuo, L. et al. (2020). Recent progress on all-small molecule organic solar cells using small-molecule nonfullerene acceptors. *InfoMat* 3 (2): 175–200.

20 de Miguel, G., Ziółek, M., Zitnan, M. et al. (2012). Photophysics of H- and J-Aggregates of Indole-Based Squaraines in Solid State. *J. Phys. Chem. C* 116 (17): 9379–9389.

21 Wurthner, F., Kaiser, T.E., Saha-Moller, C.R. (2011). J-aggregates: from serendipitous discovery to supramolecular engineering of functional dye materials. *Angew. Chem. Int. Ed.* 50 (15): 3376–3410.

22 Kim, J.H., Schembri, T., Bialas, D. et al. (2022). Slip-stacked J-aggregate materials for organic solar cells and photodetectors. *Adv. Mater.* 34 (22): e2104678.

23 Chen, G., Sasabe, H., Sasaki, Y. et al. (2014). A series of squaraine dyes: Effects of side chain and the number of hydroxyl groups on material properties and photovoltaic performance. *Chem. Mater.* 26 (3): 1356–1364.

24 Li, T.Y., Meyer, T., Meerheim, R. et al. (2017). Aza-BODIPY dyes with heterocyclic substituents and their derivatives bearing a cyanide co-ligand: NIR donor materials for vacuum-processed solar cells. *J. Mater. Chem. A* 5 (21): 10696–10703.

25 Gao, K., Li, L., Lai, T. et al. (2015). Deep absorbing porphyrin small molecule for high-performance organic solar cells with very low energy losses. *J. Am. Chem. Soc.* 137 (23): 7282–7285.

26 Mayerhoffer, U., Deing, K., Gruss, K. et al. (2009). Outstanding short-circuit currents in BHJ solar cells based on NIR-absorbing acceptor-substituted squaraines. *Angew. Chem. Int. Ed.* 48 (46): 8776–8779.

27 Deing, K.C., Mayerhoffer, U., Wurthner, F. et al. (2012). Aggregation-dependent photovoltaic properties of squaraine/PC61BM bulk heterojunctions. *Phys. Chem. Chem. Phys.* 14 (23): 8328–8334.

28 Yang, D., Jiao, Y., Huang, Y. et al. (2016). Two different donor subunits substituted unsymmetrical squaraines for solution-processed small molecule organic solar cells. *Org. Electron.* 32: 179–186.

29 Ren, Y., Hiszpanski, A.M., Whittaker-Brooks, L. et al. (2014). Structure-property relationship study of substitution effects on isoindigo-based model compounds as electron donors in organic solar cells. *ACS Appl. Mater. Interfaces* 6 (16): 14533–14542.

30 Ren, Y., Hailey, A.K., Hiszpanski, A.M. et al. (2014). Isoindigo-containing molecular semiconductors: Effect of backbone extension on molecular organization and organic solar cell performance. *Chem. Mater.* 26 (22): 6570–6577.

31 Chen, G., Si, C.F., Zhang, P.P. et al. (2017). The effect of processing solvent dependent film aggregation on the photovoltaic performance of squaraine: PC71BM bulk heterojunction solar cells. *Org. Electron.* 51: 62–69.

32 Chen, G., Sasabe, H., Lu, W. et al. (2013). J-aggregation of a squaraine dye and its application in organic photovoltaic cells. *J. Mater. Chem. C* 1 (40): 6547–6552.

33 Xiao, L., Li, Z., Hu, Q. et al. (2019). Improving the efficiencies of small molecule solar cells by solvent vapor annealing to enhance J-aggregation. *J. Mater. Chem. C* 7 (31): 9618–9624.

34 Ge, J., Hong, L., Song, W. et al. (2021). Solvent annealing enables 15.39% efficiency all-small-molecule solar cells through improved molecule interconnection and reduced non-radiative loss. *Adv. Energy Mater.* 11 (22): 2100800.

35 Li, T.Y., Benduhn, J., Qiao, Z. et al. (2019). Effect of H- and J-aggregation on the photophysical and voltage loss of boron dipyrromethene small molecules in vacuum-deposited organic solar cells. *J. Phys. Chem. Lett.* 10 (11): 2684–2691.

36 Zitzler-Kunkel, A., Lenze, M.R., Meerholz, K. et al. (2013). Enhanced photocurrent generation by folding-driven H-aggregate formation. *Chem. Sci.* 4 (5): 2071–2075.

37 Zhao, Q., Yu, X., Liu, J. et al. (2016). Increasing H-aggregation of p-DTS(FBTTh2)2 to improve photovoltaic efficiency by solvent vapor annealing. *Org. Electron.* 37: 6–13.

38 An, C., Zheng, Z., Hou, J. (2020). Recent progress in wide bandgap conjugated polymer donors for high-performance nonfullerene organic photovoltaics. *Chem. Commun.* 56 (35): 4750–4760.

39 Karki, A., Gillett, A.J., Friend, R.H. et al. (2020). The path to 20% power conversion efficiencies in nonfullerene acceptor organic solar cells. *Adv. Energy Mater.* 11 (15): 2003441.

40 Fu, H., Wang, Z., Sun, Y. (2019). Polymer donors for high-performance non-fullerene organic solar cells. *Angew. Chem. Int. Ed.* 58 (14): 4442–4453.

41 Zhao, J., Yao, C., Ali, M.U. et al. (2020). Recent advances in high-performance organic solar cells enabled by acceptor–donor–acceptor–donor–acceptor (A–DA′D–A) type acceptors. *Mater. Chem. Front.* 4 (12): 3487–3504.

42 He, K., Kumar, P., Yuan, Y. et al. (2021). Wide bandgap polymer donors for high efficiency non-fullerene acceptor based organic solar cells. *Mater. Adv.* 2 (1): 115–145.

43 Qian, D., Ye, L., Zhang, M. et al. (2012). Design, application, and morphology study of a new photovoltaic polymer with strong aggregation in solution state. *Macromolecules* 45 (24): 9611–9617.

44 Zhang, M., Guo, X., Ma, W. et al. (2015). A large-bandgap conjugated polymer for versatile photovoltaic applications with high Performance. *Adv. Mater.* 27 (31): 4655–1660.

45 Zhang, S., Qin, Y., Zhu, J. et al. (2018). Over 14% efficiency in polymer solar cells enabled by a chlorinated polymer donor. *Adv. Mater.* 30 (20): e1800868.

46 Hu, H., Chow, P.C.Y., Zhang, G. et al. (2017). Design of donor polymers with strong temperature-dependent aggregation property for efficient organic photovoltaics. *Acc. Chem. Res.* 50 (10): 2519–2528.

47 Li, Z.a., Chueh, C.-C., and Jen, A.K.Y. (2019). Recent advances in molecular design of functional conjugated polymers for high-performance polymer solar cells. *Prog. Polym. Sci.* 99: 101175.

48 Qiao, S., Li, X., Wang, H. et al. (2019). Temperature-dependent and aggregation-breaking strategy for benzodifuran-constructed organic solar cells. *Sol. RRL* 3 (10): 1900159.

49 Park, J.S., Choi, N., Lee, C. et al. (2020). Elucidating roles of polymer donor aggregation in all-polymer and non-fullerene small-molecule–polymer solar cells. *Chem. Mater.* 32 (8): 3585–3596.

50 Liu, T. and Troisi, A. (2013). What makes fullerene acceptors special as electron acceptors in organic solar cells and how to replace them. *Adv. Mater.* 25 (7): 1038–1041.

51 Liu, Y., Zhao, J., Li, Z. et al. (2014). Aggregation and morphology control enables multiple cases of high-efficiency polymer solar cells. *Nat. Commun.* 5: 5293.

52 He, Z., Xiao, B., Liu, F. et al. (2015). Single-junction polymer solar cells with high efficiency and photovoltage. *Nat. Photonics* 9 (3): 174–179.

53 Kumari, T., Lee, S.M., Kang, S.-H. et al. (2017). Ternary solar cells with a mixed face-on and edge-on orientation enable an unprecedented efficiency of 12.1%. *Energy Environ. Sci.* 10 (1): 258–265.

54 Yue, Q., Liu, W., Zhu, X. (2020). n-type molecular photovoltaic materials: Design strategies and device applications. *J. Am. Chem. Soc.* 142 (27): 11613–11628.

55 Chong, K., Xu, X., Meng, H. et al. (2022). Realizing 19.05% efficiency polymer solar cells by progressively improving charge extraction and suppressing charge recombination. *Adv. Mater.* 34 (13): e2109516.

56 Zhan, X., Facchetti, A., Barlow, S. et al. (2011). Rylene and related diimides for organic electronics. *Adv. Mater.* 23 (2): 268–84.

57 Zhang, G., Zhao, J., Chow, P.C.Y. et al. (2018). Nonfullerene acceptor molecules for bulk heterojunction organic solar cells. *Chem. Rev.* 118 (7): 3447–3507.

58 Sauve, G. and Fernando, R. (2015). Beyond fullerenes: Designing alternative molecular electron acceptors for solution-processable bulk heterojunction organic photovoltaics. *J. Phys. Chem. Lett.* 6 (18): 3770–3780.

59 Singh, R., Lee, J., Kim, M. et al. (2017). Control of the molecular geometry and nanoscale morphology in perylene diimide based bulk heterojunctions enables an efficient non-fullerene organic solar cell. *J. Mater. Chem. A* 5 (1): 210–220.

60 Cheng, P., Li, G., Zhan, X. et al. (2018). Next-generation organic photovoltaics based on non-fullerene acceptors. *Nat. Photonics* 12 (3): 131–142.

61 Han, G., Guo, Y., Song, X. et al. (2017). Terminal π–π stacking determines three-dimensional molecular packing and isotropic charge transport in an A–π–A electron acceptor for non-fullerene organic solar cells. *J. Mater. Chem. C* 5 (20): 4852–4857.

62 Zhang, J., Tan, H.S., Guo, X. et al. (2018). Material insights and challenges for non-fullerene organic solar cells based on small molecular acceptors. *Nat. Energy* 3 (9): 720–731.

63 Zhang, Z., Wang, T., Ding, Z. et al. (2019). Small molecular donor/polymer acceptor type organic solar cells: Effect of molecular weight on active layer morphology. *Macromolecules* 52 (22): 8682–8689.

64 Zhu, J., Wu, Y., Rech, J. et al. (2018). Enhancing the performance of a fused-ring electron acceptor via extending benzene to naphthalene. *J. Mater. Chem. C* 6 (1): 66–71.

65 Zhu, J., Ke, Z., Zhang, Q. et al. (2018). Naphthodithiophene-based nonfullerene acceptor for high-performance organic photovoltaics: Effect of extended conjugation. *Adv. Mater.* 30 (2): 1704713.

66 Li, T., Dai, S., Ke, Z. et al. (2018). Fused Tris(thienothiophene)-based electron acceptor with strong near-infrared absorption for high-performance as-cast solar cells. *Adv. Mater.* 30 (10): 1705969.

67 Dai, S., Li, T., Wang, W. et al. (2018). Enhancing the performance of polymer solar cells via core engineering of NIR-absorbing electron acceptors. *Adv. Mater.* 30 (15): e1706571.

68 Lin, Y., Li, T., Zhao, F. et al. (2016). Structure evolution of oligomer fused-ring electron acceptors toward high efficiency of as-cast polymer solar cells. *Adv. Energy Mater.* 6 (18): 1600854.

69 Wang, J., Zhang, J., Xiao, Y. et al. (2018). Effect of isomerization on high-performance nonfullerene electron acceptors. *J. Am. Chem. Soc.* 140 (29): 9140–9147.

70 Yuan, J., Zhang, Y., Zhou, L. et al. (2019). Single-junction organic solar cell with over 15% efficiency using fused-ring acceptor with electron-deficient core. *Joule* 3 (4): 1140–1151.

71 Zhu, L., Zhang, M., Zhou, G. et al. (2020). Efficient organic solar cell with 16.88% efficiency enabled by refined acceptor crystallization and morphology with improved charge transfer and transport properties. *Adv. Energy Mater.* 10 (18): 1904234.

72 Zhu, W., Spencer, A.P., Mukherjee, S. et al. (2020). Crystallography, Morphology, Electronic structure, and transport in non-fullerene/non-indacenodithienothiophene polymer:Y6 solar cells. *J. Am. Chem. Soc.* 142 (34): 14532–14547.

73 Zhang, G., Chen, X.K., Xiao, J. et al. (2020). Delocalization of exciton and electron wavefunction in non-fullerene acceptor molecules enables efficient organic solar cells. *Nat. Commun.* 11 (1): 3943.

74 Zhang, W., Huang, J., Xu, J. et al. (2020). Phthalimide polymer donor guests enable over 17% efficient organic solar cells via parallel-like ternary and quaternary strategies. *Adv. Energy Mater.* 10 (32): 2001436.

75 Yu, H., Qi, Z., Zhang, J. et al. (2020). Tailoring non-fullerene acceptors using selenium-incorporated heterocycles for organic solar cells with over 16% efficiency. *J. Mater. Chem. A* 8 (45): 23756–23765.

76 Zhang, Z., Li, Y., Cai, G. et al. (2020). Selenium heterocyclic electron acceptor with small Urbach energy for as-cast high-performance organic solar cells. *J. Am. Chem. Soc.* 142 (44): 18741–18745.

77 Wang, X., Lu, H., Liu, Y. et al. (2021). Simple nonfused ring electron acceptors with 3D network packing structure boosting the efficiency of organic solar cells to 15.44%. *Adv. Energy Mater.* 11 (45): 2102591.

78 Liu, Z.X., Yu, Z.P., Shen, Z. et al. (2021). Molecular insights of exceptionally photostable electron acceptors for organic photovoltaics. *Nat. Commun.* 12 (1): 3049.

79 Liu, H., Li, Y., Xu, S. et al. (2021). Emerging chemistry in enhancing the chemical and photochemical stabilities of fused-ring electron acceptors in organic solar cells. *Adv. Funct. Mater.* 31 (50): 2106735.

80 Ma, L., Zhang, S., Zhu, J. et al. (2021). Completely non-fused electron acceptor with 3D-interpenetrated crystalline structure enables efficient and stable organic solar cell. *Nat. Commun.* 12 (1): 5093.

81 Hong, L., Yao, H., Wu, Z. et al. (2019). Eco-compatible solvent-processed organic photovoltaic cells with over 16% efficiency. *Adv. Mater.* 31 (39): e1903441.

82 Zhan, L., Li, S., Li, Y. et al. (2022). Manipulating charge transfer and transport via intermediary electron acceptor channels enables 19.3% efficiency organic photovoltaics. *Adv. Energy Mater.* 12 (39): 2201076.

83 Holliday, S., Ashraf, R.S., Wadsworth, A. et al. (2016). High-efficiency and air-stable P3HT-based polymer solar cells with a new non-fullerene acceptor. *Nat. Commun.* 7: 11585.

84 Zhang, Z., Yu, J., Yin, X. et al. (2018). Conformation locking on fused-ring electron acceptor for high-performance nonfullerene organic solar cells. *Adv. Funct. Mater.* 28 (11): 1705095.

85 Li, Y., Song, J., Dong, Y. et al. (2022). Polymerized small molecular acceptor with branched side chains for all polymer solar cells with efficiency over 16.7. *Adv. Mater.* 34 (14): e2110155.

86 Liu, W., Zhang, R., Wei, Q. et al. (2022). Manipulating molecular aggregation and crystalline behavior of A-DA'D-A type acceptors by side chain engineering in organic solar cells. *Aggregate* 3: e183.

87 Cui, Y., Yao, H., Zhang, J. et al. (2020). Single-junction organic photovoltaic cells with approaching 18% efficiency. *Adv. Mater.* 32 (19): e1908205.

88 Yang, Y., Zhang, Z.G., Bin, H. et al. (2016). Side-chain isomerization on an n-type organic semiconductor ITIC acceptor makes 11.77% high efficiency polymer solar cells. *J. Am. Chem. Soc.* 138 (45): 15011–15018.

89 Mai, J., Xiao, Y., Zhou, G. et al. (2018). Hidden structure ordering along backbone of fused-ring electron acceptors enhanced by ternary bulk heterojunction. *Adv. Mater.* 30 (34): e1802888.

90 Lin, Y., Zhao, F., He, Q. et al. (2016). High-performance electron acceptor with thienyl side chains for organic photovoltaics. *J. Am. Chem. Soc.* 138 (14): 4955–4961.

91 Wang, J., Wang, W., Wang, X. et al. (2017). Enhancing performance of nonfullerene acceptors via side-chain conjugation strategy. *Adv. Mater.* 29 (35): 1702125.

92 Li, S., Ye, L., Zhao, W. et al. (2016). Energy-level modulation of small-molecule electron acceptors to achieve over 12% efficiency in polymer solar cells. *Adv. Mater.* 28 (42): 9423–9429.

93 Zhang, Z., Liu, W., Rehman, T. et al. (2017). Energy-level modulation of non-fullerene acceptors to achieve high-efficiency polymer solar cells at a diminished energy offset. *J. Mater. Chem. A* 5 (20): 9649–9654.

94 Meng, D., Zheng, R., Zhao, Y. et al. (2022). Near-infrared materials: The turning point of organic photovoltaics. *Adv. Mater.* 34 (10): e2107330.

95 He, B., Yang, B., Kolaczkowski, M.A. et al. (2018). Molecular engineering for large open-circuit voltage and low energy loss in around 10% non-fullerene organic photovoltaics. *ACS Energy Lett.* 3 (4): 1028–1035.

96 Dai, S., Zhao, F., Zhang, Q. et al. (2017). Fused nonacyclic electron acceptors for efficient polymer solar cells. *J. Am. Chem. Soc.* 139 (3): 1336–1343.

97 Li, Z., Ying, L., Zhu, P. et al. (2019). A generic green solvent concept boosting the power conversion efficiency of all-polymer solar cells to 11%. *Energy Environ. Sci.* 12 (1): 157–163.

98 Xu, X., Yu, L., Yan, H. et al. (2020). Highly efficient non-fullerene organic solar cells enabled by a delayed processing method using a non-halogenated solvent. *Energy Environ. Sci.* 13 (11): 4381–4388.

99 Lin, B., Zhou, X., Zhao, H. et al. (2020). Balancing the pre-aggregation and crystallization kinetics enables high efficiency slot-die coated organic solar cells with reduced non-radiative recombination losses. *Energy Environ. Sci.* 13 (8): 2467–2479.

100 Zhao, H., Naveed, H.B., Lin, B. et al. (2020). Hot hydrocarbon-solvent slot-die coating enables high-efficiency organic solar cells with temperature-dependent aggregation behavior. *Adv. Mater.* 32 (39): e2002302.

101 McDowell, C., Abdelsamie, M., Toney, M.F. et al. (2018). Solvent additives: Key morphology-directing agents for solution-processed organic solar cells. *Adv. Mater.*: e1707114.

102 Li, G., Shrotriya, V., Huang, J. et al. (2005). High-efficiency solution processable polymer photovoltaic cells by self-organization of polymer blends. *Nat. Mater.* 4 (11): 864–868.

103 Li, G., Yao, Y., Yang, H. et al. (2007). "Solvent Annealing" effect in polymer solar cells based on poly(3-hexylthiophene) and methanofullerenes. *Adv. Funct. Mater.* 17 (10): 1636–1644.

104 Dong, S., Zhang, K., Jia, T. et al. (2019). Suppressing the excessive aggregation of nonfullerene acceptor in blade-coated active layer by using n-type polymer additive to achieve large-area printed organic solar cells with efficiency over 15%. *EcoMat* 1 (1): e12006.

105 Wang, J., Zhu, R., Wang, S. et al. (2021). Enhancing photovoltaic performance via aggregation dynamics control in fused-ring electron acceptor. *Aggregate* 2 (3): e29.

106 Li, D., Guo, C., Zhang, X. et al. (2021). Heating-induced aggregation control for efficient sequential-cast organic solar cells. *Aggregate* 3 (3): e104.

107 He, D., Zhao, F., Xin, J. et al. (2018). A fused ring electron acceptor with decacyclic core enables over 13.5% efficiency for organic solar cells. *Adv. Energy Mater.* 8 (30): 1802050.

108 Zhao, Q., Lai, H., Chen, H. et al. (2021). H- and J-aggregation inspiring efficient solar conversion. *J. Mater. Chem. A* 9 (2): 1119–1126.

8

Molecular Aggregates for Second-order Nonlinear Optical Effect

Wenbo Wu[1], and Zhen Li[1,2,3]*

[1] *Department of Chemistry, Institute of Molecular Aggregation Science, Tianjin University, Tianjin, China*
[2] *Department of Chemistry, Wuhan University, Wuhan, China*
[3] *Joint School of National University of Singapore and Tianjin University, International Campus of Tianjin University, Binhai New City, Fuzhou, China*

*Corresponding author: wuwb@tju.edu.cn

8.1 Microscopic and Macroscopic Nonlinearities of Organic Systems

The twenty-first century we live in is the global information age, and the requirements for information technology are increasing more and more rapidly. At present, microelectronic materials with electrons as the carriers are approaching the limit in terms of transmission speed, capacity, and spatial compatibility, making it difficult to meet the needs of future communication technology. Alternatively, many advantages have been observed when transmitting information by photonics, such as high transmission speed, wide bandwidth, good parallelism, anti-jamming, high density, etc.[1,2] As early as 2010, it was pointed out by the People's Daily that "it is an irresistible historical trend to use photons instead of electrons to realize data connection."[3] Nonlinear optical (NLO) material is one of the indispensable keys to realize this "photon revolution." For instance, the phase and amplitude of light are mainly modulated by electro-optic (EO) modulator in opto-electronic technology, so as to effectively load the data signal onto the light wave, and the EO materials, which determined the performance of the modulator, are essentially the second-order NLO materials.[2,4] How to design second-order NLO materials with high performance has therefore become an important research frontier.

The first observation of NLO effects can be dated back to 1961, Franken and co-workers observed an unprecedented phenomenon that one beam of 694.3 nm ruby laser can pass through the quartz crystal to generate another beam of ultraviolet laser with the wavelength of 347.2 nm, whose frequency was exactly twice that of the original ruby laser.[5] This phenomenon has been termed as second-harmonic generation (SHG), which gave birth to this research field. Besides SHG, more and more NLO effects, such as third or even higher harmonic generation, photorefractive, two-photon absorption effect, photomixing, Pockels effect, etc., have been developed since then, making NLO gradually become an important and independent optical discipline.[6–9]

Herein, we would like to introduce some basic notions about this area first. Generally speaking, optical frequency polarization occurs in micro media (such as atoms and molecules) and macro media (such as crystals, powders, and films) after being irradiated by light, and the induced polarization intensity is a nonlinear function of the electric field, which can be described by the following two equations[10,11]:

$$p = \alpha E + \beta E^2 + \gamma E^3 + \dots \tag{8.1}$$

$$P = \chi^{(1)} E + \chi^{(2)} E^2 + \chi^{(3)} E^3 + \dots \tag{8.2}$$

The first equation is used to describe the microscopic nonlinearities, in which, p is the induced polarization intensity of the microscopic molecules, and E is the photoelectric field intensity, α is the linear optical coefficient, while β (or γ, etc.) is the second-order (or third-order, etc.) nonlinear optical coefficient, which is also termed as the first (or second, etc.) hyperpolarizability.

Similarly, the second equation is for the macroscopic nonlinearities, in which P is the induced polarization intensity of the materials, while $\chi^{(1)}$, $\chi^{(2)}$, and $\chi^{(3)}$ are the linear (or the first) polarizability, the second nonlinear polarizability, the third nonlinear polarizability, etc. From $\chi^{(1)}$ to $\chi^{(2)}$, then to $\chi^{(3)}$, etc., the polarizabilities become weaker and weaker, and their order of magnitude satisfies the following equation approximately:

$$\frac{|\chi^{(n)}|}{|\chi^{(n-1)}|} \approx \frac{1}{|E_0|} \tag{8.3}$$

where E_0 is the average electric field intensity in atoms with the order of magnitude of 10^{11} V/m. The photoelectric field intensity of E is therefore obviously much lower than E_0, so that the quadratic term and subsequent terms can be ignored. Under such conditions, the induced polarization intensity is proportional to the photoelectric field, leading to the common linear optical phenomenon, such as reflection, refraction, etc. Things changed when a laser with high power density is used. At that time, the enhanced E makes the quadratic term, cubic term, and even higher order terms cannot be ignored anymore, leading to NLO phenomenon. Accordingly, βE^2 (or γE^3, etc.) and $\chi^{(2)} E^2$ (or $\chi^{(3)} E^3$, etc.) are the physical roots of microscopic and macroscopic second-order (third-order) NLO effects, respectively. In the following part of this chapter, we will focus on the second-order NLO properties, to present their Molecular Uniting Set Identified Characteristic (MUSIC).

"Two-level" model, which was proposed by Oudar and Chemla in 1977, is a classic theory to describe the β value of 1D linear molecules.[12,13] According to "two-level" model, β value can be calculated by Equation 8.4:

$$\beta = \frac{3}{2h^2c}\frac{\nu_{eg}^2 r_{eg}^2 \Delta\mu_{eg}}{\left(\nu_{eg}^2 - \nu_L^2\right)\left(\nu_{eg}^2 - 4\nu_L^2\right)} \tag{8.4}$$

in which, h is the Planck's constant, c is the light speed, ν_{eg} is the charge transfer transition frequency, ν_L is the incident radiation frequency, r_{eg} is the dipole moment of the transition, and $\Delta\mu_{eg}$ is the difference of the dipole moment between the excited state and ground state. Therefore, it is obvious that for a centrosymmetric molecule, its $\Delta\mu_{eg}$ is 0, so that it could not demonstrate second-order NLO effect. According to Equation 8.4, if one molecule has large $\Delta\mu_{eg}$ and r_{eg} values, as well as low ν_{eg} value, it is expected to show good second-order NLO effect. Thus, the NLO chromophores were usually designed by linking an electron donor and electron acceptor to a polarizable bridge π-bridge (Figure 8.1 lists some NLO chromophores

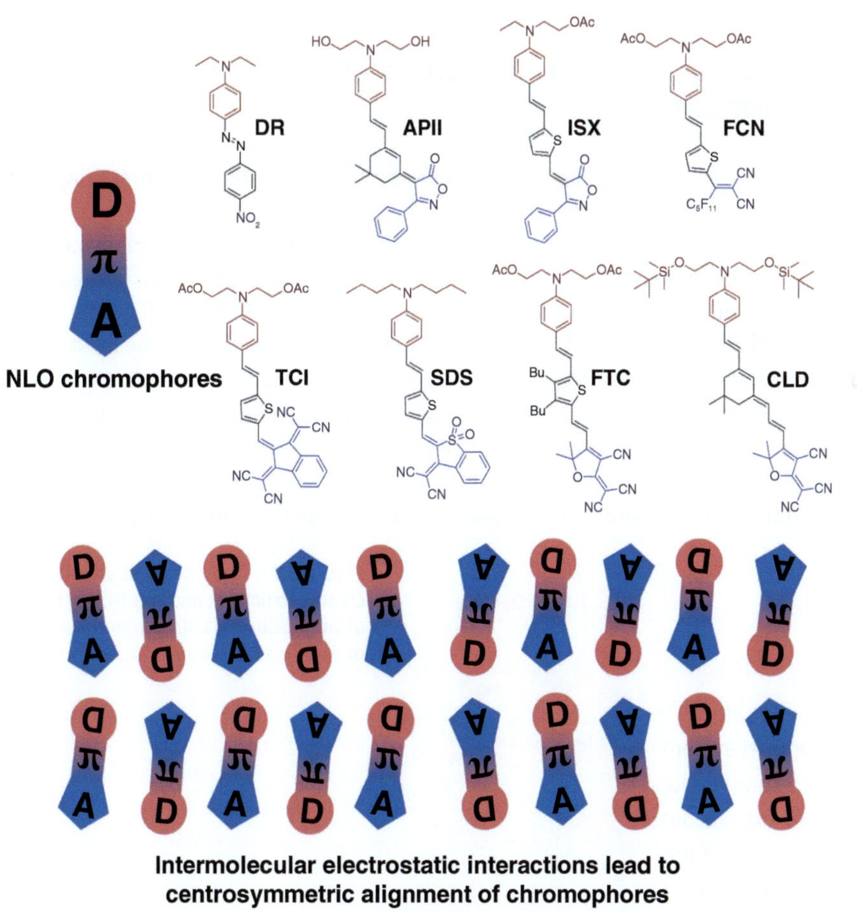

Figure 8.1 Some reported second-order NLO chromophores and their centrosymmetric alignment in macroscopic materials. *Source*: Wenbo Wu and Zhen Li.

as examples). Thanks to the efforts of scientists, the β values of NLO chromophores have been improved by two or three orders of magnitude so far.

However, in second-order NLO fields, it is a pity that large microscopic β values of molecules in fact do not mean high macroscopic nonlinearities of materials. From Figure 8.1, it is clear that all of these second-order NLO chromophores are rod-like structures with terrible charge separation, leading to intermolecular electrostatic interactions, which consequently make NLO chromophores centrosymmetric alignment in macroscopic materials.[14,15] However, the centrosymmetric macroscopic configuration of materials cannot demonstrate second-order NLO effect, similar to single molecules. From this point of view, it is one of the most important things in this field that to conquer the intermolecular electrostatic interaction for making the chromophores to form non-centrosymmetric aggregates, so that the microscopic β values of molecules can be translated into macroscopic second-order NLO effects of materials. In the following part of this chapter, we would like to discuss the recent advances in how to control the aggregates of NLO chromophores for high performance second-order NLO materials in detail, especially the molecular pack mode in aggregated states.

8.2 Organic Molecules for Second-order Nonlinear Optics

Whenever we mention regular arrangement, the crystal is usually the first thing that comes to our mind. This is also applicable to second-order NLO field, in which the crystallization is used to realize the non-centrosymmetric regular arrangement of NLO chromophores with strong intermolecular electrostatic interactions.[7,8] As early as 1977, it was found that the crystal of urea demonstrated second-order NLO effect with SHG coefficient of 1.3 pm V^{-1}, while the amorphous solid did not.[16,17] The most stable urea crystal is tetragonal system with $P\bar{4}2_1m$ space group.[18] As presented in Figure 8.2, it is clear that the urea molecules are all regularly arranged in crystals with the dipole moments pointing in the same direction, leading to macroscopic nonlinearities.

Although 1.3 pm V^{-1} is very low from the current perspective, urea crystal opened the curtain of second-order NLO organic crystals. Since then, many organic crystals

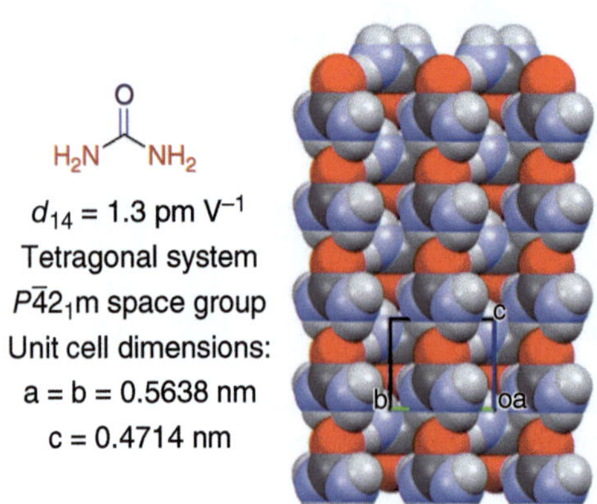

Figure 8.2 The chemical structure, crystal structure, and the NLO effect of urea. *Source*: [18]/with permission of American Chemical Society.

have been grown to provide macroscopic second-order NLO properties. In theory, ionic crystals generally show greater lattice energy than molecular crystals, and they were therefore first studied in organic NLO field, since lattice energy makes the key contribution to overcome the strong interactions among chromophores. 4-(4-(*N,N*-dimethylamino)styryl)-1-methylpyridinium 4-methylbenzenesulfonate (**DAST**) is one of the most commonly used second-order NLO crystals.[19,20] As shown in Figure 8.3a, **DAST** crystals have parallel-type packing, in which the cationic pyridinium salt acts as the acentric core, while 4-methylbenzenesulfonate is used as the counter anion. The parallel-type packing makes the chromophores orderly arranged in a crystal, with the order parameter (in the range of 0–1, which can be used to illustrate the alignment of the chromophore moieties) of 0.83, and a molecular ordering angle of 20°.

Subsequently, the hydrogen bonds were further utilized to achieve even better alignment of the chromophore moieties in parallel-type crystals, and **HMQ-TMS** (Figure 8.3b) is a typical example.[21] In comparison with **DAST**, the most significant improvement in chemical structure is that the phenolic donor with -OH groups, which can form hydrogen bonds with the counter anions or themselves for counteracting strong interactions among chromophores and stabilizing the ordered arrangement of cationic chromophores. Although the cationic chromophores in **HMQ-TMS** showed a larger dipole moment than those in **DAST** with stronger electrostatic interactions, as a result of stronger electron-withdrawing strength of quinolinium acceptor than pyridinium acceptor, the arrangement of chromophores

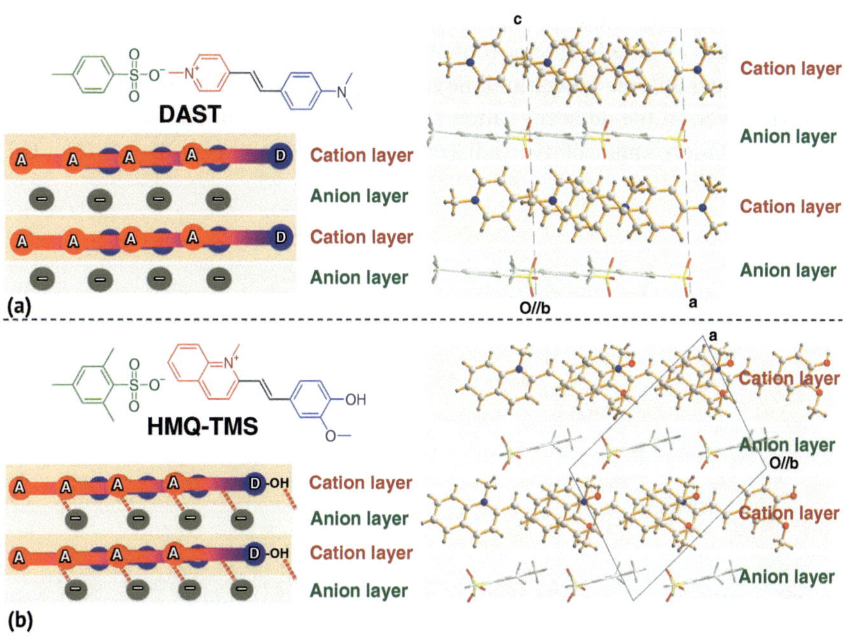

Figure 8.3 The chemical structures, schematic diagram molecular arrangements in crystals, and crystal structures of (a) **DAST** and (b) **HMQ-TMS**. *Source*: [7]/with permission of John Wiley & Sons.

in **HMQ-TMS** crystal is more regular (Figure 8.3b) with the nearly maximum value of the order parameter (≈ 1). **HMQ-TMS** was then applied in multicycle THz wave generation, another famous NLO application field, and demonstrated 20 times higher efficiency than inorganic ZnTe crystals.[22]

It should be noted that, although the introduction of phenolic donor can significantly improve the arrangement of chromophores, the enhancement in macroscopic second-order NLO coefficient was, in fact, limited, due to the weaker electron-donating strength of phenol donor than that of dialkylamino ones.[23] Therefore, the piperidin-4-ylmethanol donor, which inherited both advantages of strong electron-donating strength from dialkylamino donor and H-bond forming ability from phenol donor, was developed (Figure 8.4). Further investigation demonstrated that when this donor was used, the corresponding NLO chromophores (such as **PMB-T** in Figure 8.4) can hardly keep the parallel architecture in crystals anymore, but tend to form series-type crystals, in which the chromophores and the counter anions were linearly assembled with each other through hydrogen bonds between -OH and -SO_3^-.[24] In **PMB-T** crystals, the order parameter of chromophores was close to 1, making them to show even better performance than **HMQ-TMS** in THz generation.

In fact, not only the chromophores themselves can affect the molecular packing, but the counter anions can also do, although they are usually more or less ignored. Figure 8.5 shows a series of NLO chromophores with the same donor-acceptor structure, but different counter anions.[25] To the surprise, only two molecules of **DINS-2** and **DITMS** can form acentric crystals, while all the other seven molecules can only yield centrosymmetric crystals without second-order NLO properties. In **DINS-2** crystals (Figure 8.5), the cationic NLO chromophore sheets are interleaved with the counter anionic sheets, and the chromophores show a head-to-tail arrangement. However, in the adjacent cation sheets, the polarity orientation is opposite to the orientation angle of two adjacent chromophore sheets of 173.82°. This

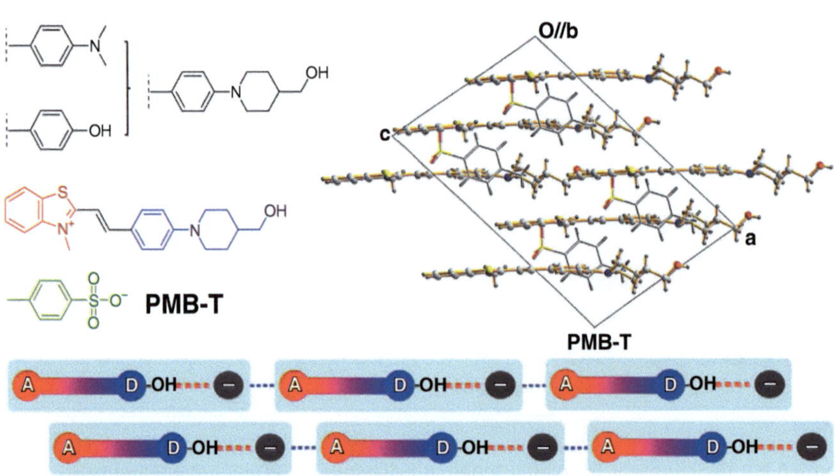

Figure 8.4 The chemical structure, schematic diagram molecular arrangement in crystal, and crystal structure of **PMB-T**. *Source*: Adapted from [24].

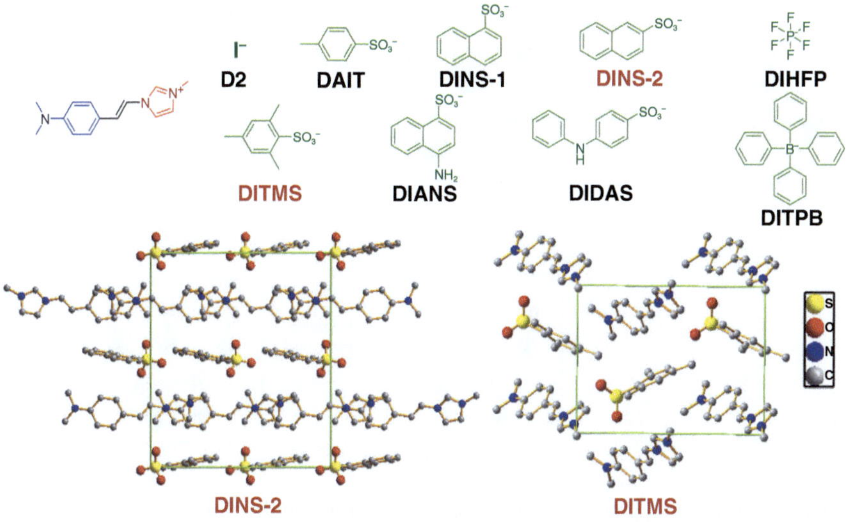

Figure 8.5 The chemical structures of **D2**, **DAIT**, **DINS-1**, **DINS-2**, **DIHFP**, **DITMS**, **DIANS**, **DIDAS**, and **DITPB**, and the crystal structures of **DINS-2** and **DITMS**. *Source*: [25]/with permission of Elsevier.

centrosymmetry-like packing makes **DINS-2** crystals show very low NLO effects, and its SHG signal is only 3% of that of **DAST**. Fortunately, in **DITMS** crystals, the orientation angle of two adjacent chromophore sheets is only 22.35°, indicating the similar dipole moment directions of the chromophores. Consequently, the NLO effect of **DITMS** crystal is more than three times higher than that of the crystal. Herein, it is noted that even a very tiny change in counter anion would lead to totally different packing of chromophores, such as **DINS-1** and **DINS-2**. Their only difference is the different linkage position of the sulfonate group. However, their packing modes are totally different in crystals: **DINS-1** is centrosymmetric without NLO effect, while **DINS-2** is opposite.

In addition to ionic chromophores, the crystallization can also realize the regular arrangement of neutral ones. However, since the lattice energies of molecular crystals are not such high as those of ionic crystals, only the chromophores with relatively low β values are suitable for this strategy, such as *N*-benzyl-2-methyl-4-nitroaniline (BNA) crystals,[26] and 2-(3-(4-hydroxystyryl)-5,5-dimethylcyclohex-2-enylidene)malononitrile (OH1),[27] etc. As a consequence, relevant studies are becoming less and less common in recent years.

8.3 Polymers for Second-order Nonlinear Optics

Alternatively, poling procedure is a more general strategy to achieve non-centrosymmetric alignment of chromophores, no matter how large the dipole moment of the chromophore. The general procedure is as following (Figure 8.6): the organic chromophores are firstly doped into polymer matrix with good

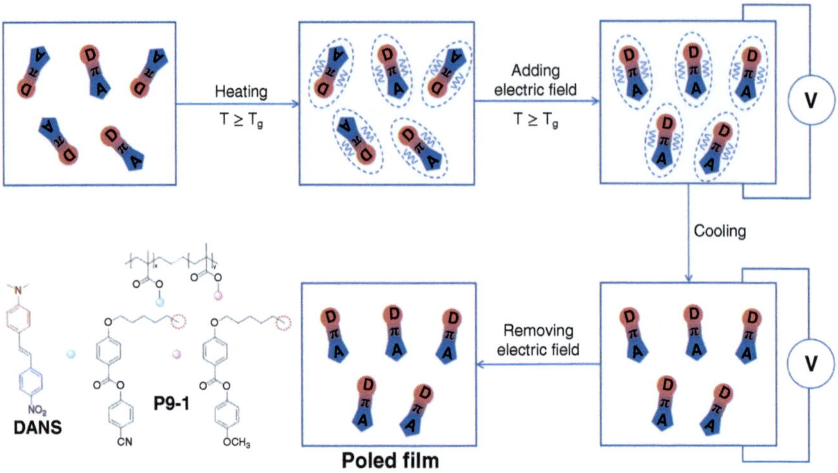

Figure 8.6 The poling procedure for NLO chromophores, and the chemical structures of **DANS** and **P9-1**. *Source*: Wenbo Wu and Zhen Li.

film-forming ability, to prepare the chromophore polymer film; after heating the film to or above the glass transition temperature under which condition the chromophores can be relatively moved freely, a strong direct current electric field of usually several kilovolts is applied to induce their reorientation; the film is then cooled to room temperature before removing the electric field, so that the chromophores can be frozen in the poled film with non-centrosymmetric alignment to demonstrate NLO characteristic.[28–30] This process was first proposed by Williams and co-workers in the early 1980s.[31] In their study, **DNAS** with a β value of 4.5×10^{-28} esu was used as the chromophore, and a liquid crystal polymer of **P9-1** as the host polymer (Figure 8.6). After poling, the film with a 3-μm-thick 2% doped **DANS** in **P9-1** film could demonstrate a $\chi^{(2)}$ value of 3×10^{-9} esu.

The poled polymers show several obvious advantages in comparison with NLO crystals: the orientation of chromophores is controlled by external electric field, which can avoid the tedious and time-consuming crystal growth process; the structures of the polymers can be easy to modify, making it easy to adjust the refractive index and meet the phase matching conditions; polymers usually have excellent processability and mechanical properties, etc.[32] These advantages of poled polymer naturally led a vigorous development of organic/polymeric second-order NLO materials, and the material type has also expanded from the initial host-guest type.[31,33] to side chain type,[34–36] main chain type,[37,38] cross-link type,[39,40] and even self-assembly type,[41] etc. To date, the polymeric NLO materials are almost exclusively referred to as poled polymers.

Despite the notable success of poled polymers in NLO materials, an unalterable fact is that the strong intermolecular electrostatic interactions among rod-like chromophores are always in existence, which could slowly destroy the non-centrosymmetric alignment of chromophores in poled films. Accordingly, the stability of the NLO effect of poled film is usually lower than that of organic crystal.

More importantly, the strong electrostatic interactions can decrease the poling efficiency. Generally speaking, the order parameters of poled films are usually only in the range of 0.05–0.3. In addition, when the loading density of NLO chromophores in the film increases, these electrostatic interactions would also further become stronger, finally leading to a decreased NLO effect.[42] Therefore, it has become the most important key issue in organic NLO field that how to effectively translate the large microscopic β values of chromophore to macroscopic NLO effects of materials with excellent stability.

Introducing some isolation groups onto the chromophores is a very direct, but effective strategy to decrease the interactions among chromophores through enlargement of their distance (Figure 8.7). This is the famous site-isolation principle proposed by Dalton and co-workers.[43,44] And they further proposed that the spherical shape is the ideal conformation to minimize chromophore-chromophore interactions, according to the experimental results and theoretical analysis. Figure 8.7 is a typical example of the site-isolation effect. In 2002, Jen's group synthesized chromophore **C9-2** by introducing three perfluorinated dendritic isolation groups onto chromophore **C9-1**.[45] As a consequence, the r_{33} value (electro-optical coefficient, an important parameter to express the macroscopic NLO effect) of spherical-shaped chromophore **C9-2** reached 30 pm V^{-1} after poling, three times to that of its prototype **C9-1** (10 pm V^{-1}). Meanwhile, its thermal stability and optical transparency have also been significantly improved.

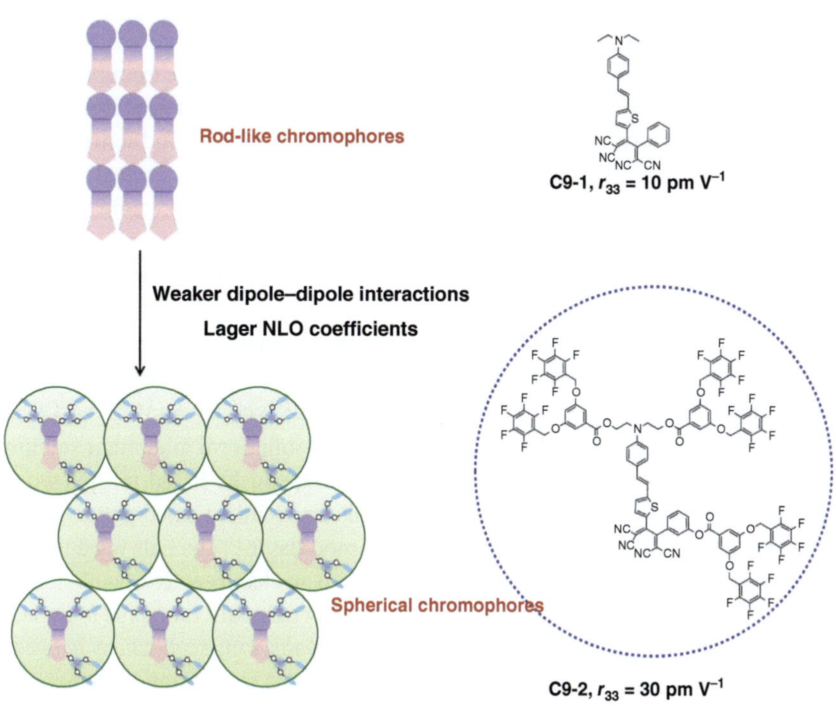

Figure 8.7 A 2D graphical illustration of site-isolation principle, and the chemical structures of **C9-1** and **C9-2**. *Source*: Wenbo Wu and Zhen Li.

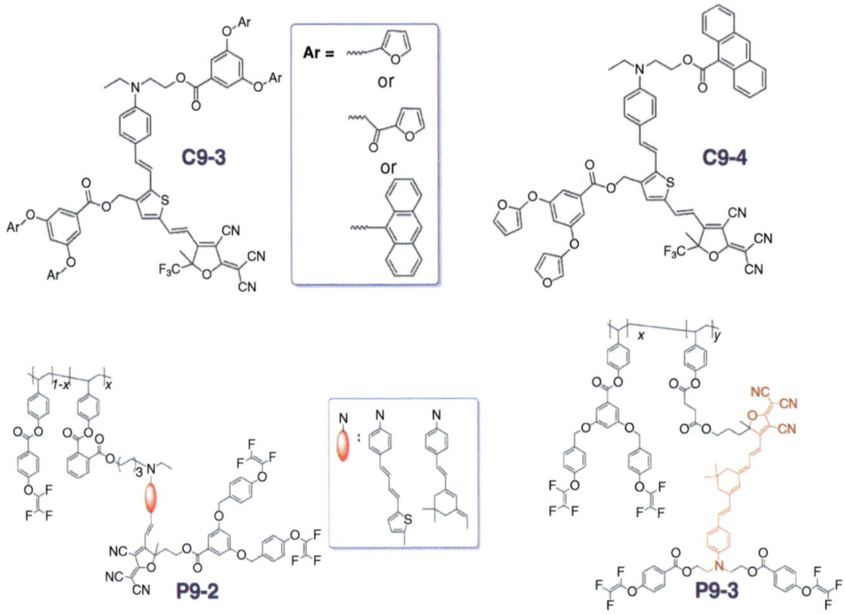

Figure 8.8 Several examples of NLO dendrimers (**C9-3** and **C9-4**) and dendronized polymers (**P9-2** and **P9-3**). *Source*: Wenbo Wu and Zhen Li.

Subsequently, according to the site-isolation principle, various kinds of NLO dendrimers.[46,47] and dendronized polymers.[48,49] (Figure 8.8 shows some examples), with similar chemical structural features that some bulky isolation groups linked onto chromophore moieties, have been developed from the early 2000s, and some of them show very good NLO performance. In addition, by changing the normal isolation groups to the cross-linkable ones, the stability of the non-centrosymmetric alignment of chromophores can also be improved, such as the example in Figure 8.9.[50] During the poling process, dendritic chromophores **C9-5** can be cross-linked with host polymer **P9-4** through in-situ Diels-Alder reaction, and the cross-linked systems demonstrated r_{33} values of 15–43 pm V^{-1}. Subsequently, chromophore **C9-6** with a very high β value was used as the guest chromophore to further enhance the loading density of the chromophores. When doping 25 wt% **C9-6** into **P9-4/C9-5b** systems, both ultrahigh r_{33} value of 263 pm V^{-1} and long-term alignment stability of chromophores at 85 °C could be achieved simultaneously. Based on this work, Peyghambarian and co-workers further mixed this cross-linked poled polymer with organic silicone gel to prepare an electro-optical modulator.[51] The half-wave voltage of the resultant modulator was 2.5 V at 1550 nm, with an electro-optical coefficient of 170 pm V^{-1}, five to six times higher than those of the benchmark inorganic materials.

In theory, the larger size of isolation groups means longer distance between two adjacent chromophores, and weaker intermolecular electrostatic interactions. However, as early as 2006, we found that the d_{33} values (SHG coefficient, another important parameter to express the macroscopic NLO effect) of polymers were in

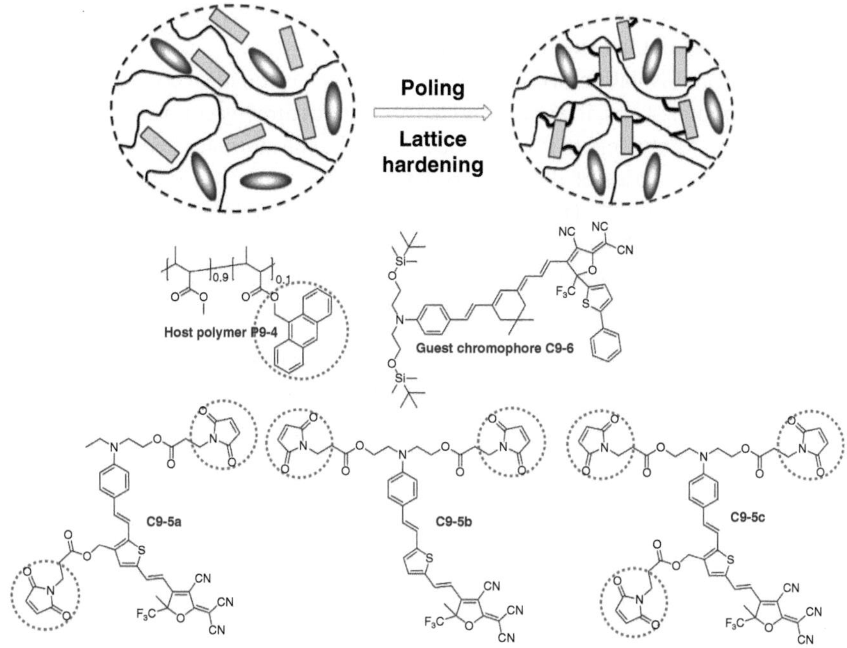

Figure 8.9 In situ poling and Diels–Alder cross-linking process of host–guest cross-linkable NLO system, and the chemical structures of host polymer **P9-4**, cross-linkable chromophores **C9-5**, and guest chromophore **C9-6**. *Source*: [50]/with permission of John Wiley & Sons.

fact not in accordance with the sizes of their isolation groups.[52] As presented in Figure 8.10, with the increasing sizes of isolation spacers, the d_{33} values of **P9-5**, with nitroazobenzene as the chromophore, were first enhanced and then decreased, and the Ph is the most suitable isolation group to make **P9-5c** show the highest d_{33} value of 74.8 pm V^{-1}. The reason is easily conceived: a) the isolation groups are not polar after all, so that they could not contribute to NLO effects directly, but decrease the loading density of NLO chromophores; b) the introduction of isolation groups can also enhance the whole size of the chromophores, and the oversized chromophores would make the alignment of NLO chromophore more difficult during the poling process. According to this consideration, lots of NLO polymers containing different sizes of isolation groups, including different polymer backbones, different polymer types, different chromophores, different linkage positions of isolation spacers, different amount of isolation groups, etc., have been prepared from 2006 by us and some other groups,[53–61] to find out how to choose the most suitable size of isolation groups, and the results have been concluded as the concept of suitable isolation group by us: for a given chromophore moiety and a given linkage position, there should be a suitable isolation group present to boost its microscopic β value to possible higher macroscopic NLO property efficiently.[6,15,30] Generally speaking, when the dipole moment of the chromophore is high, the size of corresponding suitable isolation group should also be large; for the same chromophore, when the amount of isolation group increases, the size of corresponding suitable isolation group should consequently be small; the rigid spacer group should show better isolation effect than the flexible one.

According to site-isolation principle, spherical shape is the ideal conformation of NLO chromophores. Thus, it was natural to think about directly controlling the confirmation of NLO polymers to the spherical shape by the suitable isolation groups? Considering this, a series of dendrimers **D9-1-G1** to **D9-1-G5** (Figure 8.11) were prepared by our group.[62,63] Dendrimers are highly branched defect-free monodisperse macromolecules.[64] Besides the globular shape, the perfect dendritic structures in high generation dendrimers could also endow the skeleton with more or less site

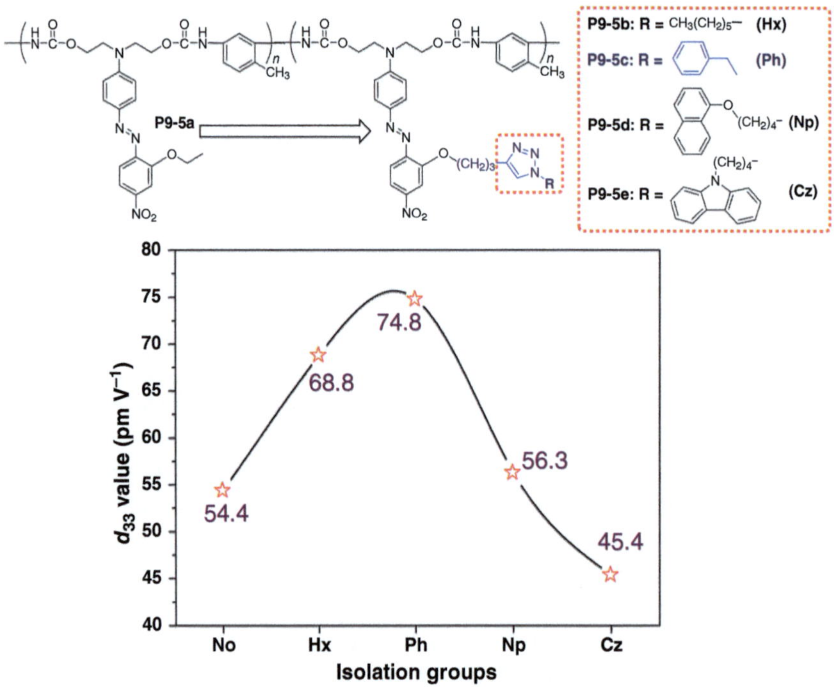

Figure 8.10 The chemical structures of NLO polymer **P9-5** with different sizes of isolation groups, as well as their d_{33} values. *Source*: Wenbo Wu and Zhen Li.

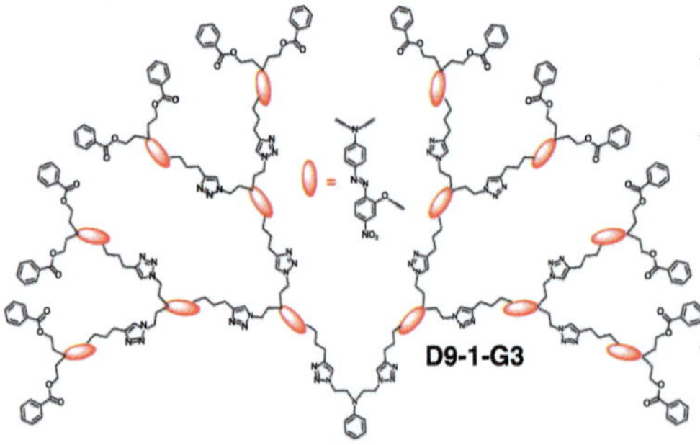

no.	d_{33} (pm V^{-1})	Loading density of chromophore
D9-1-G1	100	0.402
D9-1-G2	108	0.488
D9-1-G3	123	0.520
D9-1-G4	177	0.537
D9-1-G5	193	0.544

Figure 8.11 The chemical structures of NLO dendrimers **D9-1** (taking **D9-1-G3** as the example), as well as their d_{33} values. *Source*: Wenbo Wu and Zhen Li.

isolation effect to the NLO chromophores. In **D9-1**, there are three isolation groups surrounding each nitroazobenzene chromophore, and therefore, the small size triazole rings, which can be simply yielded from click chemistry reaction, are suitable enough to minimize the electrostatic interaction among chromophores. From the first-generation dendrimer to the fifth-generation one, the site isolation effect gradually gets better, and the loading density of chromophore also gradually becomes higher, leading to significantly enhanced NLO effects. For **D9-1-G5**, its d_{33} value is up to 193 pm V^{-1}, which is the highest value for nitroazobenzene chromophore at that time.

The subsequent study revealed that the shapes of **D9-1**, especially the low generation ones, were conical, rather than spherical. Therefore, we further utilized the concept of star-type chromophore (Figure 8.12)[65–67] to modify the topological structures of NLO dendrimers. In 2012, a series of global-like dendrimers **D9-2** (Figure 8.12), which were designed by linking the low generation NLO dendrimers of **D9-1** onto a "star-type chromophore" core, were prepared in our group.[68] To our surprise, the d_{33} value of **D9-2-G1** was already 214 pm V^{-1}, even higher than that of **D9-1-G5**.

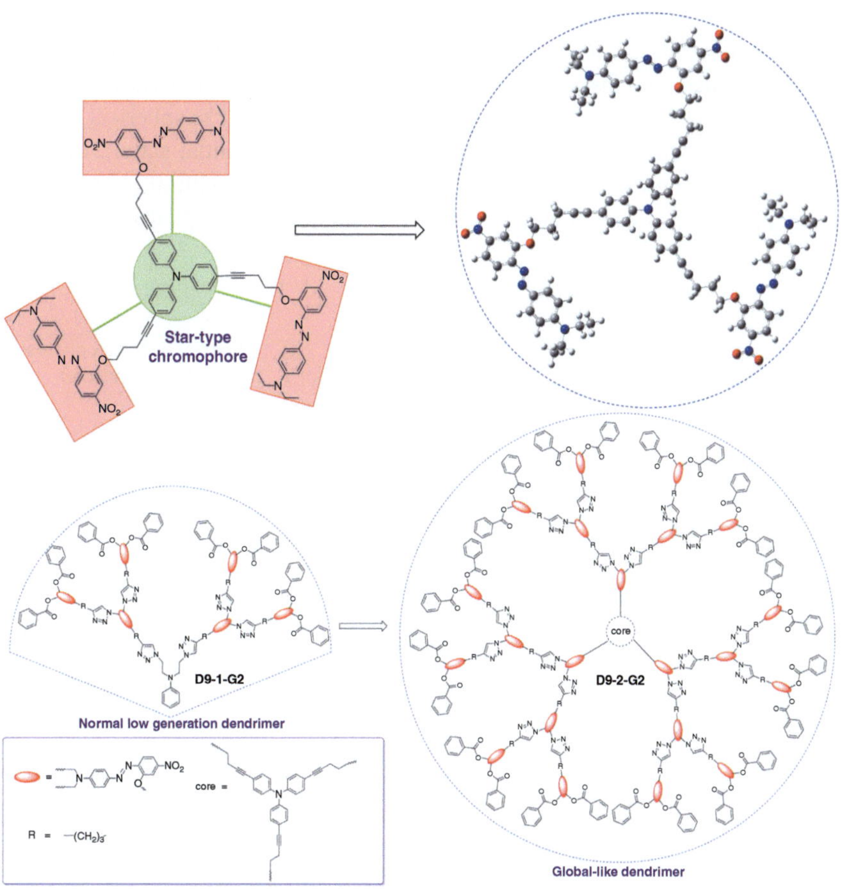

Figure 8.12 The sketch map of star-type chromophore, and the differences between traditional NLO dendrimers and the global-like ones. *Source*: Wenbo Wu and Zhen Li.

More excitingly, accompanying with the growth of the NLO dendrimers, the d_{33} value of **D9-2-G3** was further enhanced to 246 pm V^{-1}. In addition, no matter the normal NLO dendrimers or global-like ones, after the dendrimers grew larger than a critical point, the NLO effect would be significantly enhanced: from **D9-1-G3** to **D9-1-G4**, 54 pm V^{-1} was enhanced, while from **D9-2-G2** to **D9-2-G3**, 27 pm V^{-1} was enhanced. Possibly at that point, the dendritic architecture was sophisticated enough to demonstrate an excellent site-isolation effect for improving NLO coefficients.

On the other hand, some scientists attempted to control the arrangement of chromophores to be regular in short range, and found that the short-range ordered arrangement can improve the final macroscopic arrangement of chromophores after poling.[69–71] In this respect, the NLO coefficients of dendrimers could be further enhanced, because in the above dendrimers, including both conical-like and global-like ones, the chromophores demonstrated totally centrosymmetric arrangement in the initial state before poling. In 2015, an X-type chromophore, in which 5 chromophore moieties were more or less arranged, was designed by us.[72] Based on X-type chromophore, we further prepared a series of Janus dendrimers containing nitroazobenzene chromophore (**D9-3** in Figure 8.13).[72,73] In these Janus dendrimers, the NLO effects were not always improved by enhancing the generation number, and we should also pay attention to the balance of the sizes of dendrons on the two sides. The theoretical simulation indicated that if the sizes of the two dendrons are similar, the whole dendrimers would be more like spherical, and the chromophores would also show better alignment in the initial state. Therefore, **D9-3c** demonstrated a record high d_{33} value of 299 pm V^{-1} with nitroazobenzene chromophore. If the chromophore was further improved to FTC (Figure 8.1) with

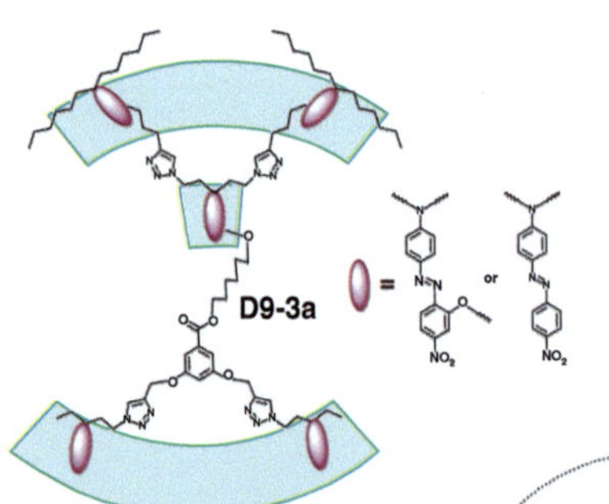

no.	d_{33} (pm V^{-1})	Number of chromophores in each dendrimer
D9-3a	157	5
D9-3b	195	9
D9-3c	299	13
D9-3d	238	17
D9-3e	261	21

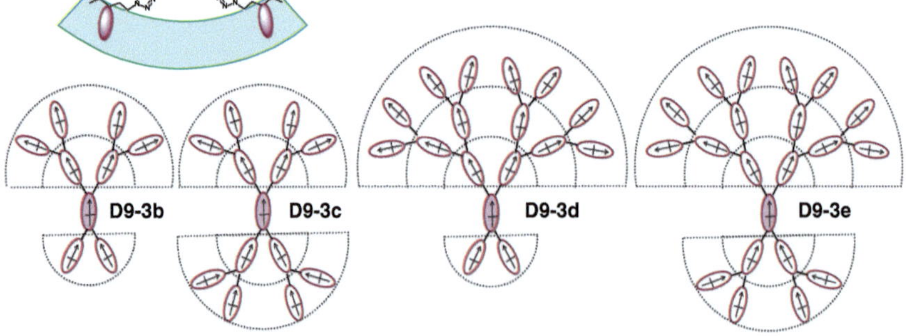

Figure 8.13 The chemical structures of NLO dendrimers **D9-3** based on X-type chromophores, as well as their d_{33} values. *Source*: Wenbo Wu and Zhen Li.

a higher β value, an even higher d_{33} value of 529 pm V^{-1} could be achieved by this type of Janus dendrimer.[74]

In all the above cases, all the isolation groups were non-polar or weakly polar ones designed by site-isolation principle and the concept of suitable isolation group, and could only decrease the strong electrostatic interactions among chromophores, but cannot directly contribute to NLO effects. One of our improvement strategies is to modify the isolation group to isolation chromophore with a relatively lower β value. As presented in Figure 8.14a, P9-6a, constructed by two different chromophore moieties of nitroazobenzene and sulfonylazobenzene ones with strictly alternative arrayment, exhibited the d_{33} value of 116.8 pm V^{-1}, higher than that of **P9-6b** with only nitroazobenzene chromophore.[75] Meanwhile, the d_{33} value of **P9-6d** with irregular structure was between those of **P9-6b** and **P9-6c**. Further deep investigation

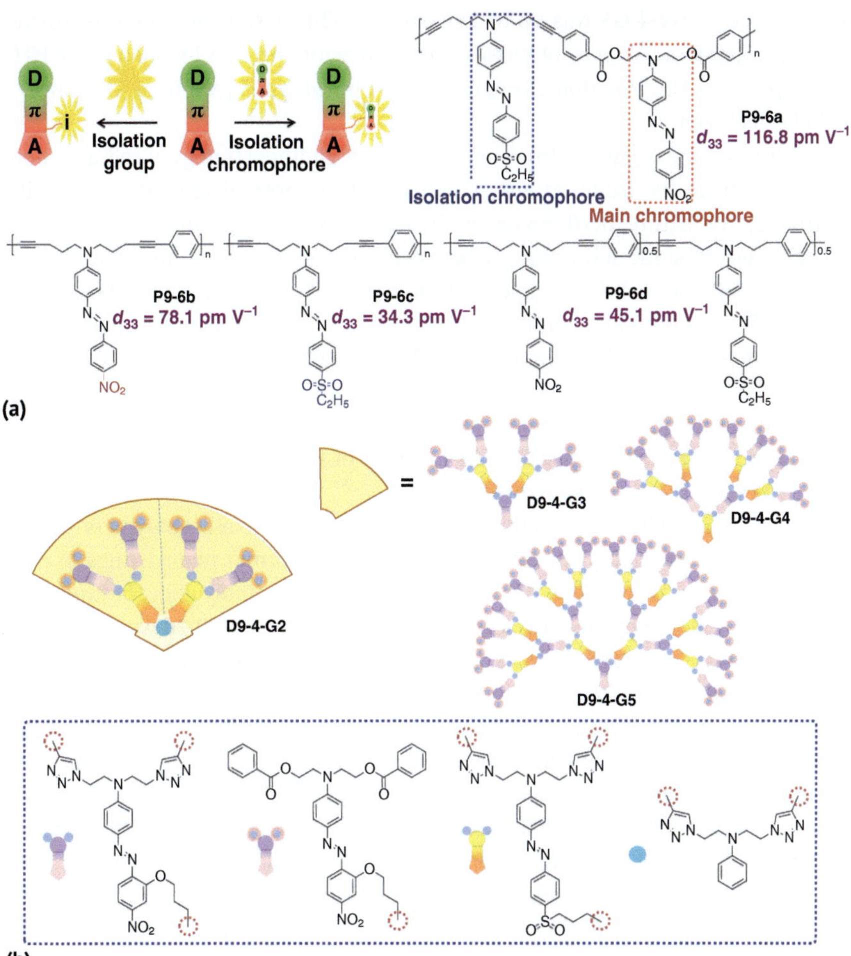

Figure 8.14 The NLO materials containing isolation chromophore. (a) The chemical structures and d_{33} values of **P9-6**. (b) The chemical structures of NLO dendrimer **D9-4**. *Source:* Wenbo Wu and Zhen Li.

showed that the size of suitable isolation group in the NLO systems containing isolation chromophore should become very small. In a typical example, the naphthalene is enough to be the suitable isolation group for two chromophores of nitroazobenzene and sulfonylazobenzene, indicating that dipole-dipole interactions really became much smaller by the introduction of isolation chromophore.[76]

Based on these results, the concept of isolation chromophore has been applied to various polymeric NLO systems, including side-chain polymers,[77] main-chain polymers,[78] H-type polymers,[79] and even hyperbranched polymers[80], etc. Among them, the dendrimers still demonstrated the best NLO performance.[75,81,82] **D9-4** is one of the typical examples, whose chemical structures are very similar to those of **D9-1**, but with orderly arranged two different kinds of chromophore moieties instead of the single nitroazobenzene chromophore (Figure 8.14b). Thanks to the isolation chromophore, **D9-4** demonstrated much higher NLO coefficients than **D9-1**, and the d_{33} value of **D9-4-G5** was up to 253 pm V^{-1}. This design strategy was further used to FTC system. When a nitroazobenzene chromophore was linked onto an FTC chromophore as the isolation chromophore, an ultrahigh d_{33} value of 384 pm V^{-1} could be achieved.[83]

In addition to the isolation chromophore, there is another strategy of self-assembly to further utilize the isolation group. Different from normal aromatic rings with electronegative feature, perfluoroaromatic rings were electropositive, leading to the reversible electrostatic self-assembly between these two types of aromatic rings.[84] As early as 2007, Jen and co-workers developed a series of NLO molecular glasses, and the FTC chromophore containing both benzene and pentafluorobenzene rings as isolation groups with self-assembly effect demonstrated a large r_{33} value of 108 pm V^{-1}, more than twice to those of FTC chromophores with only phenyl isolation group or pentafluorophenyl one.[85] If CLD chromophore with even better microscopic NLO performance was used instead of FTC chromophore, a higher r_{33} value of 318 pm V^{-1} could be realized.[86]

Then, in 2012, we first applied this self-assembly effect into NLO polymers (Figure 8.15a).[87,88] After changing the isolation groups from normal phenyl rings to pentafluorophenyl rings to endow the NLO polymers self-assembly effect between isolation groups and polymer chains, the d_{33} values of **P9-7b** and **P9-7d** were improved to more than six and three times to those of **P9-7a** and **P9-7c**, respectively. In addition, since additional disruption of these self-assembly effects is required for the relaxation of the arranged chromophore moieties in poled films **P9-7b** and **P9-7d**, their stability of NLO coefficients would also be improved.[87] After that, the NLO dendrimers with pentafluorophenyl groups on the periphery were prepared, in order to further enhance the NLO effects and stability, and the d_{33} value of **D9-5-G5** was 206 pm V^{-1} (Figure 8.15b), higher than that of **D9-1-G5** (193 pm V^{-1}, Figure 8.11) with a similar structure but normal phenyl rings on the periphery.[89] This strategy was further utilized in other types of NLO dendrimers, such as global-like ones[90] and Janus ones,[91] etc., and also achieved success.

In addition to dendrimer, a hyperbranched polymer is another common type of dendritic macromolecule. Different from dendrimers usually with tedious and costly preparation process, hyperbranched polymers are famous for their easy synthetic

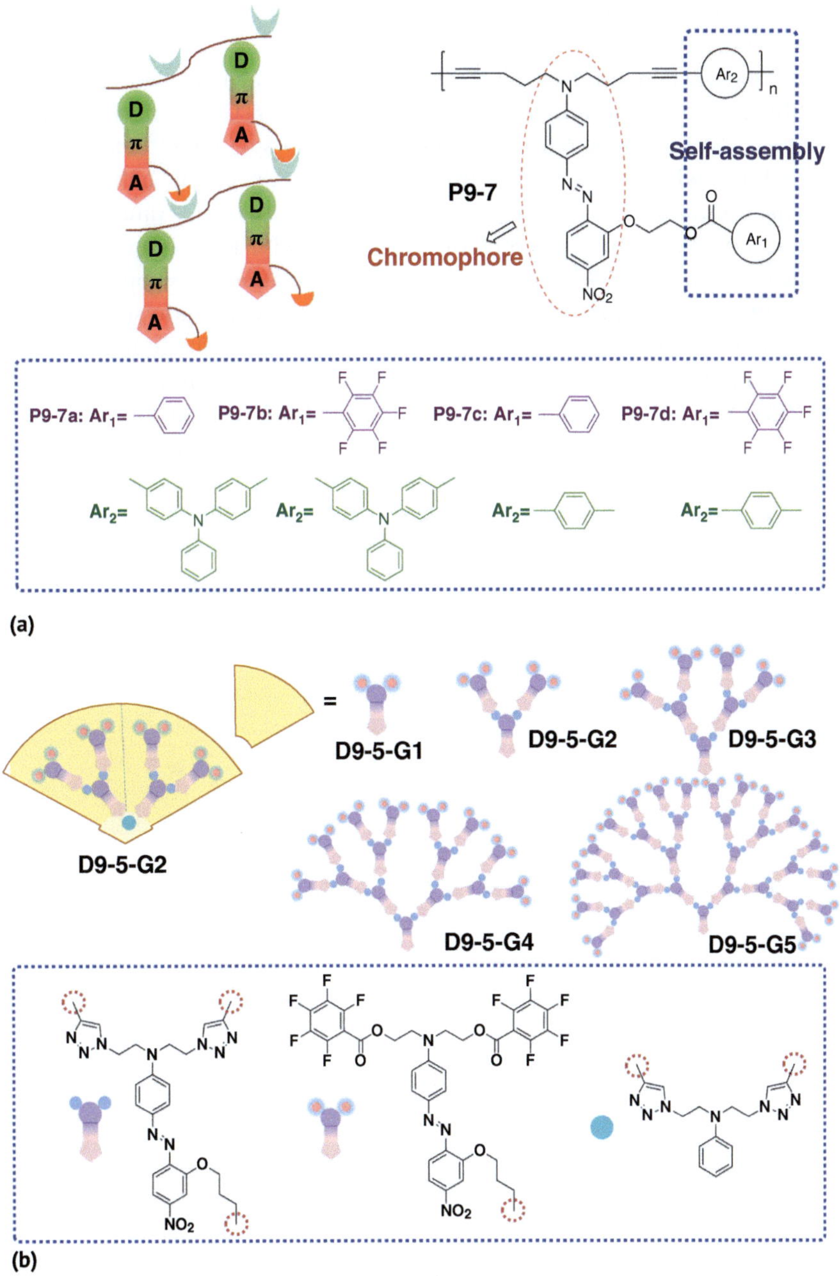

Figure 8.15 The NLO materials containing perfluoroaromatic rings as isolation groups. (a) The chemical structures and d_{33} values of **P9-7**. (b) The chemical structures of NLO dendrimer **D9-5**. *Source:* Wenbo Wu and Zhen Li.

accessibility, typically by one-pot syntheses through AB_n $(n \geq 2)$ or $A_2 + B_3(n \geq 3)$ polymerization, at the expense of their imperfect dendritic architectures, still containing some linear units.[92] Even though hyperbranched polymers still inherit many special properties from dendrimers, such as good solubility, low viscosity, and the most important site-isolation effects for NLO properties.[92]

From 2004, when the first NLO hyperbranched polymer was reported, to date, there were various of hyperbranched polymeric NLO materials have been developed to exhibit good NLO performance.[93–101] **P9-8**,[102] **P9-9**,[103] and **P9-10**[102] (Figure 8.16), which were derived from the similar AB_2 or AB_4 monomers are the typical examples. As presented in Figure 8.16, their d_{33} values are 117, 167, and 145 pm V^{-1}, respectively. In comparison with **P9-8**, it was clear that chemical structure of **P9-9** was improved by the isolation chromophore, while that of **P9-10** was improved by the aromatic/perfluoroaromatic self-assembly effect. Therefore, their enhanced d_{33} values confirmed that both design strategies remain effective in hyperbranched polymers. In addition, as compared to **D9-1-G5** (193 pm V^{-1}), **D9-4-G5** (253 pm V^{-1}), and **D9-5-G5** (206 pm V^{-1}), which are the fifth-generation dendrimers derived from the same AB_n-type monomer to **P9-6**, **P9-7**, and **P9-8**, respectively, these hyperbranched polymers represented lower NLO coefficients, because of the linear defects. On the contrary, the easy synthetic procedures usually make hyperbranched polymers have higher molecular weights (such as 6.04×10^4 g mol^{-1} for **P9-6**, 1.79×10^5 g mol^{-1} for **P9-7**, and 1.09×10^5 g mol^{-1} for **P9-8** tested by gel permeation chromatography), so that their stability of poled films is usually higher than that of dendrimers.

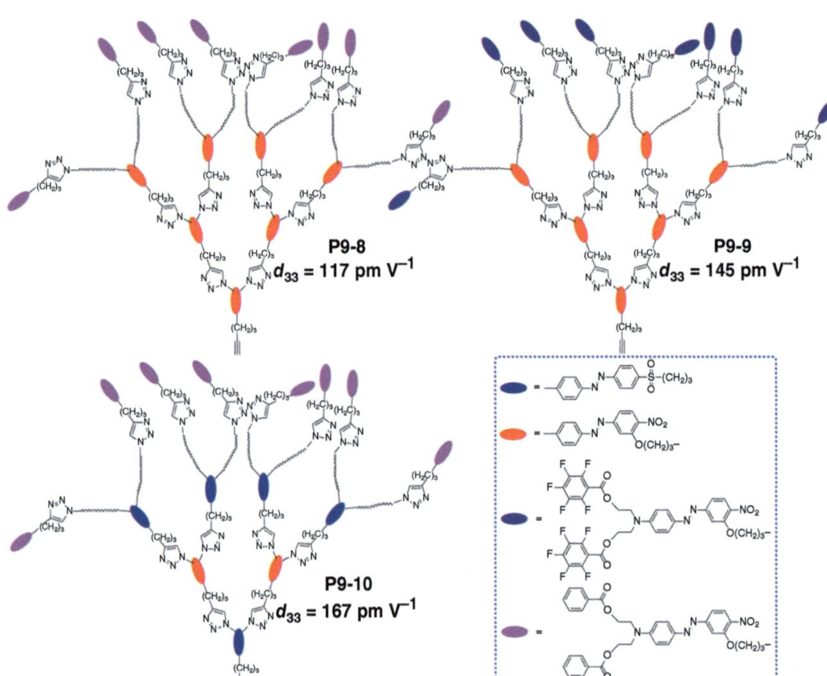

Figure 8.16 The chemical structures and NLO coefficients of hyperbranched polymers **P9-8**, **P9-9**, and **P9-10**. *Source*: Wenbo Wu and Zhen Li.

Accordingly, some scientists tried to decrease the linear defects in NLO hyperbranched polymers, in order to further improve the NLO performance on the premise of easy synthesis. Degree of branching (DB) is a physical quantity to characterize the content of branched components in macromolecules. The DB value of linear polymer is 0, while 1 for dendrimer and 0~1 for hyperbranched polymer. As for hyperbranched polymers from AB_2-type monomer, the DB value is 0.5 when the activities of two B functional groups are the same theoretically.[104] Specially, for **P9-6** derived from AB_2-type monomer, its DB value was tested as 0.457. Once prepared by AB_4-type monomer, its DB value could be enhanced to 0.691, leading to a 22% increase in d_{33} value.[105] And then, a new hyperbranched polymerization of "$AB_2 + AC_2$" was developed to further enhance the DB value of AB_2-type NLO hyperbranched polymers. As a result, the DB value could be enhanced from 0.554 to 0.791, so that the d_{33} value was also improved by 53%.[106]

However, it was found that many strategies to enhance the DB value of normal hyperbranched polymers are not applicable in the NLO field due to several reasons. Firstly, the polymerization must be carried out under mild conditions to avoid the chromophore destruction. Secondly, the sizes of functional groups and their corresponding products must be as small as possible, for ensuring the high loading density of chromophores in the obtained hyperbranched polymers. In addition, some rigid isolation groups with suitable sizes should be also introduced into the hyperbranched polymers during polymerization according to site-isolation principle and the concept of suitable isolation group. It is therefore still a big challenge to prepare NLO hyperbranched polymers with high DB values near to 1.

Alternatively, we can introduce low-generation dendrimers onto a hyperbranched polymer backbone to yield another kind of dendritic macromolecules, namely dendronized hyperbranched polymers (Figure 8.17). They usually showed much less linear structural defects than hyperbranched polymers, as well as much easier syntheses than high-generation dendrimers. The first dendronized hyperbranched

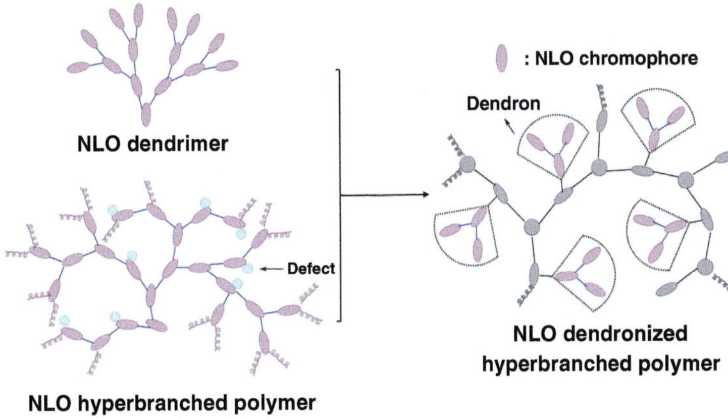

Figure 8.17 The schematic overview of different types of NLO dendritic macromolecules, including dendrimers, hyperbranched polymers, and dendronized hyperbranched polymers. *Source*: Wenbo Wu and Zhen Li.

polymer (**P9-11**, Figure 8.18 presented **P9-11-G1** as the example) was prepared through the Suzuki polycondensation between A_2-type monomer with a low-generation NLO dendrimer as the side chain and B_3-type monomer of borate ester by us in 2013.[107] The d_{33} value of **P9-11-G1** was tested as 133 pm V^{-1}, which was higher than that of hyperbranched polymer **P9-8** and between those of dendrimers **D9-1-G3** and **D9-1-G4** with the same nitroazobenzene chromophore; and the decay temperature of the NLO effect was 100°C, similar to that of the hyperbranched polymer **P9-8**, and even higher than that of **D9-1-G5**. These results indicated that the dendronized hyperbranched polymers can inherit the high NLO effects from high-generation dendrimers and the good stability and easy syntheses from hyperbranched polymers, and thus become a highly potential dendritic architecture to optimize the comprehensive performance of NLO polymers.

After that, our group paid much attention on optimizing the structures of NLO dendronized hyperbranched polymers in the following several years. **P9-12** can be termed as main chain dendronized hyperbranched polymers, since the NLO dendrimers were

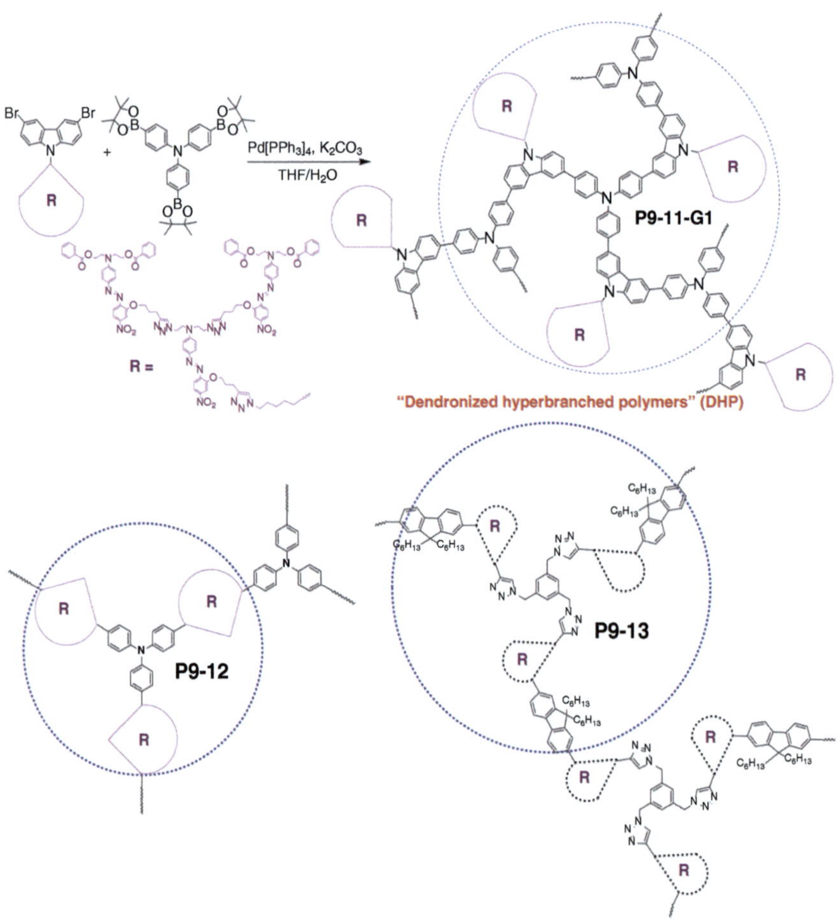

Figure 8.18 The chemical structures of dendronized hyperbranched polymers **P9-11-G1**, **P9-12**, and **P9-13**. *Source*: Wenbo Wu and Zhen Li.

embedded into the polymer chains (Figure 8.18). This change could further enhance the stability of NLO effect, but slightly decrease the NLO coefficient.[108] Once using the functional global-like dendrimer as the monomer, another type of dendronized hyperbranched polymer **P9-13** (Figure 8.18) could be yielded. Excitingly, by using nitroazobenzene as the chromophore, the d_{33} value of **P9-13-G1** was improved to 166 pm V^{-1}, with the NLO decay temperature of 117 °C, both much better than those of **P9-11-G1** and **P9-12-G1**.[109] We also used several other methods to improve the NLO performance of dendronized hyperbranched polymers, such as usage of the X-type chromophores as the dendron,[101] changing the position of NLO dendrons to the periphery,[110] introduction of isolation chromophore,[110–112] and enhance the loading density of chromophores by using another chromophore as the monomer,[112,113] etc. Thanks to the efforts of scientists, a high d_{33} value of 220 pm V^{-1} with good NLO decay temperature of 116 °C can be achieved by dendronized hyperbranched polymer with only nitroazobenzene as the chromophore.[113] Besides our groups, some other scientists have also focused on this "young" kind of polymer in recent years, and achieved some success in their syntheses, self-assemble effect, and applications in some other fields.[114–120] It can be predicted that this type of polymer will play more and more important roles in even more fields, accompanied by the deeper and deeper understanding of them in the near future.

8.4 Summary and Perspective

In summary, organic/polymeric NLO materials have achieved great success in recent years. While their structure–property relationships are becoming clearer and clearer, the comprehensive properties of NLO materials are continuously improved. In a sense, it is the most important thing in the second-order NLO field to overcome the strong electrostatic interaction among chromophores with large dipole moment to realize the non-centrosymmetric arrangement of NLO chromophores in aggregated states.

Crystallization and poling are the two most general strategies. In comparison with crystallization, which is only suitable for the chromophores with relatively lower $\mu\beta$ values, the poling process can make almost all chromophores to be non-centrosymmetric arrangement. That is to say, the external electronic field can tune the molecular packing of chromophore moieties, to modify their Molecular Uniting Set Identified Characteristic (MUSIC). Generally speaking, a higher electric field in poling process leads to a more orderly arrangement of NLO chromophores in films for better macroscopic NLO effects. However, the exorbitant electric field would also destroy the chemical structures of NLO chromophores. In order to realize more regular chromophore arrangement under relatively low electric field poling, in recent 20 years, several methods, such as site-isolation principle, the concept of suitable isolation group, isolation chromophore, etc., have therefore been developed for improving the poling efficiency of organic/polymeric NLO materials. Meanwhile, the types of NLO polymers have also been developed from the initial host-guest polymers to the recently dendritic macromolecules, since the dendritic skeleton could separate the chromophore moieties through site-isolation effect, and thus enhance the NLO effects. Relatively speaking, dendrimers, especially the high-generation ones, are difficult to prepare, and show better NLO coefficients but lower stability; while easier-to-prepare hyperbranched polymers show better stability but relatively lower NLO effects. Our newly designed dendronized hyperbranched polymer can balance these factors. By using simple nitroazobenzene as the chromophore, an ultrahigh d_{33} value of 220 pm V^{-1} with good NLO decay temperature of 116 °C can be achieved by an easy-synthesized dendronized hyperbranched polymer. It should be noted that there were only a several years since the dendronized hyperbranched polymer was proposed, and the understanding of this kind of "young" polymer is not such thorough. Accompanied by the future deep understanding of their structure–property relationship, even better NLO performance can be expected.

Acknowledgements

We are grateful for the financial support of the National Natural Science Foundation of China (52103228, 22075199), and Starting Grants from Tianjin University and the Tianjin Government.

References

1 Qiu, F. and Han, Y. (2021). Electro-optic polymer ring resonator modulators. *Chin. Opt. Lett.* 19: 041301.

2 Enami, Y., Seki, A., Masuda, S. et al. (2018). Bandwidth optimization for Mach-Zehnder polymer/sol-gel modulators. *J. Lightwave. Technol.* 36: 4181–4189.

3 Yu, J. and Huamg, Q. (2010). Photon revolution in computer field, the People's Daily. Page, 22(in November 3).

4 Zwickel, H., Wolf, S., Kieninger, C. et al. (2018). Silicon-organic hybrid (SOH) modulators for intensity-modulation/direct-detection links with line rates of up to 120 Gbit/s. *Opt. Express.* 25: 23784–23800.

5 Franken, P.A., Peters, C.W., and Weinreich, G. (1961). Generation of optical harmonics. *Phys. Rev. Lett.* 7: 118–120.

6 Li, Z., Li, Q., and Qin, J. (2011). Some new design strategies for second-order nonlinear optical polymers and dendrimers. *Polym. Chem.* 2: 2723–2740.

7 Kim, S.J., Kang, B.J., Puc, U. et al. (2021). Highly nonlinear optical organic crystals for efficient terahertz wave generation, detection, and applications. *Adv. Optical Mater.* 9: 2101019.

8 Shi, R., Han, X., Xu, J., and Bu, X.H. (2021). Crystalline porous materials for nonlinear optics. *Small* 17: 2006416.

9 Liu, J., Ouyang, C., Huo, F. et al. (2020). Progress in the enhancement of electro-optic coefficients and orientation stability for organic second-order nonlinear optical materials. *Dyes Pigm.* 181: 108509.

10 Prasad, P.N. and Williams, D.J. (1991). *Introduction to Nonlinear Optical Effects in Molecules and Polymers.* John Wiley & Sons.

11 Colombo, A., Dragonetti, C., Guerchais, V., and Roberto, D. (2021). An excursion in the second-order nonlinear optical properties of platinum complexes. *Coord. Chem. Rev.* 446: 214113.

12 Oudar, J.L. and Chemla, D.S. (1977). Hyperpolarizabilities of the nitroanilines and their relations to the excited state dipole moment. *J. Chem. Phys.* 66: 2664–2668.

13 Oudar, J.L. (1977). Optical nonlinearities of conjugated molecules. Stilbene derivatives and highly polar aromatic compounds. *J. Chem. Phys.* 67: 446–457.

14 Dalton, L.R., Harper, A.W., Ren, A. et al. (1999). Polymeric electro-optic modulators: from chromophore design to integration with semiconductor very large scale integration electronics and silica fiber optics. *Ind. Eng. Chem. Res.* 38: 8–33.

15 Wu, W., Qin, J., and Li, Z. (2013). New design strategies for second-order nonlinear optical polymers and dendrimers. *Polymer* 54: 4351–4382.

16 Bäuerle, D., Betzler, K., Hesse, H. et al. (1977). Phase-matched second harmonic generation in urea. *Phys. Status Solidi.* 42: K119–K121.

17 Halbout, J.M., Blit, S., Donaldson, W., and Tang, C. (1979). Efficient phase-matched second-harmonic generation and sum-frequency mixing in urea. *IEEE J. Quantum Electron.* 15: 1176–1180.

18 Roszak, K. and Katrusiak, A. (2017). Giant anomalous strain between high-pressure phases and the mesomers of urea. *J. Phys. Chem. C.* 121: 778–784.

19 Marder, S.R., Perry, J.W., and Schaefer, W.P. (1989). Synthesis of organic salts with large second-order optical nonlinearities. *Science* 245: 626–628.

20 Kim, P.J., Jeong, J.H., Jazbinsek, M. et al. (2011). Acentric nonlinear optical N-benzyl stilbazolium crystals with high environmental stability and enhanced molecular nonlinearity in solid state. *CrystEngComm* 13: 444–451.

21 Jeong, J.H., Kim, J.S., Campo, J. et al. (2015). N-Methylquinolinium derivatives for photonic applications: Enhancement of electron-withdrawing character beyond that of the widely-used N-methylpyridinium. *Dyes Pigm.* 113: 8–17.

22 Lu, J., Hwang, H.Y., Li, X. et al. (2015). Tunable multi-cycle THz generation in organic crystal HMQ-TMS. *Opt. Express* 23: 22723–22729.

23 Lee, S.H., Jazbinsek, M., Hauri, C.P., and Kwon, O.P. (2016). Recent progress in acentric core structures for highly efficient nonlinear optical crystals and their supramolecular interactions and terahertz applications. *CrystEngComm* 18: 7180–7203.

24 Lee, S.C., Kang, B.J., Lee, J.A. et al. (2018). Single crystals based on hydrogen-bonding mediated cation-anion assembly with extremely large optical nonlinearity and their application for intense THz wave generation. *Adv. Opt. Mater.* 6: 1701258.

25 Guo, D., Ren, Y., Liu, J., and Tao, X. (2013). Synthesis and crystal structures of imidazolium salts for quadratic optical nonlinearities. *Dyes Pigm.* 98: 384–392.

26 Miyamoto, K., Minamide, H., Fujiwara, M. et al. (2008). Widely tunable terahertz-wave generation using an N-benzyl-2-methyl-4-nitroaniline crystal. *Opt. Lett.* 33: 252–254.

27 Brunner, F.D.J., Kwon, O.P., Kwon, S.J. et al. (2008). A hydrogen-bonded organic nonlinear optical crystal for high-efficiency terahertz generation and detection. *Opt. Express* 16: 16496–16508.

28 Li, Q. and Li, Z. (2020). Molecular packing: Another key point for the performance of organic and polymeric optoelectronic materials. *Acc. Chem. Res.* 53 (4): 962–973.

29 Dalton, L.R., Sullivan, P.A., and Bale, D.H. (2010). Electric field poled organic electro-optic materials: State of the art and future prospects. *Chem. Rev.* 110: 25–55.

30 Tang, R. and Li, Z. (2017). Second-order nonlinear optical dendrimers and dendronized hyperbranched polymers. *Chem. Rec.* 17: 71–89.

31 Meredith, G.R., VanDusen, J., and Williams, D.J. (1982). 4-(dimethylamino)-4'-nitrostilbene (DANS). *Macromolecules* 15: 1385–1389.

32 Fang, C.Y.S. (1990). Polymeric nonlinear optical materials. *Polym. Bull.* 3: 95–103. (in Chinese).

33 Singer, K.D., Sohn, J.G., and Lalama, S.G. (1986). Second harmonic generation in poled polymer films. *Appl. Phys. Lett.* 49: 248–250.

34 Ye, C., Marks, T.J., Yang, J., and Wong, G.K. (1987). Synthesis of molecular arrays with nonlinear optical properties: Second-harmonic generation by covalently functionalized glassy polymers. *Macromolecules* 20: 2322–2324.

35 Ye, C., Minami, N., Marks, T.J. et al. (1988). Efficient frequency doubling by poled annealed films of a chromophore-functionalized poly(p-hydroxystyrene). *Macromolecules* 21: 2899–2901.

36 Verbiest, T., Burland, D.M., Jurich, M.C. et al. (1995). Exceptionally thermally stable polyimides for second-order nonlinear optical applications. *Science* 268: 1604–1606.

37 Xu, C., Wu, B., Becker, M.W. et al. (1993). Main-chain second-order nonlinear optical polymers: random incorporation of amino-sulfone chromophores. *Chem. Mater.* 5: 1439–1444.

38 Kang, C.S., Heldmann, C., Winkelhahn, H.J. et al. (1994). Rigid rodlike main chain polymers with conformationally restricted nonlinear optical chromophores: Synthesis and properties. *Macromolecules* 27: 6156–6162.

39 Yu, L., Chan, W., and Bao, Z. (1992). Synthesis and characterization of a thermally curable second-order nonlinear optical polymer. *Macromolecules* 25: 5609–5612.

40 Zhang, C., Wang, C., Yang, J., and Dalton, L.R. (2001). Electric poling and relaxation of thermoset polyurethane second-order nonlinear optical materials: role of cross-linking and monomer rigidity. *Macromolecules* 34: 235–243.

41 Saadeh, H., Wang, L., and Yu, L. (2000). Supramolecular solid-state assemblies exhibiting electrooptic effects. *J. Am. Chem. Soc.* 122: 546–547.

42 Harper, A.W., Sun, S., Dalton, L.R. et al. (1998). Translating microscopic optical nonlinearity into macroscopic optical nonlinearity: The role of chromophore-chromophore electrostatic interactions. *J. Opt. Soc. Am. B.* 15: 329–337.

43 Robinson, B.H. and Dalton, L.R. (2000). Monte Carlo statistical mechanical simulations of the competition of intermolecular electrostatic and poling-field interactions in defining macroscopic electro-optic activity for organic chromophore/polymer materials. *J. Phys. Chem. A*. 104: 4785–4795.

44 Robinson, B.H., Dalton, L.R., Harper, H.W. et al. (1999). The molecular and supramolecular engineering of polymeric electro-optic materials. *Chem. Phys.* 245: 35–50.

45 Luo, J., Ma, H., Haller, M. et al. (2002). Large electro-optic activity and low optical loss derived from a highly fluorinated dendritic nonlinear optical chromophore. *Chem. Commun.* 8: 888–889.

46 Shi, Z., Hau, S., Luo, J. et al. (2007). Highly efficient Diels-Alder crosslinkable electro-optic dendrimers for electric-field sensors. *Adv. Funct. Mater.* 17: 2557–2563.

47 Shi, Z., Luo, J., Huang, S. et al. (2008). Reinforced site isolation leading to remarkable thermal stability and high electrooptic activities in cross-linked nonlinear optical dendrimers. *Chem. Mater.* 20: 6372–6377.

48 Luo, J., Liu, S., Haller, M. et al. (2002). Design, Synthesis, and properties of highly efficient side-chain dendronized nonlinear optical polymers for electro-optics. *Adv. Mater.* 14: 1763–1768.

49 Luo, J., Haller, M., Ma, H. et al. (2004). Nanoscale architectural control and macromolecular engineering of nonlinear optical dendrimers and polymers for electro-optics. *J. Phys. Chem. B*. 108: 8523–8530.

50 Kim, T.D., Luo, J., Ka, J.W. et al. (2006). Ultralarge and thermally stable electro-optic activities from Diel-Alder crosslinkable polymers containing binary chromophore systems. *Adv. Mater.* 18: 3038–3042.

51 Enami, Y., Derosei, C.T., Mathinei, D. et al. (2007). Hybrid polymer/sol-gel waveguide modulators with exceptionally large electro-optic coefficients. *Nat. Photonics*. 1: 180–185.

52 Li, Z., Zeng, Q., Li, Z. et al. (2006). An attempt to modify nonlinear optical effects of polyurethanes by adjusting the structure of the chromophore moieties at the molecular level using "click" chemistry. *Macromolecules* 39: 8544–8546.

53 Li, Z., Li, Z., Di, C. et al. (2006). Structural control of the side-chain chromophores to achieve highly efficient nonlinear optical polyurethanes. *Macromolecules* 39: 6951–6961.

54 Li, Q., Li, Z., Zeng, F. et al. (2007). From controllable attached isolation moieties to possibly highly efficient nonlinear optical main-chain polyurethanes containing indole-based chromophores. *J. Phys. Chem. B*. 111: 508–514.

55 Zeng, Q., Li, Z., Li, Z. et al. (2007). Convenient attachment of highly polar azo chromophore moieties to disubstituted polyacetylene through polymer reactions by using "click" chemistry. *Macromolecules* 40: 5634–5637.

56 Li, Z., Zeng, Q., Yu, G. et al. (2008). New azo chromophore-containing conjugated polymers: facile synthesis by using "click" chemistry and enhanced nonlinear optical properties through the introduction of suitable isolation groups. *Macromol. Rapid Commun.* 29: 136–141.

57 Li, Z., Wu, W., Ye, C. et al. (2009). Two types of nonlinear optical polyurethanes containing the same isolation groups: Syntheses, optical properties and influence of binding mode. *J. Phys. Chem. B*. 113: 14943–14949.

58 Zhu, Z., Li, Q., Zeng, Q. et al. (2008). New azobenzene-containing polyurethanes: post functional strategy and second-order nonlinear optical properties. *Dyes Pigm.* 78: 199–206.

59 Li, Q., Li, Z., Ye, C., and Qin, J. (2008). New indole-based chromophore-containing main-chain polyurethanes: Architectural modification of isolation group, enhanced nonlinear optical property, and improved optical transparency. *J. Phys. Chem. B.* 112: 4928–4933.

60 Chang, P., Chen, J., Tsai, H., and Hsiue, G. (2009). Molecular design of nonlinear optical polymer based on DCM to enhance the NLO efficiency and thermal stability. *J. Polym. Sci., Part A: Polym. Chem.* 47: 4937–4949.

61 Hammond, S.R., Clot, O., Firestone, K.A. et al. (2008). Site-isolated electro-optic chromophores based on substituted 2, 2′-bis(3, 4-propylenedioxythiophene) π-conjugated bridges. *Chem. Mater.* 20: 3425–3434.

62 Li, Z., Yu, G., Wu, W. et al. (2009). Nonlinear optical dendrimers from click chemistry: convenient synthesis, new function of the formed triazole rings, and enhanced NLO effects. *Macromolecules* 42: 3864–3868.

63 Li, Z., Wu, W., Li, Q. et al. (2010). High-generation second-order nonlinear optical (NLO) dendrimers: Convenient synthesis by click chemistry and the increasing trend of NLO effects. *Angew. Chem. Int. Ed.* 49: 2763–2767.

64 Walter, M.V. and Malkoch, M. (2012). Simplifying the synthesis of dendrimers: Accelerated approaches. *Chem. Soc. Rev.* 41: 4593–4609.

65 Wu, W., Wang, C., Zhong, C. et al. (2013). Changing the shape of chromophores from "H-type" to "star-type": Increasing the macroscopic NLO effects by a large degree. *Polym. Chem.* 4: 378–386.

66 Gopalan, P., Katz, H.E., McGee, D.J. et al. (2004). Star-shaped azo-based dipolar chromophores: Design, synthesis, matrix compatibility, and electro-optic activity. *J. Am. Chem. Soc.* 126: 1741–1747.

67 Campbell, V.E., In, I., McGee, D.J. et al. (2006). Chromophore orientation dynamics, phase stability, and photorefractive effects in branched azobenzene chromophores. *Macromolecules* 39: 957–961.

68 Wu, W., Huang, L., Song, C. et al. (2012). Novel global-like second-order nonlinear optical dendrimers: Convenient synthesis through powerful click chemistry and large NLO effects achieved by using simple azo chromophore. *Chem. Sci.* 3: 1256–1261.

69 Ronchi, M., Pizzotti, M., Biroli, A.O. et al. (2009). Second-order nonlinear optical (NLO) properties of a multichromophoric system based on an ensemble of four organic NLO chromophores nanoorganized on A cyclotetrasiloxane architecture. *J. Phys. Chem. C.* 113: 2745–2760.

70 Ronchi, M., Biroli, A.O., Marinotto, D. et al. (2011). The role of the chromophore size and shape on the SHG stability of PMMA films with embebbed NLO active macrocyclic chromophores based on a cyclotetrasiloxane scaffold. *J. Phys. Chem. C.* 115: 4240–4246.

71 Chen, P., Liu, G., Zhang, H. et al. (2018). A rigid ringlike molecule: large second-order nonlinear optical performance, good temporal and thermal stability, and ideal spherical structure conforming to the 'site isolation' principle. *J. Mater. Chem. C.* 6: 6784–6791.

72 Tang, R., Zhou, S., Xiang, W. et al. (2015). New 'X-type' second-order nonlinear optical (NLO) dendrimers: Fewer chromophore moieties and high NLO effects. *J. Mater. Chem. C.* 3: 4545–4552.

73 Tang, R., Zhou, S., Cheng, Z. et al. (2016). Janus second-order nonlinear optical dendrimers: Their controllable molecular topology and corresponding largely enhanced performance. *Chem. Sci.* 8: 340–347.

74 Chen, P., Zhang, H., Han, M. et al. (2018). Janus molecules: Large second-order nonlinear optical performance, good temporal stability, excellent thermal stability and spherical structure with optimized dendrimer structure. *Mater. Chem. Front.* 2: 1374–1382.

75 Wu, W., Li, C., Yu, G. et al. (2012). High-generation second-order nonlinear optical (NLO) dendrimers that contain isolation chromophores: Convenient synthesis by using click chemistry and their increased NLO effects. *Chem. Eur. J.* 18: 11019–11028.

76 Wu, W., Ye, C., Qin, J., and Li, Z. (2013). Further enhancement of the second-order nonlinear optical (NLO) coefficient and the stability of NLO polymers that contain isolation chromophore moieties by using the 'suitable isolation group' concept and the Ar/ArF self-assembly effect. *Chem. Asian J.* 8: 1836–1846.

77 Wu, W., Xu, Z., Xiong, Y. et al. (2013). The self-assembly effect in NLO polymers containing isolation chromophores: Enhanced NLO coefficient and stability. *New J. Chem.* 37: 1789–1796.

78 Wu, W., Xin, S., Xu, Z. et al. (2013). Main-chain second-order nonlinear optical polyaryleneethynylenes containing isolation chromophores: enhanced nonlinear optical properties, improved optical transparency and stability. *Polym. Chem.* 4: 3196–3203.

79 Wu, W., Ye, C., Qin, J., and Li, Z. (2013). Introduction of an isolation chromophore into an 'H'-shaped NLO polymer: Enhanced NLO effect, optical transparency, and stability. *Chempluschem* 78: 1523–1529.

80 Wu, W., Fu, Y., Wang, C. et al. (2013). Second-order nonlinear optical hyperbranched polymer containing isolation chromophore moieties derived from both 'H'-type and star-type chromophores. *Chinese J. Polym. Sci.* 31: 1415–1423.

81 Wu, W., Huang, Q., Xu, G. et al. (2013). Using an isolation chromophore to further improve the comprehensive performance of nonlinear optical (NLO) dendrimers. *J. Mater. Chem. C.* 1: 3226–3234.

82 Wu, W., Ye, C., Qin, J., and Li, Z. (2013). Dendrimers with large nonlinear optical performance by introducing isolation chromophore, utilizing the Ar/ArF self-assembly effect, and modifying the topological structure. *ACS Appl. Mater. Interfaces* 5: 7033–7041.

83 Chen, P., Yin, X., Xie, Y. et al. (2016). FTC-containing molecules: Large second-order nonlinear optical performance and excellent thermal stability, and the key development of the "isolation chromophore" concept. *J. Mater. Chem. C.* 4: 11474–11481.

84 Coates, G.W., Dunn, A.R., Henling, L.M. et al. (1997). Phenyl-perfluorophenyl stacking interactions: A new strategy for supermolecule construction. *Angew. Chem. Int. Ed.* 36: 248–251.

85 Kim, T.D., Kang, J.W., Luo, J. et al. (2007). Ultralarge and thermally stable electro-optic activities from supramolecular self-assembled molecular glasses. *J. Am. Chem. Soc.* 129: 488–489.

86 Zhou, X., Luo, J., Huang, S. et al. (2009). Supramolecular self-assembled dendritic nonlinear optical chromophores: Fine-tuning of arene-perfluoroarene interactions for ultralarge electro-optic activity and enhanced thermal stability. *Adv. Mater.* 21: 1976–1981.

87 Wu, W., Huang, Q., Qiu, G. et al. (2012). Aromatic/perfluoroaromatic self-assembly effect: an effective strategy to improve the NLO effect. *J. Mater. Chem.* 22: 18486–18495.

88 Wu, W., Huang, Q., Ye, C. et al. (2013). Second-order nonlinear optical (NLO) polymers containing perfluoroaromatic rings as isolation groups with Ar/ArF self-assembly effect: Enhanced NLO coefficient and stability. *Polymer* 54: 5655–5664.

89 Wu, W., Yu, G., Liu, Y. et al. (2013). Using two simple methods of Ar-ArF self-assembly and isolation chromophores to further improve the comprehensive performance of NLO dendrimers. *Chem. Eur. J.* 19: 630–641.

90 Wu, W., Wang, C., Tang, R. et al. (2013). Second-order nonlinear optical dendrimers containing different types of isolation groups: Convenient synthesis through powerful 'click chemistry' and large NLO effects. *J. Mater. Chem. C.* 1: 717–728.

91 Liu, G., Liao, Q., Deng, H. et al. (2019). Janus NLO dendrimers with different peripheral functional groups: Convenient synthesis and enhanced NLO performance with the aid of the Ar-ArF self-assembly. *J. Mater. Chem. C.* 7: 7344–7351.

92 Wu, W., Tang, R., Li, Q., and Li, Z. (2015). Functional hyperbranched polymers with advanced optical, electrical and magnetic properties. *Chem. Soc. Rev.* 44: 3997–4022.

93 Zhang, Y., Wang, L., Wada, T., and Sasabe, H. (1996). One-pot synthesis of a new hyperbranched polyester containing 3, 6-di-acceptor-substituted carbazole chromophores for nonlinear optics. *Macromol. Chem. Phys.* 197: 667–676.

94 Li, Z., Qin, A., Lam, J.W.Y. et al. (2006). Facile synthesis, large optical nonlinearity, and excellent thermal stability of hyperbranched poly(aryleneethynylene)s containing azobenzene chromophores. *Macromolecules* 39: 1436–1442.

95 Bai, Y., Song, N., Gao, J. et al. (2005). A new approach to highly electrooptically active materials using cross-linkable, hyperbranched chromophore-containing oligomers as a macromolecular dopant. *J. Am. Chem. Soc.* 127: 2060–2061.

96 Li, Z., Wu, W., Ye, C. et al. (2012). New main-chain hyperbranched polymers: facile synthesis, structural control, and second-order nonlinear optical properties. *Polymer* 53: 153–160.

97 Li, Z., Yu, G., Hu, P. et al. (2009). New azo-chromophore-containing hyperbranched polytriazoles derived from AB_2 monomers via click chemistry under copper(I) catalysis. *Macromolecules* 42: 1589–1596.

98 Wu, W., Ye, C., Qin, J., and Li, Z. (2013). The utilization of isolation chromophore in an "$A_3 + B_2$" type second-order nonlinear optical hyperbranched polymer. *Macromol. Rapid Commun.* 34: 1702–1709.

99 Zang, X., Liu, H., Li, Q. et al. (2020). A TCBD-based AB_2-type second-order nonlinear optical hyperbranched polymer prepared by a facile click-type postfunctionalization. *Polym. Chem.* 11: 5493–5499.

100 Zang, X., Liu, G., Li, Q. et al. (2020). A correlation study between dendritic structure and macroscopic nonlinearity for second-order nonlinear optical materials. *Macromolecules* 53: 4012–4021.

101 Tang, R., Chen, H., Zhou, S. et al. (2015). The integration of an "X" type dendron into polymers to further improve the comprehensive NLO performance. *Polym. Chem.* 6: 6680–6688.

102 Wu, W., Ye, C., Yu, G. et al. (2012). New hyperbranched polytriazoles containing isolation chromophore moieties derived from AB_4 monomers through click chemistry under copper(I) catalysis: Improved optical transparency and enhanced NLO effects. *Chem. Eur. J.* 18: 4426–4434.

103 Wu, W., Fu, Y., Wang, C. et al. (2011). A series of hyperbranched polytriazoles containing perfluoroaromatic rings from AB_2-type monomers: Convenient syntheses by click chemistry under copper(I) catalysis and enhanced optical nonlinearity. *Chem. Asian J.* 6: 2787–2795.

104 Holter, D., Burgath, A., and Frey, H. (1997). Degree of branching in hyperbranched polymers. *Acta. Polym.* 48: 30–35.

105 Wu, W. and Li, Z. (2014). Further improvement of the macroscopic NLO coefficient and optical transparency of hyperbranched polymers by enhancing the degree of branching. *Polym. Chem.* 5: 5100–5108.

106 Wu, W. (2014). Design, synthesis and characterization of macromolecules with large second-order nonlinear optical effects. PhD Thesis. Wuhan University.

107 Wu, W., Huang, L., Fu, Y. et al. (2013). Design, synthesis and nonlinear optical properties of 'dendronized hyperbranched polymers'. *Chin. Sci. Bull.* 58: 2753–2761.

108 Wu, W., Wang, Z., Xiao, R. et al. (2015). Main chain dendronized hyperbranched polymers: Convenient synthesis and good second-order nonlinear optical performance. *Polym. Chem.* 6: 4396–4403.

109 Wu, W., Xu, Z., and Li, Z. (2014). Using low generation dendrimers as monomers to construct dendronized hyperbranched polymers with high nonlinear optical performance. *J. Mater. Chem. C.* 2: 8122–8130.

110 Yang, H., Tang, R., Wu, W. et al. (2016). A series of dendronized hyperbranched polymers with dendritic chromophore moieties in the periphery: convenient synthesis and large nonlinear optical effects. *Polym. Chem.* 7: 4016–4024.

111 Yang, H., Cheng, Z., Liu, C. et al. (2018). A second-order nonlinear optical dendronized hyperbranched polymer containing isolation chromophores: Achieving good optical nonlinearity and stability simultaneously. *Sci. China. Chem.* 61: 584–591.

112 Tang, R., Chen, H., Zhou, S. et al. (2015). Dendronized hyperbranched polymers containing isolation chromophores: Design, synthesis and further enhancement of the comprehensive NLO performance. *Polym. Chem.* 6: 5580–5589.

113 Liu, G., Chen, P., Tang, R., and Li, Z. (2016). Synthesis and characterization of dendronized hyperbranched polymers through the '$A_3 + B_2$' approach. *Sci. China. Chem.* 59: 1561–1567.

114 Hatton, F.L., Chambon, P., McDonald, T.O. et al. (2014). Hyperbranched polydendrons: A new controlled macromolecular architecture with self-assembly in water and organic solvents. *Chem. Sci.* 5: 1844–1854.

115 Hatton, F.L., Tatham, L.M., Tidbury, L.R. et al. (2015). Hyperbranched polydendrons: A new nanomaterials platform with tuneable permeation through model gut epithelium. *Chem. Sci.* 6: 326–334.

116 Rogers, H.E., Chambon, P., Auty, S.E. et al. (2015). Synthesis, nanoprecipitation and pH sensitivity of amphiphilic linear-dendritic hybrid polymers and hyperbranched-polydendrons containing tertiary amine functional dendrons. *Soft Matter* 11: 7005–7015.

117 Lederer, A., Burchard, W., Hartmann, T. et al. (2015). Dendronized hyperbranched macromolecules: Soft matter with a novel type of segmental distribution. *Angew. Chem. Int. Ed.* 54: 12578–12583.

118 Kuchkina, N.V., Zinatullina, M.S., Serkova, E.S. et al. (2015). Hyperbranched pyridylphenylene polymers based on the first-generation dendrimer as a multifunctional monomer. *RSC Advances.* 5: 99510–99516.

119 Zhang, X., Liu, T., Yan, J. et al. (2019). Multiple-responsive dendronized hyperbranched polymers. *ACS Omega.* 4: 7667–7674.

120 Morcuende-Ventura, V., Hermoso-Durán, S., Abian-Franco, N. et al. (2021). Fluorescence liquid biopsy for cancer detection is improved by using cationic dendronized hyperbranched polymer. *Int. J. Mol. Sci.* 22: 6501.

9

Other Opto-electronic Materials in Aggregate

Shaoqiang Dong[1], Meiting Zhao[1*], Zhijun Ruan[2*], Zekun Tong[1], Yutian Qin[1], Chaoyang Zheng[1], Xinfang Zhang[1], and Zhen Li[1,3]*

[1] *Institute of Molecular Aggregation Science, Tianjin University, Tianjin, China*
[2] *College of Chemistry and Chemical Engineering, Huanggang Normal University, Huanggang, China*
[3] *Department of Chemistry, Wuhan University, Wuhan, China*

*Corresponding authors: shqdong@tju.edu.cn; mtzhao@tju.edu.cn; ruanzhijun87@126.com

9.1 Magnetic Materials

9.1.1 Introduction

Molecular magnetic materials have garnered great interest owing to their diverse magnetic phenomena and important potential for applications in information storage, molecular switches, electronic devices, and sensors.[1–5] Among these materials, spin-crossover (SCO) materials and single-molecule magnet (SMM), recognized as magnetically bistable systems with highly tunable physical properties, have drawn considerable research interest and gained great progress.[6–10] Especially, for these materials, their properties are exquisitely sensitive to minute variations of the coordination environment. Changes in the local geometry of the central metal ion and the supramolecular interactions enable the targeted tuning of a variety of magnetic characteristics, such as SCO and SMM.[10–13] Therefore, different packing arrangements of molecular aggregates, such as the formation of polymorphs (differences in crystal packing, configuration, or architectures) and the single-crystal-to-single-crystal (SCSC) transformation, have an extremely important effect on their magnetic properties, providing outstanding systems for the detailed exploration of the structure–property relationships.[11]

9.1.2 Molecular Magnetic Materials

Octahedral first-row transition-metal complexes with d^4-d^7 electronic structure can take either low spin (LS, minimum number of unpaired spins) or high spin

(HS, maximum number of unpaired spins) configurations.[6] The SCO phenomenon arises from a thermodynamic balance where the LS state is enthalpically favored, and the HS state is entropically preferred. An external stimulus can tip this balance, triggering a reversible switch between the two states. For example, Fe(II) with a d^6 electronic structure (usually with 6N donors), is currently the most studied metal ion in the SCO field, due to its extreme change in magnetism, undergoing a diamagnetic LS state to the HS state that with four unpaired electrons (Figure 9.1a). Interestingly, the Fe—N bond length in the LS state is about 10% shorter than that in the HS state (Figure 9.1b). Furthermore, crystals in LS state usually have a darker color whereas a relatively lighter color in the HS state, mainly due to a reduction of the M—L bonds with a toilless metal-to-ligand charge transfer.[6]

Generally, temperature-dependent SCO can occur in one of the five different manners: gradual, abrupt, with thermal hysteresis, multi-step, or incomplete (Figure 9.1c).[14] Among these types, the abrupt (SCO happens within a few

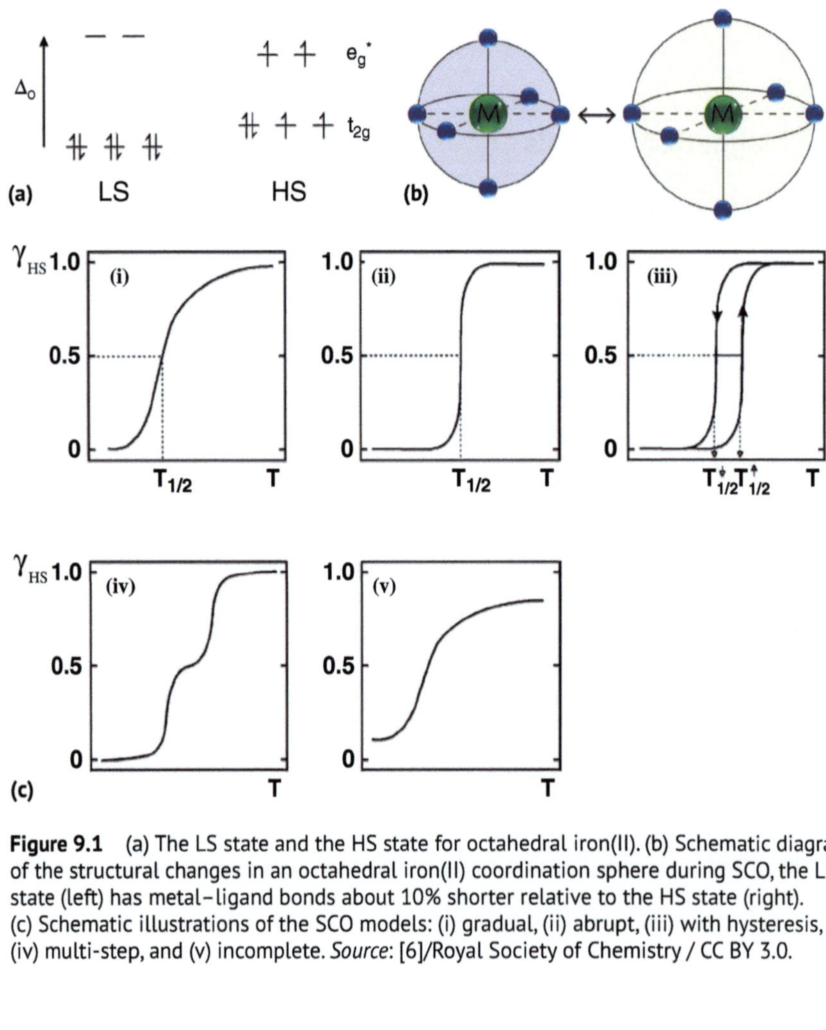

Figure 9.1 (a) The LS state and the HS state for octahedral iron(II). (b) Schematic diagram of the structural changes in an octahedral iron(II) coordination sphere during SCO, the LS state (left) has metal–ligand bonds about 10% shorter relative to the HS state (right). (c) Schematic illustrations of the SCO models: (i) gradual, (ii) abrupt, (iii) with hysteresis, (iv) multi-step, and (v) incomplete. *Source*: [6]/Royal Society of Chemistry / CC BY 3.0.

degrees) and one with hysteresis (the lag of magnetic response as the temperature is changed) are most important for practical applications, e.g. switching and memory devices. And the abrupt/hysteresis manner means strong cooperativity between the SCO centers, which is spread via intermolecular interactions, such as hydrogen bonding, π–π stacking, S–S contacts, and metallophilic interactions, etc. These interactions are heavily dependent on crystal packing, so different stacking methods of the same molecule (polymorphism) in the aggregate state can change its crystal field strength and consequently modulate the SCO properties.[11,15–17] Therefore, studying polymorphism of SCO compounds, especially the contribution of intermolecular interactions to spin transition, is crucial for better understanding of the correlation between molecular packing and the nature of SCO. Some classical polymorphism molecular systems are listed below to illustrate the influence of molecular packing on SCO properties.

The influence of polymorphism on SCO properties was first documented in $[Fe(L)_2(NCS)_2]$,[18,19] where L is phenanthroline or 2,2′-bipyridine, by König et al. Since then, polymorphism has received increasing interest. $[Fe(bt)_2(NCS)_2]$(bt = 2,2′-bithiazoline) as a typical SCO system has two different single crystals, and their magnetic behavior and crystal structures were well characterized. McGarvey and co-workers got two different single crystals through slow evaporation of $[Fe(bt)_2(NCS)_2]$ solutions in ethanol or methanol at different temperatures. Among them, one crystal exhibited a spin transition with hysteresis between approximately 170–180 K, while the other crystal remained paramagnetic from 77 to 300 K without SCO behavior.[20] Subsequently, Real et al. obtained polymorph **I** and polymorph **II** from $[Fe(bt)_2(NCS)_2]$ solutions in absolute ethanol at room temperature, and investigated their magnetic properties and crystal structures in detail.[21] The temperature-dependent $\chi_M T$ plots at various pressures were shown in Figure 9.2. At ambient pressure, polymorph **I** exhibits a spin transition with the critical temperature as $T_c\downarrow = 176$ K (cooling mode) and $T_c\uparrow = 187$ K (warming mode). Meanwhile, the spin transition in polymorph **I** exhibits a significant pressure dependence (Figure 9.2a). In sharp contrast, polymorph **II** shows no spin transition, and its magnetic properties are completely insensitive to pressure, remaining identical from ambient conditions up to 1.06 GPa (Figure 9.2b).

Crystal structure analysis revealed that polymorph **I** crystallizes in the triclinic space group $P1$, whereas polymorph **II** adopts the monoclinic $C2/c$. As illustrated in Figure 9.2c, the crystal packing of polymorph **I** viewed along the *bc* plane consists of monomeric chains of $[Fe(bt)_2(NCS)_2]$ extending along the [110] direction. Within each chain, adjacent units are connected through strong S···S interactions, with an S1···S5 distance of 3.340 Å, significantly shorter than the sum of the van der Waals radii of two sulfur atoms (3.7 Å). Weaker interchain S···S contacts are also present.

In contrast, the molecular packing of polymorph **II** differs substantially, as shown in Figure 9.2d. Each $[Fe(bt)_2(NCS)_2]$ unit engages with four neighboring molecules in the *ac* plane via six strong S···S interactions, forming a corrugated layered structure. At 293 K, the average Fe—N(NCS) bond length in polymorph **I** (2.159 Å) is slightly shorter than that in polymorph **II** (2.168 Å). A more pronounced distinction lies in

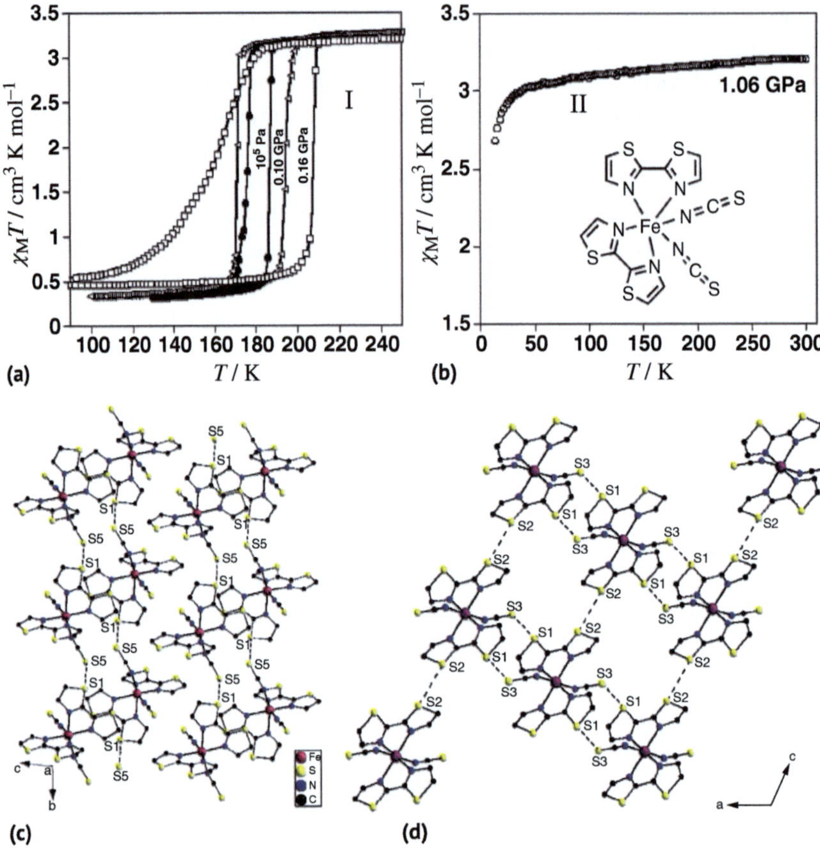

Figure 9.2 Magnetic susceptibility ($\chi_M T$) for polymorphs (a) **I** and (b) **II** of $[Fe(bt)_2(NCS)_2]$. Crystal packing of (c) polymorph **I** along the *bc* plane and (d) polymorph **II** along the *ac* plane. Intermolecular S⋯S contacts are indicated by dotted lines. *Source*: [21]/with permission of American Chemical Society.

the degree of trigonal distortion, quantified by the angle φ, which is 5.1° larger in polymorph **II** ($\varphi = 14.7°$) than in polymorph **I** ($\varphi = 9.6°$), indicating a more significant deviation from ideal octahedral geometry toward a trigonal polyhedron in the former.[21] Given that the magnetic bistability of SOC complexes is governed by the ligand field strength at the metal center, the combined structural features, shorter Fe—N bonds, and lower trigonal distortion in polymorph **I**, are consistent with a stronger ligand field relative to polymorph **II**, thereby rationalizing the occurrence of spin crossover exclusively in polymorph **I**.

Mononuclear complex *fac*-$[Fe(dppa)_2(NCS)_2]$, where dppa is (3-aminopropyl) bis(2-pyridylmethyl)amine, was synthesized and reported by Matouzenko et al. to demonstrate the influence of polymorphism on SCO. The three polymorphs of this complex with totally different magnetic behaviors are presented in this excellent work.[22] Variable-temperature magnetic susceptibility measurements and Mössbauer spectroscopy revealed distinct behaviors for the three polymorphs. Polymorph **I** undergoes a complete, yet gradual spin crossover centered at approximately 176 K, with no observable hysteresis (Figure 9.3a). In contrast, polymorph **II** remains in the HS state and is paramagnetic between 4.5 and 295 K (Figure 9.3b). Polymorph **III**, however, exhibits a highly abrupt and complete SCO with a thermal hysteresis of 8 K, characterized by transition temperatures of $T_c\downarrow = 112$ K (cooling) and $T_c\uparrow = 120$ K (warming) (Figure 9.3c).

The crystal structure investigations at 298 K revealed that polymorph **I** crystallizes in the triclinic *P*1, whereas polymorph **II** in the monoclinic $P2_1/c$ and polymorph **III** in the orthorhombic *Pbca*. Although the asymmetric units of the three polymorphs are nearly identical, each containing one chiral complex molecule with the same configuration and a similarly distorted $[FeN_6]$ octahedron, their crystal packing arrangements differ substantially. This structural divergence is reflected in the metal-ligand bond lengths: the mean Fe—N (aromatic) distances are similar in polymorphs **I** (2.196 Å) and **III** (2.205 Å), but observably longer in polymorph **II** (2.230 Å). Moreover, the dihedral angle between the two pyridine planes in polymorph **II** is much smaller than that of the other two crystals, which are 70.7° and 82° for the SCO polymorphs **I** and **III**, respectively, and 40.7° for the paramagnetic

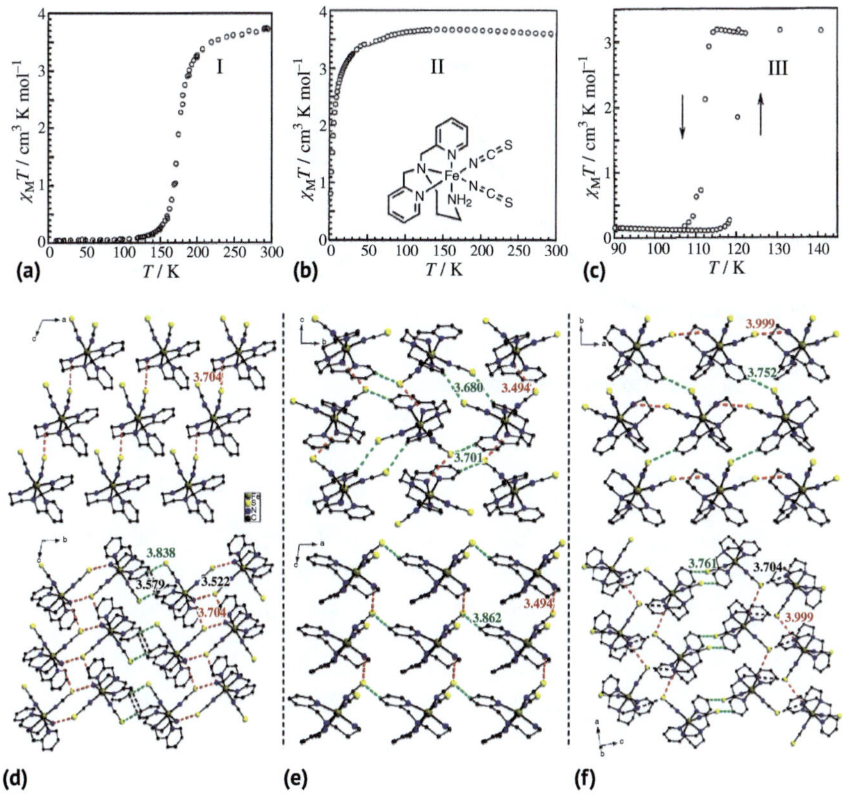

Figure 9.3 The $\chi_M T$ plots for polymorphs (a) **I**, (b) **II**, and (c) **III** of *fac*-$[Fe(dppa)_2(NCS)_2]$. *Source*: [22]/with permission of American Chemical Society. The packing diagrams of *fac*-$[Fe(dppa)_2(NCS)_2]$ for (d) **I**, (e), and (f) **III**, respectively. *Source*: [11]/with permission of Royal Society of Chemistry.

polymorph **II**.[22] Figure 9.3d–f shows the crystal packing for these three polymorphs based on the detailed structural analysis, the authors speculated that the high-spin state in polymorph **II** is stabilized by two centers of steric strain within the crystal lattice, which leads to an elongation of the Fe—N (aromatic) bonds. In contrast, the hysteretic behavior in polymorph **III** was attributed to its extensive network of intermolecular contacts, which enhances cooperativity during the spin transition relative to polymorph **I**.

In comparison with Fe^{II} SCO complexes, Co(II) SCO ones have been rarely reported.[23–25] A typical Co(II) SCO compound of $[Co(4\text{-terpyridone})_2](BF_4)_2{\cdot}(H_2O)$ that could crystallize into two different polymorphs was reported by Real et al. in 2006.[26] For polymorph **II**, it suffered a normal spin transition between 5 and 340 K, that centered at 152 K without hysteresis (Figure 9.4a). The magnetic behavior of polymorph **II** is characteristic of a non-hysteretic SOC. At 340 K, the $\chi_m T$ value of 2.48 $cm^3\,K\,mol^{-1}$ is consistent with a high-spin (S = 3/2) Co(II) ion, and the overlapping cooling-warming curves confirm the absence of hysteresis.[26]

As to polymorph **I**, its magnetic behavior is more complex and interesting. From 5 to 175 K, a constant $\chi_m T$ value of ~0.40 $cm^3\,K\,mol^{-1}$ indicates an S = 1/2 low-spin

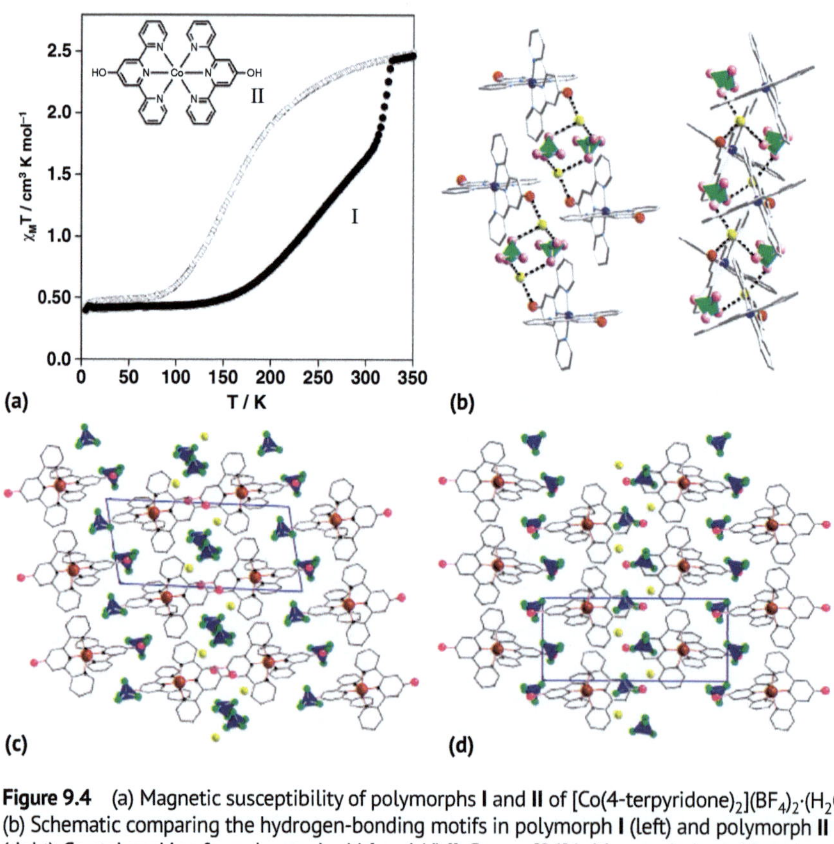

Figure 9.4 (a) Magnetic susceptibility of polymorphs **I** and **II** of $[Co(4\text{-terpyridone})_2](BF_4)_2{\cdot}(H_2O)$. (b) Schematic comparing the hydrogen-bonding motifs in polymorph **I** (left) and polymorph **II** (right). Crystal packing for polymorphs (c) **I** and (d) **II**. *Source*: [26]/with permission of American Chemical Society.

ground state. Upon being warmed to 300 K, a smooth increase to 1.53 cm^3 K mol^{-1} signals an incomplete and non-cooperative spin transition. Remarkably, further heating to 338 K triggers an abrupt transition to a full high-spin state ($\chi_m T = 2.47$ cm^3 K mol^{-1}). Subsequent cooling does not retrace the original warming path; instead, the magnetic curve mirrors that of polymorph **II** and remains irreversible over multiple cycles. This evidence strongly suggests that polymorph **I** undergoes an irreversible transformation into polymorph **II** above 340 K, a conclusion corroborated by single-crystal X-ray diffraction.[26]

Although polymorphs **I** and **II** of [Co(4-terpyridone)$_2$](BF_4)$_2$·(H_2O) share a similar distribution of primary components, their key difference lies in the arrangement of H_2O molecules and BF_4^- anions in the second coordination sphere. In polymorph **II**, these species alternate homogeneously, whereas in polymorph I, the anions form isolated "pairs" (Figure 9.4c, d). This structural divergence dictates the nature of the hydrogen-bonding network. In polymorph **I**, the interactions are discrete, linking two adjacent complex cations, two water molecules, and two anions. In contrast, polymorph **II** features a continuous, infinite helicoidal chain of hydrogen bonds connecting all components (Figure 9.4b). This extended network in polymorph **II** facilitates a more effective transmission of structural changes throughout the crystal, thereby promoting the cooperative spin crossover observed in this polymorph.[26]

Configurational polymorphism, a special class of polymorphism arising from flexible organic ligands that exist as configurational isomers, is exemplified by the complex [Fe(stpy)$_4$(NCS)$_2$].[27] This compound forms two distinct polymorphs: polymorph **I**, [Fe(*trans*-stpy)$_4$(NCS)$_2$] (orthorhombic, $Pna2_1$), and polymorph **II**, [Fe(*cis*-stpy)$_4$(NCS)$_2$] (monoclinic, $P2_1/n$). Their magnetic behaviors differ drastically; polymorph **I** undergoes an abrupt spin crossover near 108 K, whereas polymorph **II** remains in the high-spin state at all temperatures (Figure 9.5a). This contrast implies that the *trans*-stpy isomer acts as a significantly stronger field ligand than its *cis*-counterpart. The weaker ligand field strength of *cis*-stpy can be correlated with its pronounced non-planarity, a consequence of steric hindrance between the adjacent pyridine and phenyl rings, which likely disrupts optimal orbital overlap with the metal center.[27]

Polymorphic networks (polymorphism from architecture) are constructed with the same molecular modules but adopt distinct spatial arrangements. Complex {Fe(3-CNpy)[Au(CN)$_2$]$_2$}·nH_2O (3-CNpy = 3CN-pyridine) can exist in three forms, polymorph **I** (n = 2/3) as long yellow needles, polymorph **II** (n = 0) as colorless squares and polymorph **III** (n = 0) as pale-yellow parallelepipeds.[28] As shown in Figure 9.5c, polymorph **I** crystallizes in the trigonal space group *P*3, with the iron atom situated at an inversion center within an elongated octahedral coordination geometry. The average equatorial Fe—N bond length (2.151 Å), formed by four $[Au(CN)_2]^-$ groups, is shorter than the axial Fe—N bond length (2.193 Å) from the 3-CNpy ligands. This bond length distribution is characteristic of a HS Fe(II) ion at the measurement temperature. However, upon cooling below 120 K, the complex undergoes a very incomplete SOC, with only approximately 30% of the Fe(II) sites converting to the LS state.

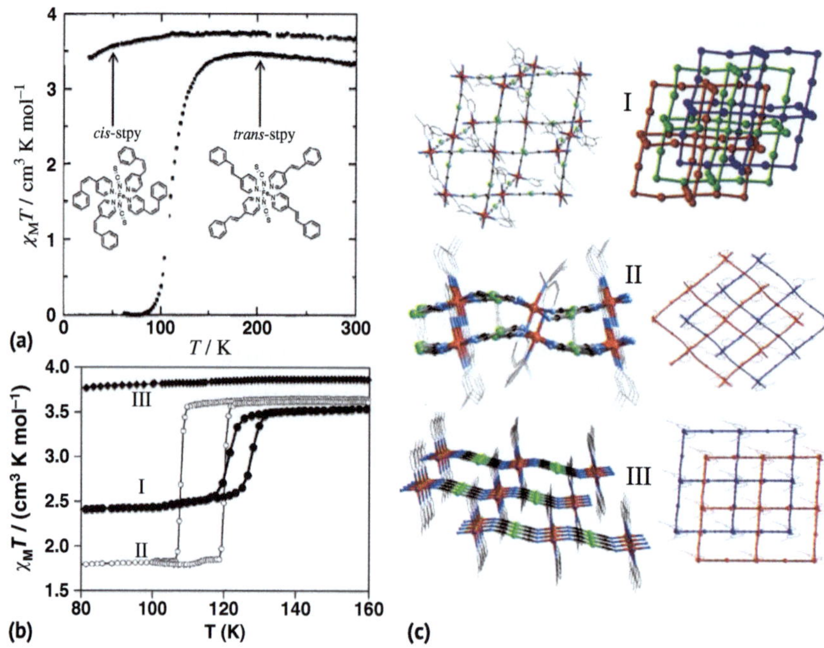

Figure 9.5 (a) The $\chi_M T$ for *trans*- and *cis*-[Fe(stpy)$_4$(NCS)$_2$]. (b) Magnetic behaviors of polymorphs I–III of {Fe(3-CNpy)[Au(CN)$_2$]$_2$}. (c) Crystal packing of I–III of {Fe(3-CNpy)[Au(CN)$_2$]$_2$}. *Source*: [28]/with permission of Royal Society of Chemistry.

Polymorph **II** crystallizes in the monoclinic space group P2$_1$/c and features two crystallographically independent Fe(II) sites within distorted octahedra. These sites are distinguished by their axial 3-CNpy ligands: the Fe1 site has the cyano groups in a *cis* conformation, while the Fe2 site adopts a *trans* conformation. Magnetic and structural analysis reveal that only the Fe1 site undergoes a complete spin transition, whereas the Fe2 site remains in the high-spin state. In contrast, polymorph **III** adopts the monoclinic C2 space group and forms a homogeneous, grid-like layer structure. These slightly corrugated layers stack uniformly along the c-axis (Figure 9.5c). The distance between planes, also represents an interlayer shortest Au···Au distance (5.154 Å), indicating no aurophilic interactions exist in this isomer. Moreover, polymorph **III** features only one type of Fe(II) site, where the 3-CNpy groups adopt a *cis* conformation. Both the equatorial and axial Fe—N bond distances remain invariant with temperature, confirming that the complex remains in the high-spin state over the measured range.[28]

Figure 9.5b shows the χ_mT plots for the three supramolecular isomers. Their magnetic behaviors are fully consistent with the structural data. Polymorph **I** exhibits a cooperative spin transition for approximately 30% of the Fe(II) sites, accompanied by a 6 K thermal hysteresis. In polymorph **II**, about 50% of the Fe(II) centers undergo a cooperative transition centered at 107 K with a wider hysteresis of 12 K. Conversely, polymorph **III** remains paramagnetic across the entire temperature range.[28]

Generally, obtaining polymorphs of SCO compounds is very difficult and fortunate, and direct polymorphic transformation is particularly rare. A remarkable exception is the classical SCO compound [Fe(tpa)(NCS)$_2$](tpa = tris(2-pyridylmethyl)amine), which exhibits an uncommon succession of polymorphic transformations. Four distinct polymorphs (**I–IV**) were obtained through different crystallization and solution-mediated conversion pathways (Figure 9.6a). Specifically, polymorph **II** undergoes a stepwise transformation to polymorph **IV** via a clearly observable intermediate phase (polymorph **III**) over one month when immersed in a DMF–H_2O (1:5) solution (Figure 9.6b). Structurally, polymorphs **I–III** crystallize in the monoclinic $P2_1/c$ space group, while polymorph **IV** adopts the triclinic $P1$ group. A key distinction lies in their asymmetric units: polymorph **I** contains one unique [Fe(tpa)(NCS)$_2$] molecule, whereas polymorphs **II–IV** each feature two crystallographically independent molecules (FeA and FeB sites, Figure 9.6c). This structural difference is directly reflected in their magnetic properties. Magnetic susceptibility measurements show that polymorphs **I** and **II** undergo a one-step incomplete spin transition. In contrast, polymorphs **III** and **IV** exhibit a two-step transition, attributed to the two distinct Fe sites undergoing SCO at different temperatures. Notably, only polymorph **III** displays cooperative SCO behavior below 200 K with a 20 K thermal hysteresis loop.[29,30]

Analysis of the crystal packing reveals that all four polymorphs are stabilized by a combination of intermolecular π–π interactions between pyridine rings and S···H—C hydrogen bonds. The specific strength and arrangement of these interactions differ among the polymorphs, accounting for their varying transition profiles. Polymorph **III**, which shows the most abrupt and cooperative transition, is characterized by the strongest intermolecular interactions, underscoring the critical role of lattice cooperativity in modulating SCO properties.

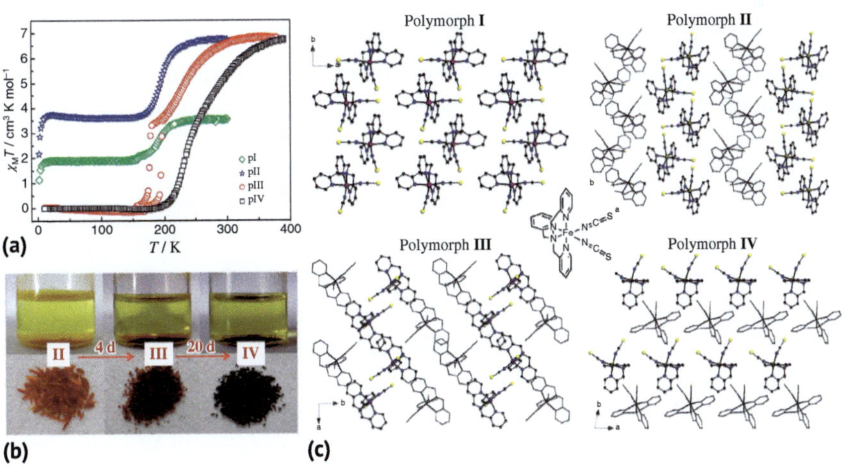

Figure 9.6 (a) Temperature-dependent magnetic susceptibility of polymorphs **I–IV** of [Fe(tpa)(NCS)$_2$]. (b) Picture of polymorph **II** transforming into polymorph **IV**. (c) Corresponding crystal structures viewed down the *c*-axis, with FeA and FeB sites shown in capped-stick and ball-and-stick modes, respectively. *Source*: [11]/with permission of Royal Society of Chemistry.

SMMs are paramagnetic molecules that can be magnetized below a characteristic blocking temperature (T_B), holding significant promise for applications in high-density information storage and spintronic devices.[10,31–34] A major research focus has been the development of SMMs with high effective energy barriers (U_{eff}) and elevated T_B values. In a notable study, Wang et al. reported the reversible on-off switching of SMM behavior via a single-crystal-to-single-crystal (SCSC) transformation. This process involves the reversible conversion between green crystals of the trinuclear complex $[Mn(L)(H_2O)]_2[Mo(CN)_7]·2H_2O$ (**Ia**) and brown crystals of the hexanuclear complex $[Mn(L)(H_2O)]_2[Mn(L)]_2[Mo(CN)_7]_2$ (**Ib**), triggered by dehydration and rehydration, respectively (Figure 9.7).[35]

Compound **Ia** (monoclinic, C2/*c*) exhibits definitive SMM properties, evidenced by pronounced frequency-dependent ac magnetic susceptibility ($\varphi = 0.14$) and magnetization hysteresis loops below 3.0 K, with an energy barrier of 44.9 cm^{-1}. In contrast, the dehydrated form **Ib** (monoclinic, C2/*m*) loses its SMM character and behaves as a simple paramagnet. Structural analysis reveals that the SCSC transformation induces a critical change: the geometry of the $\left[Mo^{III}(CN)_7\right]^{4-}$ unit distorts from an ideal pentagonal bipyramid in **Ia**. This geometric change in **Ib** alters the Mo^{III}–Mn^{II} bridging modes and disrupts the magnetic anisotropy necessary for SMM behavior. This work demonstrates that precise control over molecular geometry via chemical stimuli can effectively switch magnetic functionality.

Building upon previous work, the same research group reported a mononuclear cobalt(II) complex exhibiting reversible on–off switching of both spin-crossover (SCO) and single-molecule magnet (SMM) behaviors via a crystal-to-crystal transformation. This switching, triggered by a dehydration–rehydration process, is accompanied by a distinct color change between dark red (**IIa**, SCO-on/SMM-off) and orange (**IIb**, SCO-off/SMM-on).[13] The dark-red crystals of the hydrated phase, $\left[Co^{II}(L)_2\right](DPAS)_2·DMF·2H_2O$ (**IIa**) (L = 4′-(4-bromophenyl)-2,2′:6′,2″-terpyridine and DPAS = 4-(phenylamino)benzenesulfonate), were obtained from a DMF/H_2O solution. Heating **IIa** at 180 °C under dry N_2 yielded the orange, dehydrated

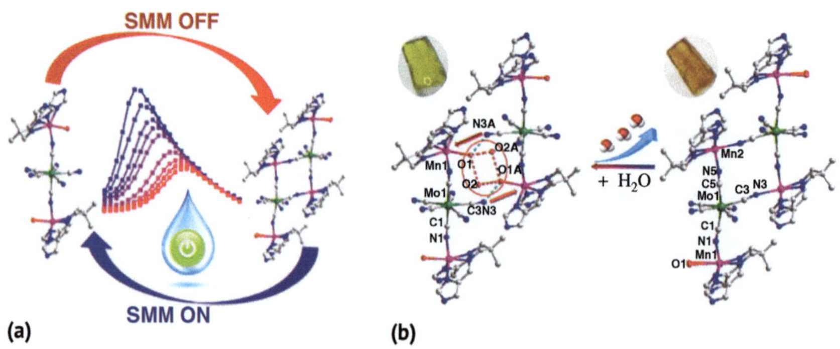

Figure 9.7 (a) View of the reversible on–off switching of a SMM through a crystal-to-crystal conversion. (b) Structures of the trinuclear Mn_2Mo (**Ia**, left) and hexanuclear Mn_4Mo_2 (**Ib**, right) clusters. *Source*: [35]/with permission of American Chemical Society.

phase **IIb** ($Co^{II}(L)_2$] $(DPAS)_2$·DMF), which reverted to **IIa** upon re-exposure to air. Both polymorphs crystallize in the triclinic space group $P1$. The Co^{II} center is coordinated in a distorted octahedral geometry by six N atoms from two meridionally coordinating L ligands, denoted L_A and L_B (Figure 9.8a). Magnetic and structural studies confirmed that **IIa** undergoes a gradual SCO between 300 and 100 K, as evidenced by a smooth decrease in the average Co—N bond length from 2.117 to 2.074 Å. In contrast, the bond lengths in **IIb** remain virtually constant, indicating it remains in the high-spin state (Figure 9.8b).[13]

The switching mechanism was elucidated through structural analysis. In the hydrated form **IIa**, an extensive supramolecular network, including hydrogen bonds (involving H_2O, $DPAS^-$, and L) and π–π interactions, rigidifies the structure (Figure 9.8c). Notably, the L_A ligands are locked within ladder-like frameworks by CH···O bonds (Figure 9.8d). Dehydration to **IIb** disrupts this network: the water-mediated hydrogen bonds are eliminated, dismantling the ladders and releasing the L_A ligands (Figure 9.8e). Concurrently, the π–π interactions are weakened due to rotations of the $DPAS^-$ anions. This leads to a looser molecular packing in **IIb**, which translates into a greater distortion of the Co^{II} coordination octahedron. This enhanced distortion stabilizes the HS state, preventing SCO and, in turn, enabling SMM behavior at low temperatures by altering the magnetic anisotropy. This work demonstrates that fine-tuning weak supramolecular interactions is a powerful strategy for designing bifunctional switchable materials with coupled SCO and SMM properties.

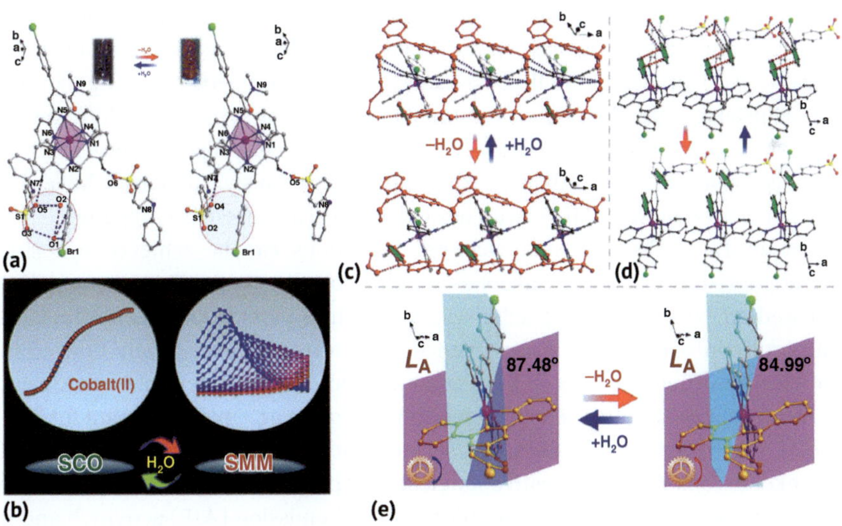

Figure 9.8 (a) Asymmetric units of **IIa** and **IIb**. The purple dashed lines indicate H-bonds. (b) Reversible switching of both SCO and SMM properties through a crystal-to-crystal conversion. (c) $DPAS^-$ connect themselves and interact with H_2O, forming porous 1D ladders along the *a*-axis via hydrogen bonds, which vanished in dehydrated **IIb**. (d) The breaking/reformation of π–π interactions between the pyridine units of L_B and the benzene units of the $DAPS^-$. (e) L_A ligand undergoes a rotation during the dehydration/rehydration cycle between **IIa** and **IIb**. *Source*: [13]/with permission of Royal Society of Chemistry.

9.1.3 Summary and Perspective

In brief, molecular magnets, especially the bistable SCO and SMM materials, have garnered significant interest as promising candidates for applications in high-density data storage, molecular switches, and spintronic devices et al. Polymorphism and the SCSC transformation exert a critical influence on the magnetic properties, e.g. the same molecule with different crystal forms could possess extremely different magnetic properties. Although abundant achievements have been obtained, more polymorphs and SCSC transformation examples are needed to better understand magneto-structural relationships. Similarly, for opto-electronic functional materials, the systematical research of structure–packing–performance relationships indicated that the properties of materials are a direct consequence of their supramolecular organization in an aggregated state, and controlling molecular packing is a practical method to optimize the performance of materials.[36–38] In addition, multifunctional molecular magnet materials are greatly attractive, thus functional modification could be cannily introduced into some excellent molecular systems to achieve multi-response and multi-functions.

9.2 Metal–Organic Frameworks

9.2.1 Introduction

Metal–organic frameworks (MOFs) are porous materials fabricated by connecting metal ions and organic ligands via coordination bonds.[39,40] As emerging organic–inorganic hybrid materials, MOFs possess several unique properties, e.g. large surface area, adjustable pore dimension, and tunable topological structure.[40] More importantly, as the connection of metal nodes and organic linkers is highly diverse in structure, it is applicable to control the packing and aggregation style of organic linkers in MOFs, thereby endowing MOFs with desired physical and chemical properties.

Particularly, MOFs with opto-electronic functions can be achieved by using chromophore-based organic linkers as building blocks. Generally, chromophores are based on extended conjugated π-systems, such as pyrene, tetraphenylethylene, and porphyrin compounds. Pyrene and its derivatives are extensively utilized in opto-electronic fields such as OLEDs, organic lasers, and organic photovoltaic cells. For instance, 1,3,6,8-tetraphenylpyrene can function as an emissive dopant for blue OLEDs.[41] Accordingly, 1,3,6,8-tetrakis(4-benzoicacid)pyrene (TBAPy) has emerged as a linker to construct MOFs with opto-electronic properties.[42] Tetraphenylethylene is a landmark molecule with aggregation-induced emission (AIE) activity,[43] and it can be simply incorporated into the organic ligand of MOFs via the attachment of carboxylic/carboxyphenyl groups, i.e. tetrakis(4-carboxyphenyl)ethylene (TCPE) and tetrakis(4′-carboxy(1,1′-biphenyl)-4-yl)ethylene (TCBPE).[44] Porphyrin compounds are also efficient light absorbers, which display intense absorption at the Soret band (400–440 nm), and their opto-electronic properties can be further tuned via functionalization at *meso*-positions and incorporation of metal ions.[45] Therefore,

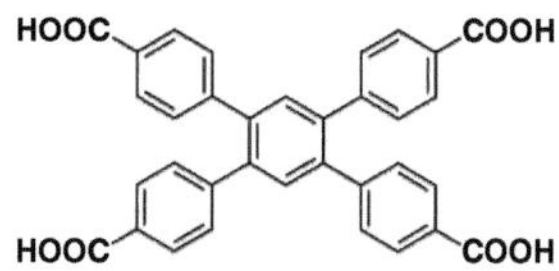

1,2,4,5-Tetrakis(4-carboxyphenyl)benzene
TCPB

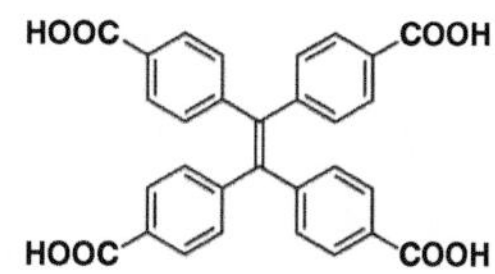

Tetrakis(4-carboxyphenyl)ethylene
TCPE

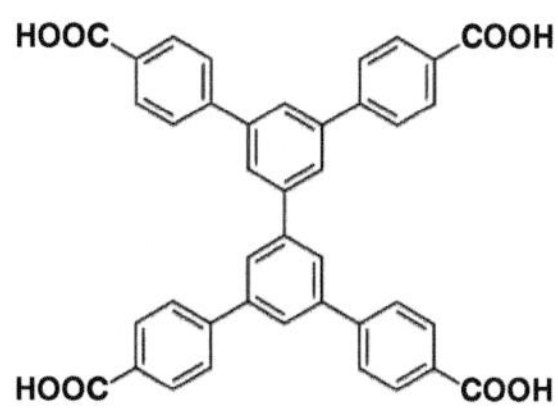

3,3',5,5'-Tetrakis(4-carboxyphenyl)biphenyl
TCPBP

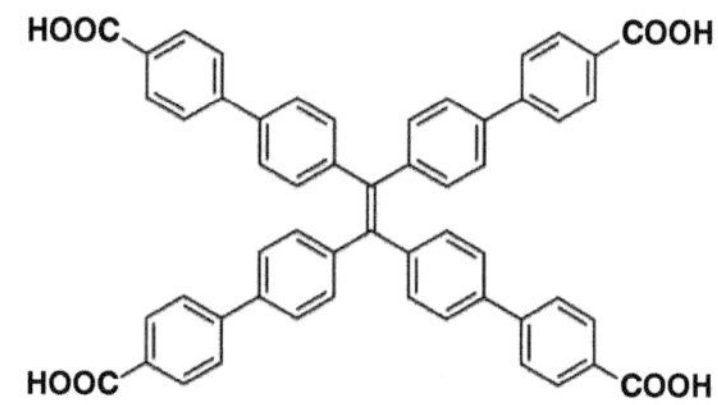

Tetrakis(4'-carboxy(1,1'-biphenyl)-4-yl)ethylene
TCBPE

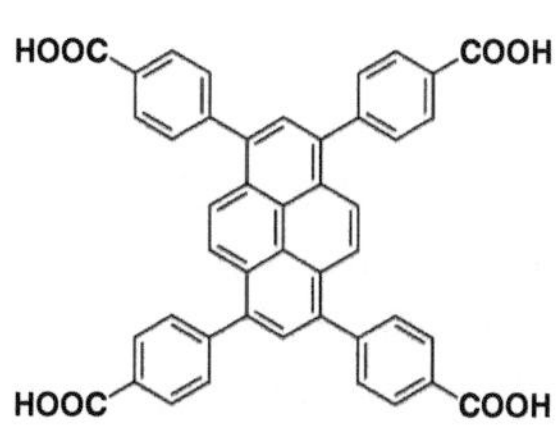

1,3,6,8-Tetrakis(4-benzoicacid)pyrene
TBAPy

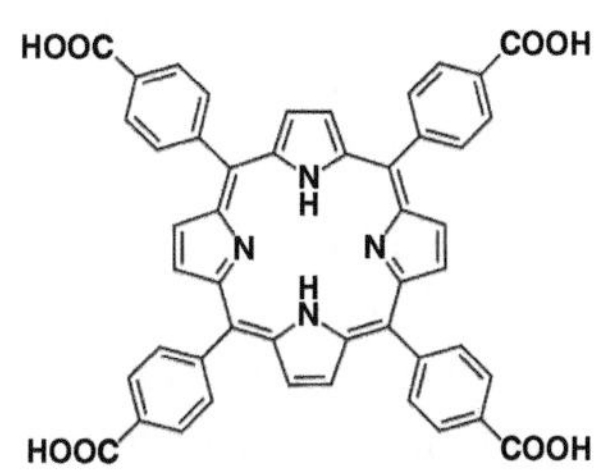

5,10,15,20-Tetrakis(4-carboxyphenyl)porphyrin
TCPP

Figure 9.9 Classic organic linkers with opto-electronic properties for the construction of MOFs.

porphyrin-based organic linkers, e.g. 5,10,15,20-tetrakis(4-carboxyphenyl)porphyrin (TCPP), have been developed for the assembly of various opto-electronic MOFs. Some typical chromophoric organic linkers are shown in Figure 9.9.

9.2.2 Structure of Opto-electronic MOFs

So far, significant research efforts have focused on the development of opto-electronic MOFs. Among the progresses that have been achieved, Zr^{IV} (or Hf^{IV}) based MOFs are remarkable due to their excellent stability. The tetravalent metal atoms, in many cases, form $M_6(\mu_3\text{-O})_4(\mu_3\text{-OH})_4$ (M = Zr, Hf) clusters (abbreviated as M_6), which act as secondary building blocks (SBUs) for MOF construction.[46] In each M_6 cluster, six metal atoms are arranged in an octahedron, and each plane of the octahedron is capped with a μ_3-O/μ_3-OH group (Figure 9.10a). The connection number of the M_6 cluster may vary from 12, 8, 6, to 4. Generally, when all twelve connection sites are combined with carboxylate groups, the M_6 cluster presents O_h symmetry (Figure 9.10b), whereas D_{6h} symmetry was also reported (Figure 9.10c). 8-Connected fashion

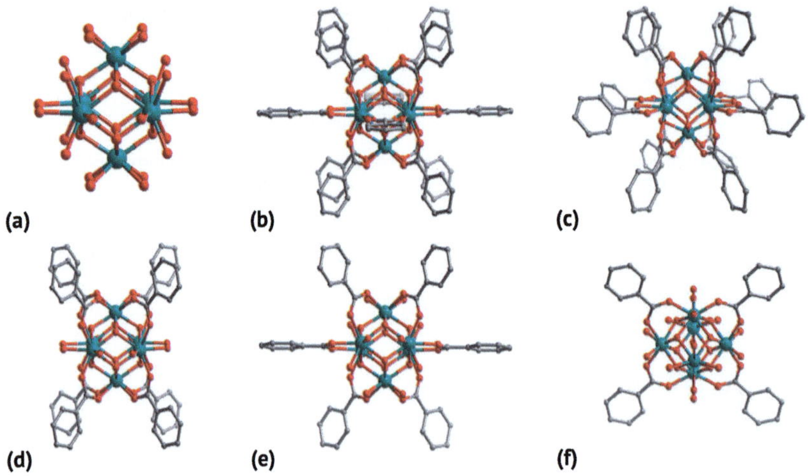

Figure 9.10 Connection modes of (a) $M_6(\mu_3\text{-O})_4(\mu_3\text{-OH})_4$ cluster: (b) 12-connected cluster with O_h symmetry; (c) 12-connected cluster with D_{6h} symmetry; (d) 8-connected cluster with D_{4h} symmetry; (e) 6-connected cluster with D_{3d} symmetry; (f) 4-connected cluster with D_{4h} symmetry.

is also common for the M_6 cluster. In a typical mode, eight edges of the octahedron are connected to carboxylate groups, and other connection sites, which are near the plane of the equator, are terminated with weakly-coordinated moieties, resulting in the cluster symmetry reduced to D_{4h} (Figure 9.10d). As for 6-connected M_6 clusters, a common mode is that carboxylate groups occupy six connection sites on a planar hexagon, thereby leading to D_{3d} symmetry (Figure 9.10e). In addition, 4-connected fashion was reported as well. For example, on the contrary of the above-mentioned D_{4h}-symmetric 8-connected M_6 cluster, four connection sites near the equatorial plane are coordinated to carboxylate linkers, while the other eight sites are occupied with weakly-coordinated moieties (Figure 9.10f). Hereinafter, several typical Zr- (or Hf-) based opto-electronic MOFs are introduced, and their aggregation modes (i.e. topologic structures) are illustrated in detail.

9.2.2.1 TBAPy-based MOFs

Two typical TBAPy-based MOFs, NU-901, and NU-1000, have been developed. Both of them are constructed with Zr_6 clusters and TBAPy linkers, but their aggregation fashions are different (Figure 9.11).

NU-901 crystallizes on the **scu** network, where 8-connected Zr_6 clusters (Figure 9.10d) and tetratopic TBAPy linkers having respective cubic and planar square configurations are arranged in *P4/mmm* space group, and other coordination sites of the Zr_6 clusters are occupied with labile molecules. The Zr_6 and TBAPy structural units are aligned respectively in the *c*-axis direction to form diamond channels.[47] Similar to NU-901, the Zr_6 nodes and TBAPy linkers in NU-1000 are also 8- and 4- connected, respectively. In contrast, however, NU-1000 adopts a **csq** aggregating mode, which belongs to *P6/mmm* space group. This network can also be considered

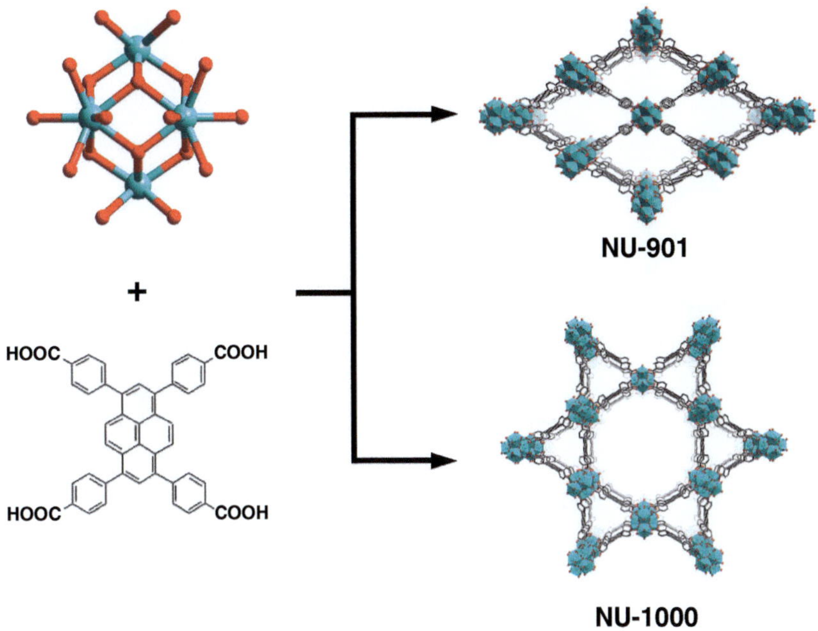

Figure 9.11 Structures of NU-901 and NU-1000.

as 2D Kagome (a plane piled with extended corner-sharing triangles) layers linked by TBAPy units.[48,49]

9.2.2.2 TCBPE-based MOFs

Three MOFs fabricated with the M_6 cluster and TCBPE linker in different aggregation modes, i.e. PCN-94 (PCN = porous coordinative network), PCN-128, and Zr-TCBPE-MOL (MOL = metal-organic layer), have been reported (Figure 9.12).

In PCN-94, the 12-connected M_6 clusters (Figure 9.10b) and TCBPE units aggregate in **ftw** fashion (in *Pm3m* space group) with cubic cages, of which the eight vertices and six faces are occupied with M_6 clusters and TCBPE units, respectively. That is, each TCBPE unit is connected with four M_6 clusters, and each M_6 cluster is linked to other 18 M_6 clusters via twelve TCBPE units.[50,51] Similar to NU-1000, PCN-128 crystallizes in **csq-a** network and belongs to *P6/mmm* space group, in which, 8-connected M_6 clusters (Figure 9.10d) having a rectangular spatial configuration are linked to each other through tetratopic TCBPE units. It is noteworthy that PCN-128 is of partial flexibility. That is, under compression/heating, the shape of TCBPE can alter reversely in a scissor manner.[51,52] In contrast to the 3D MOFs above, Zr-TCBPE-MOL is of 2D nanosheet morphology. In this structure, the TCBPE linkers and Zr_6 clusters are both 4-connected (Figure 9.10f for 4-connected M_6 cluster), and fabricate a planar network with **sql** topology. The thickness of Zr-TCBPE-MOL is 2.43 nm, which corresponds to a double-layered structure.[53]

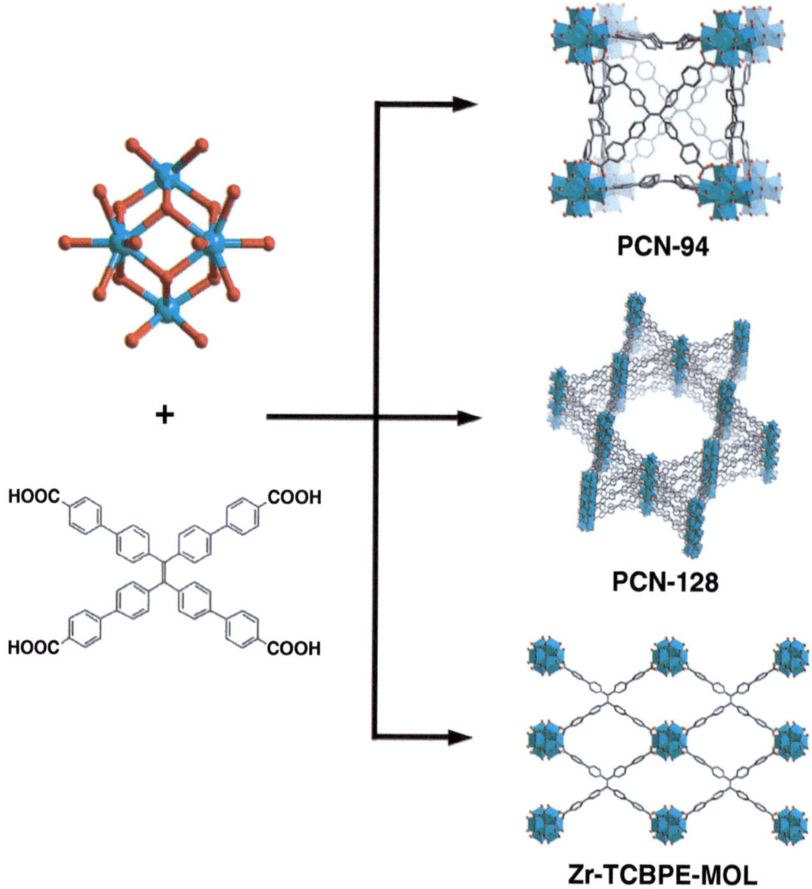

Figure 9.12 Structures of PCN-94, PCN-128, and Zr-TCBPE-MOL.

9.2.2.3 TCPP-based MOFs

Researchers have developed a number of MOFs with Zr and TCPP, such as NU-902, MOF-525, and PCN-222/223/224/225. Despite the same metal element and organic linker, their aggregation modes are quite different.

As shown in Figure 9.13, NU-902 adopts a (4,8)-connected **scu** aggregating mode, which is similar to the structure of the abovementioned NU-901, but the TBAPy ligand is replaced with TCPP.[54] PCN-222, also known as MOF-545, features a **csq** topology similar to NU-1000 and PCN-128. That is, among the 12 edges of an octahedral Zr_6 cluster, eight of them are bridged by planar square TCPP linkers (8-connected Zr_6 cluster, see Figure 9.10d), and other four edges are coordinated with terminal hydroxy groups, thus endowing the Zr_6 cluster with D_{4h} symmetry (Figure 9.13).[55,56] PCN-223 crystallizes in **shp-a** aggregating fashion, which is an edge-transitive (4,12)-connected network in $P6/m$ space group. Each Zr_6 cluster in PCN-223 is 12-connected. However, in contrast to the conventional 12-connected octahedral Zr_6 clusters, where all carboxylate

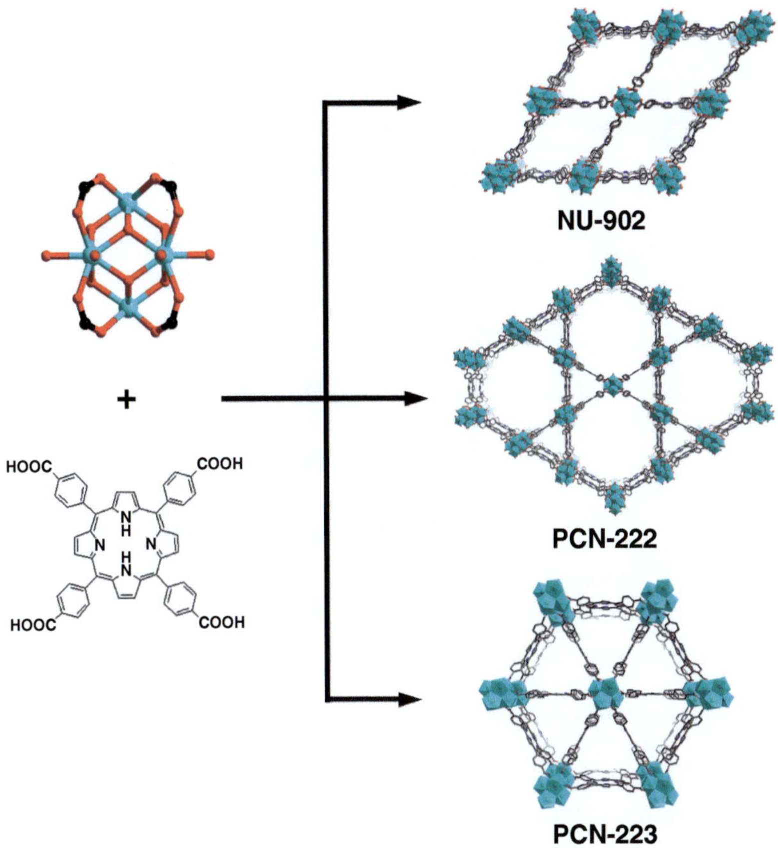

Figure 9.13 Structures of NU-902, PCN-222, and PCN-223.

groups bridge two metal atoms. In PCN-223, among the 12 COO^- connected to each Zr_6 node, eight bridge adjacent Zr atoms, while the remaining four chelate individual Zr atoms, thereby forming a hexagonal prism configuration with D_{6h} symmetry. And Figure 9.10c shows the structure of the unusual 12-connected Zr_6 cluster in PCN-223.[57]

In addition to NU-902, PCN-222, and PCN-223, the connection of Zr and TCPP can also form other MOFs, such as MOF-525 and PCN-224, indicating the powerful potential for controlling the aggregation style of molecules in MOFs. MOF-525 possesses a cubic **ftw** network in $Pm3\overline{m}$ space group. As shown in Figure 9.14, six 4-connected TCPP linkers are located at the faces of a cube, and eight 12-connected Zr_6 clusters (see Figure 9.10b for detailed cluster structure) occupy the vertices.[54,56] In PCN-224, for each Zr_6 cluster, only six of the twelve connection sites (see Figure 9.10e for detailed cluster structure) are combined with TCPP linkers, endowing the cluster with D_{3d} symmetry. When viewed from the direction of S_6 axis, the two triangles at axial vertices of each octahedral Zr_6 cluster are occupied with terminal water or hydroxy moieties,

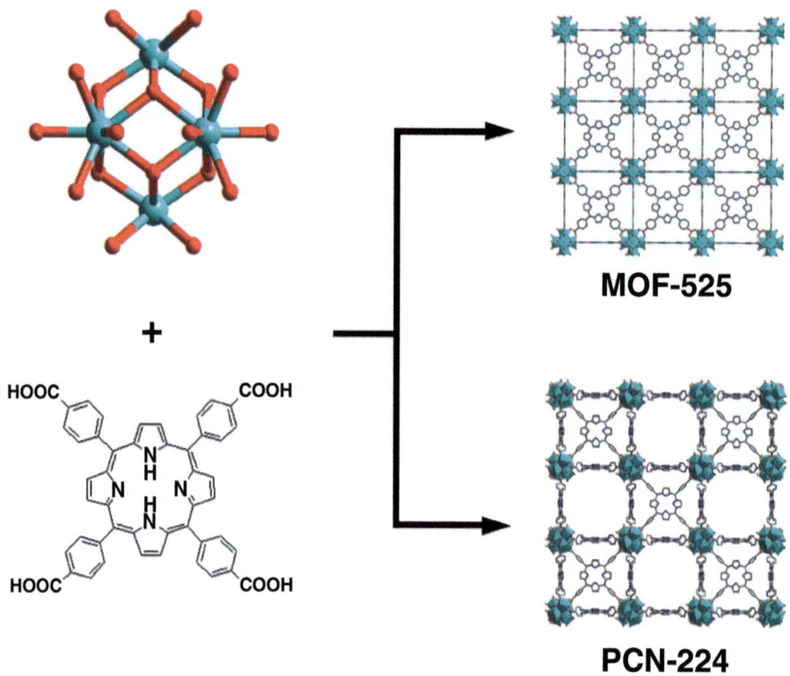

Figure 9.14 Structures of MOF-525 and PCN-224.

and the other six sites at the equatorial plane are connected to the planar square TCPP, thereby forming a 3D (4,6)-connected **she** network in $Im3\overline{m}$ space group (Figure 9.14).[58]

In some unusual cases, Zr_6 clusters in Zr-MOFs adopt different coordination models. For example, in the MOF PCN-225, there are two special types of 8-coordinated metal atoms in the Zr_6 clusters, in which one is coordinated with four μ_3-oxygen and four carboxylate O, and the other one is connected to three μ_3-oxygen, three terminal hydroxy oxygen, and two carboxylate oxygen atoms. The μ_3-oxygen atoms cap on the triangular faces of the octahedron assembled with six metal atoms, forming a highly distorted polyhedron. Hence, the Zr_6 nodes of PCN-225 have a low symmetry (D_{2d}) than common Zr_6 clusters (D_{4h}, in PCN-222). Each of the special Zr_6 cluster can be seen as a cube with eight vertices connected to TCPP linkers, and the topology of PCN-225 can be described as a (4,8)-connected **sqc** network that belongs to *I4(1)/amd* space group.[59]

9.2.2.4 Other MOFs

TCPB is similar to TBAPy, no matter if the pyrene group is replaced with a benzene ring, and it is possible to build TCPB-based MOFs with structures similar to the TBAPy-based MOFs discussed above. For example, researchers have developed a MOF isoreticular to NU-901, which is known as NU-903 (or CAU-24).[60] The Zr_6 clusters and TCPB linkers fabricate a **scu** network with

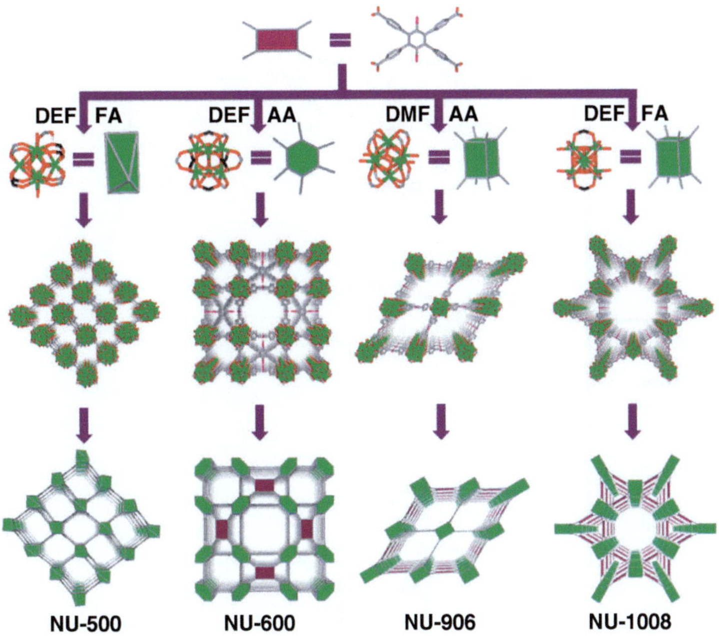

Figure 9.15 Structures of NU-500, NU-600, NU-906, and NU-1008. *Source*: Modified from [61].

diamond channels extending in the *c*-axis direction.[61] Moreover, a series of MOFs consisting of 1,4-dibromo-2,3,5,6-tetrakis(4-carboxyphenyl)benzene (TCPB-Br_2), NU-500/600/906/1008, have been developed. NU-500 crystallizes in a rare (4,4,4,5)-connected monoclinic network with *C2/m* space group, within this structure, each Zr_6 cluster is coordinated to five TCPB-Br_2 linkers.[62] NU-600 crystallizes in the space group of $Pm3\overline{m}$, where hexagonal Zr_6 nodes and square TCPB-Br_2 linkers assemble a (4,6)-connected **she** network.[62] NU-906 adopts a (4,8)-connected **scu** aggregating mode that belongs to *P4/mmm* space group; each Zr_6 cluster features a nearly cubic configuration, with eight vertices coordinated to TCPB-Br_2 units.[63] NU-1008 crystallizes in **csq** topology with *P6/mmm* space group, which is isoreticular to NU-1000 (Figure 9.15).[61]

Similarly, Zr_6 clusters can be connected by TCPBP (or its derivatives) linkers to assemble MOFs with various structures, e.g. PCN-605/606/608. PCN-605 crystallizes in (4,8)-connected **flu** topology, in which, the tetratopic TCPBP-based linker has an approximate T_d symmetry due to the rotation of the C—C bond between the two phenyl rings of the central biphenyl group. In contrast, the central biphenyl group in PCN-606 is coplanar, endowing the tetratopic linker with a D_{2h} symmetry, and an **scu** network is assembled once it is linked with the 8-connected Zr_6 cluster. Besides, in PCN-608, the TCPBP-based linker is presented in a C_{2v} symmetry, and thus leading to a (4,8)-connected **csq** framework that is isoreticular to NU-1000 (Figure 9.16).[64]

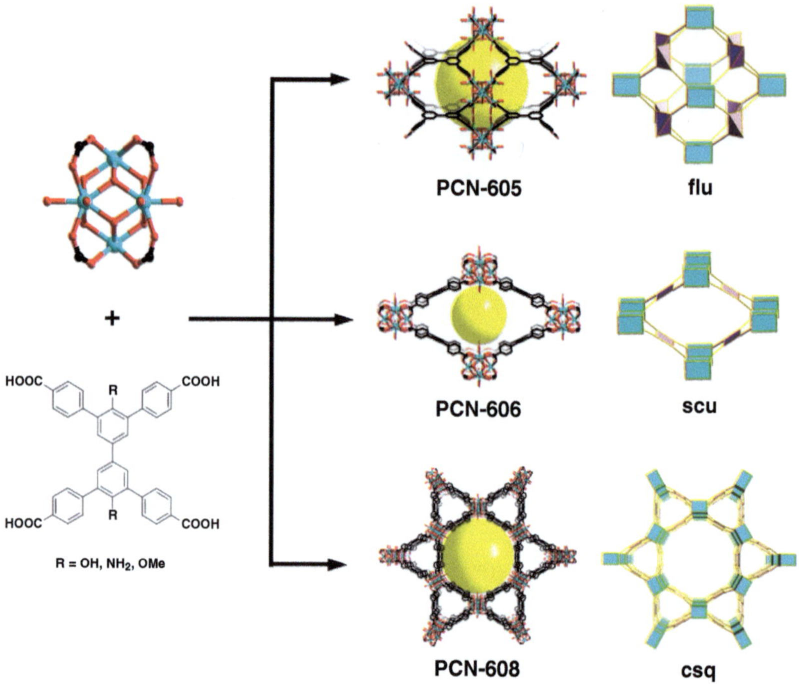

Figure 9.16 Structures of PCN-605, PCN-606 and PCN-608. *Source*: [64]/with permission of American Chemical Society.

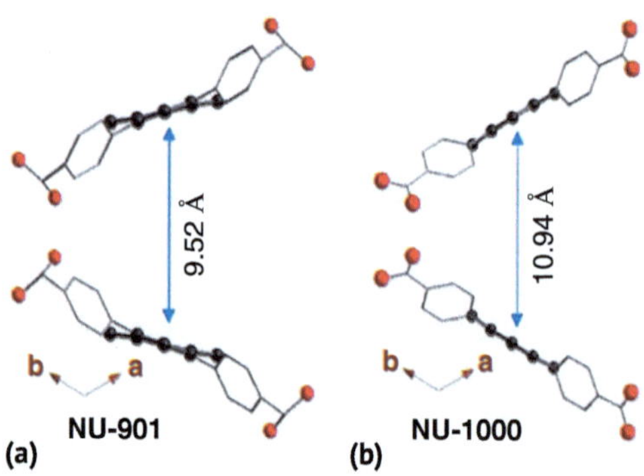

Figure 9.17 Arrangement of adjacent TBAPy in (a) NU-901 and (b) NU-1000. *Source*: [42]/with permission of American Chemical Society.

9.2.3 Properties of Opto-electronic MOFs

The aggregating fashions of metal nodes and organic linkers have significantly impacted the photoluminescent behaviors of MOFs. For example, among the TBAPy-based MOFs of NU-901 and NU-1000, the center-to-center distance between two adjacent TBAPy units in NU-901 (9.52 Å) is smaller than that in NU-1000 (10.94 Å), and the intermolecular pyrene-pyrene torsional angle in NU-901 (20°) is smaller than that in NU-1000 (60°) (Figure 9.17). Appropriate distance and orientation between the pyrene chromophores may lead to an intense inter-chromophore interaction, which reduces the energy difference between ground state and first excited singlet state. In this case, compared with NU-1000, such inter-chromophore interaction is stronger in NU-901. Therefore, in fluorescence emission spectra, NU-1000 exhibits a slight red-shift maximum emission wavelength of 8 nm compared with free TBAPy, while NU-901 shows a much more significant red-shift of 71 nm.[42]

The correlation between aggregating fashions and fluorescence properties in porphyrin MOFs has been studied as well. Taking MOF-525 and NU-902 as examples, owing to the different topological networks (MOF-525, **ftw**; NU-902, **scu**), their porphyrin–porphyrin torsional angles are 90° and 60°, and center-to-center distances between porphyrin units are 13.5 and 10.5 Å, respectively (Figure 9.18a, b). The smaller inter-porphyrin angle and distance in NU-902 lead to a stronger inter-chromophoric coupling, thus reducing the energy gap between the lowest excited singlet state and the ground state. As a result, NU-902 exhibits a longer maximum emission wavelength (~728 nm) and shorter fluorescence lifetime (4.6 ns) in comparison with MOF-525 (~718 nm; 5.6 ns) (Figure 9.18c, d).[65]

Moreover, the aggregating modes of structural units may determine the rigidity of MOFs, which further impacts the opto-electronic performance. For example, in the rigid (4,12)-connected **ftw** network of PCN-94, the TCBPE ligand is highly twisted and constrained. Hence, the torsional relaxation process, which may result in fluorescence quenching, is suppressed. As a result, an enhanced quantum yield (99.9%) is achieved in PCN-94 compared with free TCBPE (~30%).[50] In another TCBPE-based

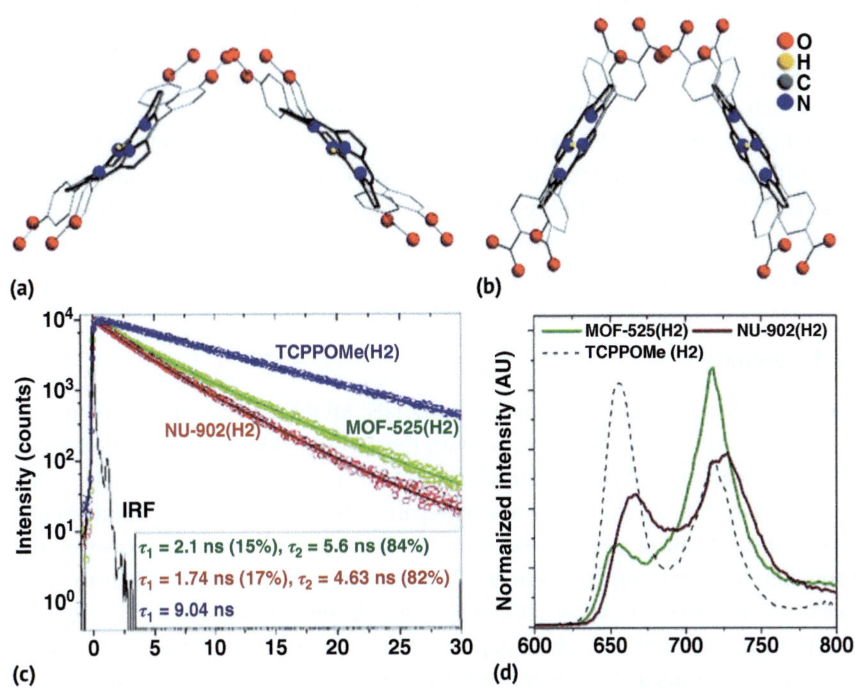

Figure 9.18 Schematic views of two positioned TCPP dimers in (a) MOF-525 and (b) NU-902; (c) transient emission decay profiles and (d) steady state fluorescence emission curves of MOF-525 and NU-902. *Source*: [65]/with permission of Royal Society of Chemistry.

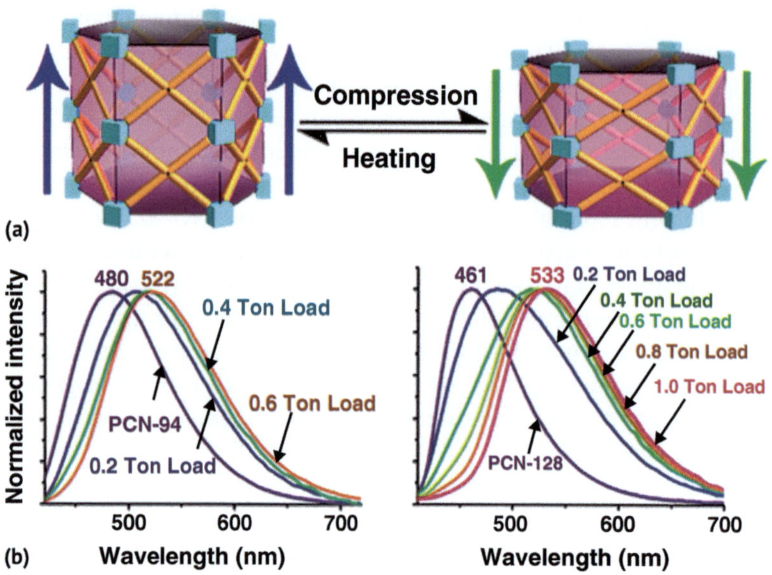

Figure 9.19 (a) Schematic illustration of the structural flexibility of PCN-128; *Source*: [52]/with permission of American Chemical Society. (b) pressure-induced fluorescence (λ_{ex} = 398 nm) of PCN-94 and PCN-128. *Source*: [66]/with permission of John Wiley & Sons.

MOF, PCN-128, the partially flexible **csq-a** network leads to significant piezofluorochromic behavior. The conformation of TCBPE is allowed to change like a pair of scissors. That is, the two sides of the "scissors" may "close" under external compaction, which leads to the size and shape alternation of the TCBPE segment from 18.9 × 11.7 to 19.7 × 10.0 Å (Figure 9.19a).[52] Since the tetraphenylethylene group in TCBPE is of AIE character, the compressed PCN-128 exhibits an obvious red-shift ($\Delta\lambda$ = 72 nm) in the fluorescence emission peak. In contrast, PCN-94 only shows minor piezofluorochromic characters ($\Delta\lambda$ = 42 nm) due to the relatively rigid structure that limits the deformation of TCBPE units (Figure 9.19b).[66]

9.2.4 Summary

As emerging organic–inorganic hybrid materials, MOFs provide an appealing platform for opto-electronic applications. With the combination of polydentate ligands with opto-electronic properties, it is feasible to fabricate MOFs with opto-electronic functions. Benefiting from the diverse aggregation modes of metal nodes and organic linkers, the structural parameters, such as the distance and angle between the chromophores, can be tuned effectively. The structural differences are likely to influence the opto-electronic properties by impacting the energy of excimers. Besides, the rigidity/flexibility of MOFs, which depends on the topological structures, is relevant to the opto-electronic behaviors as well. Therefore, it is feasible to build MOFs with desired opto-electronic performance via precise design of structure. Here are some key issues on the fabrication of MOFs with certain topologic structures.

1. Precise selection of metal nodes and organic linkers. Since metal ions and clusters with different radii, charges, and electron configurations may have different coordination geometries, and organic linkers may be various in shape, length, rigidity, and conjugation degree, the combination of metal nodes and organic linkers may have a great influence on the connective modes and symmetry of MOFs.
2. Templated methods. The crystal lattices of a template may match a certain topologic structure of MOFs, thereby guiding the metal nodes and organic linkers to arrange in corresponding symmetry via metal-template or linker-template interactions.
3. Topologic structure transformation. The stability of MOFs in certain topologic structures may differ in various physical and chemical conditions. With the treatment of solvent, heating, and mechanical forces, MOFs in sub-stable topologic structures can be transformed into more stable ones, extending the routes to synthesis MOFs in desired topologic structures.
4. Theoretical computation. With the help of calculation tools and crystal databases, researchers can design and predict the topological structures of MOFs more efficiently. Particularly, the utilization of artificial intelligence (AI) is appealing in the elevation of precision and accuracy.

9.3 Covalent Organic Frameworks

9.3.1 Introduction

Similar to MOFs, covalent organic frameworks (COFs) are another kind of porous material with powerful potential for regulating the aggregation style of organic molecular. As is known, COFs are crystalline porous materials formed by linking organic building blocks through strong covalent bonds, such as B—O, C=N, C—N, C=C, etc. Since the synthesis of the first COF materials (COF-1 and COF-5) in 2005,[67] this type of functional material has aroused extensive research interest. Connected by lightweight elements through strong covalent bonds and possessing periodic and ordered porous, COFs are characterized by high specific surface area, low density, exceptional thermal stability, and long-range ordered structure. Therefore, COFs with tailorable structures and functions have demonstrated considerable potential for diverse applications, particularly in gas adsorption and storage,[68–71] molecular separation,[72–75] and heterogeneous catalysis.[76–79] Importantly, the diversity of the functional structures of COFs materials lies in the covalent bonds in varied forms (also known as linkages, such as B—O, C=N, and C=C) and the widely selective combination of organic building units. In addition, the stacking mode and distance between layers in 2D COFs are other factors that affect the structural and functional diversity.

Herein, we summarize the construction and performance of COFs materials with different aggregation states. We will discuss the role of single molecular and their aggregation behaviors in COFs and introduce the internal aggregation structures that affect their optical and semiconductor properties.

9.3.2 Construction of COFs with Different Aggregation States

As one of the most promising functional materials, COFs offer many opportunities for designing opto-electronic materials by using organic building units with different optical and electronic functions. Up to now, researchers have developed a variety of linkages to construct COFs, and the linkages also control the intralayer electronic interactions and conjugations in the whole framework system. The structures of B—O, C=N, and C=C linkages, as well as the corresponding model reactions, are shown in Figure 9.20a. And some typical organic linkers with opto-electronic properties are shown in Figure 9.20b. The as-formed COFs known as boronate ester-linked COFs, imine-linked COFs, and vinylene-linked COFs have shown an irreplaceable important role in opto-electronic applications. These linkages and the conjugated degree have a profound effect on the framework, especially their π structure, electronic interaction, frontier orbital energy level, and bandgap, which have an impact on the properties and functions.

Generally speaking, boronate ester- and boroxine ring-linked COFs are built via the highly reversible condensation reactions of boric acid and o-diphenol to form boronate ester, or boric acid self-condensation to form boroxine rings. The imine-linked COFs are formed by condensation of amine and aldehyde. Compared to boronate ester-linked COFs, imine-linked COFs show better stability in water, organic solvents, and even acid solutions. Therefore, imine-linked COFs are the most important subclass of COFs in the whole COF family. However, the rotation of the imine bond is the direct cause of the reduction of conjugation and the most important reason for limiting the photoelectric performance of the imine-linked COFs. Thereby, chemical reactions of cyclization can be a promising method to lock the imine bond.[80,81] In view of the fact that B—O bond COFs and imine COFs have certain limitations in opto-electronic applications due to their structure and electronic properties, it is urgent to develop a novel linkage with both conjugation and stability. Therefore, fully π-conjugated COFs with C=C linkage (COF-701 and V-COFs as typical representatives) were established by Knövenagel polycondensation reaction and aldol condensation. This linkage visually magnifies the opto-electronic effects of corresponding COFs.[80]

From the viewpoint of molecular aggregates, another factor that affects the opto-electronic properties of COFs, especially in 2D COFs, is the stacking modes. Generally, the stacking of 2D COFs can be categorized mainly into two modes, i.e. eclipsed and staggered ones, according to the stacking angle and symmetry between layers.[81] Taking COF-5 constructed from hexahydroxytriphenylene (HHTP) and diboronic acid (BDBA) as an example (Figure 9.21), in the eclipsed stacking mode (AA stacking), there is no offset between two adjacent layers, and the line of atoms in the two adjacent layers is perpendicular to the plane where the two layers are located. While in the staggered stacking mode (AB stacking), there is a complete interleaving between layers, and the three-connected intersection of the atoms in the upper layer is directly above the geometric center of the lower six-membered ring. The other two stacking modes are based on the AA stacking model with different shift styles between adjacent layers. The alternate layers of serrated stacking are eclipsed, while the layers shift along the same directions in the inclined stacking mode.

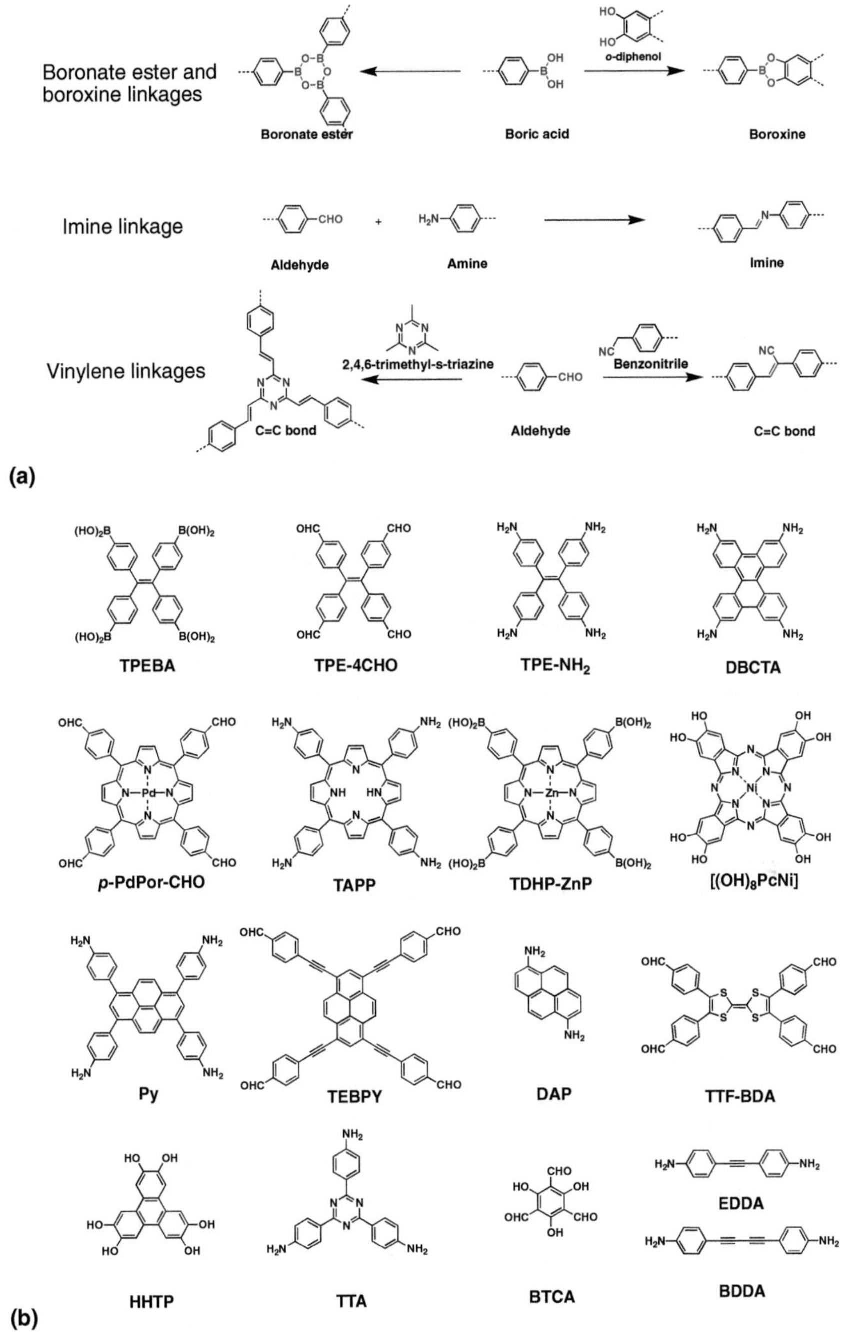

Figure 9.20 Schematic representation of model reactions of (a) B—O, C=N, and C=C linkages and (b) the structures of some organic building units with photoelectric functions.

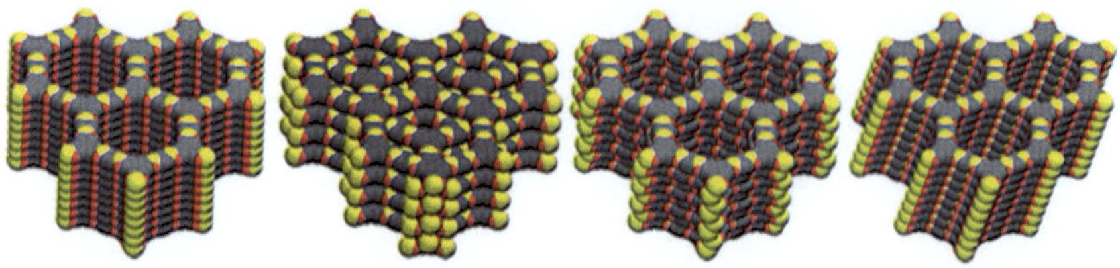

Figure 9.21 Schematic representation of the modes of eclipsed stacking, staggered stacking, serrated stacking, and inclined stacking in COF-5 (from left to right). *Source*: [81]/Beilstein-Institut/CC BY 4.0.

9.3.3 Properties of COFs with Different Aggregation States

9.3.3.1 Optical Properties of COFs

Due to the non-radiative transition generated via the rotation of the molecule or the heat loss produced via the π–π stacking, most of the optically active linkers lose their functions after the formation of COFs. Tang et al. reported the seminal discovery of aggregation-induced emission (AIE) and proposed the restriction of intramolecular bond rotations (RIR) mechanism for this phenomenon.[82] Inspired by this, organic linkers with AIE characteristics, such as tetraphenylethylene (TPE), have been widely a research hotspot.

In 2016, Jiang et al. reported an extremely emissive COF constructed by linking AIE-active TPE-cored boronic acids (TPEBA) and 1,2,4,5-tetrahydroxybenzene (THB).[83] The covalent bonds in the layer, the π–π interactions between the layers, and the regular columnar AIE π-arrays endow the TPE-Ph COF with excellent quantum yield, which could be used as an excellent probe for ammonia detection. Loh et al. integrated 1,3,6,8-tetrakis(4-aminophenyl)-pyrene (Py) with non-planar 1,1,2,2-tetrakis(4-formylphenyl)ethane (TPE-4CHO) to weaken the π–π reactions to inhibit the fluorescence quenching process.[84] The imine-linked Py-TPE-COF showed high photoluminescence quantum yields (PLQY) after being dispersed in acetonitrile (21.1%), acetone (18.0%), DMF (13.7%), and THF (15.7%).

However, most of the imine-linked COFs containing TPE units always exhibit low emission or no emission due to the rotatability of the TPE and imine linkage. To overcome this issue, construction of TPE-based COFs using organic building units with functional groups to restrict the rotation of TPE has been successfully utilized to enhance the optical properties of TPE-COFs. As typical examples, OH-functionalized TPE-COF, TPE-COF-OH, was complexed with boron through post-synthetic modification method to lock the imine bond, and the corresponding TPE-COF-BF_2 exhibited the turn-on fluorescence and aggregation-induced emission (Figure 9.22a).[85] The rotatable nature of the vinylene group in TPE molecular leads to its propeller structure. So as to eliminate the limitation of space accumulation generated via the propeller structure of TPE, researchers have proposed a strategy to lock the rotation by cyclization reaction at the vinylene group.[86] For example, Bein and co-workers designed a new rigid tetradentate linker dibenzochrysenetetraamine (DBCTA) by cyclization reaction of the TPE molecular (Figure 9.22b). As expected, DBC-COFs successfully synthesized from the DBCTA with better planarity exhibit longer photoluminescence (PL) lifetimes than TPE-COFs, since the tight

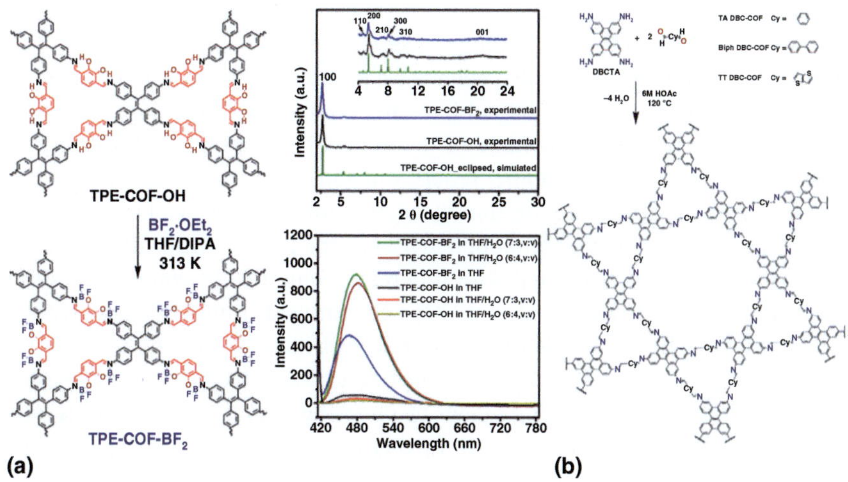

Figure 9.22 Schematic representation of (a) synthetic routes, structures, powder X-ray diffraction patterns, and fluorescence emission spectra of TPE-COF-BF$_2$. *Source*: [85]/with permission of American Chemical Society. (b) Synthetic routes and structures of DBC-COFs. *Source*: Reproduced with permission from [87], The Royal Society of Chemistry.

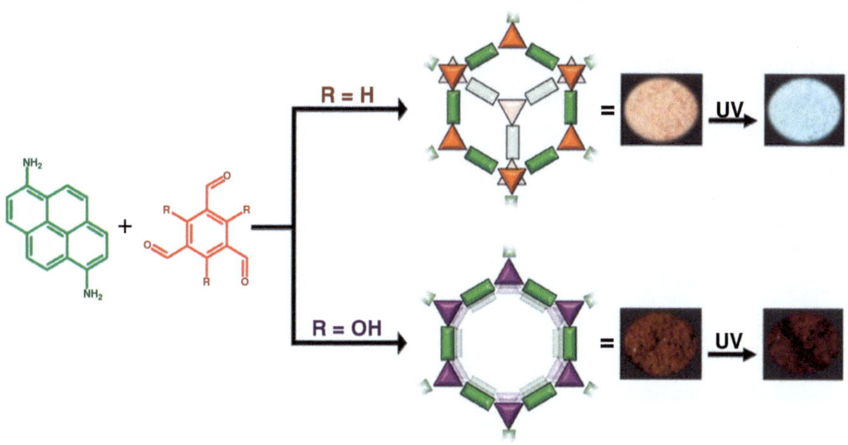

Figure 9.23 Schematic representation of synthetic routes, stacking modes, and luminescent behaviors of IMDEA-COF-1 (R=H) and IMDEA-COF-2 (R=OH). *Source*: [88]/with permission of American Chemical Society.

stacking of molecular in DBC-COF enhanced the delocalization of the photoexcited states across the π-stacked system.[87]

The stacking modes of optical molecular in COFs also play a crucial role in determining the properties of COFs. Zamora group utilized the tritopical 1,3,5-benzenetricarbaldehyde (BTCA) and 1,6-diaminopyrene (DAP) to synthesize IMDEA-COF-1 with staggered stacking mode (Figure 9.23).[88] Interestingly, the stacking mode was changed to eclipsed stacking, when BTCA was replaced by 2,4,6-triformylphloroglucinol(TP) in IMDEA-COF-2, possibly due to the coexistence of enol-imine tautomeric and keto-amine in IMDEA-COF-2. Importantly, the obtained COFs with different stacking modes showed different optical properties. The green-emitting IMDEA-COF-1 achieved a photoluminescence quantum yield of 3.5%, while IMDEA-COF-2 did not have any fluorescence emission. This study confirmed that the stacking method of materials had an irreplaceable influence on the optical properties. In another report, Cui and co-workers deeply researched the effect of the spatial configuration of organic units on the stacking modes of COFs.[89] Three types of tris(4-aminophenyl)benzene (TAPB) with different alkyl chain substitutions were reacted with different geometrically symmetrical units to construct COFs. It was found that COFs constructed from TAPB adopted AA stacking mode, whereas ethyl- and isopropyl-substituted TAPB tend to adopt AB or ABC stacking modes. This study establishes a reasonable tactic for constructing COFs with different stacking modes by choosing organic building units with different spatial configurations, which further affects the electronic, optical, and catalytic properties of materials.

The interlayer distance of 2D COFs is another influencing factor for their optical properties. In the study by Wang and collaborators, after removing the dioxane (Dio) solvent molecules between the COF layers, the interlayer distance of the produced eclipsed COF reduced from 3.7 Å in Dio ⊂ BZL-COF to 3.4 Å in BZL-COF (Figure 9.24a).[90] At the same time, the phosphorescent emission behavior of the COFs changed significantly. For Dio ⊂ BZL-COF, it exhibited strong yellow phosphorescence at 545 nm with a lifetime of 1.27 ms at 77 K (Figure 9.24b). On the contrary, the activated BZL-COF was less luminescent, and the resolved emission bands were at 540 nm with a lifetime of 1.02 ms at 77 K. The possible reasons for the difference in phosphorescence originated from the lone electron pairs of the C=O in benzil segments in Dio ⊂ BZL-COF, which could elevate the

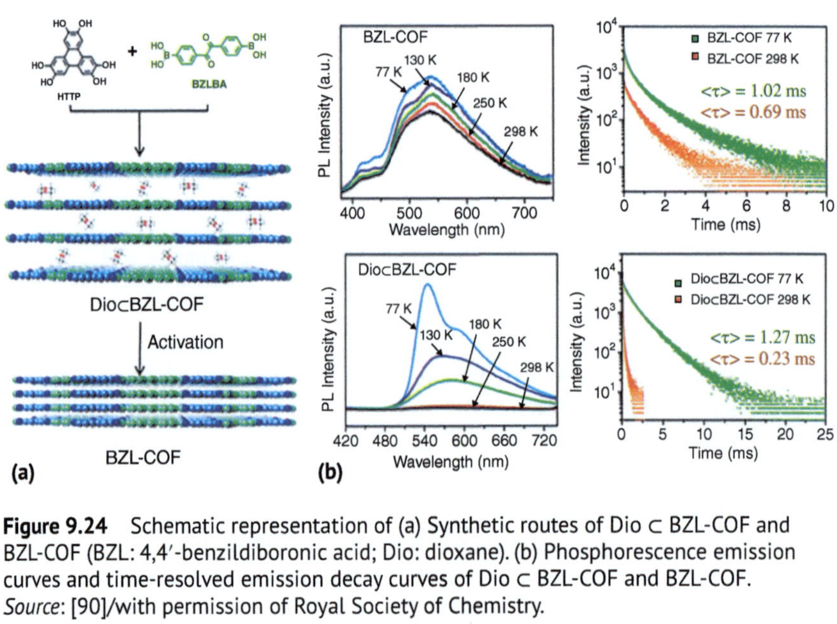

Figure 9.24 Schematic representation of (a) Synthetic routes of Dio ⊂ BZL-COF and BZL-COF (BZL: 4,4′-benzildiboronic acid; Dio: dioxane). (b) Phosphorescence emission curves and time-resolved emission decay curves of Dio ⊂ BZL-COF and BZL-COF. *Source*: [90]/with permission of Royal Society of Chemistry.

intersystem crossing (ISC) transition. This research strongly proved that the distance of interlayers had certain effects on the photoelectric performance of the COFs, and it was not that the shorter the interlayer distance, the better the photoelectric performance.

9.3.3.2 Semiconducting Properties of COFs

Organic linkers composed of aromatic organic monomers with unique structural characteristics have become the perfect choice for designing and synthesizing semiconductor COFs. In 2008, Jiang's group reported the first semiconducting COF using HHTP and pyrene-2,7-diboronic acid (PDBA).[91] The author indicated that the properties of this p-type semiconductor are mainly due to the stacking of large π-conjugated units. This research opened the prelude to the exploration, design, and synthesis of semiconductor COFs materials constructed by organic units of pyrene and triphenylene. Also, porphyrin and phthalocyanine organic building units have become the key building units for designing and synthesizing semiconductor COF materials, owing to the ability to coordinate with metal cations (e.g. Ni^{2+}, Cu^{2+}, Zn^{2+}, and Co^{2+}) to form metalloporphyrins and metallophthalocyanines to adjust the electron cloud density on the macrocyclic rings. For example, the research groups of Yaghi and Jiang respectively reported plenty of 2D COFs (NiPc COF, COF-366, COF-66, and ZnP-COF) with excellent semiconducting properties based on (metallo)porphyrin and metallophthalocyanine.[92–94] In another study of metalloporphyrin-based COFs, the authors used induced self-complementary π-electron interactions to control the distance of the interlayer.[95] They integrated the fluorine-substituted linear unit (2,3,5,6-tetrafluoroterephthalaldehyde) with the unsubstituted unit (terephthalaldehyde) in different molar ratios to construct a variety of 2D porphyrin-COFs. With the control of the layer spacing, this strategy effectively enhanced the π–π interactions, improved the crystallinity, shortened the gap between the HOMO–LUMO, and accelerated the electron transport.

Tetrathiafulvalene (TTF) and salts of its oxidized derivatives are also popular raw materials for the research of synthetic semiconductors. Because of the high symmetry and planarity of TTF, they can effectively promote charge delocalization and π–π stacking in the COFs. In 2014, Jiang's group prepared two TTF-based COFs, i.e. TTF-Ph-COF and TTF-Py-COF constructed from 2,3,6,7-tetra(4-formylphenyl) tetrathiafulvalene and 1,4-phenyldiamine and 1,3,6,8-tetra(4-aminophenyl)pyrene, respectively (Figure 9.25a).[96] The monomers of TTF can also form organic salts through redox reaction in COFs, so that the mono-oxidized cation radicals could exert excellent conductivity. Liu and co-workers deeply explored the changes in conductivity of TTF-COF after dopants were added, and proposed that the crystalline TTF-COF film showed higher conductivity after doping with acceptors e.g. I_2 and tetracyanoquinodimethane (TCNQ). The I_2 doped TTF-COF showed a conductivity of 0.28 S m^{-1}, which was three orders of magnitude higher than TTF-COF and the highest value reported at that time (Figure 9.25b).[97]

Owing to their designable structures and functions, semiconductor COFs have emerged as a focal point of research in the field of photocatalysis. Until now, researchers have reported a variety of COFs materials for photocatalysis with greatly

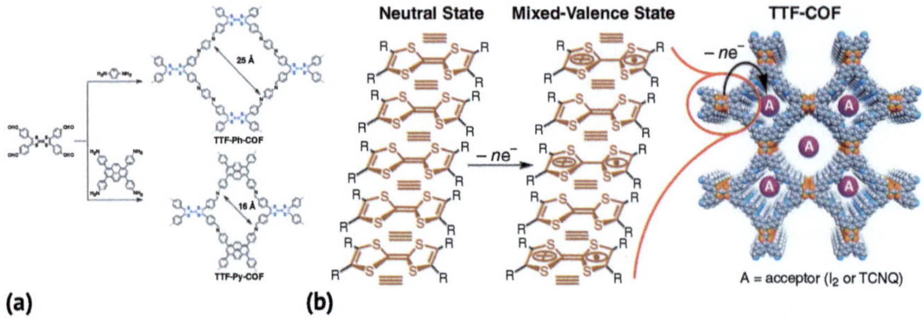

Figure 9.25 Schematic representation of (a) synthetic routes and structures of TTF-Ph-COF and TTF-Py-COF. *Source:* [96]/with permission of John Wiley & Sons. (b) Illustration of the transition from neutral state to mixed-valence state. *Source*: [97]/with permission of Royal Society of Chemistry.

improved selectivity of chemical reactions. For example, triazine-based COFs are widely used in photocatalytic water splitting and other photoelectric fields,[98,99] due to the planarity of the triazine units in COFs, which promotes the complete delocalization of π electrons and the strong π–π interaction with the peripheral aromatic rings. An example of triazine-based COF with a big surface area prepared by 1,3,5-tris-(4-formyl-phenyl)triazine (TFPT) and 2,5-diethoxy-terephthalohydrazide (DETH) was reported for photocatalytic water splitting by Lotsch's group.[100] The obtained stable hydrazone-linked TFPT-COF can catalyze the decomposition of H_2O to generate hydrogen under daylight when coupled with Pt as a proton reduction catalyst (PRC). In another work, a 2D triazine-based COF synthesized from 4,4′,4″-(1,3,5-triazine-2,4,6-triyl) trianiline (TTA) and 2,2-bipyridyl-5,5-dialdehyde (BPDA) exhibits an efficiency of 98% in catalyzing the reduction of CO_2 to form CO after combining with tricarbonylchloro(bipyridyl) Re complex (Re(bpy)$(CO)_3$Cl).[101]

As a "star molecular," porphyrin and metalloporphyrin-based COFs have also attracted lots of attention in photocatalysis. Their applications in the fields of accelerating the generation of singlet oxygen and the selective oxidation of sulfides have been widely developed. For example, Jiang's group reported a new kind of porphyrin-based 2D COF from Copper(II) 5,10,15,20-tetrakis(4-aminophenyl)porphyrin (TAP-CuP) and squaraine (SQ).[102] They confirmed CuP-SQ COF with a conjugated structure could boost the photon absorption and generate the triplet excited state that induces the activation of molecular oxygen. Wang et al. expanded the structural dimension of porphyrin-based COFs to 3D by introducing 3D building block.[103] Both of the as-synthesized 3D-Por-COF and 3D-CuPor-COF are photosensitive materials, which can be utilized to catalyze the production of singlet oxygen under light irradiation (Figure 9.26a). 3D-Por-COF showed superior photocatalytic activity under the same conditions as 3D-CuPor-COF, suggesting the metallization of porphyrins would decrease the photophysical and redox properties. This experimental result was consistent with the theory put forward in previous work that the photosensitivity of porphyrin decreases after integrating paramagnetic metal ions.[104] Importantly, compared with 2D porphyrin-based COFs, the 3D COFs with highly accessible porphyrin units can provide more active sites in the catalytic process. Moreover, they

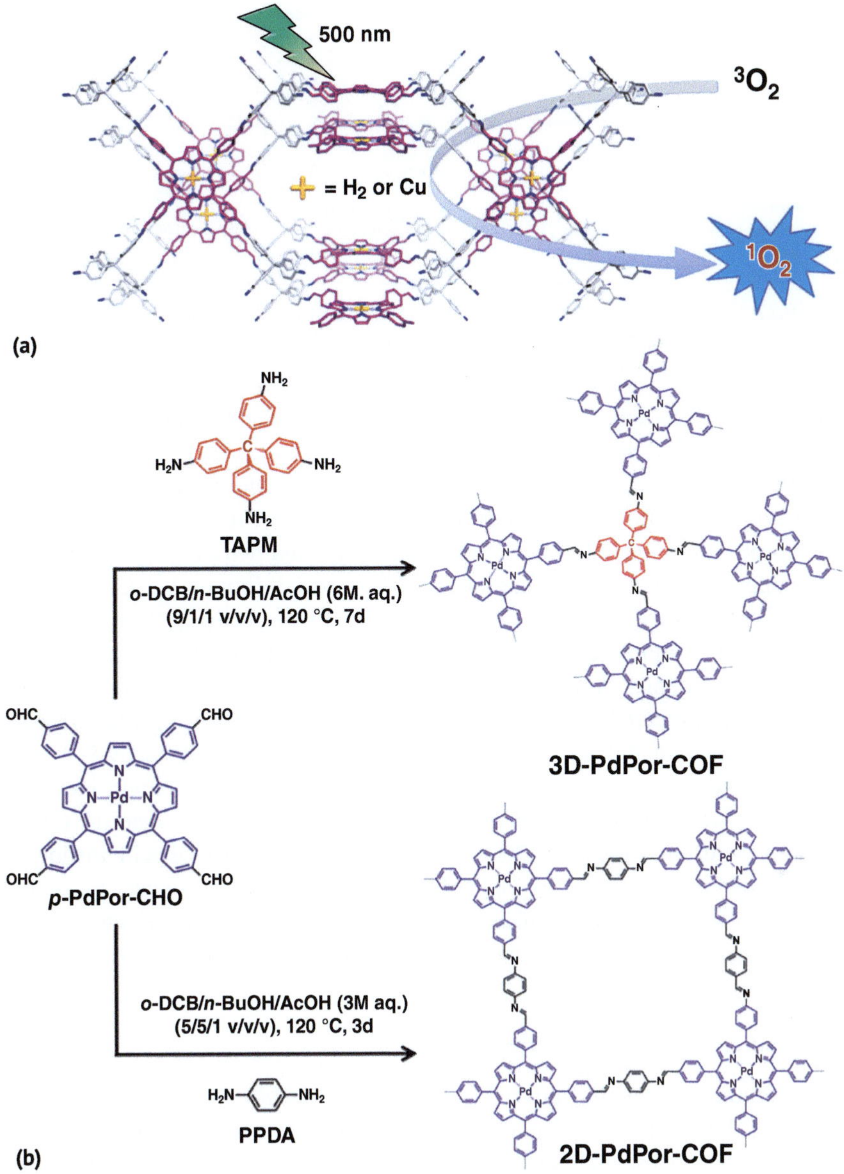

Figure 9.26 Schematic representation of (a) 3D-Por-COF and 3D-CuPor-COF catalyze the production of singlet oxygen under light irradiation. *Source*: [103]/with permission of American Chemical Society. (b) Synthetic routes and structures of 2D-PdPor-COF and 3D-PdPor-COF. *Source*: Reproduced with permission from [105], Wiley-VCH.

also chose [5,10,15,20-tetrakis(4-benzaldehyde)porphyrin]palladium (*p*-PdPor-CHO) as the organic unit to explore the difference between 2D-PdPor-COF and 3D-PdPor-COF (Figure 9.26b).[62] Through various experimental characterizations, they found that compared with 2D-PdPor-COF, 3D-PdPor-COF displayed better photocatalytic properties and size-selective photocatalysis. Therefore, the

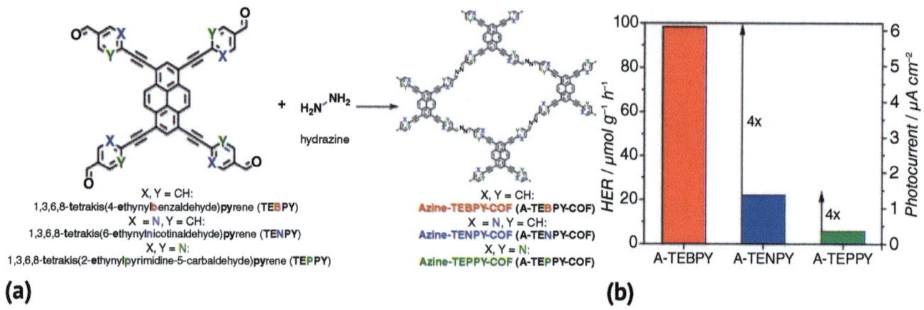

Figure 9.27 Schematic representation of (a) synthetic routes and structures of azine-linked COFs. (b) Comparison of the HER rates and photocurrents of azine-linked COFs. *Source:* [107]/with permission of John Wiley & Sons.

dimensionality of COFs was a significant factor that affected the functions and performance of COF materials.

In 2017, Thomas and co-workers found that COFs materials constructed from organic units containing acetylene group (—C ≡ C—) had outstanding performance in photocatalytic water splitting.[105] The authors reported two COFs synthesized by connecting 4′-(ethyne-1,2-diyl)dianiline (EDDA) and 4,4′-(buta-1,3-diyne-1,4-diyl)-dianiline (BDDA) with triformylphloroglucinol (TP), respectively. TP-BDDA COF displayed higher photocatalytic water splitting efficiency, due to its functional diacetylene diad unit, high porosity, and considerable chemical stability. One year later, Lotsch's group prepared a group of azine-linked COFs from an organic unit containing both pyrene and acetylene groups (Figure 9.27a).[106] Photocatalysis experiments showed that the efficiency of light-induced hydrogen evolution increases significantly with the decrease of nitrogen atom content in COFs. And the rate of hydrogen evolution reaction (HER) is the highest for A-TEBPY-COF with the lowest nitrogen content (Figure 9.27b). The enhanced HER activity with lower N-content is consistent with a more favorable proton reduction thermodynamics, which is further confirmed via photocurrent data and theoretical calculations.

9.3.4 Summary and Perspectives

In this section, we systematically analyze the main strategies for designing and synthesizing COFs with excellent photoelectric functions from the viewpoints of molecules and aggregates, and focus on the important roles of organic building units, linkage motifs, and stacking of π-systems in the construction of COFs. We provided reasonable guidance for the design of opto-electronic COFs displaying photoluminescent, semiconductive, and photocatalytic effects. We also summarized the construction of COFs using different linkages, the stacking modes of 2D COFs, as well as the important roles of stacking on the opto-electronic properties of COFs.

It has been proposed that the molecular structures of organic linkers have a great impact on the stacking modes of corresponding COFs, and further influence their opto-electronic properties. Only linkers with certain high symmetries are capable of forming well-defined stacking modes. The functional groups on linkers may also influence the stacking modes. For example, bulky groups like alkyl chains can affect the tight stacking, thus favoring staggered rather than eclipsed stacking, while substituents with hydrogen bond donors/acceptors may promote the generation of highly ordered stacking because of relatively tough interlayer hydrogen bond interactions. Moreover, conjugated degree is another key factor, as linkers in higher conjugated degree tend to form stronger interlayer π–π interactions, which can stabilize certain stacking modes. This is particularly essential in opto-electronic COFs since the interlayer π–π interactions are highly relevant to the opto-electronic properties like bandgap. Despite progresses have been made, more study is required to promote the development of molecular design of opto-electronic COFs.

Actually, opto-electronic COFs are one of the focuses and hotspots of current research in the field, and there are still some challenges and problems that need to be solved by researchers. For example, it is a long-standing problem to prepare COFs with good crystallinity and regular morphology, and the method of scalable synthesis of COFs remains a significant challenge. Furthermore, by using COFs as templates, the synthesis of novel COF-based composite materials is essential to further broaden the applications of opto-electronic COFs.

9.4 Organic Radical Materials

9.4.1 Introduction

Organic radical materials are also called organic open-shell materials, which involve organic small molecules and macrocycles, oligomers, dendrimers, or polymers with subvalent bonds consisting of unpaired or weakly bonded electrons. Accordingly, these species are highly reactive and easy to decompose via different pathways, such as dimerization, hydrogen addition, oxidation, or disproportionations, which is thermodynamically favorable for their low or negligible energy barriers.[107–129]

On the other hand, depending on different numbers of unpaired electrons, these organic radical materials usually exhibit special magnetic properties, which may play an important role in future flexible organic electronics,[118,123,129] organic spintronics,[107–124] organic quantum devices,[128] and organic photonics.[120–123] In this section, we will summarize different classes of stable or persistent stable organic radical materials and their recent applications on organic optical, electronic, and magnetic devices in the last decades.

9.4.2 Triarylmethyl-based Fluorescent Radical Materials

Up to now, almost all fluorescent radicals are designed based on triarylmethyl radical with an unpaired electron. Actually, when the first triphenylmethyl radical **1** (Figure 9.28) was discovered as early as 1900 by Gomberg,[130] it is a quite unstable radical due to its thermodynamic equilibrium between the monoradical and its dimer **2** even under an inert atmosphere. Furthermore, monoradical **1** decomposed quickly to peroxide compound **3** once it was exposed to air. Theoretically, monoradical **1** usually possesses a magnetic moment. Its spin quantum number (S) is 1/2, and magnetic components (m_s) are +1/2 and −1/2. Thus, the monoradical's spin

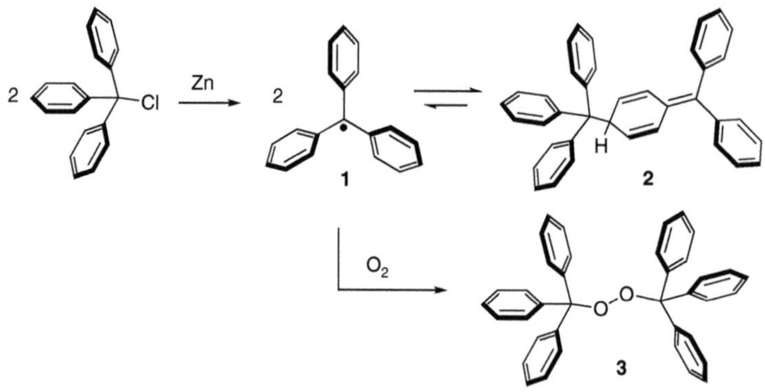

Figure 9.28 Triphenylmethyl radical and its decomposing pathway.

multiplicity is a doublet (since $2S+1 = 2$), which is why it is referred to as a doublet monoradical.[107,109]

Early work on triarylmethyl radicals mainly focuses on high-spin polyradicals despite the stability problem. Usually, the overall spin quantum number (S) of a pair of electrons is either 0 or 1, depending on their spin directions. Consequently, the spin multiplicity ($2S+1 = 2$) of diradicaloid is either singlet or triplet, accordingly.[129] As shown in Figure 9.29, the calculated $\Delta E_{\mathrm{S-T}}$ of *m*-benzoquinodimethane by ab initio calculations was about 10 kcal/mol, indicating its open-shell triplet ground state due to its non-Kekulé structure with strong ferromagnetic coupling via 1,3-linked benzene and the absence of formation of double bond between the two radical centers.[131–133] From 1990s A. Rajca synthesized a series of high-spin triphenylmethyl radicals with multiple radical centers.[134–138] **4** with two polyarylmethyl radicals (PAMRs) at *meso* positions also shows a triplet diradical ground state.[134] And its extended structure **5a** with four PAMRs shows eight symmetrical ESR signals in $\Delta m_s = 1$ region and a significant half-field $\Delta m_s = 2$ resonance peak, revealing its quintet ground state ($S = 2$).[134] This star-branched topology allows for the synthesis of heptaradical **5b** and decaradical **5c**. Further superconducting quantum interference device (SQUID) measurements on the high-spin π-conjugated polyradicals revealed the ground state of octet for **5b** ($S = 7/2$) and undecet for **5c** ($S = 5$), respectively.[135]

Later, calixarene based macrocylic octaradical **6a** with nonet ($S = 4$) ground state was synthesized.[136] Based on this calixarene structure, a ladder-type π-conjugated polymer magnet **7** with alternative macrocylic tetraradical ($S = 2$) and cross-linked diradical ($S = 1/2$) blocks were prepared. The structure caused large resultant S values arising from the combination of (anti)ferromagnetic interactions along the polymer backbones. The S value of the polymer could reach about 5000. Thus, the slow magnetic relaxation induced by a small magnetic field can be found at 10 K. Therefore, the polymer magnet could be regarded as an intermediate state between insulating spin glasses and blocked superparamagnets.[137] The further designed dendritic macrocyclic polyradicals, **6b** and **8** with 8 and 24 radical centers, were demonstrated to show high-spin states with S values of 10 and 3.6–3.8, respectively. Actually, compound **8** could be regarded as a molecular nanomagnet with small magnetic anisotropic barrier for the magnetization inversal.[138]

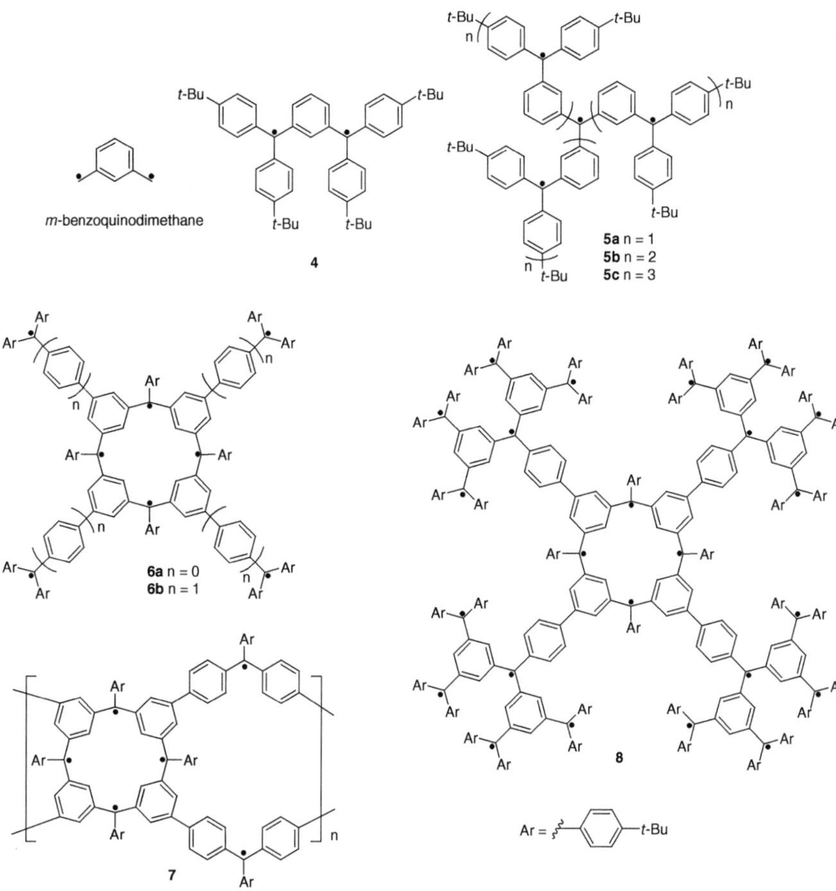

Figure 9.29 Chemical structures of high-spin triphenylmethyl radicals.

Meanwhile, these polyradicals are only transiently stable in air, which hampers their further possible applications.[107] Scientists tried to protect these triarylmethyl radicals by introducing steric groups to design and synthesize air-persistent open-shell molecules. As illustrated in Figure 9.30, in 1994, hexamethoxyl substituted monoradical **9** was prepared with good stability in the solid state. However, it is gradually oxidized in solution by oxygen in air to give a peroxide dimer.[139] Kubo et al. in 2018 disclosed a novel monoradical **12a**, protected by highly steric hindered tri(9-anthryl)methyl groups. It exhibits localized spin-electron densities at the central C atom and corresponding remarkable air-stability. In contrast, the absence of a mesityl group at the 10-position of one anthracene unit (**12b**) results in the formation of **13** as a tail-to-tail dimer, unlike the formation of head-to-tail dimer (**2**) in unprotected triphenylmethyl monoradical.[140]

The benchmark molecule, multi-chlorinated monoradicals **10**-**11** (Figure 9.30), was developed in the 1970s to 1980s, which are air-stable and thermal (in air) stable for a couple of months.[141] From 2014, Li, Nishihara, and Friend have pioneered research on the applications of these radicals in next-generation organic

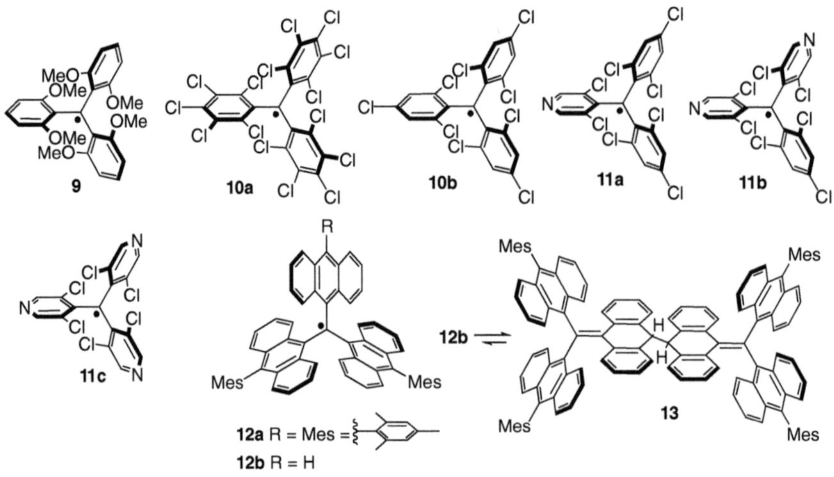

Figure 9.30 Steric protected stable triphenylmethyl radicals.

light-emitting diode (OLED) materials. Among them, monoradicals exhibit doublet fluorescence from their first excited D_1 state to ground-state D_0 state, diradicaloids show spin-allowed fluorescent transition from their lowest triplet excited states to triplet ground states. In contrast to the 25% internal quantum efficiency limit imposed by spin statistics on conventional closed-shell fluorophores, monoradicals or triplet diradicaloids possess a theoretical IQE of 100%.[123] However, the advancement of this field has been limited primarily by the low photoluminescent quantum yield (PLQY) of these radical compounds. For instance, **10b** displays a maximum emission peak (λ_{em}) at 563 nm (cyclohexane) with a low PLQY of 0.03. Actually, Juliá et al. have prepared monoradical **14a** by the coupling of **10b** and carbazole in 2006. Excitedly, the fluorescence quantum yield of **14a** is 0.53 with a λ_{em} at 563 nm in cyclohexane, almost 18 times higher than that of **10b,** and with much red-shifted and enhanced fluorescence emission.[142]

In 2015, Li et al. pioneered the fabrication of OLED devices using **14a** (Figure 9.31) by doping the monoradical into a 4,4′-bis(carbazol-9-yl)biphenyl matrix. The maximum external quantum efficiency (EQE, η_{EQE}) of **14a** is 2.4%.[143] They further synthesized **15a** and **15b**, through device engineering, the OLED performance could be improved to 17 and 27%, respectively.[144] The improved luminescent efficiency is attributed to the formation of charge-transfer states. These states modify the symmetry of TTM molecule (**10b**), thereby altering the degeneracy of the first excited state. Further fine-tuning of the imine nitrogen position on pyridoindolyl groups afforded **14b-14d** with the photoluminescence quantum yield (PLQY) of 91, 89, 32 and 99%, respectively. Consequently, OLED devices with enhanced efficiency could be realized, exhibiting maximum EQE of 9.6% for **14b**, 12.2% for **14c**, 2.9% for **14c,** and 9.5% for **14d**, respectively.[145] **16** is a nontypical monoradical, which can violate the Aufbau principle, exhibiting a SOMO orbital located below its HOMO and HOMO-1 orbitals. This unique electronic structure contributes to the monoradical's stability by suppressing the reactivity of the singly occupied electron. Thus, the half-life of **16** is as long as several months under UV

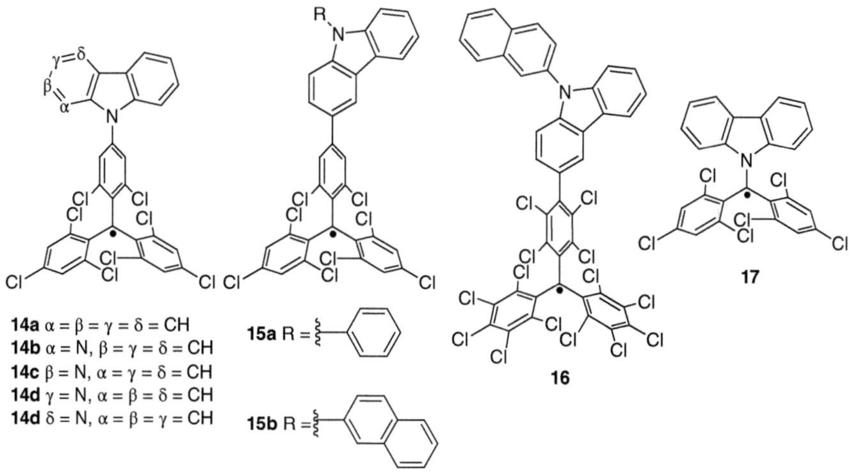

Figure 9.31 Chemical structures of multi-chlorinated organic luminescent monoradicals.

exposure. Furthermore, OLED devices of **16** show a deep-red emissive peak at 700 nm with 5.3% EQE.[146] By linking carbazole' N atoms directly to the carbon radical center, Li successfully synthesized a novel luminescent biphenylmethyl radical **17**.[147] EPR spectra demonstrated its monoradical character with a g value of 2.0035. Moreover, compound **17** exhibits photoluminescence at 697 nm in cyclohexane (PLQY = 2.0 %), and its corresponding OLED device displays an EQE of 0.66 %.

In recent years, a series of diradical emitters were developed and their unique high-spin emission phenomena were detailed examined, which may open a new era of luminescent radical materials. In 2023, Perpichka et al. (Figure 9.32) reported a series of thermal, oxidative, and photostable Thieles's hydrocarbons **18a-c** with singlet diradical characters (y_0) around 0.35 (Figure 9.32).[148] The emission spectra of **18a-c** show strong solvent dependence and a large Stokes shift from 601 nm up to NIR region. The PLQY of **18a-c** decreased from 100% in cyclohexane to a negligible value in polar solvent, which could be ascribed to their zwitterionic excited states. In 2024, Delius et al. synthesized multi-chlorinated Chichibabin's diradical **19** by linking two TTM radicals together,[149] which has a singlet diradical ground state with a y_0 value of 0.58 and a singlet–triplet bandgap (ΔE_{ST}) of -3.11 kcal mol^{-1}. **19** exhibits a λ_{em} of 780 nm and a relatively low PLQY of 0.8%.

Recently, Zeng et al. presented a new class of sulfone-locked Chichibabin's diradicals **20a-c** with y_0 values of 0.44, 0.58, and 0.7, respectively (Figure 9.32).[150] Embedding sulfone into Chichibabin's backbone induced their molecular symmetry breaking, which further leads to an excited-state charge transfer character. Therefore, the λ_{em} of **20a-c** appears in NIR region of 764, 840, 918 nm, respectively. **20a-c** shows moderate PLQY of 5.2, 2.0, and 2.4%, respectively. Further extended Müller's hydrocarbon diradical **21** synthesized by Abaurahman et al., also has a singlet diradical ground state with a y_0 value of 0.90 and a ΔE_{S-T} of -1.51 kcal mol^{-1}, indicating its thermally accessible triplet state.[151] **21** is also emissive with a λ_{em} at 671 nm, but its PLQY is only 0.4%.

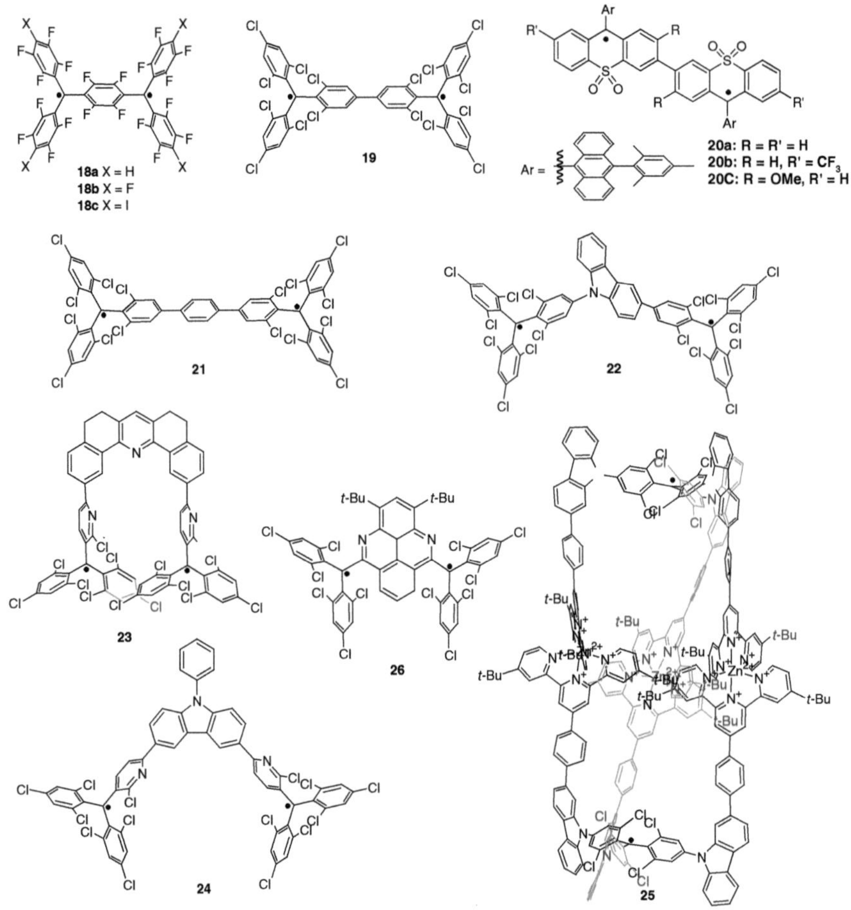

Figure 9.32 Molecular structures of luminescent diradicals.

In 2024, Abaurahman further synthesized diradical **22** by replacing benzene of **21** with carbazole.[152] **22** is also an singlet diradicaloid with a larger y_0 value (0.95) and a smaller ΔE_{S-T} (-0.051 kcal mol^{-1}). **22** shows much higher luminescent character with a PLQY of 16% in solution and unusually strong magnetoluminescence (ML) phenomena by increasing of external magnetic field, indicating its possible thermally activated triplet emission. Similarly, Kusamoto et al. synthesized a spatially confined diradical emitter **23**, which possesses a small ΔE_{S-T} of 0.02 kcal mol^{-1}.[153] Therefore, a single-molecule ML could be observed, which could be ascribed to the manipulation of ground state spin populations and spin-related excited-state distributions by adjusting the intensity of the external magnetic field. Very recently, Kusamoto et al. reported another singlet ground-state diradical **24**, which possesses a singlet ground state and a ΔE_{S-T} of -0.05 kcal mol^{-1}.[154] **24** in PMMA with 0.01 w% shows clear ML phenomena with enhanced luminescence intensity by increasing the magnetic field at 4.2 K. Meanwhile, it also shows temperature dependent increased emission spectra as the temperature increases from 3.6 to 50 K. DFT calculation reveals their unique ground state and excited state, which is inversed to common closed-shell emitters. The heavy-atom solvent effect on photophysical measurements indicates the nonemissive character of the single-ground state of **24**. Furthermore, temperature dependent ML phenomena confirmed the unique thermally activated triplet emission of this diradicaloid. For the diradical cage **25** reported by Wang et al., due to the long distance between the two carbon radical centers, no spin-spin interaction can be generated. Thus, the magnetic field's effect on the emissions of **25** is almost negligible.[155] Zhao et al. developed a novel non-Kekulé (diaza)pyrene-4,10-diydiradical **26**, which features a triplet ground state (ΔE_{S-T} of 1.1 kcal mol^{-1}) and demonstrates excellent stability with a half-life of 31 hrs. Furthermore, **26** shows intramolecular ferromagnetic spin coupling and localized non-disjoint frontier orbitals.[156]

Actually, for these multi-chlorinated triphenylmethayl monoradicals, the stability problem under photoexcitation is still a hindrance for their further development. As shown in Figure 9.33, Nishihara et al. disclosed (3,5-dichloro-4-pyridyl)-bis(2,4,6-trichlorophenyl)methyl radical **11a**, which exhibits a photostability 115 times higher than that of **10b**, owing to the pyridyl substitution and reduced molecular orbital energies. Compounds **11a** is also fluorescent, exhibiting PLQY of 0.03 in solution, 0.26 in PMMA film at room temperature, and 0.81 in a diethyl ether:isopentane: ethanol matrix at 77 K.[157] Subsequently, Nishihara et al. synthesized **27a** and **27b** by replacing Cl atom on pyridyl with Br and F atoms. Their study revealed that the halogen's electronegativity significantly modulated the HOMO energy, whereas the

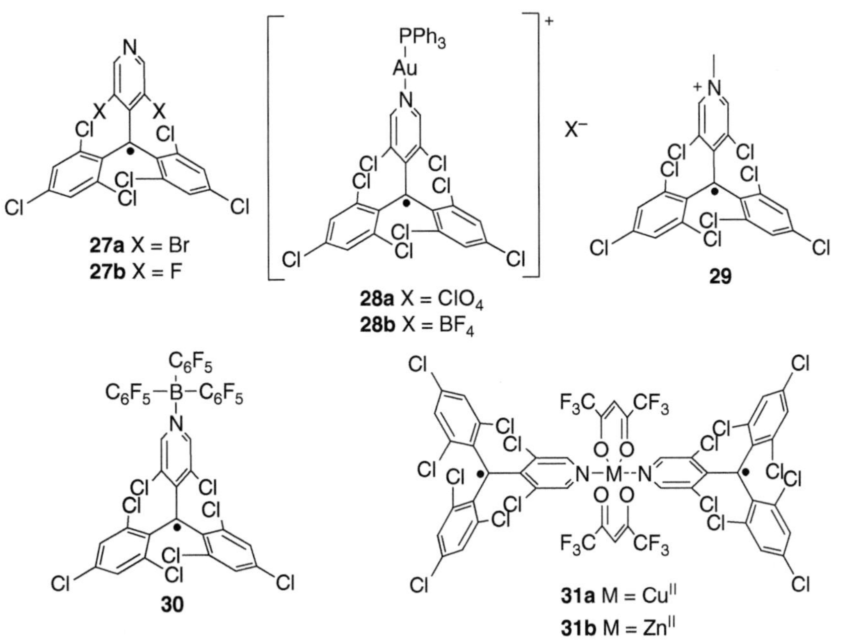

Figure 9.33 Molecular structures of chlorinated pyridyl monoradical derivatives.

SOMO remained largely unaffected. Accordingly, both the absorption and emission bands of these compounds are red-shifted, following the order F < Cl < Br.[158]

Furthermore, Nishihara et al. developed a new class of luminescent monoradicals based on pyridine's coordination chemistry. By coordinating **11a** with Au^I, they synthesized **28a** and **28b.**[159] Comparison to **10b** (λ_{em} = 585 nm), the Au^I complex **28b** displayed red-shifted fluorescent emission at 653 nm. This shift may be attributed to the excitation transition from the D_1 state to the D_0 state involving the pyridyl radical center. Nishihara et al. reported the synthesis of two other luminescent monoradicals **29** and **30**, via methylation and coordinating $B(C_6F_5)_3$ to the pyridyl imine N atom.[160] Further investigations illustrated that the functionalization of the imine nitrogen atom significantly alters the β-SOMO, a key determinant of the electrochemical and optical features. Consequently, compounds **29** and **30** displayed red-shifted fluorescent bands (712 and 660 nm, respectively) relative to **11a**. Likewise, the authors described complexes **31a** and **31b**, which are the Cu^{II} and Zn^{II} complexes coordinated by the bis(pyridyl radical) ligand **11a**. Structural analysis reveals **31a** and **31b** adopt a hexa-coordinated octahedral geometry.[161] The observed elongation of the Zn—N bonds relative to the Cu—N bonds is consistent with a structural model, where the ligand adopts an axial coordination in the Zn^{II} complex but an equatorial one in the Cu^{II} complex. The observation of intramolecular ferromagnetic coupling between the two carbon radicals in **31a**, mediated by the Cu^{II} center, can be rationalized by the orthogonality of the magnetic orbitals involved.

Later in 2019, Nishihara et al. presented a 1D polymer chain **32a** consisting Cu^{II} ions and polychlorinated di(pyridyl)phenylmethyl monoradical **11b.**[162] As illustrated in Figure 9.34, the structure self-assembles into 1D – $Cu^{II}(hfac)_2$–**11b**–zigzag chains extending along the a + c direction. Subsequently, these chains stack to generate a layered architecture. The system exhibited a temperature dependent Jahn–Teller (JT) distortion, which involves a rotation of the Jahn-Teller axis from being aligned with the Cu—N bond at 298 K to the Cu—O bond at 93 K. Thus, the JT distortion at low temperature enhanced the intramolecular ferromagnetic interaction by reorienting the $Cu^{II}d_{x2-y2}$ orbital. In 2021, Nishihara et al. developed the tris(3,5-dichloro-4-pyridyl)methyl radical (**11c**), which exhibits remarkably enhanced photostability, possessing a half-life of 2.2×10^4 s. Compared to TTM **10b**, mono-, di-, and tri-pyridyl radicals (**11a, 11b,** and **11c**) exhibit a dramatic increase in half-life by factors of 4, 160, and 10 000, respectively, resulting from the increased number of imine N atoms and progressive decrease in frontier orbital energy levels.[163] Consequently, the solution emission (λ_{em}) red-shifts steadily across the series **10b** (570 nm), **11a** (585 nm), **11b** (650 nm), and **11c** (700 nm), as the number of imine N atoms on the TTM skeleton increases. Monoradical **11c**, exhibiting a blue-shifted solid-state emission at 665 nm, was employed to construct a 2D coordination polymer (**33**) upon reaction with $Zn^{II}(hfac)_2$. As illustrated in Figure 9.34c, 2D polymer **33** featured honeycomb spin-lattices with a graphene-like spin topology. Notably, it exhibited luminescence at 79 K (λ_{max} = 695 nm), representing a rare example of a luminescent open-shell 2D material. Magnetic studies further verified that the monoradical character of each **11c** remained within the 2D polymer **33**. They further

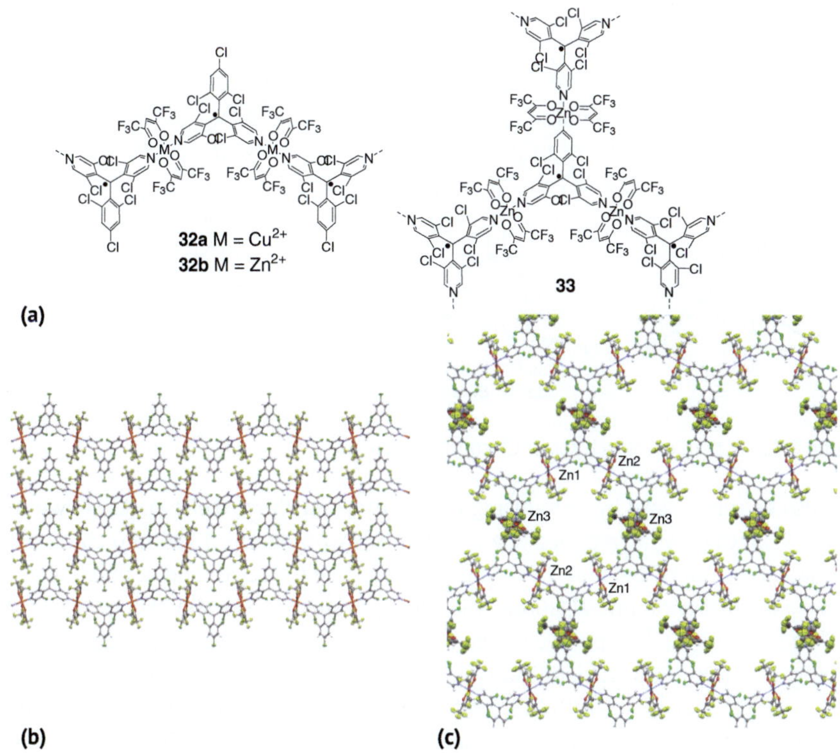

Figure 9.34 (a) Molecular structures of 1D and 2D magnetic molecules, X-ray crystal structure of (b) **32a** and (c) **33**. *Source*: [162]/Royal Society of Chemistry (a), [164]/ American Chemical Society.

investigated the luminescence behavior in response to an external magnetic field and temperature for monoradical **11b**, **11c**, and coordination polymers **32b** (1D) and **33** (2D). External magnetic field shows no effects on the emissions of two monoradicals. Whereas, **32b** and **33** exhibit remarkable ML effects. 1D polymer **32b** was almost non-emissive at 4.2 K and 0 T. However, upon the application of an external magnetic field, emission peaks at 626 and 675 nm gradually appeared. Similarly, a 25% enhancement in the emission intensity of **33** was observed at 18 T, a phenomenon arising from the suppression of ground-state radical-radical interactions by the Zn^{II} centers, which allows for their effective modulation by the magnetic field.[164]

Up to now, circularly polarized luminescent radical materials have been rarely reported. Actually, multiple chlorinated triphenylmethyl monoradicals **10a** and **10b** are both composed of racemic *M* and *P* enantiomers, defined by the opposite helical chirality (left- or right-handed) of the three protecting phenyl rings. However, the enantiomers of **10a** and **10b** were separated until 2019 with rapid racemization even at room temperature.[165] The luminescence asymmetry factors (g_{lum}) of **10a** and **10b** are 8×10^{-4} and 5×10^{-4}, respectively. Later, Veciana et al. further synthesized brominated analogue **34**.[166] Due to the introduction of a bulky Br atom, **34** has a stronger ability against racemization ($t_{1/2}$ = 18 s at 60 °C) with a g_{lum} of 8×10^{-4}. Chiroptical radical

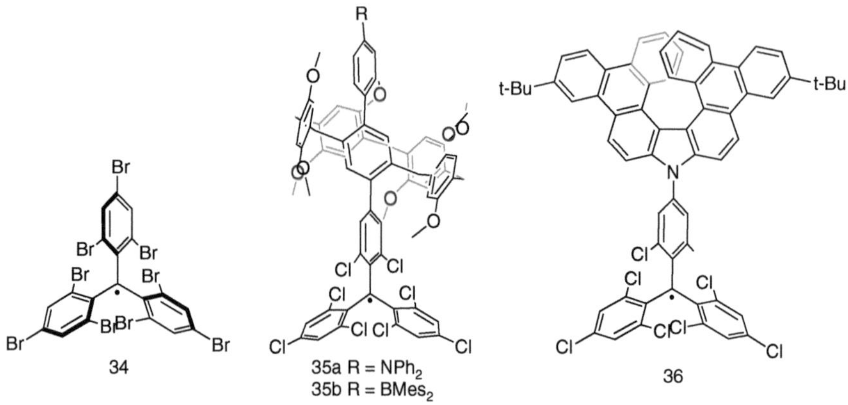

Figure 9.35 Chemical structures of Chiroptical radicals.

can also be synthesized by introducing radical units on chiral skeleton. As shown in Figure 9.35, Chen et al. reported **35a** and **35b** by incorporating **TTM** into chiral pillar[5]arenes.[167] Steric bulk on the macrocycles enabled the separation of their enantiomers and chiroptical investigations, and the g_{lum} values of **35a** and **35b** are -8.43×10^{-4} and -1.51×10^{-3}, respectively. Meanwhile, the photostability and configurational stability of **35a** and **35b** are significantly improved in comparison with bare **TTM** radicals. In 2023, Chen et al. synthesized π-conjugated chiral radical **36** by incorporating **TTM** into heteroatom-doped aza[7]helicenes.[168] **36** can be resolved into *P*/*M* enantiomers with a g_{lum} value of 3.57×10^{-3}. Due to the strong donor-acceptor effect between aza[7]helicene and **TTM** radical, the emission peak of **36** appears in the NIR region. Meanwhile, the PLQY of two chiral isomers is 44 and 50%, respectively.

9.4.3 π-conjugated Conductive Radical Materials

Usually, π-conjugated compounds with radical electrons delocalizing on the backbone would show good stability, benefiting from the spin dilution effect. Meanwhile, the large conjugated planes and good solid-state packing may satisfy the demand for charge transport of organic conductive or semiconductive materials.

Similar to triphenylmethyl radicals, early work on these polycyclic aromatic hydrocarbon radicals mainly focuses on their stability issues. Compounds **37a**, **37b,** and **37c** (Figure 9.36) are characteristic pentaphenylcyclopentadienyl monoradicals, whose unpaired electrons are delocalized on the cyclopentadienyl ring.[169–171] Notably, the presence of TIPS-ethynyl substituents on **37c** (Figure 9.36) provides a stabilizing effect through delocalization, making it the only air-stable monoradical in this family of these compounds.[170] DFT calculations on the fluorenyl radical—which can be considered a dibenzo-annulated cyclopentadienyl radical—showed that the unpaired electron is delocalized over the entire π-system of the fluorene unit. Therefore, by strategically protecting the reactive 3, 6, and 9 positions with bulky substituents, Wu et al. successfully developed air stable fluorenyl radical **38**, which remained isolable even through a standard silica gel column.[172,173]

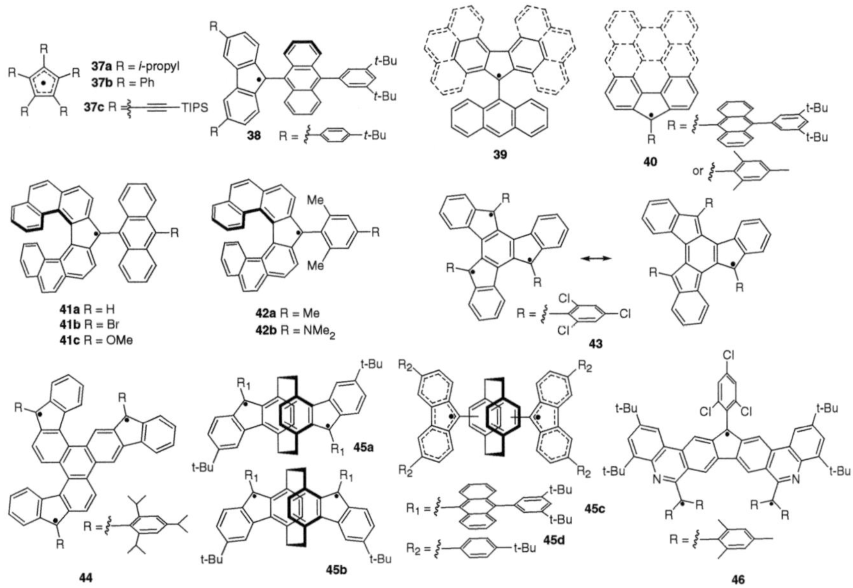

Figure 9.36 Fluorenyl-based monoradicals.

Fluorenyl radicals can also be thermodynamically stabilized by spin delocalization of the fluorenyl skeleton with π-extension of two more aromatic benzene rings. As illustrated in Figure 9.36, Kubo et al. developed three long-lived conjugated extended fluorenyl radicals **39** with moderate stability and a half-life time of 43 days.[174] Sun et al. reported two stable serrated edge extended fluorenyl radicals **40**, one of which exhibited excellent stability in an ambient saturated solution with a half-life of 260 days, providing their possible applications in electronics.[175] Gu et al. developed a group of highly stable, helical chiral fluorenyl radicals (**41a-c**, **42a-b**), which are separable on a conventional silica gel column. These monoradicals exhibit stability toward oxygen, acid, and base.[176] Zhao et al. reported a triradical molecule **43** with intramolecular strong spin couplings. Since the quinoid form significantly promotes antiferromagnetic coupling between two spin centers than the triradical form, which makes the quinoid resonance form feasible. However, it did not exhibit sufficient environmental stability and therefore was not separated.[177] Stępień et al. reported spin frustrated doublet ground-state triradical **44** with doublet–quartet energy gap $\left(\Delta E_{\text{D-Q}}\right)$ of −0.14 kcal mol^{-1}. Although possessing open-shell properties, **44** are stable enough to be stored under environmental conditions for several days.[178] In 2021, Zhao et al. reported two pairs of [2.2]paracyclophane (CP) coupled polycyclic aromatic hydrocarbons diradical regio-isomers **45a-b** and **45c-d** with antiferromagnetic and ferromagnetic couplings.[179] Two indeno fused fluorenyl radicals **45a** and **45b** have singlet ground state with diradical indices of 0.8 and 0.9, respectively. Remarkably, both **45a** and **45b** exhibit antiferromagnetic coupling regardless of the fusing positions, due to the significant orbital overlapping beneficial from the short π–π stacking distances of CP units. On the contrary, when two fluorenyl radicals are directly connected to CP through a single C—C bond via C9 position, both **45c** and **45d** exhibit ferromagnetic coupling, which could be ascribed to the limited spin delocalization on the CP units. In addition, Zhao also reported a remarkably stable quartet ground-state triradical **46** with a ΔE_{D-Q} of 0.78 kcal mol^{-1}.[180] Two nitrogen atoms embedded in specific spin-sharing positions of the aromatic skeleton have been shown to be crucial for the stability of radical in their previous studies.[156] Therefore, this triradical exhibits excellent photochemical stability with an ultra-long half-life of 90 days ($\tau_{1/2}$) in ambient conditions and excellent thermal stability without significant thermal decomposition or phase transition at temperatures as high as 175 °C under nitrogen flow. Notably, selective excitations between different Zeeman splitting energy levels can be realized by X-band pulsed ESR measurements, clearly revealing the magnetic anisotropy of the quartet ground state in compound **46** in a frozen toluene solution.

Phenalenyl **47a** (Figure 9.37) is the archetypal monoradical hydrocarbon containing an odd number of C atoms.[181–183] Owing to its high reactivity, it exhibits instability in air. However, phenalenyl **47b** with three *tert*-butyl substitutes is stable and readily forms a π–π stacked dimer in its crystal state and in solution state at low temperature.[181] Kubo et al. developed **48** with two phenalenyls fused by cyclopentalienyl group, which showcases remarkable air stability ($t_{1/2}$ = 60 h). Notably, although **48** carries only one *n*-butyl substituent (of modest size), its stability arises from the effective delocalization throughout the π framework.[184] In 2020, Sun et al. developed the air-stable olympicenyl monoradical **49**, stabilized by a TIPS-ethynyl substituent and exhibiting a long half-life (seven days). Subsequent single-crystal analysis revealed that **49** forms a head-to-tail dimer, featuring a unique 20-center-2-electron structure. Notably, this π-dimer has a singlet diradical ground state.[185] By further lateral extension, Sun et al. also synthesized benzo[*c*]anthanthrenyl radical **50a** and **50b** with half-life times of 36 and 48 h, respectively. X-ray crystallographic analysis revealed the 1D stacking of **50a**, with a balanced π–π stacking distance of 3.56 Å, probably due to its twisted backbone structure for the introduction of two bulky TIPS ethynyl substituents.[186]

Through further lateral π-extension, Sun synthesized previously undiscovered nine-ring hydrocarbon π-radical **51**. Both thermodynamic spin delocalization and kinetically protection of bulky substituents enhance the stability of **51** with

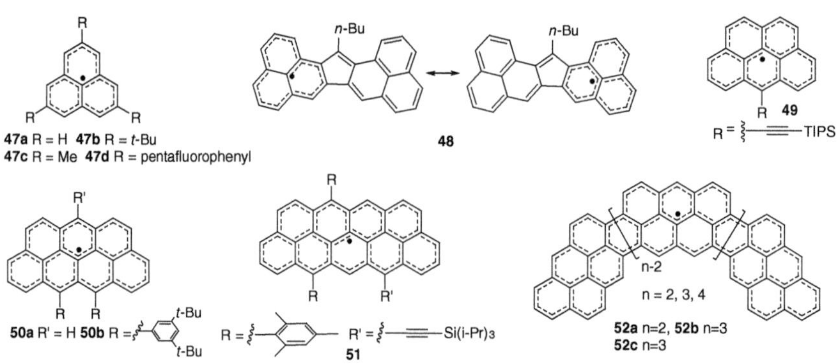

Figure 9.37 Phenlenyl-based monoradicals.

a half-life of 13 days in ambient toluene solution. Similarly, a π-dimer in the single crystal structure with an average distance of 3.175 Å could be observed. The π-dimers can further form 1D chains, featuring an average intermolecular distance of 3.365 Å.[187] Later, Wu and Sun reported olympicenyl radical-based graphene fragments **52a-c** based on radical-mediated intramolecular coupling strategy.[188] When n=2, **52a** forms a dimer in a solid state, exhibiting a slipped stacked arrangement with a π–π distance of 3.42 Å. When n = 3, dimer structures of **52b** could be observed in both solid and solution, which is fundamentally different from the dimer structure of **52a**. This interaction can be named as a "32 centered-2 electron" pancake bonding, similar to the "12 centered-2 electron," and "20 centered-2 electron" pancake bonding in the π-dimer of **47** and **49**. The interlayer distance of the π-dimer is 3.22 Å, slightly shorter than **47** and **49** (3.20-3.32 Å and 3.30 Å, respectively).[161,185] In addition, **52a** exhibits global aromatic ring currents and **52c** could be regarded as the dimer structure of **52a**.

As shown in Figure 9.38, [n]triangulene could be viewed as triangular zigzag-edged extended graphene fragments, where n represents the zigzag carbon numbers. When n is an even number, for example, [2]triangulene (**47a**), also called phenalenyl radical, has total odd odd-numbered carbon atoms. Thus, no Kekulé-type resonance structures can be drawn for this molecule. However, even n is an even number, such as [3]triangulene, which is also a typical Clar's hydrocarbon, no Kekulé-type structure can be drawn either. Moreover, the total spin of these monoradicals can be calculated based on the Ovchinnikov rule.[189] Simply, the ground-state spin quantum number S is (n−1)/2. Therefore, for high-order [n]triangulenes, ferromagnetic high-spin states would be the great interest for scientists.

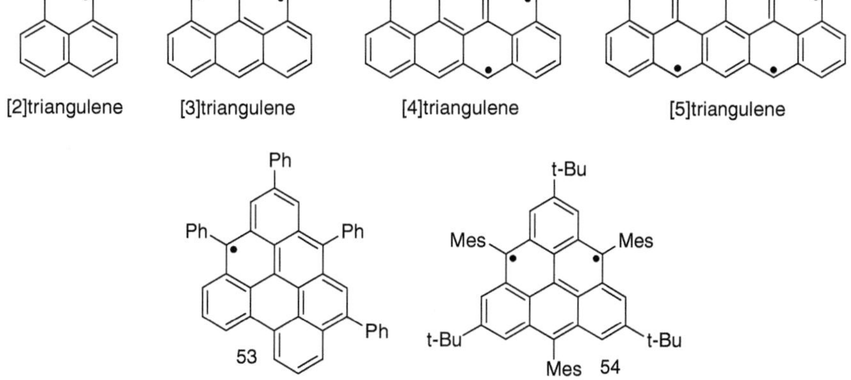

Figure 9.38 Chemical structures of [n]triangulene and their derivatives.

However, synthesis and purification of unprotected high-order [n]triangulenes is almost impossible due to their intrinsic instability. As shown in Figure 9.38, recent studies demonstrated that the synthesis of high-order [n]triangulene up to [7]triangulene could be realized by on-surface synthesis with a delicately molecular design of precursors.[190–194] Actually, [3]triangulene has been pursued for almost 70 years.[195] In 2020, Juríček reported benzo[*cd*]triangulene **53** as a persistent stable radical compound. Different from [3]triangulene, **53** had a spin quantum number *S* of 1/2, with doublet ground state. Even though **53** was still quite sensitive to air and would quickly convert to its oxidation product.[196] In 2021, Shintani synthesized and isolated [3]triangulene compound **54** by multiple kinetically protecting t-butyl groups and mesityl groups, and its triplet ground state could be revealed by ESR and SQUID measurements.[197] Although the high-spin characters of [n]triangulenes are fascinating, their applications are still limited by stability problems.

Diradicaloids are molecules featuring two unpaired electrons in three situations. When the exchange interaction (*J*) between two unpaired electrons in one molecule becomes nearly negligible, this system can be described as a biradical. When two unpaired electrons are in close proximity, a spin-spin (exchange) interaction occurs, which results in the formation of two electronic states with different spin multiplicities: a singlet state ($S = 0$) and a triplet state ($S = 1$).[107,109,129] Therefore, compared to biradicals and high-spin triplet diradicals, diradicaloids with singlet ground states usually possess moderate air and light stability, making their possible applications in organic electronics, such as organic field effect transistors (OFETs).

As reported by Haley et al. early in 2012, a single-crystal OFET device of pentafluorophenyl substituted indofluorene **55** (Figure 9.39) exhibited hole and electron carrier mobilities of 7×10^{-4} and 3.0×10^{-3} $cm^2\ V^{-1}\ s^{-1}$, respectively.[198] In 2016, they developed an extended indofluorene derivative **56**, with a higher y_0 value

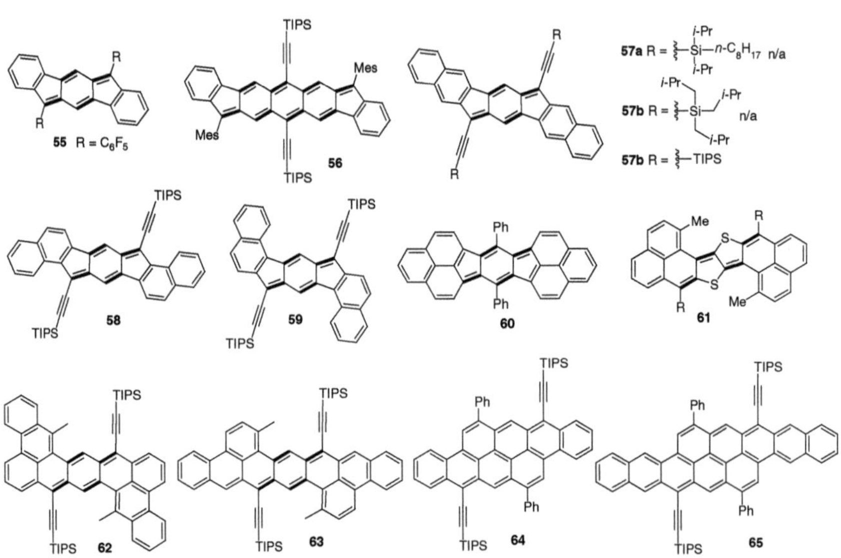

Figure 9.39 Singlet diradicaloids for OFETs.

of 0.381 at the PUHF level. **56** exhibited balanced ambipolar carrier transport with hole and electron carrier mobilities of 2.0×10^{-3} and 4.0×10^{-3} $cm^2 V^{-1} s^{-1}$, respectively.[199] It is obvious that bulky substituents of **56**, such as two mesityl groups, are harmful for effective π–π stacking. To increase the π–π overlap of adjacent molecules, Haley et al. further developed lateral extended indenofluorene diradicaloids **57-59**.[200] Single-crystal structural analysis revealed that the fusion modes on the indenofluorenes directly influence their crystal packing. The presence of larger bulky substituents in **57a** and **58** leads to disconnected 1D π–π packing motifs, **59** arranged in dimer form without π–π or C—H···π interactions between adjacent dimers. Therefore, **57a**, **57b**, and **59** did not exhibit measurable OFET properties. In contrast, like TIPS-pentacene, **35c** and **58** adopt a 2D π–π packing motif, a structure known to facilitate superior electron transport. Accordingly, OFET devices based on crystalline **57c** and **58** exhibit remarkable p-type hole mobilities of 1.04 ± 0.68 and 4.72 ± 1.97 $cm^2 V^{-1} s^{-1}$, respectively.

As illustrated in Figure 9.39, Kubo developed an open-shell singlet diradicaloid **60** with a y_0 value of 0.68 (at CASSCF(2,2)/6-31G level) and a small optical bandgap of 1.1 eV.[201] The high-lying HOMO and low-lying LUMO energy levels of **60** enable its ambipolar carrier transport character. Compound **60** forms 1D slipped π-stacks with an average π–π distance of 3.317 Å, which is less than the sum of the van der Vaals radius of sp^2 C atoms, implying its potential in efficient charge transport. Whereas, the conductivity of **60** is only 5.0×10^{-5} $S\,cm^{-1}$. Moreover, OFET devices of **60** demonstrate balanced hole and electron mobilities of 2.8×10^{-4} and 3.0×10^{-4} $cm^2 V^{-1} s^{-1}$, respectively.[202]

Chi's group prepared bisphenaleno-thieno[3,2-b]thiophene **61**.[203] X-ray single-crystal analysis discloses its unexpected wave-shape structure with a typical quinoidal resonance form, arising from strong intermolecular π–π interactions between two adjacent phenalenyls. DFT calculations at UCAM-B3LYP/6-31G level suggest its singlet diradical character with a y_0 of 0.186, which can be attributed to the thermodynamic stabilization conferred by the delocalized phenalenyls. Consequently, solution-processed OFET devices of **61** exhibit p-type charge-transport mobility of 0.26 $cm^2 V^{-1} s^{-1}$. Two isomeric dibenzo-extended heptazethrenes (**62** and **63**) were later disclosed by Sun and coworkers. These isomers demonstrated moderate diradical character, with calculated y_0 values of 0.11 and 0.13 (at the UCAM-B3LYP/6-31G level), respectively.[204] The dimeric structures of **62** and **63**, as determined by single-crystal X-ray diffraction, are facilitated by a periodic alignment in solid states. Despite the absence of distinct 1D or 2D π–π stacking, single crystal OFETs of **62** and **63**, with top-contact and bottom-gate configuration, exhibited maximum hole mobilities of 0.15 and 0.017 $cm^2 V^{-1} s^{-1}$, respectively.

Limited by their poor stability of extended acenes with significant diradical characters, their charge-transport properties based on OFETs are rarely reported.[205] Frigoli et al. developed slipped-bisacene compounds **64** and **65**, which exhibit high diradical character (0.50 and 0.64 at UHF/6-31G(d, p) level) and moderate half-lives of 6.5 and 4.9 days, respectively. Notably, hole-transport mobilities of 0.77 and 1.4 $cm^2 V^{-1} s^{-1}$ were achieved in the thin-film OFET devices of **42** and **43**, respectively.

9.4.4 Charged Radical Materials

Thiele's hydrocarbon is the oxidative product of the dimer **2** of tripenylmethyl monoradical, which is a quinoidal closed-shell structure.[206] Its extended structure **67**, also named as Chichibabin's hydrocarbon, is a typical singlet diradicaloid. The ground state of **45** was demonstrated by its X-ray structural analysis in 1986.[207,208] Further extended structure **68** shows a triplet ground state, although theoretical calculations indicated its singlet ground state with very small ΔE_{S-T}.[209,210] The intrinsic stability problem of these extended quinoidal structures makes the result still controversial.

As shown in Figure 9.40, Wang developed a series of bis(triarylamine) dications as the analogue of these typical hydrocarbons. Compared to their all-carbon counterparts, the moderate stability of these charged species in the crystalline state enables the deep research of their solid-state structures and electronic structure possible. They first synthesized dication **69**, **71**, and SQUID measurements reveal the closed-shell feature of **69** and **70**, which is similar to Thiele's hydrocarbon. However, due to the steric repulsion effect of the central anthrancene moiety, **71** showed a singlet diradical ground state with a small ΔE_{S-T} and a thermally excited triplet ESR signal.[211] Later, Wang reported compound **50** with magnetic bistability due to its intramolecular electron-exchange interaction. Temperature-dependence of the magnetic susceptibility experiment of **72** demonstrated a spin state conversion between two singlets at the temperature window of 118 to 131 K.[212] The analogues of Chichibabin's hydrocarbon **73a**, **73b**, and **73c** showed gradually increased diradical character and decreased ΔE_{S-T} with obvious substituent dependence.[213] By further extension, Wang synthesized dication **74** as the analogue of Muller's hydrocarbon. It is found that **72** exhibited

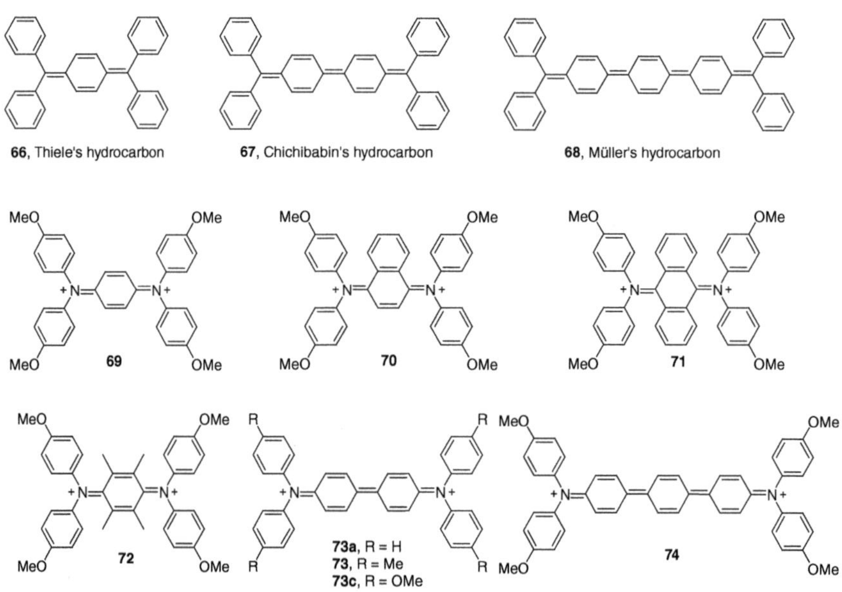

Figure 9.40 Thiele's hydrocarbon, Chichibabin's hydrocarbon, Müller's hydrocarbon and their N analogues.

two singlet ground states with different spin coupling constants depending on temperature, due to the conformation change of the terphenyl bridge.[214]

Meso position nitrogen-bridged macrocycle **75** (Figure 9.41) is an air-stable diradical dication with a triplet ground state (ΔE_{S-T} = 0.5 kcal/mol). In solid state, it can form a Heisenberg chain with antiferromagnetic coupling and a small interchain ferromagnetic coupling.[215] Chi reported a global antiaromatic all-benzenoid macrocycle **76** with a total 32 π electrons. By two-electron and four-electron oxidation, **76**$^{2+}$ and **76**$^{4+}$ exhibit global aromaticity and antiaromaticity with 30 π and 28 π electrons, respectively. Both **76**$^{2+}$ and **76**$^{4+}$ possess open-shell singlet ground state with the ΔE_{S-T} of −1.07 and −2.96 kcal mol^{-1}, respectively.[216] By linking benzene rings together with nitrogen atoms, Ito synthesized a series of macrocycles **77a**-**77c**, and these electron-rich compounds are easy to be oxidized into their radical cation species.[217,218] Among them, **77a** could be oxidized to **77a**$^{2+}$. Interestingly, although the square-type macrocycle possesses large intramolecular tension, **77a**$^{2+}$ shows much planar structure compared to its neutral counterpart. Thus, **77a**$^{2+}$ exhibits 18 π electrons global aromatic character with a singlet ground state.

In 2024, Chi reported a tetraazasupernaphthalene compound **78** (Figure 9.41).[219] Furthermore, **78** could be oxidized to its dication **78**$^{2+}$, tetracation **78**$^{4+}$ and hexacation **78**$^{6+}$ with two, four, and six equivalents of oxidant, respectively. **78** showed

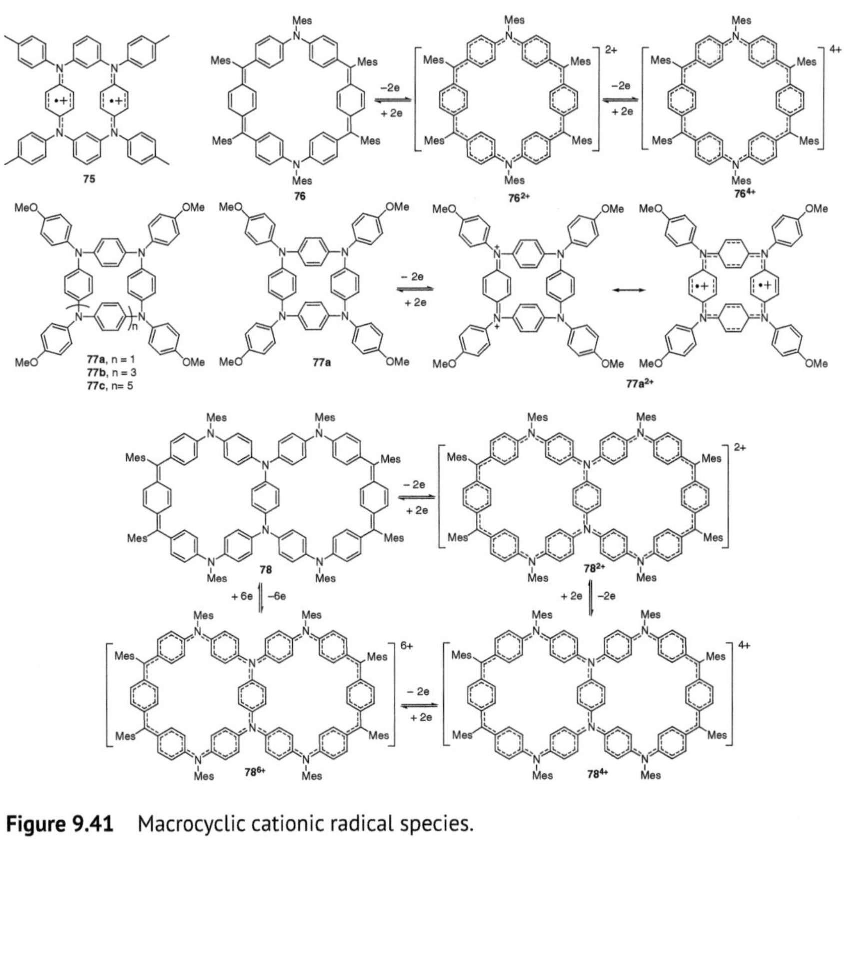

Figure 9.41 Macrocyclic cationic radical species.

localized aromaticity due to the weak conjugations between neighboring phenyl rings linked by N atoms. Therefore, even though it was oxidized to its dication $\mathbf{78}^{2+}$, which possesses a singlet diradical ground state, the dication only showed localized aromaticity. In contrast, ^{1}H NMR of tetracation $\mathbf{78}^{4+}$ clearly revealed its globally antiaromatic character, which is also confirmed by ACID plot and NICS values. Moreover, hexacation $\mathbf{78}^{6+}$ showed clear ESR signals, demonstrating its singlet diradical character. ACID plots show ditropic ring currents along the dual macrocycle, indicating its global aromaticity with 52 π electrons following Hückel's rule.

Photoinduced organic radicals play a pivotal role in the emission process of close-shell triphenylamine or triphenylphosphine following prolonged light exposure. The proposed mechanism and potential applications were explored through several representative examples. Giuseppone found the triarylamine derivative to be photoactive, forming radical species under mild white light. For instance, in chloroform solutions of compounds **79** and **80** (Figure 9.42), the proportion of triaryammonium cationic radical reached up to 44% upon long-time halogen light irradiation.[220,221] While later studies reveal that the self-assembled nanostructure of **80** displays supramolecular self-healable feature and metallic behavior, its emission characteristics remained unexplored.[221]

Compound **81**, reported by Chi et al., is characterized by its photoinduced radical character and an unexpected, rapid, reversible luminescent color change from blue to pinkish-purple in single crystals upon irradiation under ambient conditions.[222] The intensity could be significantly enhanced as the irradiative time is prolonged. Moreover, the absence of this phenomenon in the amorphous phase highlights the essential role of specific molecular packing and the accompanying intermolecular interactions. The measured lifetime of 4.2 ns for 580 nm emission is consistent with fluorescence. Single-crystal analysis confirmed no π–π interactions in its solid state. However, in the crystalline state, C—H···π interactions effectively separate the photoinduced radical cations. This spatial isolation can facilitate the rapid release of energy from its excited state to the ground state. Later in 2022, Tang reported the phosphorus-containing photoinduced radicals with bright red emissions from its crystalline state.[223] After being irradiated **82** (**i-82**), **i-82** exhibited two new absorption bands (496 and 520 nm), originating from radical species and confirmed by ESR spectroscopy. The fluorescence character of **i-82** was established by its 3.9 ns PL lifetime. Like **82**, the emission of **i-82** was grinding-sensitive, demonstrating the lattice's role as a protective cage against ambient oxygen and moisture. Notably, the

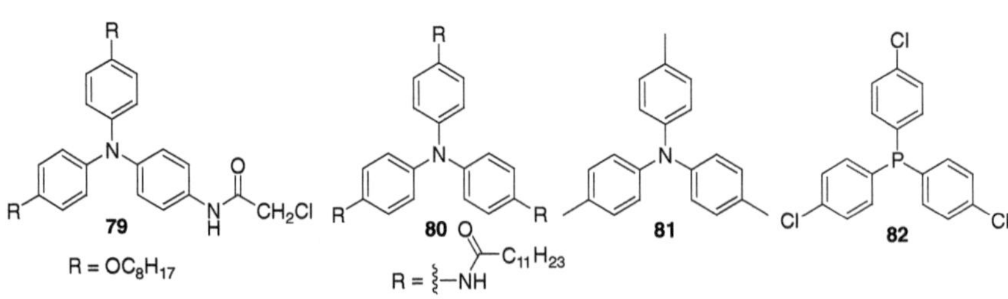

Figure 9.42 Triphenylamine and tripenylphosphine derivatives.

red emission of **i-82** persisted for over seven days in air when in the crystalline state. DFT and single-crystal analyses revealed that a symmetry-breaking phenomenon in the crystal induces conformational changes, modifying the photoredox properties. The resultant asymmetric packing, with a 3.29 Å π–π distance, may promote exciton separation and stabilization through a charge-hopping mechanism. The work presents a feasible strategy for generating in situ stable radicals upon visible-light irradiation, highlighting their potential as photosensitive materials.

9.4.5 Room-temperature Ferromagnetic Radical Materials

Magnetism of traditional magnets originates from unpaired electrons of the *d* or *f* orbitals, in contrast, the magnetic moments of organic molecular magnets could be ascribed to unpaired electrons of *s* or *p* orbitals. Up to now, studies on plastic organic magnets mainly focus on their solid-state study of small molecules and 2D materials.

In 2018, J. Baek reported the facile synthesis of 2D ferromagnetic material **83** (Figure 9.43) via self-polymerization of tetracyanoquinodimenthane in trifluoromethanesulfonic acid at 155 °C.[224] The radical character of **83** originates from the rotation of cyclohexadiene rings. **83** can form a 120 ferromagnetic order with high glass transition temperature larger than 400 K. Furthermore, it exhibits magnetic response under an external strong permanent magnet and excellent stability and can be stored in ambient conditions for two years. Later, Wu reported a porous organic radical anion framework **84** as a room-temperature (RT) organic magnet.[225] X-ray structure of **84** revealed its close distance between two radical layers of 2.57 Å, which is much shorter than common π–π distances. Furthermore, **84** shows a critical temperature as high as 465 K but a relatively low magnetic susceptibility of 0.33 emu·K·mol^{-1}, which could be ascribed to the partial degradation of the radical

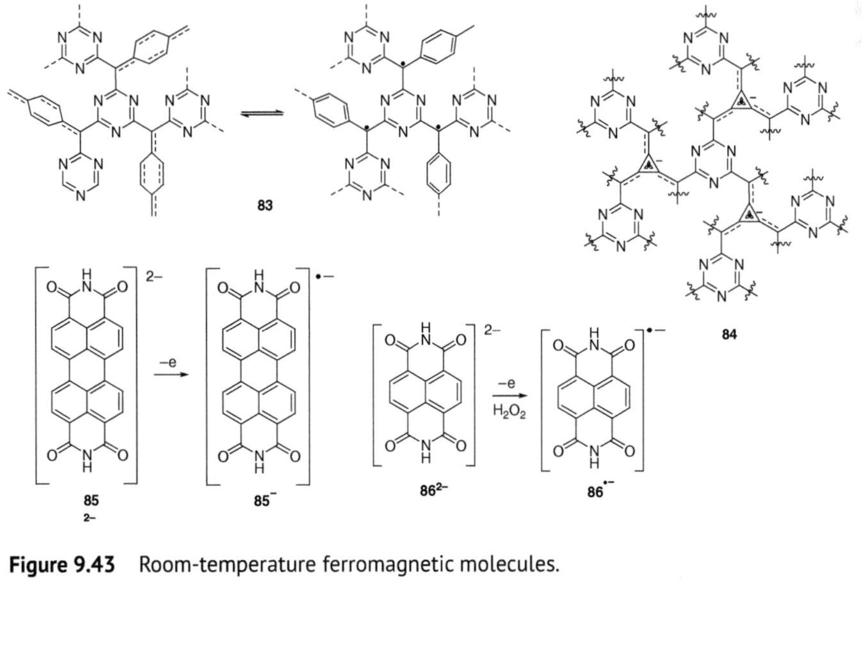

Figure 9.43 Room-temperature ferromagnetic molecules.

anions. In 2022, Ma reported a small molecule RT organic magnet **85**, which was prepared by the oxidation of perylene diimide (PDI) dianion.[226] X-ray diffraction displays a π–π distance of 3.30 Å, which is important for the strong spin interaction and the formation of ferromagnetic packing. The magnetic susceptibility of **85** was measured to be 0.8 emu·K·mol^{-1} in air condition. Later, they reported a RT ferromagnetism of naphthalene diimide molecule **86**$^{\cdot-}$, which could be reduced to **86**$^{2-}$ by H_2O_2 in a quantitative yield.[227] X-ray diffraction exhibits a π–π distance of 3.26 Å, and RT ferromagnetism measurements of **86**$^{\cdot-}$ indicate the highest saturated magnetization of 1.1 emu·K·mol^{-1} in pure organic ferromagnets. The RT ferromagnets represent a potential candidate for the next generation of magnetic materials.

9.4.6 Summary and Perspective

In last decades, based on the progress and fundamental understanding of organic radicals, scientists have gradually transferred their attentions from the stabilization of organic radicals to the exploration of their applications. From 2000 (Figure 9.44), various of applications of radical materials based on their electronic, optical, and magnetic properties were developed. Especially, luminescent radical materials have made great progress due to the breakthrough of double emissive TTM monoradicals and their applications in OLEDs. Meanwhile, some new emission phenomena and concepts which may change the present cognition were developed, such as photoinduced radical emission, ML, triplet fluorescence, and so on. Although several conductive radicals were designed, the relatively low conductivity still lags behind their closed-shell counterparts. Similarly, semiconductive radical compounds also showed relatively low mobility, and their stability would be a huge obstacle in practice compared to the closed-shell compounds. In recent years, several room-temperature ferromagnets have been prepared, which represent a new possible area of organic plastic magnets.

In the future, radical molecules would still be an emerging force in material and chemical world for their unique chemical structures, especially in luminescent, conductive, semiconductive, ferromagnetic, and spintronic areas. For luminescent

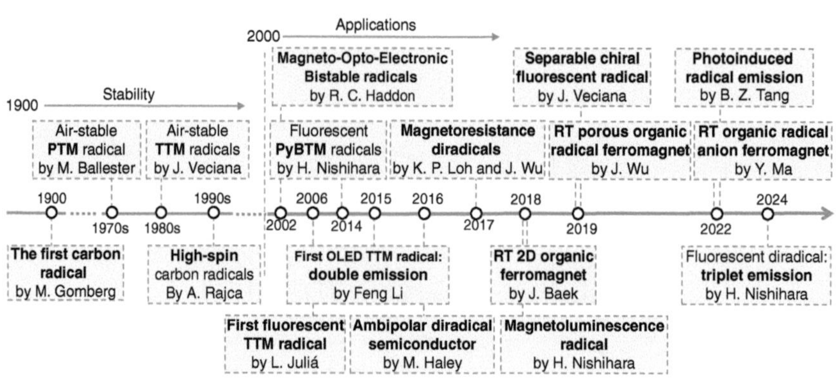

Figure 9.44 Timeline for radical materials.

radical materials, the first challenging task is how to develop novel stable luminescent radicals based on the mechanism and design principle of now available luminescent radical materials; the second one is how to design high-spin luminescent radicals, which is predicted to exhibit higher emission quantum yield and photon utilization efficiency compared to traditional closed-shell luminescent materials; the third one is how to design and synthesize chiral luminescent radical materials, which is far from explored. For conductive and semiconductive radicals, these molecules are also expected to exhibit good conducting properties for their unique structure with unpaired electrons or weakly bonded electrons, which are theoretically much easier to move compared to their closed-shell counterparts under the driving force of electric field. However, the past progress mainly focuses on the stabilization of these radicals, and bulky groups would be a kinetic obstacle for their tight packing and charge transport. Therefore, the critical difficulty is how to form a uniform conductive or semiconductive channel through both molecular design and device engineering. Organic plastic ferromagnetic materials are promising next generation of plastic magnets. Less reports on these materials make their structure design principle and origin of ferromagnetism still controversial, which also would attract more interest of material and chemical scientists.

References

1 Huang, W., Ma, X., Sato, O. et al. (2021). Controlling dynamic magnetic properties of coordination clusters via switchable electronic configuration. *Chem. Soc. Rev.* 50 (12): 6832–6870.

2 Liu, F., Hou, Y., and Gao, S. (2014). Exchange-coupled nanocomposites: Chemical synthesis, characterization and applications. *Chem. Soc. Rev.* 43 (23): 8098–8113.

3 Kumar, K.S. and Ruben, M. (2021). Sublimable spin-crossover complexes: from spin-state switching to molecular devices. *Angew. Chem. Int. Ed.* 60 (14): 7502–75211.

4 Wang, X.-Y., Avendaño C., and Dunbar, K.R. (2011). Molecular magnetic materials based on 4d and 5d transition metals. *Chem. Soc. Rev.* 40 (6): 3213–3238.

5 Feng, M., Ruan, Z.-Y., Chen, Y.-C. et al. (2020). Physical stimulus and chemical modulations of bistable molecular magnetic materials. *Chem. Commun.* 56 (89): 13702–13718.

6 Brooker, S. (2015). Spin crossover with thermal hysteresis: Practicalities and lessons learnt. *Chem. Soc. Rev.* 44 (10): 2880–2892.

7 Miller, J.S. (2011). Magnetically ordered molecule-based materials. *Chem. Soc. Rev.* 40 (6): 3266–3296.

8 Sanvito, S. (2011). Molecular spintronics. *Chem. Soc. Rev.* 40 (6): 3336–3355.

9 Cornia, A., Mannini, M., Sainctavit, P. et al. (2011). Chemical strategies and characterization tools for the organization of single molecule magnets on surfaces. *Chem. Soc. Rev.* 40 (6): 3076–3091.

10 Shao, D. and Wang, X.-Y. (2020). Development of single-molecule magnets. *Chin. J. Chem.* 38 (9): 1005–1018.

11 Tao, J., Wei, R.-J., Huang, R.-B. et al. (2012). Polymorphism in spin-crossover systems. *Chem. Soc. Rev.* 41 (2): 703–737.

12 Sato, O., Tao, J., and Zhang, Y.-Z. (2007). Control of magnetic properties through external stimuli. *Angew. Chem. Int. Ed.* 46 (13): 2152–2187.

13 Shao, D., Shi, L., Yin, L. et al. (2018). Reversible on–off switching of both spin crossover and single-molecule magnet behaviours via a crystal-to-crystal transformation. *Chem. Sci.* 9 (41): 7986–7991.

14 Gütlich, P. and Goodwin, H.A. (2004). Spin crossover-an overall perspective. *Top. Curr. Chem.* 233: 1–47.

15 Wei, R.-J., Huo, Q., Tao, J. et al. (2011). Spin-crossover Fe^{II}_4 squares: Two-step complete spin transition and reversible single-crystal-to-single-crystal transformation. *Angew. Chem. Int. Ed.* 50 (38): 8940–8943.

16 Chun, H., Bill, E., Weyhermulller, T. et al. (2003). $S = {}^3/_2 \leftrightarrow S = {}^1/_2$ spin crossover behavior in five-coordinate halido- and pseudohalido-bis(*o*-iminobenzosemiquinonato)iron(III) complexes. *Inorg. Chem.* 42 (18): 5612–5620.

17 Shao, D., Shi, L., Shen, F.-X. et al. (2019). Reversible on—off switching of the hysteretic spin crossover in a cobalt(II) complex via crystal to crystal transformation. *Inorg. Chem.* 58 (17): 11589–11598.

18 König, E. and Madeja, K. 5T2-1A1 (1967). Equilibriums in some iron(II)-bis(1,10-phenanthroline) complexes. *Inorg. Chem.* 6 (1): 48–55.

19 König, E., Madeja, K., and Watson, K.J. (1968). Reversible quintet-singlet transition in dithiocyanato-bis (2, 2'-dipyridyl) iron (II). *J. Am. Chem. Soc.* 90 (5): 1146–1153.

20 Ozarowski, A., McGarvey, B.R., Sarker, A.B. et al. (1988). EPR study of manganese(I1) in two crystalline forms of $Fe(C_6H_8N_2S_2)_2(NCS)_2$ and the high-spin-low-spin transition that occurs in only one form, X-ray structure determination of both forms. *Inorg. Chem.* 27 (4): 628–635.

21 Galet, A., Gaspar, A.B., Muñoz, M.C. et al. (2006). Pressure effect and crystal structure reinvestigations on the spin crossover system: $[Fe(bt)_2(NCS)_2]$ (bt = 2,2′-Bithiazoline) polymorphs A and B. *Inorg. Chem.* 45 (24): 9670–9679.

22 Matouzenko, G.S., Bousseksou, A., Lecocq, S. et al. (1997). Polymorphism in spin transition systems. crystal structure, magnetic properties, and Mo1ssbauer spectroscopy of three polymorphic modifications of $[Fe(DPPA)(NCS)_2]$ [DPPA = (3-Aminopropyl)bis(2-pyridylmethyl)amine]. *Inorg. Chem.* 36 (25): 5869–5879.

23 Tiwary, S. and Vasudevan, S. (1998). Void geometry driven spin crossover in zeolite-encapsulated cobalt tris(bipyridyl) complex ion. *Inorg. Chem.* 37 (20): 5239–5246.

24 Clérac, R., Cotton, F.A., Dunbar, K.R. et al. (2001). Tuning the metal—metal bonds in the linear tricobalt compound $Co_3(dpa)_4Cl_2$: Bond-stretch and spin-state isomers. *Inorg. Chem.* 40 (6): 1256–1264.

25 Hayami, S., Hashiguchi, K., Juhász, G. et al. (2004). 1-D Cobalt(II) spin transition compound with strong interchain interaction: $[Co(pyterpy)Cl_2]\cdot X$. *Inorg. Chem.* 43 (14): 4124–4126.

26 Galet, A., Gaspar, A.B., Muñoz, M.C. et al. (2006). Influence of the counterion and the solvent molecules in the spin crossover system $[Co(4\text{-terpyridone})_2]X_p{\cdot}nH_2O$. *Inorg. Chem.* 45 (11): 4413–4422.

27 Roux, C., Zarembowitch, J., Gallois, B. et al. (1994). Toward ligand-driven light-induced spin change, influence of the configuration of 4-Styrylpyridine (stpy) on the magnetic properties of $Fe^{II}(stpy)_4(NCS)_2$ complexes. crystal structures of the spin-crossover species Fe(*trans*-stpy)$_4$(NCS)$_2$ and of the high-spin species Fe(*cis*-stpy)$_4$(NCS)$_2$. *Inorg. Chem.* 33 (10): 2273–2279.

28 Galet, A., Muñoz, M.C., Martínez, V. et al. (2004). Supramolecular isomerism in spin crossover networks with aurophilic interactions. *Chem. Commun.* (20):2268–2269.

29 Wei, R.-J., Li, B., Tao, J. et al. (2011). Making spin-crossover crystals by successive polymorphic transformations. *Inorg. Chem.* 50 (4): 1170–1172.

30 Li, B., Wei, R.J., Tao, J. et al. (2010). Solvent-induced transformation of single crystals of a spin-crossover (SCO) compound to single crystals with two distinct SCO centers. *J. Am. Chem. Soc.* 132 (5): 1558–1566.

31 Bogani, L. and Wernsdorfer, W. (2008). Molecular spintronics using single-molecule magnets. *Nat. Mater.* 7: 179–186.

32 Serrano, G., Poggini, L., Briganti, M. et al. (2020). Quantum dynamics of a single molecule magnet on superconducting Pb(111). *Nat. Mater.* 19: 546–551.

33 Aulakh, D., Pyser, J.B., Zhang, X. et al. (2015). Metal—organic frameworks as platforms for the controlled nanostructuring of single-molecule magnets. *J. Am. Chem. Soc.* 137 (29): 9254—9257.

34 Aulakh, D., Liu, L., Varghese, J.R. et al. (2019). Direct imaging of isolated single-molecule magnets in metal—organic frameworks. *J. Am. Chem. Soc.* 141 (7): 2997—3005.

35 Wu, D.-Q., Shao, D., Wei, X.-Q. et al. (2017). Reversible on—off switching of a single-molecule magnet via a crystal-to-crystal chemical transformation. *J. Am. Chem. Soc.* 139 (34): 11714—11717.

36 Li, Q. and Li, Z. (2020). Molecular packing: Another key point for the performance of organic and polymeric optoelectronic materials. *Acc. Chem. Res.* 53 (4): 962—973.

37 Gu, J., Li, Z., and Li, Q. (2023). From single molecule to molecular aggregation science. *Coord. Chem. Rev.* 475: 214872.

38 Huang, A., Li, Q., and Li, Z. (2022). Molecular uniting set identified characteristic (MUSIC) of organic optoelectronic materials. *Chin. J. Chem.* 40 (19): 2356—2370.

39 Zhou, H.-C., Long, J.R., and Yaghi, O.M. (2012). Introduction to metal–organic frameworks. *Chem. Rev.* 112: 673–674.

40 Furukawa, H., Cordova, K.E., O'keeffe, M., and Yaghi, O.M. (2013). The chemistry and applications of metal-organic frameworks. *Science* 341: 1230444.

41 Figueira-Duarte, T.M. and Müllen, K. (2011). Pyrene-based materials for organic electronics. *Chem. Rev.* 111: 7260–7314.

42 Deria, P., Yu, J., Smith, T., and Balaraman, R.P. (2017). Ground-state versus excited-state interchromophoric interaction: Topology dependent excimer contribution in metal–organic framework photophysics. *J. Am. Chem. Soc.* 139: 5973–5983.

43 Hu, R., Leung, N.L.C., and Tang, B.Z. (2014). AIE macromolecules: Syntheses, structures and functionalities. *Chem. Soc. Rev.* 43: 4494–4562.

44 Zhu, L., Zhu, B., Luo, J., and Liu, B. (2021). Design and property modulation of metal–organic frameworks with aggregation-induced emission. *ACS Mater. Lett.* 3: 77–89.

45 Rajasree, S. S., Li, X., and Deria, P. (2021). Physical properties of porphyrin-based crystalline metal–organic frameworks. *Commun. Chem.* 4: 47.

46 Webber, T.E., Liu, W.-G., Desai, S.P. et al. (2017). Role of a modulator in the synthesis of phase-pure NU-1000. *ACS Appl. Mater. Interfaces* 9: 39342–39346.

47 Kung, C.-W., Wang, T.C., Mondloch, J.E. et al. (2013). Metal–organic framework thin films composed of free-standing acicular nanorods exhibiting reversible electrochromism. *Chem. Mater.* 25: 5012–5017.

48 Mondloch, J.E., Bury, W., Fairen-Jimenez, D. et al. (2013). Vapor-phase metalation by atomic layer deposition in a metal–organic framework. *J. Am. Chem. Soc.* 135: 10294–10297.

49 Li, P., Klet, R.C., Moon, S.-Y. et al. (2015). Synthesis of nanocrystals of Zr-based metal–organic frameworks with csq-net: Significant enhancement in the degradation of a nerve agent simulant. *Chem. Commun.* 51: 10925–10928.

50 Wei, Z., Gu, Z.-Y., Arvapally, R.K. et al. (2014). Rigidifying fluorescent linkers by metal–organic framework formation for fluorescence blue shift and quantum yield enhancement. *J. Am. Chem. Soc.* 136: 8269–8276.

51 Medishetty, R., Nemec, L., Nalla, V. et al. (2017). Multi-photon absorption in metal–organic frameworks. *Angew. Chem. Int. Ed.* 56: 14743–14748.

52 Zhang, Q., Su, J., Feng, D. et al. (2015). Piezofluorochromic metal–organic framework: A microscissor lift. *J. Am. Chem. Soc.* 137: 10064–10067.

53 Hu, X., Wang, Z., Lin, B. et al. (2017). Two-dimensional metal-organic layers as a bright and processable phosphor for fast white-light communication. *Chem. Eur. J.* 23: 8390–8394.

54 Deria, P., Gómez-Gualdrón, D.A., Hod, I. et al. (2016). Framework-topology-dependent catalytic activity of zirconium-based (porphinato)zinc(II) MOFs. *J. Am. Chem. Soc.* 138: 14449–14457.

55 Feng, D., Gu, Z.-Y., Li, J.-R. et al. (2012). Zirconium-metalloporphyrin PCN-222: Mesoporous metal–organic frameworks with ultrahigh stability as biomimetic catalysts. *Angew. Chem. Int. Ed.* 51: 10307–10310.

56 Morris, W., Volosskiy, B., Demir, S. et al. (2012). Synthesis, structure, and metalation of two new highly porous zirconium metal–organic frameworks. *Inorg. Chem.* 51: 6443–6445.

57 Feng, D., Gu, Z.-Y., Chen, Y.-P. et al. (2014). A highly stable porphyrinic zirconium metal–organic framework with shp-a topology. *J. Am. Chem. Soc.* 136: 17714–17717.

58 Feng, D., Chung, W.-C., Wei, Z. et al. (2013). Construction of ultrastable porphyrin Zr metal–organic frameworks through linker elimination. *J. Am. Chem. Soc.* 135: 17105–17110.

59 Jiang, H.-L., Feng, D., Wang, K. et al. (2013). An exceptionally stable, porphyrinic Zr metal–organic framework exhibiting pH-dependent fluorescence. *J. Am. Chem. Soc.* 135: 13934–13938.

60 Lammert, M., Reinsch, H., Murray, C.A. et al. (2016). Synthesis and structure of Zr(IV)- and Ce(IV)-based CAU-24 with 1,2,4,5-tetrakis(4-carboxyphenyl)benzene. *Dalton Trans.* 45: 18822–18826.

61 Lyu, J., Zhang, X., Otake, K.-I. et al. (2019). Topology and porosity control of metal–organic frameworks through linker functionalization. *Chem. Sci.* 10: 1186–1192.

62 Meng, Y., Luo, Y., Shi, J.L. et al. (2020). 2D and 3D porphyrinic covalent organic frameworks: The influence of dimensionality on functionality. *Angew. Chem. Int. Ed.* 59: 3624–3629.

63 Lyu, J., Gong, X., Lee, S.-J. et al. (2020). Phase transitions in metal–organic frameworks directly monitored through in situ variable temperature liquid-cell transmission electron microscopy and in situ X-ray diffraction. *J. Am. Chem. Soc.* 142: 4609–4615.

64 Pang, J., Yuan, S., Qin, J. et al. (2017). Control the structure of Zr-tetracarboxylate frameworks through steric tuning. *J. Am. Chem. Soc.* 139: 16939–16945.

65 Deria, P., Yu, J., Balaraman, R.P. et al. (2016). Topology-dependent emissive properties of zirconium-based porphyrin MOFs. *Chem. Commun.* 52: 13031–13034.

66 Chen, C.-X., Yin, S.-Y., Wei, Z.-W. et al. (2019). Pressure-induced multiphoton excited fluorochromic metal–organic frameworks for improving MPEF properties. *Angew. Chem. Int. Ed.* 58: 14379–14385.

67 Côté, A.P., Benin, A.I., Ockwig, N.W. et al. (2005). Porous, crystalline, covalent organic frameworks. *Science* 310: 1166–1170.

68 Alahakoon, S.B., Thompson, C.M., Nguyen, A.X. et al. (2016). An azine-linked hexaphenylbenzene based covalent organic framework. *Chem. Commun.* 52: 2843–2845.

69 Li, Z., Feng, X., Zou, Y. et al. (2014). A 2D azine-linked covalent organic framework for gas storage applications. *Chem. Commun.* 50: 13825–13828.

70 Kang, Z., Peng, Y., Qian, Y. et al. (2016). Mixed matrix membranes (MMMs) comprising exfoliated 2D covalent organic frameworks (COFs) for efficient CO_2 separation. *Chem. Mater.* 28: 1277–1285.

71 Gao, Q., Li, X., Ning, G.-H. et al. (2018). Covalent organic framework with frustrated bonding network for enhanced carbon dioxide storage. *Chem. Mater.* 30: 1762–1768.

72 Khan, N.A., Zhang, R., Wu, H. et al. (2020). Solid-vapor interface engineered covalent organic framework membranes for molecular separation. *J. Am. Chem. Soc.* 142: 13450–13458.

73 Li, Y., Wu, Q., Guo, X. et al. (2020). Laminated self-standing covalent organic framework membrane with uniformly distributed subnanopores for ionic and molecular sieving. *Nat. Commun.* 11: 599.

74 Liu, J., Han, G., Zhao, D. et al. (2020). Self-standing and flexible covalent organic framework (COF) membranes for molecular separation. *Sci. Adv.* 6: eabb1110.

75 Ma, Y., Wang, Y., Li, H. et al. (2020). Three-dimensional chemically stable covalent organic frameworks through hydrophobic engineering. *Angew. Chem. Int. Ed.* 59: 19633–19638.

76 Ding, S.Y., Gao, J., Wang, Q. et al. (2011). Construction of covalent organic framework for catalysis: Pd/COF-LZU1 in Suzuki–Miyaura coupling reaction. *J. Am. Chem. Soc.* 133: 19816–19822.

77 Hao, W., Chen, D., Li, Y. et al. (2019). Facile synthesis of porphyrin based covalent organic frameworks via an A_2B_2 monomer for highly efficient heterogeneous catalysis. *Chem. Mater.* 31: 8100–8105.

78 Lin, S., Diercks, C.S., Zhang, Y.-B. et al. (2015). Covalent organic frameworks comprising cobalt porphyrins for catalytic CO_2 reduction in water. *Science* 349: 1208–1213.

79 Ma, Y.X., Li, Z.J., Wei, L. et al. (2017). A dynamic three-dimensional covalent organic framework. *J. Am. Chem. Soc.* 139: 4995–4998.

80 Zhuang, X., Zhao, W., Zhang, F. et al. (2016). A two-dimensional conjugated polymer framework with fully sp^2-bonded carbon skeleton. *Polym. Chem.* 7: 4176–4181.

81 Lukose, B., Kuc, A., Frenzel, J., and Heine, T. (2010). On the reticular construction concept of covalent organic frameworks. *Beilstein J. Nanotechnol.* 1: 60–70.

82 Luo, J., Xie, Z., Lam, J.W. et al. (2001). Aggregation-induced emission of 1-methyl-1,2,3,4,5-pentaphenylsilole. *Chem. Commun.* 1740–1741.

83 Dalapati, S., Jin, E., Addicoat, M. et al. (2016). Highly emissive covalent organic frameworks. *J. Am. Chem. Soc.* 138: 5797–5800.

84 Gao, Q., Li, X., Ning, G.H. et al. (2018). Highly photoluminescent two-dimensional imine-based covalent organic frameworks for chemical sensing. *Chem. Commun.* 54: 2349–2352.

85 Peng, Y., Li, L., Zhu, C. et al. (2020). Intramolecular hydrogen bonding-based topology regulation of two-dimensional covalent organic frameworks. *J. Am. Chem. Soc.* 142: 13162–13169.

86 Xie, Z., Wang, B., Yang, Z. et al. (2019). Stable 2D heteroporous covalent organic frameworks for efficient ionic conduction. *Angew. Chem. Int. Ed.* 58: 15742–15746.

87 Keller, N., Sick, T., Bach, N.N. et al. (2019). Dibenzochrysene enables tightly controlled docking and stabilizes photoexcited states in dual-pore covalent organic frameworks. *Nanoscale* 11: 23338–23345.

88 Albacete, P., Martinez, J.I., Li, X. et al. (2018). Layer-stacking-driven fluorescence in a two-dimensional imine-linked covalent organic framework. *J. Am. Chem. Soc.* 140: 12922–12929.

89 Wu, X., Han, X., Liu, Y. et al. (2018). Control interlayer stacking and chemical stability of two-dimensional covalent organic frameworks via steric tuning. *J. Am. Chem. Soc.* 140: 16124–16133.

90 Wang, S., Ma, L., Wang, Q. et al. (2018). Covalent organic frameworks: A platform for the experimental establishment of the influence of intermolecular distance on phosphorescence. *J. Mater. Chem. C* 6: 5369–5374.

91 Wan, S., Guo, J., Kim, J. et al. (2008). A belt-shaped, blue luminescent, and semiconducting covalent organic framework. *Angew. Chem. Int. Ed.* 47: 8826–8830.

92 Ding, X., Guo, J., Feng, X. et al. (2011). Synthesis of metallophthalocyanine covalent organic frameworks that exhibit high carrier mobility and photoconductivity. *Angew. Chem. Int. Ed.* 50: 1289–1293.

93 Feng, X., Chen, L., Dong, Y., and Jiang, D. (2011). Porphyrin-based two-dimensional covalent organic frameworks: Synchronized synthetic control of macroscopic structures and pore parameters. *Chem. Commun.* 47: 1979–1981.

94 Wan, S., Gándara, F., Asano, A. et al. (2011). Covalent organic frameworks with high charge carrier mobility. *Chem. Mater.* 23: 4094–4097.

95 Chen, X., Addicoat, M., Irle, S. et al. (2013). Control of crystallinity and porosity of covalent organic frameworks by managing interlayer interactions based on self-complementary pi-electronic force. *J. Am. Chem. Soc.* 135: 546–549.

96 Jin, S., Sakurai, T., Kowalczyk, T. et al. (2014). Two-dimensional tetrathiafulvalene covalent organic frameworks: Towards latticed conductive organic salts. *Chem. Eur. J.* 20: 14608–14613.

97 Cai, S.-L., Zhang, Y.-B., Pun, A.B. et al. (2014). Tunable electrical conductivity in oriented thin films of tetrathiafulvalene-based covalent organic framework. *Chem. Sci.* 5: 4693–4700.

98 Ishi-I, T., Yaguma, K., Thiemann, T. et al. (2004). High electron drift mobility in an amorphous film of 2,4,6,-tris[4-(1-naphthyl)phenyl]-1,3,5-triazine. *Chem. Lett.* 33: 1244–1245.

99 Sun, D., Ma, S., Ke, Y. et al. (2006). An interweaving MOF with high hydrogen uptake. *J. Am. Chem. Soc.* 128: 3896–3897.

100 Stegbauer, L., Schwinghammer, K., and Lotsch, B.V. (2014). A hydrazone-based covalent organic framework for photocatalytic hydrogen production. *Chem. Sci.* 5: 2789–2793.

101 Yang, S., Hu, W., Zhang, X. et al. (2018). 2D covalent organic frameworks as intrinsic photocatalysts for visible light-driven CO_2 reduction. *J. Am. Chem. Soc.* 140: 14614–14618.

102 Nagai, A., Chen, X., Feng, X. et al. (2013). A squaraine-linked mesoporous covalent organic framework. *Angew. Chem. Int. Ed.* 52: 3770–3774.

103 Lin, G., Ding, H., Chen, R. et al. (2017). 3D porphyrin-based covalent organic frameworks. *J. Am. Chem. Soc.* 139: 8705–8709.

104 Cannistraro, S., Van De Vorst, A., and Jori, G. (1978). EPR studies on singlet oxygen production by porphyrins. *Photochem. Photobiol.* 28: 257–259.

105 Pachfule, P., Acharjya, A., Roeser, J. et al. (2018). Diacetylene functionalized covalent organic framework (COF) for photocatalytic hydrogen generation. *J. Am. Chem. Soc.* 140: 1423–1427.

106 Stegbauer, L., Zech, S., Savasci, G. et al. (2018). Tailor-made photoconductive pyrene-based covalent organic frameworks for visible-light driven hydrogen generation. *Adv. Energy Mater.* 8: 1703278.

107 Rajca, A. (1994). Organic diradicals and polyradicals: From spin coupling to magnetism? *Chem. Rev. 94*: 871–893.

108 Zhang, B. and Zhu, D. (2012). Molecular-based conducting magnet. *Sci. China Chem.* 55: 883–892.

109 Abe, M. (2013). Diradicals. *Chem. Rev.* 113: 7011–7088.

110 Winkler, M. and Sander, W. (2014). Triradicals. *Acc. Chem. Res.* 47: 31–44.

111 Zeng, Z., Shi, X., Chi, C. et al. (2015). Pro-aromatic and anti-aromatic π-conjugated molecules: an irresistible wish to be diradicals. *Chem. Soc. Rev.* 44: 6578–6596.

112 Kubo, T. (2015). Recent progress in quinoidal singlet biradical molecules. *Chem. Lett.* 44: 111–122.

113 Frederickson, C.K., Rose, B.D., and Haley, M.M. (2017). Explorations of the indenofluorenes and expanded quinoidal analogues. *Acc. Chem. Res.* 50: 977–987.

114 Shi, X. and Chi, C. (2017). Heterocyclic quinodimethanes. *Top. Curr. Chem.* 375: 68.

115 Konishi, A. and Kubo, T. (2017). Benzenoid quinodimethanes. *Top. Curr. Chem.* 375: 83.

116 Burrezo, P.M., Zafra, J.L., López Navarrete, J.T., and Casado, J. (2017). Quinoidal/aromatic transformations in π-conjugated oligomers: Vibrational Raman studies on the limits of rupture for π-bonds. *Angew. Chem. Int. Ed.* 56: 2250–2259.

117 Gopalakrishna, T.Y., Zeng, W., Lu, X., and Wu, J. (2018). From open-shell singlet diradicaloids to polyradicaloids. *Chem. Commun.* 54: 2186–2199.

118 Hu, X., Wang, W., Wang, D., and Zheng, Y. (2018). The electronic applications of stable diradicaloids: present and future. *J. Mater. Chem. C* 6: 11232–11242.

119 Kato, K. and Osuka, A. (2019). Platforms for stable carbon-centered radicals. *Angew. Chem. Int. Ed.* 58: 8978–8986.

120 Dong, S., Ong, A., and Chi, C. (2019). Photochemistry of various acene based molecules. *J. Photoch. Photobio. C: Photoch. Rev.* 38: 27–46.

121 Ji, L., Shi, J., Wei, J. et al. (2020). Air-stable organic radicals: New-generation materials for flexible electronics? *Adv. Mater.* 32: 1908015.

122 Teki, Y. (2020). Excited-state dynamics of non-luminescent and luminescent π-radicals. *Chem. Eur. J.* 26: 980–996.

123 Cui, Z., Abdurahman, A., Ai, X., and Li, F. (2020). Stable luminescent radicals and radical-based LEDs with doublet emission. *CCS Chem.* 2: 1129–1145.

124 Chen, Z.X., Li, Y., and Huang, F. (2021). Persistent and stable organic radicals: Design, synthesis, and applications. *Chem.* 7: 288–332.

125 Yuan, D., Liu, W., and Zhu, X. (2021). Design and applications of single-component radical conductors. *Chem.* 7: 333–357.

126 Zeng, W. and Wu, J. (2021). Open-shell graphene fragments. *Chem* 7: 358–386.

127 Slota, M., Keerthi, A., Myers, W.K. et al. (2018). Magnetic edge states and coherent manipulation of graphene nanoribbons. *Nature* 557: 691–696.

128 Kuhn, R., Neugebauer, F.A., and Trischmann, H. (1965). A stable triradical. *Angew. Chem. Int. Ed.* 4: 72.

129 Dong, S. and Li, Z. (2022). Recent progress in open-shell organic conjugated materials and their aggregated states. *J. Mater. Chem. C* 10: 2431–2449.

130 Gomberg, M. (1900). An instance of trivalent carbon: Triphenylmethyl. *J. Am. Chem. Soc.* 22: 757–771.

131 Kato, S., Morokuma, K., Feller, D. et al. (1983). Ab initio study of m-benzoquinodimethane. *J. Am. Chem. Soc.* 105: 1791–1795.

132 Goodman, J.L. and Berson, J.A. (1985). m-Quinodimethane, parent hydrocarbon of the m-quinonoid non-Kekule series. Low-temperature isolation and solution-phase chemical reactivity. *J. Am. Chem. Soc.* 107: 5409–5424.

133 Fort, R.C., Jr., Getty, S.J., Hrovat, D.A. et al. (1992). Ab initio calculations on m-quinone. The ground state is a triplet. *J. Am. Chem. Soc.* 114: 7549–7552.

134 Rajca, A. (1990). A polyarylmethyl quintet tetraradical. *J. Am. Chem. Soc.* 112: 5890–5892.

135 Rajca, A., Utamapanya, S., and Thayumanavan, S. (1992). Poly(arylmethyl) octet (S=7/2) heptaradical and undecet (S=5) decaradical. *J. Am. Chem. Soc.* 114: 1884–1885.

136 Rajca, A., Rajca, S., and Padmakumar, R. (1994). Calixarene-based macrocyclic nonet (S=4) octaradical and its acyclic sextet (S=5/2) pentaradical analogue. *Angew. Chem. Int. Ed.* 33: 2091–2093.

137 Rajca, A., Wongsriratanakul, J., and Rajca, S. (2001). Magnetic ordering in an organic polymer. *Science* 294: 1503–1505.

138 Rajca, S., Rajca, A., Wongsriratanakul, J. et al. (2004). Organic spin clusters. A dendritic-macrocyclic poly(arylmethyl) polyradical with very high spin of S=10 and its derivatives: Synthesis, magnetic studies, and small-angle neutron scattering. *J. Am. Chem. Soc.* 126: 6972–6986.

139 Jang, S.-H., Gopalan, P., Jackson, J.E., and Kahr, B. (1994). Jacobson and Heintschel peroxides. *Angew. Chem. Int. Ed.* 33: 775–777.

140 Nishiuchi, T., Aibara, S., and Kubo, T. (2018). Synthesis and properties of a highly congested tri(9-anthryl)methyl radical. *Angew. Chem. Int. Ed.* 57: 16516—16519.

141 Ballester, M. (1985). Inert free radicals (IFR): a unique trivalent carbon species. *Acc. Chem. Res.* 18: 380–387.

142 Gamero, V., Velasco, D., Latorre, S. et al. (2006). [4-(N-Carbazolyl)-2,6-dichlorophenyl]bis(2,4,6-trichlorophenyl)methyl radical an efficient red light-emitting paramagnetic molecule. *Tetrahedron Lett.* 47: 2305–2309.

143 Peng, Q., Obolda, A., Zhang, M., and Li, F. (2015). Organic light-emitting diodes using a neutral π radical as emitter: The emission from a doublet. *Angew. Chem. Int. Ed.* 54: 7091–7095.

144 Ai, X., Evans, E.W., Dong, S. et al. (2018). Efficient radical-based light-emitting diodes with doublet emission. *Nature* 563: 536–540.

145 Abdurahman, A., Hele, T.J.H., Gu, Q. et al. (2020). Understanding the luminescent nature of organic radicals for efficient doublet emitters and pure-red light-emitting diodes. *Nat. Mater.* 19: 1224–1229.

146 Guo, H., Peng, Q., Chen, X.-K. et al. (2019). High stability and luminescence efficiency in donor–acceptor neutral radicals not following the Aufbau principle. *Nat. Mater.* 18: 977–984.

147 Ai, X., Chen, Y., Feng, Y., and Li, F. (2018). A stable room-temperature luminescent biphenylmethyl radical. *Angew. Chem. Int. Ed.* 57: 2869–2873.

148 Liu, C.-H., He, Z., Ruchlin, C. et al. (2023). Thiele's fluorocarbons: Stable diradicaloids with efficient visible-to-near-infrared fluorescence from a zwitterionic excited state. *J. Am. Chem. Soc.* 145: 15702–15707.

149 Chang, X., Arnold, M.E., Blinder, R. et al. (2024). A stable chichibabin diradicaloid with near-infrared emission. *Angew. Chem. Int. Ed.* 63: e202404853.

150 Zhou, Z., Yang, K., He, L. et al. (2024). Sulfone-functionalized chichibabin's hydrocarbons: Stable diradicaloids with symmetry breaking charge transfer contributing to NIR emission beyond 900 nm. *J. Am. Chem. Soc.* 146: 6763–6772.

151 Abdurahman, A., Wang, J., Zhao, Y. et al. (2023). A highly stable organic luminescent diradical. *Angew. Chem. Int. Ed.* 62: e202300772.

152 Abdurahman, A., Shen, L., Wang, J. et al. (2023). A highly efficient open-shell singlet luminescent diradical with strong magnetoluminescence properties. *Light. Sci. Appl.* 12: 272.

153 Matsuoka, R., Kimura, S., Miura, T. et al. (2023). Single-molecule magnetoluminescence from a spatially confined persistent diradical emitter. *J. Am. Chem. Soc.* 145: 13615–13622.

154 Mizuno, A., Matsuoka, R., Kimura, S. et al. (2024). Spin-correlated luminescence of a carbazole-containing diradical emitter: Single-molecule magnetoluminescence and thermally activated emission. *J. Am. Chem. Soc.* 146: 18470–18483.

155 Zhu, Z., Zhang, D., Xiao, T. et al. (2023). Rational design of an air-stable, High-spin diradical with diazapyrene. *Angew. Chem. Int. Ed.* 62: e202314900.

156 Shi, J., Xu, W., Yu, H. et al. (2023). A highly luminescent metallo-supramolecular radical cage. *J. Am. Chem. Soc.* 145: 24081–24088.

157 Hattori, Y., Kusamoto, T., and Nishihara, H. (2014). Luminescence, stability, and proton response of an open-shell (3,5-Dichloro-4-pyridyl)bis(2,4,6-trichlorophenyl)methyl radical. *Angew. Chem. Int. Ed.* 53: 11845–11848.

158 Hattori, Y., Kusamoto, T., and Nishihara, H. (2015). Highly photostable luminescent open-shell (3,5-dihalo-4-pyridyl)bis(2,4,6-trichlorophenyl)methyl radicals: Significant effects of halogen atoms on their photophysical and photochemical properties. *RSC Adv.* 5: 64802–64805.

159 Hattori, Y., Kusamoto, T., and Nishihara, H. (2015). Enhanced luminescent properties of an open-shell (3,5-Dichloro-4-pyridyl)bis(2,4,6-trichlorophenyl) methyl radical by coordination to gold. *Angew. Chem. Int. Ed.* 54: 3731–3734.

160 Kusamoto, T., Kimura, S., Ogino, Y. et al. (2016). Modulated luminescence of a stable open-shell triarylmethyl radical: Effects of chemical modification on its electronic structure and physical properties. *Chem. Eur. J.* 22: 17725–17733.

161 Kusamoto, T., Hattori, Y., Tanushi, A., and Nishihara, H. (2015). Intramolecular ferromagnetic radical–CuII coupling in a CuII complex ligated with pyridyl-substituted triarylmethyl radicals. *Inorg. Chem.* 54: 4186—4188.

162 Kimura, S., Uchida, H., Kusamoto, T., and Nishihara, H. (2019). One-dimensional magnetic chain composed of CuII and polychlorinated dipyridylphenylmethyl radical: temperature-dependent Jahn-Teller distortion correlated to π-conjugation and magnetic properties. *Dalton Trans.* 48: 7090–7093.

163 Kimura, S., Uejima, M., Ota, W. et al. (2021). An open-shell, Luminescent, Two-dimensional coordination polymer with a honeycomb lattice and triangular organic radical. *J. Am. Chem. Soc.* 143: 4329–4338.

164 Kimura, S., Matsuoka, R., Kimura, S. et al. (2021). Radical-based coordination polymers as a platform for magnetoluminescence. *J. Am. Chem. Soc.* 143: 5610–5615.

165 Mayorga Burrezo, P., Jiménez, V.G., Blasi, D. et al. (2019). Organic free radicals as circularly polarized luminescence emitters. *Angew. Chem. Int. Ed.* 58: 16282–16288.

166 Mayorga-Burrezo, P., Jiménez, V.G., Blasi, D. et al. (2020). An enantiopure propeller-like trityl-brominated radical: Bringing together a high racemization barrier and an efficient circularly polarized luminescent magnetic emitter. *Chemistry* 26: 3776–3781.

167 Chen, J.-F., Tian, G., Liu, K. et al. (2022). Pillar[5]arene-based neutral radicals with doublet red emissions and stable chiroptical properties. *Org. Lett.* 24: 1935–1940.

168 Duan, J., Shi, Y., Zhao, F. et al. (2023). Chiral luminescent aza[7]helicenes functionalized with a triarylborane acceptor and near-infrared-emissive doublet-state radicals. *Inorg. Chem.* 62: 15829–15833.

169 Janiak, C., Weimann, R., and Görlitz, F. (1997). Structures of the pentaphenylcyclopentadienyl radical, (Pentaphenylcyclopentadienyl)tin(IV) trichloride, and bromo(chloro)bis(tetrahydrofuran)tin(II): Products of the reaction of C_5Ph_5X with SnX_2 (X=Cl,Br). *Organometallics* 16: 4933–4936.

170 Jux, N., Rubin, Y., and Holczer, K. (1996). An unusually stable pentaethynylcyclopentadienyl radical. *Angew. Chem. Int. Ed.* 35: 1986–1990.

171 Ogawa, K., Komatsu, K., and Kitagawa, T. (2011). Allyl radical nature of a phenylcyclopentadienyl radical annelated with two homoadamantene frameworks. *J. Org. Chem.* 76: 6095–6100.

172 Zeng, Z., Sung, Y.M., Bao, N. et al. (2012). Stable tetrabenzo-chichibabin's hydrocarbons: Tunable ground state and unusual transition between their closed-shell and open-shell resonance forms. *J. Am. Chem. Soc.* 134: 14513–14525.

173 Lu, X., Lee, S., Kim, J.O. et al. (2016). Stable 3,6-linked fluorenyl radical oligomers with intramolecular antiferromagnetic coupling and polyradical characters. *J. Am. Chem. Soc.* 138: 13048–13058.

174 Tian, Y., Uchida, K., Kurata, H. et al. (2014). Design and synthesis of new stable fluorenyl-based radicals. *J. Am. Chem. Soc.* 136: 12784–12793.

175 Gong, Z., Xiang, Q., Li, K. et al. (2022). Pentagon-containing doublet graphene fragments with edge-dependent spin/charge distribution. *Chin. J. Chem.* 40: 2525–2530.

176 Duan, L., Xue, X., Hong, B., and Gu, Z. (2023). Conjugation-induced spin delocalization in helical chiral carbon radicals via through-bond and through-space effects. *Adv. Sci.* 10: 2304563.

177 Yang, X., Zhang, D., Liao, Y., and Zhao, D. (2020). Toward an air-stable triradical with strong spin coupling: Synthesis of substituted truxene-5,10,15-triyl. *J. Org. Chem.* 85: 5761–5770.

178 Borissov, A., Chmielewski, P.J., Valdivia, A.C. et al. (2024). Triindenotriphenylenes: Spin-frustrated triskelion triradicals with excellent ambient stability. *Angew. Chem. Int. Ed.* 63: e202408510.

179 Han, H., Zhang, D., Zhu, Z. et al. (2021). Aromatic stacking mediated spin-spin coupling in cyclophane-assembled diradicals. *J. Am. Chem. Soc.* 143: 17690–17700.

180 Zhang, D., Zhu, Z., Xiao, X. et al. (2024). An air-stable carbon-centered triradical with a well-addressable quartet ground state. *J. Am. Chem. Soc.* 146: 21752–21761.

181 Goto, K., Kubo, T., Yamamoto, K. et al. (1999). A stable neutral hydrocarbon radical: Synthesis, Crystal structure, and physical properties of 2,5,8-Tri-tert-butyl-phenalenyl. *J. Am. Chem. Soc.* 121: 1619–1620.

182 Small, D., Zaitsev, V., Jung, Y. et al. (2004). Intermolecular π-to-π bonding between stacked aromatic dyads. Experimental and theoretical binding energies and near-IR optical transitions for phenalenyl radical/radical versus radical/cation dimerizations. *J. Am. Chem. Soc.* 126: 13850–13858.

183 Mou, Z., Uchida, K., Kubo, T., and Kertesz, M. (2014). Evidence of σ- and π-dimerization in a series of phenalenyls. *J. Am. Chem. Soc.* 136: 18009–18022.

184 Kubo, T., Katada, Y., Shimizu, A. et al. (2011). Synthesis, Crystal structure, and physical properties of sterically unprotected hydrocarbon radicals. *J. Am. Chem. Soc.* 133: 14240–14243.

185 Xiang, Q., Guo, J., Xu, J. et al. (2020). Stable olympicenyl radicals and their π-dimers. *J. Am. Chem. Soc.* 142: 11022–11031.

186 Guo, Y., Ding, S., Zhang, N. et al. (2022). π-extended doublet open-shell graphene fragments exhibiting one-dimensional chain stacking. *J. Am. Chem. Soc.* 144: 2095–2100.

187 Pu, Z., Xu, Z., Zhang, X. et al. (2024). Unlocking the multistage redox property of graphenic radicals by π-extension. *Angew. Chem. Int. Ed.* 63: e202406078.

188 Jiao, T., Ni, Y., Xu, T. et al. (2023). Synthesis of monolayer and persistent bilayer graphene fragments by using a radical-mediated coupling approach. *Nat. Syn.* 2: 1104–1115.

189 Ovchinnikov, A.A. (1978). Multiplicity of the ground state of large alternant organic molecules with conjugated bonds. *Theor. Chim. Acta.* 47: 297–304.

190 Su, J., Telychko, M., Song, S., and Lu, J. (2020). Triangulenes: From precursor design to on-surface synthesis and characterization. *Angew. Chem. Int. Ed.* 59: 7658–7668.

191 Pavliček, N., Mistry, A., Majzik, Z. et al. (2017). Synthesis and characterization of triangulene. *Nat. Nanotechnol.* 12: 308–312.

192 Rizzo, D.J., Veber, G., Cao, T. et al. (2018). Topological band engineering of graphene nanoribbons. *Nature* 560: 204–208.

193 Su, J., Telychko, M., Hu, P. et al. (2019). Atomically precise bottom-up synthesis of π-extended [5]triangulene. *Sci. Adv.* 5: eaav7717.

194 Mishra, S., Xu, K., Eimre, K. et al. (2021). Synthesis and characterization of [7]triangulene. *Nanoscale* 13: 1624–1628.

195 Clar, E. and Stewart, D.G. (1953). Aromatic hydrocarbons. LXV. Triangulene derivatives1. *J. Am. Chem. Soc.* 75: 2667–2673.

196 Ravat, P., Blacque, O., and Juríček, M. (2020). Benzo[cd]triangulene: A spin 1/2 graphene fragment. *J. Org. Chem.* 85: 92–100.

197 Arikawa, S., Shimizu, A., Shiomi, D. et al. (2021). Synthesis and isolation of a kinetically stabilized crystalline triangulene. *J. Am. Chem. Soc.* 143: 19599–19605.

198 Chase, D.T., Fix, A.G., Kang, S.J. et al. (2012). 6,12-diarylindeno[1,2-b]fluorenes: Syntheses, Photophysics, and ambipolar OFETs. *J. Am. Chem. Soc.* 134: 10349–10352.

199 Rudebusch, G.E., Zafra, J.L., Jorner, K. et al. (2016). Diindeno-fusion of an anthracene as a design strategy for stable organic biradicals. *Nat. Chem.* 8: 753–759.

200 Zeidell, A.M., Jennings, L., Frederickson, C.K. et al. (2019). Organic semiconductors derived from dinaphtho-fused s-indacenes: How molecular structure and film morphology influence thin-film transistor performance. *Chem. Mater.* 31: 6962–6970.

201 Kubo, T., Shimizu, A., Sakamoto, M. et al. (2005). Synthesis, intermolecular interaction, and semiconductive behavior of a delocalized singlet biradical hydrocarbon. *Angew. Chem. Int. Ed.* 44: 6564–6568.

202 Chikamatsu, M., Mikami, T., Chisaka, J. et al. (2007). Ambipolar organic field-effect transistors based on a low bandgap semiconductor with balanced hole and electron mobilities. *Appl. Phys. Lett.* 91: 043506.

203 Shi, X., Lee, S., Son, M. et al. (2015). Pro-aromatic bisphenaleno-thieno[3,2-b]thiophene versus anti-aromatic bisindeno-thieno[3,2-b]thiophene: Different ground-state properties and applications in field-effect transistors. *Chem. Commun.* 51: 13178–13180.

204 Zong, C., Zhu, X., Xu, Z. et al. (2021). Isomeric dibenzoheptazethrenes for air-stable organic field-effect transistors. *Angew. Chem. Int. Ed.* 60: 16230–16236.

205 Jousselin-Oba, T., Mamada, M., Marrot, J. et al. (2019). Excellent semiconductors based on tetracenotetracene and pentacenopentacene: From stable closed-shell to singlet open-shell. *J. Am. Chem. Soc.* 141: 9373–9381.

206 Thiele, J. and Balhorn, H. (1904). Ueber einen chinoïden Kohlenwasserstoff. *Chem. Ber.* 37: 1463.

207 Tschitschibabin, A.E. (1907). Über einige phenylierte Derivate des p, p-Ditolyls. *Chem. Ber.* 40: 1810.

208 Montgomery, L.K., Huffman, J.C., Jurczak, E.A., and Grendze, M.P. (1986). The molecular structures of Thiele's and Chichibabin's hydrocarbons. *J. Am. Chem. Soc.* 108: 6004.

209 Müller, E. and Pfanz, H. (1941). Über biradikaloide Terphenylderivate. *Ber. Dtsch. Chem. Ges.* 74: 1051.

210 Schmidt, R. and Brauer, H.-D. (1971). The energetic positions of the lowest singlet and triplet state of the Schlenk and of the Müller hydrocarbon. *Angew. Chem., Int. Ed.* 10: 506.

211 Su, Y., Wang, X., Li, Y. et al. (2015). Nitrogen analogues of thiele's hydrocarbon. *Angew. Chem., Int. Ed.* 54: 1634–1637.

212 Li, T., Tan, G., Shao, D. et al. (2016). Magnetic bistability in a discrete organic radical. *J. Am. Chem. Soc.* 138: 10092–10095.

213 Su, Y., Wang, X., Zheng, X. et al. (2013). Tuning ground states of bis(triarylamine) dications: From a closed-shell singlet to a diradicaloid with an excited triplet state. *Angew. Chem., Int. Ed.* 53: 2857–2861.

214 Su, Y., Wang, X., Wang, L. et al. (2016). Thermally controlling the singlet–triplet energy gap of a diradical in the solid state. *Chem. Sci.* 7: 6514–6518.

215 Wang, W., Chen, C., Shu, C. et al. (2018). S = 1 tetraazacyclophane diradical dication with robust stability: A case of low-temperature one-dimensional antiferromagnetic chain. *J. Am. Chem. Soc.* 140: 7820–7826.

216 Dong, S., Gopalakrishna, T.Y., Han, Y., and Chi, C. (2019). Cyclobis(7,8-(para-quinodimethane)-4,4′-triphenylamine) and its cationic species showing annulene-like global (Anti)aromaticity. *Angew. Chem., Int. Ed.* 58: 11742–11746.

217 Ito, A., Yokoyama, Y., Aihara, R. et al. (2010). Preparation and characterization of N-anisyl-substituted hexaaza[16]paracyclophane. *Angew. Chem., Int. Ed.* 49: 8205–8208.

218 Sakamaki, D., Ito, A., Tsutsui, Y., and Seki, S. (2017). Tetraaza[14]- and Octaaza[18]paracyclophane: Synthesis and characterization of their neutral and cationic states. *J. Org. Chem.* 82: 13348–13358.

219 Ma, Y., Han, Y., Hou, X. et al. (2024). Facile synthesis and global aromaticity of aza-superbenzene and aza-supernaphthalene at different oxidation states. *Angew. Chem., Int. Ed.* 63: e202407990.

220 Moulin, E., Niess, F., Maaloum, M. et al. (2010). The hierarchical self-assembly of charge nanocarriers: A highly cooperative process promoted by visible light. *Angew. Chem. Int. Ed.* 49: 6974–6978.

221 JArmao, J.J.I.V., Maaloum, M., Ellis, T. et al. (2014). Healable supramolecular polymers as organic metals. *J. Am. Chem. Soc.* 136: 11382–11388.

222 Mu, Y., Liu, Y., Tian, H. et al. (2021). Sensitive and repeatable photoinduced luminescent radicals from a simple organic crystal. *Angew. Chem. Int. Ed.* 60: 6367–6371.

223 Zhao, X., Gong, J., Alam, P. et al. (2022). A simple approach to achieve organic radicals with unusual solid-state emission and persistent stability. *CCS Chem.* 4: 1912–1920.

224 Mahmood, J., Park, J., Shin, D. et al. (2018). Organic ferromagnetism: Trapping spins in the glassy state of an organic network structure. *Chem* 4: 2357–2369.

225 Phan, H., Herng, T.S., Wang, D. et al. (2019). Room-temperature magnets based on 1,3,5-triazine-linked porous organic radical frameworks. *Chem* 5: 1223–1234.

226 Jiang, Q., Zhang, J., Mao, Z. et al. (2022). Room-temperature ferromagnetism in perylene diimide organic semiconductor. *Adv. Mater.* 34: 2108103.

227 He, X., Zhao, D., Yao, Y. et al. (2024). Magnetic properties of self-assemble naphthalene diimide radical aggregates. *Small* 20: 2311766.

10

Conclusions and Outlook

*Zhen Li**

Department of Chemistry, Wuhan University, Wuhan, China

*Corresponding author: lizhen@whu.edu.cn

In the previous nine chapters, we have demonstrated the importance of molecular aggregates, which bridges the gap between single-molecule properties and the macroscopic opto-electronic performance of materials, providing more objective and perfect analysis methods. Really, in most cases, the modification and optimization of the molecular aggregate structure is still just like opening a blind box, but more and more successful examples disclosed the inherent clues for the rational single molecular design and external stimuli for the realization of the optimized molecular aggregate structure as partially confirmed in this book. Especially, the old odd–even effect in the melting point of alkane such as methane, ethane, propane et al., demonstrates the regularity of molecular packing, which has been successfully applied for the regulation of room temperature phosphorescence performance of luminogens in 2018; the substituent position could control the molecular conformation and thus lead to the different molecular packing and the related intermolecular interactions; the intramolecular charge transfer alters the packing density; and the electronic field aligns the orderly arrangement of chromophore moieties. Figure 10.1 summarizes the adjusting approaches for the molecular aggregate structure, which, surely, should be expanded with the continued efforts of scientists.

For further development, there are mainly three issues that should be paid much attention. First, the methods to provide accurate information of molecules in aggregated states especially including the molecular packing and intermolecular interactions, since the different molecular packing can not only result in different intermolecular interactions but also change the normally recognized intrinsic parameter of chemical bonds; second, the deeply understanding of the inherent mechanism of the difference including the molecule itself and adjacent molecules caused by the different molecular packing; and third but not the last one, the accurate structure-packing-performance relationship. Although

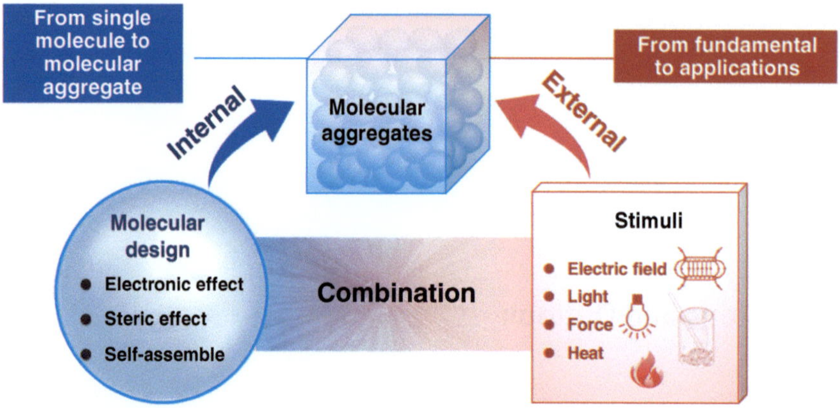

Figure 10.1 The summary of the crucial factors to the molecular aggregate structure, and the key role of molecular aggregates in the research from single molecules to molecular aggregates.

not easy, the increasing attention attracted from academic community and the industrial sector will certainly accelerate the pace of progress, and the proposed concept of molecular uniting set identified characteristic (MUSIC) can vividly describe the evolution from single molecule to molecular aggregate just like the real music in our lives: basic atoms mean notes, and melody with alignment of notes is similar with molecules constructed by atoms with specific sequences. Correspondingly, the MUSICs, which are heavily dependent on aggregated states with various packing modes, resemble the symphony with the coming together of music voiced by different instruments, and bring out the new research field of Molecular Aggregated Science.

Index

Note: Page numbers in *italics* refers to figures respectively.

d